T0336525

Linear Programming and Resource Allocation Modeling

Linear Programming and Resource Allocation Modeling

Michael J. Panik

This edition first published 2019

Registered Office(s)
John Wiley & Sons, Inc., 111 River Street, Hoboken, NJ 07030, USA

Editorial Office
111 River Street, Hoboken, NJ 07030, USA

For details of our global editorial offices, customer services, and more information about Wiley products visit us at www.wiley.com.

Wiley also publishes its books in a variety of electronic formats and by print-on-demand. Some content that appears in standard print versions of this book may not be available in other formats.

Library of Congress Cataloging-in-Publication Data

Names: Panik, Michael J., author.
Title: Linear programming and resource allocation modeling / Michael J. Panik.
Description: 1st edition. | Hoboken, NJ : John Wiley & Sons, Inc., [2018] | Includes bibliographical references and index.
Identifiers: LCCN 2018016229 | ISBN 9781119509448 (hardcover)
Subjects: LCSH: Linear programming. | Resource allocation–Mathematical models.
Classification: LCC T57.77 .P36 2018 | DDC 519.7/2–dc23
LC record available at https://lccn.loc.gov/2018016229

Cover Design: Wiley
Cover Image: © whiteMocca/Shutterstock

Set in 10/12pt Warnock by SPi Global, Pondicherry, India

Printed in the United States of America

V10004749_092118

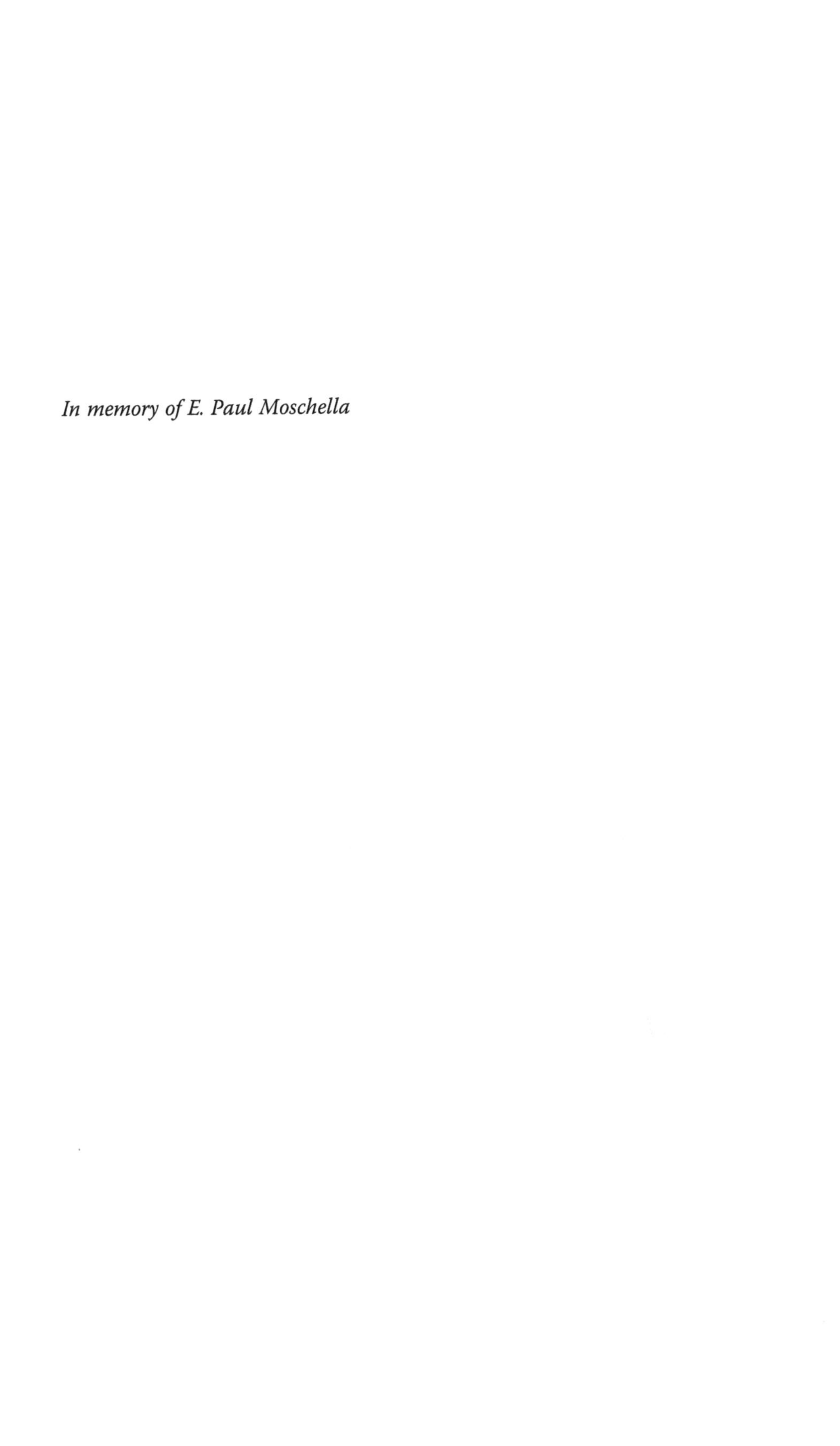

In memory of E. Paul Moschella

Contents

Preface *xi*
Symbols and Abbreviations *xv*

1 **Introduction** *1*

2 **Mathematical Foundations** *13*
2.1 Matrix Algebra *13*
2.2 Vector Algebra *20*
2.3 Simultaneous Linear Equation Systems *22*
2.4 Linear Dependence *26*
2.5 Convex Sets and n-Dimensional Geometry *29*

3 **Introduction to Linear Programming** *35*
3.1 Canonical and Standard Forms *35*
3.2 A Graphical Solution to the Linear Programming Problem *37*
3.3 Properties of the Feasible Region *38*
3.4 Existence and Location of Optimal Solutions *38*
3.5 Basic Feasible and Extreme Point Solutions *39*
3.6 Solutions and Requirement Spaces *41*

4 **Computational Aspects of Linear Programming** *43*
4.1 The Simplex Method *43*
4.2 Improving a Basic Feasible Solution *48*
4.3 Degenerate Basic Feasible Solutions *66*
4.4 Summary of the Simplex Method *69*

5 **Variations of the Standard Simplex Routine** *71*
5.1 The M-Penalty Method *71*
5.2 Inconsistency and Redundancy *78*
5.3 Minimization of the Objective Function *85*

5.4 Unrestricted Variables *86*
5.5 The Two-Phase Method *87*

6 Duality Theory *95*
6.1 The Symmetric Dual *95*
6.2 Unsymmetric Duals *97*
6.3 Duality Theorems *100*
6.4 Constructing the Dual Solution *106*
6.5 Dual Simplex Method *113*
6.6 Computational Aspects of the Dual Simplex Method *114*
6.7 Summary of the Dual Simplex Method *121*

7 Linear Programming and the Theory of the Firm *123*
7.1 The Technology of the Firm *123*
7.2 The Single-Process Production Function *125*
7.3 The Multiactivity Production Function *129*
7.4 The Single-Activity Profit Maximization Model *139*
7.5 The Multiactivity Profit Maximization Model *143*
7.6 Profit Indifference Curves *146*
7.7 Activity Levels Interpreted as Individual Product Levels *148*
7.8 The Simplex Method as an Internal Resource Allocation Process *155*
7.9 The Dual Simplex Method as an Internalized Resource Allocation Process *157*
7.10 A Generalized Multiactivity Profit-Maximization Model *157*
7.11 Factor Learning and the Optimum Product-Mix Model *161*
7.12 Joint Production Processes *165*
7.13 The Single-Process Product Transformation Function *167*
7.14 The Multiactivity Joint-Production Model *171*
7.15 Joint Production and Cost Minimization *180*
7.16 Cost Indifference Curves *184*
7.17 Activity Levels Interpreted as Individual Resource Levels *186*

8 Sensitivity Analysis *195*
8.1 Introduction *195*
8.2 Sensitivity Analysis *195*
8.2.1 Changing an Objective Function Coefficient *196*
8.2.2 Changing a Component of the Requirements Vector *200*
8.2.3 Changing a Component of the Coefficient Matrix *202*
8.3 Summary of Sensitivity Effects *209*

9 Analyzing Structural Changes *217*
9.1 Introduction *217*
9.2 Addition of a New Variable *217*

9.3 Addition of a New Structural Constraint *219*
9.4 Deletion of a Variable *223*
9.5 Deletion of a Structural Constraint *223*

10 Parametric Programming *227*
10.1 Introduction *227*
10.2 Parametric Analysis *227*
10.2.1 Parametrizing the Objective Function *228*
10.2.2 Parametrizing the Requirements Vector *236*
10.2.3 Parametrizing an Activity Vector *245*
10.A Updating the Basis Inverse *256*

11 Parametric Programming and the Theory of the Firm *257*
11.1 The Supply Function for the Output of an Activity (or for an Individual Product) *257*
11.2 The Demand Function for a Variable Input *262*
11.3 The Marginal (Net) Revenue Productivity Function for an Input *269*
11.4 The Marginal Cost Function for an Activity (or Individual Product) *276*
11.5 Minimizing the Cost of Producing a Given Output *284*
11.6 Determination of Marginal Productivity, Average Productivity, Marginal Cost, and Average Cost Functions *286*

12 Duality Revisited *297*
12.1 Introduction *297*
12.2 A Reformulation of the Primal and Dual Problems *297*
12.3 Lagrangian Saddle Points *311*
12.4 Duality and Complementary Slackness Theorems *315*

13 Simplex-Based Methods of Optimization *321*
13.1 Introduction *321*
13.2 Quadratic Programming *321*
13.3 Dual Quadratic Programs *325*
13.4 Complementary Pivot Method *329*
13.5 Quadratic Programming and Activity Analysis *335*
13.6 Linear Fractional Functional Programming *338*
13.7 Duality in Linear Fractional Functional Programming *347*
13.8 Resource Allocation with a Fractional Objective *353*
13.9 Game Theory and Linear Programming *356*
13.9.1 Introduction *356*
13.9.2 Matrix Games *357*
13.9.3 Transformation of a Matrix Game to a Linear Program *361*
13.A Quadratic Forms *363*

13.A.1 General Structure *363*
13.A.2 Symmetric Quadratic Forms *366*
13.A.3 Classification of Quadratic Forms *367*
13.A.4 Necessary Conditions for the Definiteness and Semi-Definiteness of Quadratic Forms *368*
13.A.5 Necessary and Sufficient Conditions for the Definiteness and Semi-Definiteness of Quadratic Forms *369*

14 Data Envelopment Analysis (DEA) *373*
14.1 Introduction *373*
14.2 Set Theoretic Representation of a Production Technology *374*
14.3 Output and Input Distance Functions *377*
14.4 Technical and Allocative Efficiency *379*
14.4.1 Measuring Technical Efficiency *379*
14.4.2 Allocative, Cost, and Revenue Efficiency *382*
14.5 Data Envelopment Analysis (DEA) Modeling *385*
14.6 The Production Correspondence *386*
14.7 Input-Oriented DEA Model under CRS *387*
14.8 Input and Output Slack Variables *390*
14.9 Modeling VRS *398*
14.9.1 The Basic BCC (1984) DEA Model *398*
14.9.2 Solving the BCC (1984) Model *400*
14.9.3 BCC (1984) Returns to Scale *401*
14.10 Output-Oriented DEA Models *402*

References and Suggested Reading *405*
Index *411*

Preface

Economists, engineers, and management scientists have long known and employed the power and versatility of linear programming as a tool for solving resource allocation problems. Such problems have ranged from formulating a simple model geared to determining an optimal product mix (e.g. a producing unit seeks to allocate its limited inputs to a set of production activities under a given linear technology in order to determine the quantities of the various products that will maximize profit) to the application of an input analytical technique called data envelopment analysis (DEA) – a procedure used to estimate multiple-input, multiple-output production correspondences so that the productive efficiency of decision making units (DMUs) can be compared. Indeed, DEA has now become the subject of virtually innumerable articles in professional journals, textbooks, and research monographs.

One of the drawbacks of many of the books pertaining to linear programming applications, and especially those addressing DEA modeling, is that their coverage of linear programming fundamentals is woefully deficient – especially in the treatment of duality. In fact, this latter area is of paramount importance and represents the "bulk of the action," so to speak, when resource allocation decisions are to be made.

That said, this book addresses the aforementioned shortcomings involving the inadequate offering of linear programming theory and provides the foundation for the development of DEA. This book will appeal to those wishing to solve linear optimization problems in areas such as economics (including banking and finance), business administration and management, agriculture and energy, strategic planning, public decision-making, health care, and so on. The material is presented at the advanced undergraduate to beginning graduate level and moves at an unhurried pace. The text is replete with many detailed example problems, and the theoretical material is offered only after the reader has been introduced to the requisite mathematical foundations. The only prerequisites are a beginning calculus course and some familiarity with linear algebra and matrices.

Looking to specifics, Chapter 1 provides an introduction to the primal and dual problems via an optimum product mix problem, while Chapter 2 reviews the rudiments of vector and matrix operations and then considers topics such as simultaneous linear equation systems, linear dependence, convex sets, and some n-dimensional geometry. Specialized mathematical topics are offered in chapter appendices.

Chapter 3 provides an introduction to the canonical and standard forms of a linear programming problem. It covers the properties of the feasible region, the existence and location of optimal solutions, and the correspondence between basic feasible solutions and extreme point solutions.

The material in Chapter 4 addresses the computational aspects of linear programming. Here the simplex method is developed and the detection of degeneracy is presented.

Chapter 5 considers variations of the standard simplex theme. Topics such as the M-penalty and two-phase methods are developed, along with the detection of inconsistency and redundancy.

Duality theory is presented in Chapter 6. Here symmetric, as well as unsymmetric, duals are covered, along with an assortment of duality theorems. The construction of the dual solution and the dual simplex method round out this key chapter.

Chapter 7 begins with a basic introduction to the technology of a firm via activity analysis and then moves into single- and multiple-process production functions, as well as single- and multiple-activity profit maximization models. Both the primal and dual simplex methods are then presented as internal resource allocation mechanisms. Factor learning is next introduced in the context of an optimal product mix. All this is followed by a discussion of joint production processes and production transformation functions, along with the treatment of cost minimization in a joint production setting.

The discussion in Chapter 8 deals with the sensitivity analysis of the optimal solution (e.g. changing an objective function coefficient or changing a component of the requirements vector) while Chapter 9 analyzes structural changes (e.g. addition of a new variable or structural constraint). Chapter 10 focuses on parametric programming and consequently sets the stage for the material presented in the next chapter. To this end, Chapter 11 employs parametric programming to develop concepts such as the demand function for a variable input and the supply function for the output of an activity. Notions such as the marginal and average productivity functions along with marginal and average cost functions are also developed.

In Chapter 12, the concept of duality is revisited; the primal and dual problems are reformulated and re-examined in the context of Lagrangian saddle points, and a host of duality and complementary slackness theorems are offered. This treatment affords the reader an alternative view of duality theory and, depending on the level of mathematical sophistication of the reader, can be considered as optional or can be omitted on a first reading.

Chapter 13 deals with primal and dual quadratic programs, the complementary pivot method, primal and dual linear fractional functional programs, and (matrix) game theory solutions via linear programming.

Data envelopment analysis (DEA) is the subject of Chapter 14. Topics such as the set theoretic representation of a production technology, input and output distance functions, technical and allocative efficiency, cost and revenue efficiency, the production correspondence, input-oriented models under constant and variable returns to scale, and output-oriented models are presented. DEA model solutions are also discussed.

A note of thanks is extended to Bharat Kolluri, Rao Singamsetti, and Jim Peta. I have benefited considerably from conversations held with these colleagues over a great many years. Additionally, Alice Schoenrock accurately and promptly typed the entire manuscript. Her efforts are greatly appreciated.

I would also like to thank Mindy Okura-Marszycki, editor, Mathematics and Statistics, and Kathleen Pagliaro, assistant editor, at John Wiley & Sons, for their professionalism and encouragement.

Symbols and Abbreviations

■	Denotes end of example
$\mathcal{E}^n$	n-dimensional Euclidean space
$\mathcal{E}^n_+$	$\{\boldsymbol{x} \in \mathcal{E}^n \mid \boldsymbol{x} \geq \boldsymbol{O}\}$
$\mathcal{D}(\boldsymbol{x}_o)$	Tangent support cone
$\mathcal{K}$	Region of admissible solutions
$\mathcal{D}(\boldsymbol{x}_o)^+$	Polar support cone
$\mathcal{D}(\boldsymbol{x}_o)^*$	Dual support cone
A'	Transpose of a matrix $\boldsymbol{A}$
$\mathcal{J}$	Index set of binding constraints
∇	Del operator
$\boldsymbol{O}$	Null matrix (vector)
$\boldsymbol{I}_n$	Identity matrix of order n
$(m \times n)$	Order of a matrix (with m rows and n columns)
$A \to B$	Matrix $\boldsymbol{A}$ is transformed into matrix $\boldsymbol{B}$
$\lvert A \rvert$	Determinant of a square matrix $\boldsymbol{A}$
$\mathcal{M}$	Set of all square matrices
$\boldsymbol{A}^{-1}$	Inverse of matrix $\boldsymbol{A}$
$\mathcal{V}_n$	Vector space
$\lVert \boldsymbol{x} \rVert$	Norm of $\boldsymbol{x}$
$\boldsymbol{e}_i$	ith unit column vector
$\rho(\boldsymbol{A})$	Rank of a matrix $\boldsymbol{A}$
dim	Dimension of a vector space
$\delta(\boldsymbol{x}_o)$	Spherical δ-neighborhood of $\boldsymbol{x}_o$
$\boldsymbol{x}_c$	Convex combination
$\mathcal{H}$	Hyperplane
$(\mathcal{H}^+)$, $(\mathcal{H}^-)$	Open half-planes
$[\mathcal{H}^+]$, $[\mathcal{H}^-]$	Closed half-planes
$\mathcal{C}$	Cone
$\mathcal{L}$	Ray or half-line
$\underline{\lim}$	Lower limit

$\overline{\lim}$	Upper limit
AE	Allocative efficiency
BCC	Banker, Charnes, and Cooper
CCR	Charnes, Cooper, and Rhodes
CE	Cost efficiency
CRS	Constant returns to scale
DBLP	Dual of PBLP (multiplier form of (primal) linear program)
DEA	Data envelopment analysis
DLP	Dual of PLP
DMU	Decision making unit
EDLP	Extension of DLP
Eff	Efficient
IPF	Input distance function
Isoq	Isoquant
LCP	Linear complementarity problem
ODF	Output distance function
P1	Phase 1
P2	Phase 2
PBLP	Envelopment form of the (primal) linear program
PLP	Primal linear program
RE	Revenue efficiency
TE	Technical efficiency
VRS	Variable returns to scale

1

Introduction

This book deals with the application of linear programming to firm decision making. In particular, an important resource allocation problem that often arises in actual practice is when a set of inputs, some of which are limited in supply over a particular production period, is to be utilized to produce, using a given technology, a mix of products that will maximize total profit. While a model such as this can be constructed in a variety of ways and under different sets of assumptions, the discussion that follows shall be limited to the linear case, i.e. we will consider the short-run static profit-maximizing behavior of the multiproduct, multifactor competitive firm that employs a fixed-coefficients technology under certainty (Dorfman 1951, 1953; Naylor 1966).

How may we interpret the assumptions underlying this **profit maximization model**?

1) All-around **perfect competition** – the prices of the firm's product and variable inputs are given.
2) The firm employs a **static model** – all prices, the technology, and the supplies of the fixed factors remain constant over the production period.
3) The firm operates under conditions of **certainty** – the model is deterministic in that all prices and the technology behave in a completely systematic (nonrandom) fashion.
4) All factors and products are **perfectly divisible** – fractional (noninteger) quantities of factors and products are admissible at an optimal feasible solution.
5) The character of the firm's **production activities,** which represent specific ways of combining fixed and variable factors in order to produce a unit of output (in the case where the firm produces a single product) or a unit of an individual product (when the number of activities equals or exceeds the number of products), is determined by a set of technical decisions internal to the firm. These input activities are:
   a) **independent** in that no interaction effects exist between activities;
   b) **linear**, i.e. the input/output ratios for each activity are constant along with returns to scale (if the use of all inputs in an activity increases by

Linear Programming and Resource Allocation Modeling, First Edition. Michael J. Panik.

a fixed amount, the output produced by that activity increases by the same amount);

c) **additive**, e.g. if two activities are used simultaneously, the final quantities of inputs and outputs will be the arithmetic sums of the quantities that would result if these activities were operated separately. In addition, total profit generated from all activities equals the sum of the profits from each individual activity; and
d) **finite** – the number of input activities or processes available for use during any production period is limited.

6) All structural relations exhibit **direct proportionality** – the objective function and all constraints are linear; unit profit and the fixed-factor inputs per unit of output for each activity are directly proportional to the level of operation of the activity (thus, marginal profit equals average profit).
7) The firm's objective is to **maximize total profit** subject to a set of structural activities, fixed-factor availabilities, and nonnegativity restrictions on the activity levels. Actually, this objective is accomplished in two distinct stages. First, a technical optimization problem is solved in that the firm chooses a set of production activities that requires the minimum amount of the fixed and variable inputs per unit of output. Second, the firm solves the aforementioned constrained maximum problem.
8) The firm operates in the **short run** in that a certain number of its inputs are fixed in quantity.

Why is this linear model for the firm important? It is intuitively clear that the more sophisticated the type of capital equipment employed in a production process, the more inflexible it is likely to be relative to the other factors of production with which it is combined. That is, the machinery in question must be used in fixed proportions with regard to certain other factors of production (Dorfman 1953, p. 143). For the type of process just described, no factor substitution is possible; a given output level can be produced by one and only one input combination, i.e. the inputs are **perfectly complementary**. For example, it is widely recognized that certain types of chemical processes exhibit this characteristic in that, to induce a particular type of chemical reaction, the input proportions (coefficient) must be (approximately) fixed. Moreover, mechanical processes such as those encountered in cotton textile manufacturing and machine-tool production are characterized by the presence of this **limitationality**, i.e. in the latter case, constant production times are logged on a fixed set of machines by a given number of operators working with specific grades of raw materials.

For example, suppose that a firm produces three types of precision tools (denoted x_1, x_2, and x_3) made from high-grade steel. Four separate production operations are used: casting, grinding, sharpening, and polishing. The set of input–output coefficients (expressed in minutes per unit of output), which describe the firm's technology (the firm's stage one problem, as alluded to above, has been solved) is presented in Table 1.1. (Note that each of the three columns represents a separate input activity or process.)

Additionally, capacity limitations exist with respect to each of the four production operations in that upper limits on their availability are in force. That is, per production run, the firm has at its disposal 5000 minutes of casting time, 3000 minutes of grinding time, 3700 minutes of sharpening time, and 2000 minutes of polishing time. Finally, the unit profit values for tools x_1, x_2, and x_3 are \$22.50, \$19.75, and \$26.86, respectively. (Here these figures each depict unit revenue less unit variable cost and are computed before deducting fixed costs. Moreover, we are tacitly assuming that what is produced is sold.) Given this information, it is easily shown that the optimization problem the firm must solve (i.e. the stage-two problem mentioned above) will look like (1.1):

$$\begin{aligned} max f &= 22.50x_1 + 19.75x_2 + 26.86x_3 \; s.t. \text{ (subject to)} \\ & 13x_1 + 10x_2 + 16x_3 \leq 5000 \\ & 12x_1 + 8x_2 + 20x_3 \leq 3000 \\ & 8x_1 + 4x_2 + 9x_3 \leq 3700 \\ & 5x_1 + 4x_2 + 6x_3 \leq 2000 \\ & x_1, x_2, x_3 \geq 0. \end{aligned} \tag{1.1}$$

How may we rationalize the structure of this problem? First, the **objective function f represents total profit**, which is the sum of the individual (gross) profit contributions of the three products, i.e.

$$\begin{aligned} \text{total profit} &= \sum_{j=1}^{3} (\text{total profit from } x_j \text{ sales}) \\ &= \sum_{j=1}^{3} (\text{unit profit from } x_j \text{ sales})\,(\text{number of units of } x_j \text{ sold}) \end{aligned}$$

Table 1.1 Input–output coefficients.

Tools			
x_1	x_2	x_3	**Operations**
13	10	16	Casting
12	8	20	Grinding
8	4	9	Sharpening
5	4	6	Polishing

Next, if we consider the first **structural constraint** inequality (the others can be interpreted in a similar fashion), we see that total casting time used per production run cannot exceed the total amount available, i.e.

$$\begin{aligned}\text{total casting time used} &= \sum_{j=1}^{3}(\text{total casting time used by } x_j)\\ &= \sum_{j=1}^{3}(\text{casting time used per unit of } x_j)\\ &\quad(\text{number of units of } x_j \text{ produced}) \le 5000.\end{aligned}$$

Finally, the **activity levels** (product quantities) x_1, x_2, and x_3 are nonnegative, thus indicating that the production activities are **nonreversible**, i.e. the fixed inputs cannot be created from the outputs.

To solve (1.1) we shall employ a specialized computational technique called the *simplex method.* The details of the simplex routine, as well as its mathematical foundations and embellishments, will be presented in Chapters 2–5. Putting computational considerations aside for the time being, the types of information sets that the firm obtains from an optimal solution to (1.1) can be characterized as follows. The **optimal product mix** is determined (from this result management can specify which product to produce in positive amounts and which ones to omit from the production plan) as well as the **optimal activity levels** (which indicate the exact number of units of each product produced). In addition, **optimal resource utilization** information is also generated (the solution reveals the amounts of the fixed or scarce resources employed in support of the optimal activity levels) along with the **excess (slack) capacity** figures (if the total amount available of some fixed resource is not fully utilized, the optimal solution indicates the amount left idle). Finally, the **optimal dollar value of total profit** is revealed.

Associated with (1.1) (hereafter called the **primal problem)** is a symmetric problem called its **dual.** While Chapter 6 presents duality theory in considerable detail, let us simply note without further elaboration here that the dual problem deals with the internal valuation (pricing) of the firm's fixed or scarce resources. These (nonmarket) prices or, as they are commonly called, **shadow prices** serve to signal the firm when it would be beneficial, in terms of recouping *forgone profit* (since the capacity limitations restrict the firm's production and thus profit opportunities) to acquire additional units of the fixed factors. Relative to (1.1), the dual problem appears as

$$\begin{aligned}&\min g = 5000u_1 + 3000u_2 + 3700u_3 + 2000u_4 \quad s.t.\\ &\quad 13u_1 + 12u_2 + 8u_3 + 5u_4 \ge 22.50\\ &\quad 10u_1 + 8u_2 + 4u_3 + 4u_4 \ge 19.75\\ &\quad 16u_1 + 20u_2 + 9u_3 + 6u_4 \ge 26.86\\ &\quad u_1, u_2, u_3, u_4 \ge 0,\end{aligned} \tag{1.2}$$

where the dual variables $u_1, \ldots, u_4$ are the shadow prices associated with the primal capacity constraints.

What is the interpretation of the form of this dual problem? First, the objective g depicts the **total imputed (accounting) value of the firm's fixed resources**, i.e.

$$\begin{aligned}&\text{total imputed value of all fixed resources}\\&=\sum_{i=1}^{4}(\text{total imputed value of the } i\text{th resource})\\&=\sum_{i=1}^{4}(\text{number of units of the } i\text{th resource available})\\&\quad(\text{shadow price of the } i\text{th resource}).\end{aligned}$$

Clearly, the firm must make the value of this figure as small as possible. That is, it must *minimize forgone profit*. Next, looking to the first **structural constraint** inequality in (1.2) (the rationalization of the others follows suit), we see that the total imputed value of all resources going into the production of a unit of x_1 cannot fall short of the profit per unit of x_1, i.e.

$$\begin{aligned}&\text{total imputed value of all resources per unit of } x_1\\&=\sum_{i=1}^{4}(\text{imputed value of the } i\text{th resource per unit of } x_1)\\&=\sum_{i=1}^{4}(\text{number of units of the } i\text{th resource per unit of } x_1)\\&\quad(\text{shadow price of the } i\text{th resource})\geq 22.50.\end{aligned}$$

Finally, as is the case for any set of prices, the shadow prices $u_1, \ldots, u_4$ are all nonnegative.

As will become evident in Chapter 6, the dual problem does not have to be solved explicitly; its optimal solution is obtained as a byproduct of the optimal solution to the primal problem (and vice versa). What sort of information is provided by the optimal dual solution? The **optimal (internal) valuation of the firm's fixed resources** is exhibited (from this data the firm can discern which resources are in excess supply and which ones are "scarce" in the sense that total profit could possibly be increased if the supply of the latter were augmented) along with the **optimal shadow price configuration** (each such price indicates the increase in total profit resulting from a one unit increase in the associated fixed input). Moreover, the **optimal (imputed) value of inputs** for each product is provided (the solution indicates the imputed value of all fixed resources entering into the production of a unit of each of the firm's outputs) as well as the **optimal accounting loss figures** (here, management is provided with information pertaining to the amount by which the imputed value of all resources used to produce a unit of some product exceeds the unit profit level for the same). Finally, the **optimal imputed value of all fixed resources** is determined. Interestingly enough, this quantity equals the optimal dollar value of total profit obtained from the primal problem, as it must at an optimal feasible solution to the primal-dual pair of problems.

In the preceding model we made the assumption that the various production activities were technologically independent. However, if we now assume that they are **technologically interdependent** in that each product can be produced by employing more than one process, then we may revise the firm's objective to one where a set of production quotas are to be fulfilled at minimum cost. By invoking this assumption we may construct what is called a **joint production model.**

As far as a full description of this type of production program is concerned, let us frame it in terms of the short-run static cost-minimizing behavior of a multiproduct, multifactor competitive firm that employs a fixed-coefficients technology. How can we interpret the assumptions given in support of this model?

1) **Perfect competition** in the factor markets – the prices of the firm's primary and shadow inputs are given.
2) The firm employs a **static model** – all prices, the technology, and the output quotas remain constant over the production period.
3) The firm operates under conditions of **certainty** – the model is deterministic in that all prices and the technology behave in a completely systematic (nonrandom) fashion.
4) All factors and products are **perfectly divisible** – fractional quantities of factors and products are admissible at an optimal feasible solution.
5) The character of the firm's **production activities,** which now represent ways of producing a set of outputs from the application of one unit of a primary input, is determined by a set of technical decisions internal to the firm. These output activities are:
   a) **independent** in that no interaction effects exist among activities;
   b) **linear,** i.e. the output/input ratios for each activity are constant along with the input response to an increase in outputs (if the production of all outputs in an activity increases by a fixed amount, then the input level required by the process must increase by the same amount);
   c) **additive,** e.g. if two activities are used simultaneously, the final quantities of inputs and outputs will be the arithmetic sums of the quantities which would result if these activities were operated separately. Moreover, the total cost figure resulting from all output activities equals the sum of the costs from each individual activity; and
   d) **finite** – the number of output activities or processes available for use during any production period is limited.
6) All structural relations exhibit **direct proportionality** – the objective function and all constraints are linear; unit cost and the fixed-output per unit of

input values for each activity are directly proportional to the level of operation of the activity. (Thus marginal cost equals average cost.)

7) The firm's objective is to **minimize total cost** subject to a set of structural activities, fixed output quotas, and nonnegativity restrictions on the activity levels. This objective is also accomplished in two stages, i.e. in stage one a technical optimization problem is solved in that the firm chooses a set of output activities which yield the maximum amounts of the various outputs per unit of the primary factors. Second, the firm solves the indicated constrained minimization problem.
8) The **short-run** prevails in that the firm's minimum output requirements are fixed in quantity.

For the type of output activities just described, no output substitution is possible; producing more of one output and less of another is not technologically feasible, i.e. the outputs **are perfectly complementary** or **limitational** in that they must all change together.

As an example of the type of model just described, let us assume that a firm employs three grades of the primary input labor (denoted x_1, x_2, and x_3) to produce four separate products: chairs, benches, tables, and stools. The set of output–input coefficients (expressed in units of output per man-hour) which describe the firm's technology appears in Table 1.2. (Here each of the three columns depicts a separate output activity.) Additionally, output quotas exist with respect to each of the four products in that lower limits on the number of units produced must not be violated, i.e. per production run, the firm must produce at least eight chairs, four benches, two tables, and eight stools. Finally, the unit cost coefficients for the labor grades x_1, x_2, and x_3 are \$8.50, \$9.75, and \$9.08, respectively. (Each of these latter figures depicts unit primary resource cost plus unit

Table 1.2 Output–input coefficients.

Grades of Labor			
x_1	x_2	x_3	**Outputs**
$\frac{1}{16}$	$\frac{1}{14}$	$\frac{1}{18}$	Chairs
$\frac{1}{4}$	$\frac{1}{4}$	$\frac{1}{6}$	Benches
$\frac{1}{20}$	$\frac{1}{25}$	$\frac{1}{30}$	Tables
$\frac{1}{4}$	$\frac{1}{3}$	$\frac{1}{6}$	Stools

shadow input cost.) Given this information, the firm's optimization problem may be written as:

$$\begin{aligned}
&\min f = 8.50x_1 + 9.75x_2 + 9.08x_3 \quad s.t. \\
&\frac{1}{16}x_1 + \frac{1}{14}x_2 + \frac{1}{18}x_3 \geq 8 \\
&\frac{1}{4}x_1 + \frac{1}{4}x_2 + \frac{1}{6}x_3 \geq 4 \\
&\frac{1}{20}x_1 + \frac{1}{25}x_2 + \frac{1}{30}x_3 \geq 2 \\
&\frac{1}{4}x_1 + \frac{1}{3}x_2 + \frac{1}{6}x_3 \geq 8 \\
&x_1, x_2, x_3 \geq 0.
\end{aligned} \tag{1.3}$$

How may we interpret the structure of this problem? First, the **objective function f represents total cost,** expressed as the sum of the individual cost contributions of the various labor grades and shadow factors, i.e.

$$\begin{aligned}
\text{total cost} &= \sum_{j=1}^{3} (\text{total cost of all resources associated with the } j\text{th output activity}) \\
&= \sum_{j=1}^{3} (\text{total cost of all resources per unit of the } j\text{th output activity}) \\
&\quad (\text{number of units of } x_j \text{ used}).
\end{aligned}$$

Next, if we concentrate on the first **structural constraint** inequality (the others are interpreted in like fashion), we see that the total number of chairs produced per production run cannot fall short of the total number required, i.e.

$$\begin{aligned}
&\text{total number of chairs produced} \\
&= \sum_{j=1}^{3} (\text{number of chairs produced using } x_j) \\
&= \sum_{j=1}^{3} (\text{number of chairs produced per unit of } x_j) \\
&\quad (\text{number of units of } x_j \text{ used}) \geq 8.
\end{aligned}$$

Finally, the **activity levels** (units of the various primary-input grades employed) x_1, x_2, and x_3 are all nonegative.

The information content of an optimal feasible solution to (1.3) can be characterized as follows. The **optimal primary-factor or labor-grade mix** is defined (from this result management can resolve the problem of which grades of labor to use in positive amounts and which ones not to employ) as well as the **optimal output activity levels** (the exact number of units of each labor grade utilized is indicated). Moreover, the **optimal output configuration** is decided (the solution reveals the amounts of each of the outputs produced) along with the set of **overproduction figures** (which give the amounts by which any of the production quotas are exceeded). Finally, the decision makers are provided with the **optimal dollar value of total cost.**

As was the case with (1.1), the primal problem (1.3) has associated with it a symmetric dual problem which deals with the assessment of the opportunity costs associated with fulfilling the firm's output quotas. These costs or, more properly, **marginal (imputed or shadow) costs,** are the dual variables which serve to inform the firm of the "potential" cost reduction resulting from a unit decrease in the ith (minimum) output requirement (since these production quotas obviously limit the firm's ability to reduce total cost by employing fewer inputs). Using (1.3), the dual problem has the form

$$\begin{aligned}
&max\, g = 8u_1 + 4u_2 + 2u_3 + 8u_4 \qquad s.t.\\
&\qquad \frac{1}{16}u_1 + \frac{1}{4}u_2 + \frac{1}{20}u_3 + \frac{1}{4}u_4 \le 8.50\\
&\qquad \frac{1}{14}u_1 + \frac{1}{4}u_2 + \frac{1}{25}u_3 + \frac{1}{3}u_4 \le 9.75\\
&\qquad \frac{1}{18}u_1 + \frac{1}{6}u_2 + \frac{1}{30}u_3 + \frac{1}{6}u_4 \le 9.08\\
&\qquad u_1, u_2, u_3, u_4 \ge 0,
\end{aligned} \tag{1.4}$$

where the dual variables $u_1, \ldots, u_4$ are the marginal (imputed) costs associated with the set of (minimum) output structural constraints.

What is the economic meaning of the form of this dual problem? First, the objective g represents **the total imputed cost of the firm's minimum output requirements,**

$$\begin{aligned}
&\text{total imputed cost of all output quotas}\\
&\quad = \sum_{i=1}^{4} (\text{total imputed cost of the } i\text{th output quota})\\
&\quad = \sum_{i=1}^{4} (i\text{th output quota})\,(\text{marginal cost of the } i\text{th output}).
\end{aligned}$$

Clearly the firm must make the value of this figure as large as possible, i.e. the firm seeks to maximize its *total potential cost reduction.* Next, upon examining the first **structural constraint** inequality in (1.4) (the other two are interpreted in a similar fashion) we see that the total imputed cost of the outputs produced by operating the jth activity at the unit level cannot exceed the total cost of all inputs per unit of the jth activity, i.e.

$$\begin{aligned}&\text{total imputed cost of all outputs per unit of } x_1\\&=\sum_{i=1}^{4}(\text{imputed value of the } i\text{th output per unit of } x_1)\\&=\sum_{i=1}^{4}(\text{number of units of the } i\text{th output per unit of } x_1)\\&(\text{marginal cost of the } i\text{th output}) \leq 8.50.\end{aligned}$$

Finally, the **marginal cost** figures $u_1, \ldots, u_4$ are all required to be nonnegative.

How may we interpret the data sets provided by an optimal feasible solution to this dual problem?

The set of **optimal imputed costs of the output quotas** is rendered, and from this information the firm can determine which production quotas are fulfilled and which ones are exceeded.

The **marginal (imputed) cost** configuration is determined. Each such figure reveals the potential cost reduction to the firm if the associated output quota is reduced by one unit.

Furthermore, the **optimal (imputed) value of outputs produced** for each primary factor grade is computed. Here we obtain data on the imputed cost of all outputs produced by each primary factor.

The **optimal accounting loss figures** are calculated. Here management is apprised of the amount by which the total of all resources per unit of activity j exceeds the total imputed cost of the outputs produced by running activity j at the unit level.

The **total imputed cost of all output requirements** is determined. Here, too, the optimal values of the primal and dual objectives are equal.

While it is important to obtain the information contained within an optimal solution to the primal and dual problems, additional sets of calculations that are essential for purposes of determining the *robustness* of, say, the optimal primal solution are subsumed under the heading of *postoptimality analysis*. For example, we can characterize the relevant types of postoptimality computations as follows:

a) *Sensitivity analysis* (Chapter 8) involves the introduction of discrete changes in any of the unit profit, input–output, or capacity values, i.e. these quantities are altered (increased or decreased) in order to determine the extent to which the original problem may be modified without violating the feasibility or optimality of the original solution

b) *Analyzing structural changes* (Chapter 9) determines the effect on an optimal basic feasible solution to a given linear programming problem of the addition or deletion of certain variables or structural constraints
c) *Parametric analysis* (Chapter 10) generates a sequence of basic solutions, which, in turn, become optimal, one after the other, as any or all of the unit profit coefficients or capacity restrictions or components of a particular activity vary continuously in some prescribed direction.

Once the reader has been exposed to parametric programming techniques, it is but a short step to the application of the same (see Chapter 11) in the derivation of the following:

- Supply function for the output of an activity
- Demand function for a variable input
- Marginal (net) revenue productivity function for a fixed input
- Marginal cost function for an activity
- Marginal and average productivity functions for a fixed input along with the marginal and average cost functions for the firm's output

Next, with reference to the cost minimization objective within the joint production model, we shall again employ the technique of parametric programming to derive the total, marginal, and average cost functions for a joint product. In addition, the supply function for the same is also developed.

In order to set the stage for the presentation of the theory and economic applications of linear programming, Chapter 2 discusses the rudiments of matrix algebra, the evaluation of determinants, elementary row operations, matrix inversion, vector algebra and vector spaces, simultaneous linear equation systems, linear dependence and rank, basic solutions, convex sets, and n-dimensional geometry and convex cones.

2

Mathematical Foundations

2.1 Matrix Algebra

We may define a **matrix** as an ordered set of elements arranged in a rectangular array of rows and columns. Thus a matrix $\boldsymbol{A}$ may be represented as

$$\boldsymbol{A} = \begin{bmatrix} a_{11} & a_{12} & \cdots & a_{1n} \\ a_{21} & a_{22} & \cdots & a_{2n} \\ \vdots & \vdots & \vdots & \vdots \\ a_{m1} & a_{m2} & \cdots & a_{mn} \end{bmatrix},$$

where a_{ij}, $i = 1, \ldots, m$; $j = 1, \ldots, n$, is the (representative) element in the ith row and jth column of $\boldsymbol{A}$. Since there are m rows and n columns in $\boldsymbol{A}$, the matrix is said to be "of order m by n" (denoted $(m \times n)$). When $m = n$, the matrix is square and will simply be referred to as being "an nth order" matrix. To economize on notation, we may represent $\boldsymbol{A}$ in the alternative fashion

$$\boldsymbol{A} = \left[a_{ij}\right], i = 1, \ldots, m; j = 1, \ldots, n.$$

Oftentimes we shall need to utilize the notion of a matrix within a matrix, i.e. a **submatrix** is the $(k \times s)$ matrix $\boldsymbol{B}$ obtained by deleting all but k rows and s columns of an $(m \times n)$ matrix $\boldsymbol{A}$.

Let us now examine some fundamental matrix operations. Specifically, the **sum** of two $(m \times n)$ matrices $\boldsymbol{A} = [a_{ij}]$, $\boldsymbol{B} = [b_{ij}]$ is the $(m \times n)$ matrix $\boldsymbol{A} + \boldsymbol{B} = \boldsymbol{C} = [c_{ij}]$, where $c_{ij} = a_{ij} + b_{ij}$, $i = 1, \ldots, m$; $j = 1, \ldots, n$, *i.e*, we add corresponding elements. Next, to multiply an $(m \times n)$ matrix $\boldsymbol{A}$ by a scalar λ we simply multiply each element of the matrix by the scalar or

$$\lambda \boldsymbol{A} = \left[\lambda a_{ij}\right], i = 1, \ldots, m; j = 1, \ldots, n.$$

Linear Programming and Resource Allocation Modeling, First Edition. Michael J. Panik.

(In view of these operations it is evident that $\boldsymbol{A} - \boldsymbol{B} = \boldsymbol{A} + (-1)\,\boldsymbol{B} = \boldsymbol{D} = [d_{ij}]$, $d_{ij} = a_{ij} - b_{ij}$.) The properties of scalar multiplication and matrix addition may be summarized as:

a) $\boldsymbol{A} + \boldsymbol{B} = \boldsymbol{B} + \boldsymbol{A}$ (**commutative law**);
b) $(\boldsymbol{A} + \boldsymbol{B}) + \boldsymbol{C} = \boldsymbol{A} + (\boldsymbol{B} + \boldsymbol{C})$ (**associative law**);
c) $\lambda\boldsymbol{A} = \boldsymbol{A}\lambda$;
d) $\lambda(\boldsymbol{A} + \boldsymbol{B}) = \lambda\boldsymbol{A} + \lambda\boldsymbol{B}$
e) $(\lambda_1 + \lambda_2) = \lambda_1\boldsymbol{A} + \lambda_2\boldsymbol{B}$ } (**distributive laws**);
f) $\boldsymbol{A} + \boldsymbol{O} = \boldsymbol{A}$, where the **null matrix** $\boldsymbol{O}$ (which contains only zero elements) is the **additive identity**.

The **transpose** of an $(m \times n)$ matrix $\boldsymbol{A}$, denoted $\boldsymbol{A}'$, is the $(n \times m)$ matrix obtained from $\boldsymbol{A}$ by interchanging its rows and columns, i.e. row i of $\boldsymbol{A}$ becomes column i of the transposed matrix. The essential properties of matrix transposition are:

a) $(\boldsymbol{A}')' = \boldsymbol{A}$;
b) $(\lambda_1\boldsymbol{A} + \lambda_2\boldsymbol{B})' = \lambda_1\boldsymbol{A}' + \lambda_2\boldsymbol{B}'$;
c) $(\boldsymbol{AB})' = \boldsymbol{B}'\boldsymbol{A}'$ (where the product of $\boldsymbol{A}$ and $\boldsymbol{B}$, $\boldsymbol{AB}$, is assumed to exist)

An nth order matrix $\boldsymbol{A}$ is said to be **symmetric** if it equals its transpose, i.e. $\boldsymbol{A} = \boldsymbol{A}'$ or $a_{ij} = a_{ji}$, $i \neq j$.

The **principal diagonal** of an nth order matrix $\boldsymbol{A}$ is the set of elements running from the upper left to the lower right corner of $\boldsymbol{A}$, namely $a_{11}, a_{22}, \ldots, a_{nn}$. A **diagonal matrix** of order n is one for which all elements off the principal diagonal are zero, i.e. $a_{ij} = 0$, $i \neq j$. An important special type of diagonal matrix is the **identity matrix** of order n, denoted

$$\boldsymbol{I}_n = \begin{bmatrix} 1 & 0 & \cdots & 0 \\ 0 & 1 & \cdots & 0 \\ \vdots & \vdots & \vdots & \vdots \\ 0 & 0 & \cdots & 1 \end{bmatrix}.$$

Thus $\boldsymbol{I}_n$ has ones along its principal diagonal and zeros elsewhere. A **triangular matrix** of order n is one for which all elements on one side of the principal diagonal are zero, e.g. $\boldsymbol{A}$ is **upper triangular** if $a_{ij} = 0$, $i > j$.

Given an $(m \times n)$ matrix $\boldsymbol{A}$ and an $(n \times p)$ matrix $\boldsymbol{B}$, the **product** $\boldsymbol{AB}$ is the $(m \times p)$ matrix $\boldsymbol{C}$ whose elements are computed from the elements of $\boldsymbol{A}$, $\boldsymbol{B}$ according to the rule

$$c_{ij} = \sum_{k=1}^{n} a_{ik}b_{kj}, i = 1, \ldots, m; j = 1, \ldots, p. \tag{2.1}$$

The product $\boldsymbol{AB}$ exists if and only if the matrices $\boldsymbol{A}, \boldsymbol{B}$ are **conformable for multiplication**, i.e. if and only if the number of columns in $\boldsymbol{A}$ equals the number of rows in $\boldsymbol{B}$. Moreover $\boldsymbol{AB} \neq \boldsymbol{BA}$ (generally), i.e. matrix multiplication is not commutative. Matrix multiplication possesses the following properties:

a) $(\boldsymbol{AB})\boldsymbol{C} = \boldsymbol{A}(\boldsymbol{BC})$ (**associative law**);
b) $\boldsymbol{A}(\boldsymbol{B}+\boldsymbol{C}) = \boldsymbol{AB}+\boldsymbol{AC}$ } (**distributive laws**);
c) $(\boldsymbol{A}+\boldsymbol{B})\boldsymbol{C} = \boldsymbol{AC}+\boldsymbol{BC}$ }
d) $\lambda(\boldsymbol{AB}) = \boldsymbol{A}(\lambda\boldsymbol{B}) = (\lambda\boldsymbol{A})\boldsymbol{B}$;
e) $\boldsymbol{I}_n\boldsymbol{A} = \boldsymbol{A}\boldsymbol{I}_n = \boldsymbol{A}$ ($\boldsymbol{I}_n$ is the **multiplicative identity**); and
f) $\boldsymbol{OA} = \boldsymbol{AO} = \boldsymbol{O}$.

Example 2.1 Given the matrices

$$\boldsymbol{A} = \begin{bmatrix} 1 & -2 & 3 \\ 4 & 1 & 0 \\ 1 & 5 & -1 \end{bmatrix}, \boldsymbol{B} = \begin{bmatrix} 2 & 0 & 1 \\ 3 & 1 & 1 \\ 2 & 0 & 6 \end{bmatrix},$$

it is easily show that:

$$\boldsymbol{A}+\boldsymbol{B} = \begin{bmatrix} 3 & -2 & 4 \\ 7 & 2 & 1 \\ 3 & 5 & -5 \end{bmatrix}; \boldsymbol{A}-\boldsymbol{B} = \begin{bmatrix} -1 & -2 & 2 \\ 1 & 0 & -1 \\ -1 & 5 & -2 \end{bmatrix}; 3\boldsymbol{A} = \begin{bmatrix} 3 & -6 & 9 \\ 12 & 3 & 0 \\ 3 & 15 & -3 \end{bmatrix};$$

and

$$\boldsymbol{AB} = \begin{bmatrix} 1(2)-2(3)+3(2) & 1(0)-2(1)+3(0) & 1(1)-2(1)+3(6) \\ 4(2)+1(3)+0(2) & 4(0)+1(1)+0(0) & 4(1)+1(1)+0(6) \\ 1(2)+5(3)-1(2) & 1(0)+5(1)-1(0) & 1(1)+5(1)-1(6) \end{bmatrix} = \begin{bmatrix} 2 & -2 & 17 \\ 11 & 1 & 5 \\ 15 & 5 & 0 \end{bmatrix} = \boldsymbol{C}.$$

That is, to obtain the first row of the product matrix $\boldsymbol{C}$, take the first row of the **premultiplier matrix** $\boldsymbol{A}$ and multiply it into each column of the **postmultiplier matrix** $\boldsymbol{B}$, where multiplication is executed in a pairwise fashion and the relevant products of elements are summed (see (2.1). Next, to obtain the second row of $\boldsymbol{C}$, take the second row of $\boldsymbol{A}$ and multiply it into each column of $\boldsymbol{B}$, again by multiplying appropriate pairs of elements and summing. Finally, to get the third row of $\boldsymbol{C}$, the third row of $\boldsymbol{A}$ is multiplied into each column of $\boldsymbol{B}$ in "the usual fashion." (As we shall see later on, we are actually determining what is called the set of "inner products" between the rows of $\boldsymbol{A}$ and the appropriate columns of $\boldsymbol{B}$ in order to get the elements of $\boldsymbol{C}$.)

Next $B' = \begin{bmatrix} 2 & 3 & 2 \\ 0 & 1 & 0 \\ 1 & 1 & 6 \end{bmatrix}$; $D = \begin{bmatrix} -2 & 3 \\ 1 & 0 \end{bmatrix}$ is a (2×2) submatrix of A; and

$E = \begin{bmatrix} 3 & -1 & 4 \\ 0 & 2 & 1 \\ 0 & 0 & 4 \end{bmatrix}$ is an upper-triangular matrix. ■

In what follows we shall introduce an extremely versatile computational device for performing a wide assortment of matrix calculations. To this end, an **elementary row operation** may be performed on a matrix A by:

a) interchanging any two rows of A; [TYPE I]
b) multiplying a row of A by a scalar $\lambda \neq 0$; [TYPE II]
c) adding to the ith row of A a scalar λ times any other row. [TYPE III]

For instance, given

$$A = \begin{bmatrix} 3 & 1 & 4 \\ 2 & 5 & 7 \\ 1 & 0 & 1 \end{bmatrix},$$

let us transform this matrix (where transformation is indicated by an arrow with an appropriate Roman numeral beneath it, e.g. by, say, "$\underset{I}{\rightarrow}$") into the following matrices:

$$A = \begin{bmatrix} 3 & 1 & 4 \\ 2 & 5 & 7 \\ 1 & 0 & 1 \end{bmatrix} \underset{I}{\rightarrow} \begin{bmatrix} 2 & 5 & 7 \\ 3 & 1 & 4 \\ 1 & 0 & 1 \end{bmatrix} = B \text{ (rows one and two are interchanged);}$$

$$A = \begin{bmatrix} 3 & 1 & 4 \\ 2 & 5 & 7 \\ 1 & 0 & 1 \end{bmatrix} \underset{II}{\rightarrow} \begin{bmatrix} 3 & 1 & 4 \\ 2 & 5 & 7 \\ 6 & 0 & 6 \end{bmatrix} = C \text{ (row three is multiplied by } \lambda = 6\text{);}$$

$$A = \begin{bmatrix} 3 & 1 & 4 \\ 2 & 5 & 7 \\ 1 & 0 & 1 \end{bmatrix} \underset{III}{\rightarrow} \begin{bmatrix} 3 & 1 & 4 \\ 8 & 7 & 15 \\ 1 & 0 & 1 \end{bmatrix} = D \text{ (} \lambda = 2 \text{ times row one is added to row two).}$$

Oftentimes we shall be faced with computing the determinant of a matrix A, denoted $|A|$. In general, a **determinant** is a scalar-valued function D defined on the set of all square matrices $\mathcal{M}$. So if the nth order matrix $A \in \mathcal{M}$ then $D(A)$ is the scalar $|A|$. The determinant of a (2×2) matrix is defined as follows:

$$\begin{vmatrix} a_{11} & a_{12} \\ a_{21} & a_{22} \end{vmatrix} = a_{11}a_{22} - a_{12}a_{21}.$$

Example 2.2 Given the matrices

$$\boldsymbol{A} = \begin{bmatrix} 2 & 3 \\ 5 & 1 \end{bmatrix}, \boldsymbol{B} = \begin{bmatrix} 1 & -1 \\ 2 & 4 \end{bmatrix}, \text{ and } \boldsymbol{C} = \begin{bmatrix} 3 & -2 \\ -4 & 7 \end{bmatrix},$$

it is easily shown that $|\boldsymbol{A}| = -13$, $|\boldsymbol{B}| = 6$, and $|C| = 13$. ■

Before examining a computational routine for evaluating determinants of larger matrices, let us note some of their important properties:

a) Interchanging any two rows or columns of a matrix $\boldsymbol{A}$ changes the sign of $|\boldsymbol{A}|$.
b) If the matrix $\boldsymbol{A}$ has two rows or columns which are identical, $|\boldsymbol{A}| = 0$.
c) If a row or column of a matrix $\boldsymbol{A}$ has all zero elements, $|\boldsymbol{A}| = 0$.
d) If every element of a row or column of a matrix $\boldsymbol{A}$ is multiplied by a nonzero scalar λ to yield a new matrix $\boldsymbol{B}$, $|\boldsymbol{B}| = \lambda|\boldsymbol{A}|$.
e) If every element of an nth order matrix $\boldsymbol{A}$ is multiplied by a nonzero scalar λ to yield a new matrix

$$\boldsymbol{B} = \lambda\boldsymbol{A} = \begin{bmatrix} \lambda a_{11} & \cdots & \lambda a_{1n} \\ \vdots & & \vdots \\ \lambda a_{n1} & \cdots & \lambda a_{nn} \end{bmatrix},$$

then $|\boldsymbol{B}| = |\lambda\boldsymbol{A}| = \lambda^n|\boldsymbol{A}|$.
f) If a row or column for a matrix $\boldsymbol{A}$ is a multiple of any other row or column, $|\boldsymbol{A}| = 0$.
g) For an nth order matrix $\boldsymbol{A}$, $|\boldsymbol{A}| = |\boldsymbol{A}'|$.
h) If A is a diagonal or triangular matrix, $|\boldsymbol{A}| = a_{11}, a_{22}, \ldots, a_{nn}$.
i) $|\boldsymbol{I}_n| = 1$, $|\boldsymbol{O}| = 0$.
j) If $\boldsymbol{A}$, $\boldsymbol{B}$ are of order n so that $\boldsymbol{AB}$ exists, $|\boldsymbol{AB}| = |\boldsymbol{A}| \cdot |\boldsymbol{B}|$.
k) If $\boldsymbol{A}$ is $(m \times n)$ and $\boldsymbol{B}$ is $(n \times m)$, $|\boldsymbol{AB}| = 0$ if $m > n$.
l) In general, $|\boldsymbol{A} + \boldsymbol{B}| \neq |\boldsymbol{A}| + |\boldsymbol{B}|$.
m) If $\boldsymbol{A}$, $\boldsymbol{B}$ are of order n, $|\boldsymbol{AB}| = |\boldsymbol{A}'\boldsymbol{B}| = |\boldsymbol{AB}'| = |\boldsymbol{A}'\boldsymbol{B}'|$.

Given an nth order matrix $\boldsymbol{A}$, how do these elementary row operations affect $|\boldsymbol{A}|$? The answer may be indicated as:

a) TYPE I operation – see property (a) above.
b) TYPE II operation – see property (d) above.

c) TYPE III operation – if in $\boldsymbol{A}$ we add any nonzero multiple of one row to a different row to obtain a new matrix $\boldsymbol{B}$, $|\boldsymbol{B}| = |\boldsymbol{A}|$.

Thus, only TYPE III elementary row operations leave $|\boldsymbol{A}|$ invariant. To evaluate the determinant of an nth order matrix $\boldsymbol{A}$ we now state, in terms of elementary row operations, the **SWEEP-OUT PROCESS.** If $|\boldsymbol{A}|$ can be transformed, by elementary row operations, into the product of a constant k and the determinant of an upper-triangular matrix $\boldsymbol{B}$,

$$|\boldsymbol{A}| = k|\boldsymbol{B}| = k\begin{vmatrix} b_{11} & b_{12} & \cdots & b_{1n} \\ 0 & b_{22} & \cdots & b_{2n} \\ 0 & 0 & \cdots & b_{3n} \\ \vdots & \vdots & & \vdots \\ 0 & 0 & \cdots & b_{nn} \end{vmatrix},$$

then $|\boldsymbol{A}| = kb_{11}b_{22}...b_{nn}$. See property (h) above, where k is chosen to compensate for the cumulative effects on $|\boldsymbol{A}|$ of successive TYPE I, II elementary row operations.

Example 2.3 Find $|\boldsymbol{A}|$ when

$$\boldsymbol{A} = \begin{bmatrix} 1 & 2 & 0 & 1 \\ 1 & 3 & 4 & 2 \\ 0 & 2 & 1 & 2 \\ 4 & 1 & 1 & 0 \end{bmatrix}.$$

$$|\boldsymbol{A}| = \begin{vmatrix} 1 & 2 & 0 & 1 \\ 1 & 3 & 4 & 2 \\ 0 & 2 & 1 & 2 \\ 4 & 1 & 1 & 0 \end{vmatrix} = \begin{vmatrix} 1 & 2 & 0 & 1 \\ 0 & 1 & 4 & 1 \\ 0 & 2 & 1 & 2 \\ 0 & -7 & 1 & -4 \end{vmatrix} = \begin{vmatrix} 1 & 2 & 0 & 1 \\ 0 & 1 & 4 & 1 \\ 0 & 0 & -7 & 0 \\ 0 & 0 & 29 & 3 \end{vmatrix} = -7\begin{vmatrix} 1 & 2 & 0 & 1 \\ 0 & 1 & 4 & 1 \\ 0 & 0 & 1 & 0 \\ 0 & 0 & 29 & 3 \end{vmatrix}$$

$$= -7\begin{vmatrix} 1 & 2 & 0 & 1 \\ 0 & 1 & 4 & 1 \\ 0 & 0 & 1 & 0 \\ 0 & 0 & 0 & 3 \end{vmatrix} = -7(1\cdot 1\cdot 1\cdot 3) = k|\boldsymbol{B}| = -21.$$

■

In what follows, we shall frequently need to determine the reciprocal of an nth order matrix $\boldsymbol{A}$. In this regard, if there exists an nth order matrix $\boldsymbol{A}^{-1}$ that satisfies the relation $\boldsymbol{A}\boldsymbol{A}^{-1} = \boldsymbol{A}^{-1}\boldsymbol{A} = \boldsymbol{I}_n$, then $\boldsymbol{A}^{-1}$ is termed the **inverse** of $\boldsymbol{A}$. Under what circumstances will $\boldsymbol{A}^{-1}$ exist? To answer this question, let us first consider some terminology. Specifically, an nth order matrix $\boldsymbol{A}$ is said to be *singular* if $|\boldsymbol{A}| = 0$ and **nonsingular** if $|\boldsymbol{A}| \neq 0$. In this regard, $\boldsymbol{A}^{-1}$ exists if and only if $\boldsymbol{A}$ is nonsingular. Moreover, every nonsingular matrix has an inverse; and if $\boldsymbol{A}^{-1}$ exists, it is unique. The essential properties of matrix inversion are:

a) $\left(\boldsymbol{A}^{-1}\right)^{-1} = \boldsymbol{A}$;
b) $\left|\boldsymbol{A}^{-1}\right| = 1/|\boldsymbol{A}|$;
c) $(\boldsymbol{A}')^{-1} = (\boldsymbol{A}^{-1})'$; and
d) given that $\boldsymbol{A}$, $\boldsymbol{B}$ are nonsingular and $\boldsymbol{AB}$ exists, $(\boldsymbol{AB})^{-1} = \boldsymbol{B}^{-1}\boldsymbol{A}^{-1}$.

From a computational viewpoint, how do we actually obtain $\boldsymbol{A}^{-1}$? Again relying on elementary row operations, let us initially form the $(n \times n + n)$ partitioned matrix $[\boldsymbol{A} \vdots \boldsymbol{I}_n]$.Then if a sequence of elementary row operations applied to $[\boldsymbol{A} \vdots \boldsymbol{I}_n]$ reduces $\boldsymbol{A}$ to $\boldsymbol{I}_n$, that same sequence of operation transforms $\boldsymbol{I}_n$ to $\boldsymbol{A}^{-1}$.

Example 2.4 Find $\boldsymbol{A}^{-1}$ given

$$\boldsymbol{A} = \begin{bmatrix} 2 & 5 & 1 \\ 3 & 0 & 4 \\ 2 & 8 & 3 \end{bmatrix}.$$

As a first step, let us check to see if $\boldsymbol{A}$ is nonsingular. Since $|\boldsymbol{A}| = -45 \neq 0$, $\boldsymbol{A}^{-1}$ exists. Then

$$\left[\boldsymbol{A} \vdots \boldsymbol{I}_3\right] = \begin{bmatrix} 2 & 5 & 1 & \vdots & 1 & 0 & 0 \\ 3 & 0 & 4 & \vdots & 0 & 1 & 0 \\ 2 & 8 & 3 & \vdots & 0 & 0 & 1 \end{bmatrix} \rightarrow \begin{bmatrix} 1 & 5/2 & 1/2 & \vdots & 1/2 & 0 & 0 \\ 0 & -15/2 & 5/2 & \vdots & -3/2 & 1 & 0 \\ 0 & 3 & 2 & \vdots & -1 & 0 & 1 \end{bmatrix} \rightarrow$$

$$\begin{bmatrix} 1 & 0 & 4/3 & \vdots & 0 & 1/3 & 0 \\ 0 & 1 & -1/3 & \vdots & 1/5 & -2/15 & 0 \\ 0 & 0 & 3 & \vdots & -8/5 & 2/5 & 1 \end{bmatrix} \rightarrow \begin{bmatrix} 1 & 0 & 0 & \vdots & 32/45 & 7/45 & -4/9 \\ 0 & 1 & 0 & \vdots & 1/45 & -4/45 & 1/9 \\ 0 & 0 & 1 & \vdots & -8/45 & 2/15 & 1/3 \end{bmatrix} = \left[\boldsymbol{I}_3 \vdots \boldsymbol{A}^{-1}\right].$$

As a check on whether A^{-1} has been computed correctly, it is easily verified that $AA^{-1} = I_3$. ■

2.2 Vector Algebra

A **vector** is an ordered n-type of (real) numbers expressed as a column ($n \times 1$) or row ($1 \times n$) matrix

$$X = \begin{bmatrix} x_1 \\ x_2 \\ \vdots \\ x_n \end{bmatrix}, X' = (x_1, x_2, \ldots, x_n)$$

respectively. Here the x_i's, $i = 1, \ldots, n$, are called the **components** of the vector. Geometrically, we may think of a vector as a point in n-dimensional coordinate space. To fully understand the rudiments of vector algebra, it is important to identify the properties of the space within which the vectors are defined. In this regard, a **vector space** $\mathcal{V}_n$ is a collection of n-components vectors that is closed under the operations of addition and scalar multiplication, i.e. if $X_1, X_2 \in \mathcal{V}_n$, then $X_1 + X_2 \in \mathcal{V}_n, \lambda X_1 \in \mathcal{V}_n$, where λ is a real scalar. These operations of addition and scalar multiplication obey the following rules:

a) $X_1 + X_2 = X_2 + X_1$ (**commutative law**).
b) $X_1 + (X_2 + X_3) = (X_1 + X_2) + X_3$ (**associative law**).
c) There is a unique element $O \in \mathcal{V}_n$ such that $X + O = X$ for each $X \in \mathcal{V}_n$ (the **null vector** O is the **additive identity**).
d) For each $X \in \mathcal{V}_n$ there is a unique element $-X \in \mathcal{V}_n$ such that $X + (-X) = O$ ($-X$ is the **additive inverse**).
e) $\lambda(X_1 + X_2) = \lambda X_1 + \lambda X_2$
f) $(\lambda_1 + \lambda_2)X = \lambda_1 X + \lambda_2 X$ (e, f: **distributive laws**).
g) $(\lambda_1 \lambda_2)X = \lambda_1(\lambda_2 X)$ (**associative law**).
h) $1X = X$ for every $X \in \mathcal{V}_n$ (1 is the **multiplicative identity**).

For vectors $X, Y \in \mathcal{V}_n$, the **scalar (inner) product** of X, Y is the scalar

$$X'Y = \sum_{i=1}^{n} x_i y_i$$

Here, x_i is the ith component of X and y_i is the ith component of (Y, $i = 1, \ldots, n$), where

a) $X'X \geq 0$ and $X'X = 0$ if and only if $X = O$;
b) $X'Y = Y'X$ (**commutative law**);
c) $(X' + Y')Z = X'Z + Y'Z$(**distributive law**); and
d) $\lambda(X'Y) = (\lambda X')Y$.

We note briefly that the vectors $X, Y \in \mathcal{V}_n$, with $X, Y \neq O$, are **orthogonal** (mutually perpendicular) if $X'Y = 0$.

Let us next define on $\mathcal{V}_n$ the concept of "distance." In particular, the distance between a vector X and the origin O is called the **norm** of X; it is a function that assigns to each $X \in \mathcal{V}_n$ some number

$$\|X\| = (X'X)^{1/2} = \left[\sum_{i=1}^{n} |x_i|^2\right]^{1/2}$$

such that:

a) $\|X\| \geq 0$ and $\|X\| = 0$ if and only if $X = O$;
b) if $X, Y \in \mathcal{V}_n$, $\|X + Y\| \leq \|X\| + \|Y\|$ (**the triangular inequality**);
c) $\|\lambda X\| = |\lambda| \|X\|$ (**homogeneity**); and
d) $|X'Y| \leq \|X\| \|Y\|$ (**the Cauchy-Schwarz inequality**).

In addition, for $X, Y \in \mathcal{V}_n$, the **distance between** X, Y is

$$\|X - Y\| = \left[(X - Y)'(X - Y)\right]^{1/2} = \left[\sum_{i=1}^{n} |x_i - y_i|^2\right]^{1/2}.$$

Once the concept of a norm is defined on $\mathcal{V}_n$, we obtain what is called **n-dimensional Euclidean space, $\mathcal{E}^n$**.

Quite often, two specialized types of vectors will be used in our calculations. First, the **sum vector 1** is an n-component vector whose elements are all 1's. The inner product between **1** and any vector X is the sum of the elements within X or $1'X = \sum_{i=1}^{n} x_i$. Next, the ith **unit column vector** e_i, $i = 1, \ldots, n$, is an n-component vector with a 1 as the ith component and zeros elsewhere. For any $(n \times 1)$ vector X, $e_1'X = x_i$, the ith component of X. And for any $(n \times n)$ matrix A, $Ae_i = a_i$, the ith column of A; while $e_i'A = \alpha_i$, the ith row of A, $i = 1, \ldots, n$. Note also that e_i is the ith column of I_n, i.e. $I_n = [e_1, e_2, \ldots, e_n]$, $i = 1, \ldots, n$.

Example 2.5 For the vectors

$$X = \begin{bmatrix} 3 \\ -1 \\ 2 \end{bmatrix}, Y = \begin{bmatrix} 1 \\ 4 \\ 1 \end{bmatrix},$$

it is easily shown that

$X'Y = 3 + (-4) + 2 = 1$ (these vectors are not orthogonal);

$\|X\| = (9 + 1 + 4)^{1/2} = \sqrt{14}$; $\|Y\| = (1 + 16 + 1)^{1/2} = \sqrt{18}$;

$\|X - Y\| = (4 + 25 + 1)^{1/2} = \sqrt{30}$; and the sum of the components within X is $1'X = 3 + (-1) + 2 = 4$. ■

2.3 Simultaneous Linear Equation Systems

The **simultaneous system of *n* linear equations in *n* unknowns**

$$a_{11}x_1 + a_{12}x_2 + \cdots + a_{1n}x_n = c_1$$
$$a_{21}x_1 + a_{22}x_2 + \cdots + a_{2n}x_n = c_2$$
$$\cdots\cdots\cdots\cdots\cdots\cdots\cdots\cdots$$
$$a_{n1}x_1 + a_{n2}x_2 + \cdots + a_{nn}x_n = c_n$$

may be written in matrix form as

$$\begin{bmatrix} a_{11} & a_{12} & \cdots & a_{1n} \\ a_{21} & a_{22} & \cdots & a_{21} \\ \vdots & \vdots & & \vdots \\ a_{n1} & a_{n2} & \cdots & a_{nn} \end{bmatrix} \begin{bmatrix} x_1 \\ x_2 \\ \vdots \\ x_n \end{bmatrix} = \begin{bmatrix} c_1 \\ c_2 \\ \vdots \\ c_n \end{bmatrix} \quad \text{or } AX = C, \tag{2.2}$$

where A is an $(n \times n)$ matrix of (constant) coefficients a_{ij}, X is an $(n \times 1)$ matrix of (unknown) variables x_i, and C is an $(n \times 1)$ matrix of constants c_i, $i, j = 1, \ldots, n$. Under what conditions will a solution to this system exist? Moreover, provided that at least one solution does exist, how may one go about finding it? To answer these questions, let us first develop some terminology. A system of equations is **consistent** if it has at least one solution; it is **inconsistent** if it does not possess a solution. Additionally, two systems of linear equations are **equivalent** if every particular solution of either one is also a solution of the other. Next, the **rank** of an $(n \times n)$ matrix $A, \rho(A)$, is the order of the largest nonsingular submatrix of A. In this light, an nth order matrix A is said to be of **full rank** if $\rho(A) = n$, i.e. $|A| \neq 0$. Here are the important properties of the rank of a matrix:

a) Given that A is of order $(m \times n)$, $\rho(A) \leq min\ \{m,n\}$.
b) $\rho(I_n) = n, \rho(0) = 0$.

c) $\rho(A') = \rho(A)$.
d) If A is a diagonal matrix, $\rho(A)$ = number of nonzero elements on the principal diagonal of A.
e) If A, B are both of order $(m \times n)$, $\rho(A + B) \le \rho(A) + \rho(B)$.
f) If AB exists, $\rho(AB) \le \min\{\rho(A), \rho(B)\}$.
g) If A is of order n, $\rho(A) = n$ if and only if A is nonsingular; $\rho(A) < n$ if and only if A is singular.
h) If a sequence of elementary row operations transforms a matrix A into a new matrix B, $\rho(A) = \rho(B)$.

In what follows, the discussion on the existence of a solution to (2.2) will be developed in terms of its **augmented matrix** $[A \vdots C]$. To this end, let us examine Theorem 2.1.

Theorem 2.1 Given the system $AX = C$, where A is of order $(m \times n)$, if

1) $\rho[A \vdots C] > \rho(A)$, the system is inconsistent;
2) $\rho[A \vdots C] = \rho(A)$ = number of unknowns n, the system is consistent and possesses a unique solution; and
3) $\rho[A \vdots C] = \rho(A) = k <$ number of unknowns n, the system is consistent and possesses an infinity of solutions, where arbitrary values may be assigned to $n - k$ of the variables.

To apply this theorem in actual practice, let us work with a highly specialized form of $[A \vdots C]$. Specifically, an $(n \times n)$ matrix E (obtained from a matrix of the same order by a series of elementary row operations) is an **echelon matrix** if:

a) the first k rows $(k \ge 0)$ are nonzero while the last $n - k$ rows contain only zero elements;
b) the first nonzero element in the ith row $(i = 1, \ldots, k, k \ge 1)$ equals one;
c) c_1 denotes the column in which the unity element appears, we required that $c_1 < c_2 < \cdots < c_k$; lastly, (a), (b), and (c) together imply that
d) the lower triangle of elements e_{ij}, $i > j$, are all zero, e.g. an echelon matrix may typically appear as

$$E = \begin{bmatrix} 1 & e_{12} & e_{13} & e_{14} & e_{15} \\ 0 & 1 & e_{23} & e_{24} & e_{25} \\ 0 & 0 & 1 & e_{34} & e_{35} \\ 0 & 0 & 0 & 1 & e_{45} \end{bmatrix}.$$

One convenient application of the concept of an echelon matrix is that it enables us to formulate an alternative specification of the rank of a matrix, i.e. the **rank of an $(m \times n)$ matrix A** is the number of nonzero rows in the echelon form of A.

Let us now look to the specification of a computational routine that will directly utilize the preceding theorem. To this end, we have the *Gauss elimination technique*. If the augmented matrix $[A\vdots C]$ can be transformed, by a series of elementary row operations, into an echelon matrix E, the system of equations corresponding to E is equivalent to that represented by $[A\vdots C]$. Hence, any solution of the system associated with E is a solution of the system associated with $[A\vdots C]$ and conversely.

Example 2.6 Determine (in two separate ways) the rank of the matrix

$$A = \begin{bmatrix} 3 & 1 & 4 & 2 \\ 2 & 0 & 1 & 0 \\ -1 & 2 & 2 & 4 \\ 5 & 1 & 0 & 2 \end{bmatrix}.$$

1) Since $|A| = 0, \rho(A) < 4$. Can we find at least one third-order nonsingular submatrix within A? The answer is yes, since

$$\begin{vmatrix} 3 & 1 & 4 \\ 2 & 0 & 1 \\ -1 & 2 & 2 \end{vmatrix} = 5 \neq 0.$$

Thus, $\rho(A) = 3$.

2) Since the echelon form of A is

$$E = \begin{bmatrix} 1 & 1/3 & 4/3 & 2/3 \\ 0 & 1 & 5/2 & 2 \\ 0 & 0 & 1 & 0 \\ 0 & 0 & 0 & 0 \end{bmatrix}$$

and E has three nonzero rows, it follows that $\rho(A) = 3$. ■

Example 2.7 Using the Gauss elimination technique, determine if the following systems possess a solution.

(a)

$$\begin{aligned} 4x_1 + x_2 + 3x_3 &= 5 \\ x_1 + x_2 - x_3 &= 8 \\ 2x_1 - x_2 &= 6 \end{aligned}$$

(b)

$$\begin{aligned} 2x_1 + 5x_2 &= 9 \\ \frac{16}{5}x_1 + 8x_2 &= 6 \end{aligned}$$

(c)

$$\begin{aligned} x_1 - x_2 + x_3 &= 4 \\ x_1 - 3x_2 - x_3 &= 8 \end{aligned}$$

a) Given that

$$\left[A\vdots C\right] \rightarrow \boldsymbol{E} = \begin{bmatrix} 1 & 1 & -1 & 8 \\ 0 & 1 & -7/3 & 9 \\ 0 & 0 & 1 & -17/5 \end{bmatrix},$$

it is easily seen that $\rho[A\vdots C] = \rho(A) = 3$ = number of unknowns (here $\rho(A)$ also corresponds to the number of nonzero rows in the submatrix)

$$\begin{bmatrix} 1 & 1 & -1 \\ 0 & 1 & -7/3 \\ 0 & 0 & 1 \end{bmatrix}$$

i.e. this submatrix represents the echelon form of $\boldsymbol{A}$) and thus a unique solution to the original system emerges. In fact, the said solution can readily be obtained by the process **of back-substitution**. Thus, from the last row of $\boldsymbol{E}$, we obtain $x_3 = -17/5$. Given x_3, row two of $\boldsymbol{E}$ renders $x_2 = 16/15$. Finally, given x_2 and x_3, row one yields $x_1 = 53/15$.

b) Since

$$\left[A\vdots C\right] \rightarrow \boldsymbol{E} = \begin{bmatrix} 1 & 5/2 & 9/2 \\ 0 & 0 & -42/5 \end{bmatrix},$$

it follows that $\rho[A\vdots C] = 2 > \rho(A) = 1$. Hence the original system is inconsistent and thus does not possess a solution. (In fact, the inconsistency is readily apparent from row two.)

c) For this equation system

$$\left[A\vdots C\right] \rightarrow \boldsymbol{E} = \begin{bmatrix} 1 & -1 & 1 & 4 \\ 0 & 1 & 1 & 2 \end{bmatrix}$$

and thus $\rho[A\vdots C] = \rho(A) = 2 < 3$ = number of unknowns. Hence, the original system possesses an infinity of particular solutions, where arbitrary values may be assigned to $n - k = 3 - 2 = 1$ of the variables. For instance, x_1 and x_2 may each be expressed in terms of x_3, where x_3 is deemed arbitrary, i.e. from row two of $\boldsymbol{E}$, $x_2 = 2 - x_3$; and from row one of the same $x_1 = 6 - 2x_3$. ■

The system of equations represented by (c) above has more variables than equations. A system such as this is called **underdetermined** and, as we have just seen, if a solution exists, it will not be unique. One method of generating **determinate solutions** to underdetermined systems of the form $AX = C$, where A is of order $(m \times n)$ and $m < n$, is to set $n - m$ of the variables equal to zero and then obtain a unique solution to the resulting system involving m equations in m unknowns. Returning to system (c), i.e. to

$$\begin{aligned} x_1 - x_2 + x_3 &= 4 \\ x_1 - 3x_2 - x_3 &= 8 \end{aligned}$$

it is readily seen that the three possible determinate solutions emerging via this technique are:

1) for $x_1 = 0$, it follows that $x_2 = -3$, $x_3 = 1$;
2) for $x_2 = 0$, we obtain $x_1 = 6$, $x_3 = -2$; and
3) for $x_3 = 0$, we get $x_1 = 2$, $x_2 = -2$.

At times we may be required to solve a linear simultaneous equation system of the form $AX = O$. Such a system is termed **homogeneous**. By inspection it is easily seen that this system can never be inconsistent since $\rho[A \vdots O] = \rho(A)$. Moreover, $X = O$ is always a (trivial) solution. What about the existence of nontrivial solutions? The answer to this question is provided by Theorem 2.2.

Theorem 2.2 Given the system $AX = O$, where A is of order $(m \times n)$, if

1) $\rho(A)$ = number of unknowns = n, the system has a unique (trivial) solution $X = O$;
2) $\rho(A) = k <$ number of unknowns = n, the system has an infinity of nontrivial solutions, where arbitrary values may be assigned to $n - k$ of the variables.

2.4 Linear Dependence

The vector $X \in \mathcal{E}^m$ is a **linear combination** of the vectors $X_j \in \mathcal{E}^m$, $j = 1, \ldots, n$, if there exist scalars λ_j such that

$$\sum_{j=1}^{n} \lambda_j X_j = X \tag{2.3}$$

or, for X_j the jth column of the $(m \times n)$ matrix A and λ_j the jth component of the $(n \times 1)$ vector λ, (2.3) may be rewritten as

$$A\lambda = X. \tag{2.3.1}$$

So to express X as a linear combination of the X_j's, we need only solve the simultaneous linear equation system (2.3.1) for the λ_j's. Next, a set of vectors

$\{X_j \in \mathcal{E}^m, j = 1, \ldots, n\}$ is **linearly dependent** if there exist scalars $\lambda_j, j = 1, \ldots, n$, not all zero such that

$$\sum_{j=1}^{n} \lambda_j X_j = O, \tag{2.4}$$

i.e. the null vector is a linear combination of the X_j's. If the only set of scalars λ_j for which (2.4) holds is $\lambda_j = 0, j = 1, \ldots, n$, the vectors X_j are **linearly independent**. Here the trivial combination $0X_1 + \cdots + 0X_n$ is the only linear combination of X_j's which equals the null vector. If A and λ are defined as above, then (2.4) may be rewritten as

$$A\lambda = O, \tag{2.4.1}$$

a **homogeneous equation system.** A test procedure for determining whether the set of vectors $\{X_j \in \mathcal{E}^m, j = 1, \ldots, n\}$ is linearly dependent can be outlined by applying the preceding theorem for homogeneous equation systems. That is:

1) if $m \geq n$ and $\rho(A) = n$, the set of vectors $\{X_j \in \mathcal{E}^m, j = 1, \ldots, n\}$ is linearly independent; but
2) if $m \geq n$ and $\rho(A) < n$, or if $m < n$, the set of vectors $\{X_j \in \mathcal{E}^m, j = 1, \ldots, n\}$ is linearly dependent.

A set of vectors $\{X_j \in \mathcal{E}^m, j = 1, \ldots, n\}$ is a **spanning set** for $\mathcal{E}^m$ if every vector $X \in \mathcal{E}^m$ can be written as a linear combination of the vectors X_j, i.e. if $X \in \mathcal{E}^m$, then $\sum_{j=1}^{n} \lambda_j X_j = X$, or $A\lambda = X$, where X_j is the jth column of the $(m \times n)$ matrix A and λ_j is the jth component of the $(n \times 1)$ vector λ, $j = 1, \ldots, n$. Here the set $\{X_j \in \mathcal{E}^m, j = 1, \ldots, n\}$ is said to **span** or **generate** $\mathcal{E}^m$ since every vector in the space is (uniquely) linearly dependent on the spanning set. We may note further that:

a) The vectors that span $\mathcal{E}^m$ need not be linearly independent; but
b) any set of vectors spanning $\mathcal{E}^m$ which contains the smallest possible number of vectors is linearly independent.

A **basis** for $\mathcal{E}^m$ is a linearly independent subset of vectors from $\mathcal{E}^m$ which spans $\mathcal{E}^m$. A basis for $\mathcal{E}^m$ exhibits the following characteristics:

a) A basis for $\mathcal{E}^m$ is not unique; but the vectors in the basis are.
b) Every basis for $\mathcal{E}^m$ contains the same number of basis vectors; and there are precisely m vectors in every basis for $\mathcal{E}^m$.
c) Any set of m linearly independent vectors from $\mathcal{E}^m$ forms a basis for $\mathcal{E}^m$.
d) Any set of $m + 1$ vectors from $\mathcal{E}^m$ is linearly dependent.

This discussion leads us to the specification of an alternative way to interpret the concept of the rank of an $(m \times n)$ matrix A, i.e. $\rho(A)$ is the maximum number of linearly independent vectors $\{X_j \in \mathcal{E}^m, j = 1, \ldots, n\}$, which span the columns of A; it is the number of vectors in a basis for $\mathcal{E}^m$.

We noted at the end of the preceding section that one way to generate a determinate solution to an underdetermined equation system $AX = C$, A and $(m \times n)$ coefficient matrix with $m < n$, is to set $n - m$ variables equal to zero and ultimately obtain a unique solution to the resulting system of m equations in m unknowns. If the columns of the resulting $(m \times m)$ coefficient matrix are linearly independent, i.e. form a basis for $\mathcal{E}^m$, then the particular determinate solution emerging from this process is called a **basic solution**, a concept that will be explored further in the next chapter.

Finally, the **dimension** of a vector space $\mathcal{V}_n$, dim $(\mathcal{V}_n)$, is the maximum number of linearly independent vectors that span the space, i.e. it is the number of vectors in a basis for $\mathcal{V}_n$. Thus, dim $(\mathcal{E}^n) = n$, where n is to be interpreted as the number of vectors in a basis for $\mathcal{E}^n$.

Example 2.8 Find scalars λ_1, λ_2 such that $X' = (5, 3)$ is a linear combination of $X_1' = (3,4), X_2' = (-6,8)$. From (2.3) we obtain the system

$$3\lambda_1 - 6\lambda_2 = 5$$

$$4\lambda_1 + 8\lambda_2 = 3$$

with solution $\lambda_1 = 29/24$, $\lambda_2 = -11/48$. Thus $\frac{29}{24}X_1 - \frac{11}{48}X_2 = X$. In addition, with $\lambda_1, \lambda_2 \neq 0$, the set vectors $\{X, X_1, X_2\}$ is linearly dependent. Next, let us determine whether the vectors

$$X_1 = \begin{bmatrix} 1 \\ 2 \\ 1 \end{bmatrix}, X_2 = \begin{bmatrix} 4 \\ 3 \\ 8 \end{bmatrix}, \text{and } X_3 = \begin{bmatrix} 0 \\ -1 \\ 4 \end{bmatrix}$$

provide a basis for $\mathcal{E}^3$. To this end, we must check to see if these vectors are linearly independent. Using (2.4.1) we may construct the homogeneous system

$$\begin{aligned} \lambda_1 + 4\lambda_2 \qquad &= 0 \\ 2\lambda_1 + 3\lambda_2 - \lambda_3 &= 0 \\ \lambda_1 + 8\lambda_2 + 4\lambda_3 &= 0 \end{aligned}$$

and subsequently conclude that, since $\rho(\boldsymbol{A}) = 3$, the only solution is the trivial solution $\boldsymbol{\lambda} = \boldsymbol{O}$ so that the given set of vectors is linearly independent and thus constitutes a basis for $\mathcal{E}^3$. ■

2.5 Convex Sets and *n*-Dimensional Geometry

Our discussion in this section begins with the presentation of certain fundamental concepts from the area of point-set theory. A **spherical δ-neighborhood** about the point $\boldsymbol{X}_0 \in \mathcal{E}^n$ is the set of points $\delta(\boldsymbol{X}_0) = \{\boldsymbol{X} \mid \|\boldsymbol{X} - \boldsymbol{X}_0\| < \delta\}$. Hence the distance between $\boldsymbol{X}_0$ and any other point $\boldsymbol{X}$ within the hypersphere is strictly less than (the radius) δ. For S a nonempty subset of $\mathcal{E}^n$, a point $\boldsymbol{X}_0 \in \mathcal{E}^n$ is an **interior point** of S if there exists a spherical δ-neighborhood about $\boldsymbol{X}_0$ that contains only points of S; while $\bar{\boldsymbol{X}}$ is a **boundary point** of S if every spherical δ-neighborhood about $\bar{\boldsymbol{X}}$ encompasses points in S and in the complement of S, $\bar{S}$. (A boundary point of a set S need not be a member of S.) A set $S \subseteq \mathcal{E}^n$ is termed **open** if it contains only interior points; it is termed **closed** if it contains all of its boundary points. A set $S \subseteq \mathcal{E}^n$ is **strictly bounded** if and only if there exists a real number N such that for all points $\boldsymbol{X} \in S$, $\|\boldsymbol{X}\| \leq N$.

A set $S \subseteq \mathcal{E}^n$ is **convex** if for any two points, $\boldsymbol{X}_1, \boldsymbol{X}_2 \in S$, the line segment joining them is also in S, i.e. the **convex combination** of $\boldsymbol{X}_1, \boldsymbol{X}_2$, namely $\boldsymbol{X}_c = \theta \boldsymbol{X}_2 + (1-\theta)\boldsymbol{X}_1, 0 \leq \theta \leq 1$, is also a member of S, where the ith component of $\boldsymbol{X}_c$ is $x_i^c = \theta x_i^2 + (1-\theta) x_i^1, i = 1, \ldots, n$. Here $\boldsymbol{X}_c$ represents, for a given θ, a point on the line segment joining $\boldsymbol{X}_1$, $\boldsymbol{X}_2$, i.e. for $\theta = 0$, $\boldsymbol{X}_c = \boldsymbol{X}_1$; and for $\theta = 1$, $\boldsymbol{X}_c = \boldsymbol{X}_2$ (see Figure 2.1 for $S \subseteq \mathcal{E}^2$). To generalize this definition to any number of points within S we state the following: The set $S \subseteq \mathcal{E}^n$ is convex if for points $\boldsymbol{X}_j \in \mathcal{E}^n, j = 1, \ldots, m$, the convex combination of the $\boldsymbol{X}_j$s,

$$\boldsymbol{X}_c = \sum_{j=1}^{m} \theta_j \boldsymbol{X}_j, \theta_j \geq 0, \sum_{j=1}^{m} \theta_j = 1 \tag{2.5}$$

is also a member of S.

A **linear form** or **hyperplane** in $\mathcal{E}^n$ is the set $\mathcal{H}$ of all points $\boldsymbol{X}$ such that $\sum_{i=1}^{n} c_i x_i = \boldsymbol{C}'\boldsymbol{X} = \alpha$, where $\boldsymbol{C}\ (\neq \boldsymbol{O})$ is an $(n \times 1)$ vector and α is a scalar, i.e. $\mathcal{H} = \{\boldsymbol{X} \mid \boldsymbol{C}'\boldsymbol{X} = \alpha;\ \boldsymbol{C}(\neq \boldsymbol{O}), \boldsymbol{X} \in \mathcal{E}^n\}$. Any hyperplane divides all of $\mathcal{E}^n$ into the two **closed half-planes (–spaces)**

$$[\mathcal{H}^+] = \{\boldsymbol{X} \mid \boldsymbol{C}'\boldsymbol{X} \geq \alpha; \boldsymbol{C}(\neq \boldsymbol{O}), \boldsymbol{X} \in \mathcal{E}^n\};$$
$$[\mathcal{H}^-] = \{\boldsymbol{X} \mid \boldsymbol{C}'\boldsymbol{X} \leq \alpha; \boldsymbol{C}(\neq \boldsymbol{O}), \boldsymbol{X} \in \mathcal{E}^n\}.$$

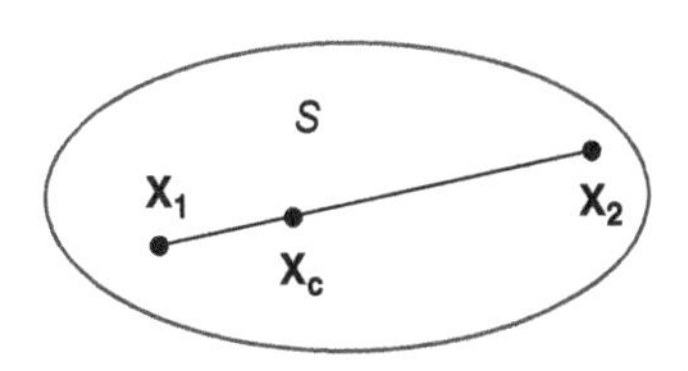

Figure 2.1 S is a convex set.

If $C'X > \alpha(<\alpha)$, we obtain the two **open half-planes (–spaces)**

$$(\mathcal{H}^+) = \{X | C'X > a; C(\neq O), X \in \mathcal{E}^n\};$$

$$(\mathcal{H}^-) = \{X | C'X < a; C(\neq O), X \in \mathcal{E}^n\}.$$

Note that $\mathcal{H}$, $[\mathcal{H}^+]$, and $[\mathcal{H}^-]$ are closed sets; $(\mathcal{H}^+)$, $(\mathcal{H}^-)$ are open sets. Additionally, each of these five sets is convex. In view of these definitions, we may now specify two specialized types of hyperplanes. First, the hyperplane $\mathcal{H}$ is a separating hyperplane for the nonempty sets $\mathcal{A}, \mathcal{B} \subseteq \mathcal{E}^n$ if, for $X \in \mathcal{A}$, $C'X \geq \alpha$, while for $X \in \mathcal{B}$, $C'X \leq \alpha$. Note that if $\mathcal{H}$ exists and $C'X > \alpha$ for $X \in \mathcal{A}$ and $C'X < \alpha$ for $X \in \mathcal{B}$, then sets $\mathcal{A}$, $\mathcal{B}$ are said to be **strictly separable**. Moreover, separability does not imply that $\mathcal{A} \cap \mathcal{B} = \emptyset$ since these sets may have boundary points in common. This definition simply states that all points in $\mathcal{A}$ lie in one of the closed half-planes determined by $\mathcal{H}$ and all points in $\mathcal{B}$ lie in the other. Second, the hyperplane $\mathcal{H}$ is a **supporting hyperplane** for a nonempty set $\mathcal{A} \subseteq \mathcal{E}^n$ if $\mathcal{H}$ contains at least one boundary point of $\mathcal{A}$ and $C'X \leq \alpha(\geq \alpha)$ for all $X \in \mathcal{A}$. In this instance, if $\mathcal{H}$ exists, all points in $\mathcal{A}$ lie in one of the closed half-planes determined by $\mathcal{H}$. Under what conditions will a separating (supporting) hyperplane exist? The answer to this question is provided by Theorems 2.2 and 2.3:

Theorem 2.3 (Weak Separation Theorem). If $\mathcal{A}$, $\mathcal{B} \subseteq \mathcal{E}^n$ are two nonempty disjoint convex sets, then there exists a hyperplane $\mathcal{H}$ that separates them, i.e. there exists a nonnull vector C and a scalar α such that, for $X \in \mathcal{A}$, $C'X \geq \alpha$, and for $X \in \mathcal{B}$, $C'X \leq \alpha$.

If both $\mathcal{A}$, $\mathcal{B}$ are closed and at least one of them is bounded, then $\mathcal{H}$ strictly separates $\mathcal{A}$, $\mathcal{B}$, i.e. for $X \in \mathcal{A}$, $C'X > \alpha$, and for $X \in \mathcal{B}$, $C'X < \alpha$.

Theorem 2.4 (Plane of Support Theorem). If $\mathcal{A} \subseteq \mathcal{E}^n$ is a nonempty convex set with a boundary point $Y \in \mathcal{E}^n$, then there exists a supporting hyperplane $C'Y = \alpha$, $C(\neq O) \in \mathcal{E}^n$, at Y.

A **cone** $\mathcal{C} \subseteq \mathcal{E}^n$ is a set of points such that if $X \in \mathcal{C}$, then so is every nonnegative scalar multiple of X, i.e. if $X \in \mathcal{C}$, then $\lambda X \in \mathcal{C}$, $\lambda \geq 0$ (see Figure 2.2a for the illustration of a cone in $\mathcal{E}^2$). The point $O \in \mathcal{E}^n$ (the null vector) is termed the **vertex** of a cone and is an element of every cone. A cone $\mathcal{C} \subseteq \mathcal{E}^n$ is termed a **convex cone** if and only if it is closed under the operations of addition and multiplication by a nonnegative scalar, i.e. $\mathcal{C}$ is a convex cone if and only if: (i) for X, Y, $\in \mathcal{C}$, $X + Y \in \mathcal{C}$; and (ii) for $X \in \mathcal{C}$, $\lambda X \in \mathcal{C}$, $\lambda \geq 0$, where X, $Y \in \mathcal{E}^n$. (Note that not every cone is convex, as Figure 2.2b attests.) Clearly a hyperplane $\mathcal{H}$ passing through the origin is a convex cone as are the closed half-planes $[\mathcal{H}^+]$, $[\mathcal{H}^-]$ determined by $\mathcal{H}$. Continuing in this vein, a convex cone $\mathcal{C} \subset \mathcal{E}^n$ is termed a **finite cone** if it consists

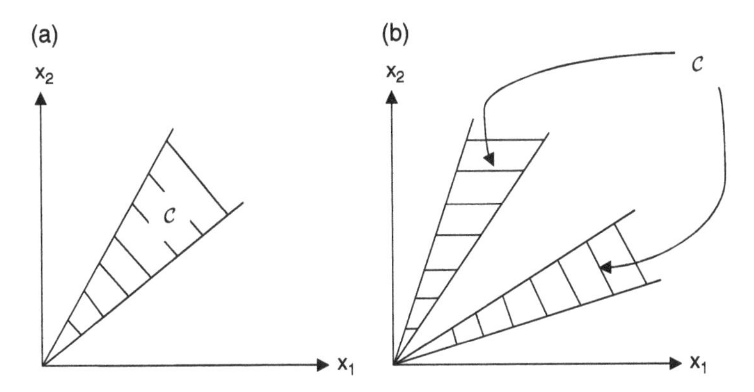

Figure 2.2 (a) $\mathcal{C}$ is a convex cone. (b) $\mathcal{C}$ is not a convex cone.

of the set of all nonnegative linear combinations of a finite set of vectors, i.e. for points $X_j \in \mathcal{E}^n$, $j = 1, \ldots, m$,

$$\mathcal{C} = \left\{ X | X = \sum_{j=1}^{m} \lambda_j X_j, \lambda_j \geq 0 \right\}.$$

Here $\mathcal{C}$ is **spanned** or **generated** by the points X_j. Hence, any vector that can be expressed as a nonnegative linear combination of a finite set of vectors X_j, $j = 1, \ldots, m$, lies in the finite convex cone spanned by these vectors. An alternative way of looking at a finite cone is: a convex cone $\mathcal{C} \subset \mathcal{E}^m$ is termed a finite cone if for some $(m \times n)$ matrix $A = [a_1, \ldots, a_n]$, every $X \in \mathcal{C}$ is a nonnegative linear combination of the $(m \times 1)$ column vectors a_j, $j = 1, \ldots, n$, of A, i.e.

$$\mathcal{C} = \left\{ X | X = A\lambda = \sum_{j=1}^{m} \lambda_j a_j, \lambda \geq O, X \in \mathcal{E}^m \right\}, \tag{2.6}$$

where λ_j is the jth component of the $(n \times 1)$ vector λ. A finite convex cone is sometimes referred to as a **convex polyhedral cone**, i.e. a convex cone is polyhedral if it is generated by a matrix A.

Example 2.9 Let us express the vector $X' = (4, 4)$ as a convex combination of the vectors $X_1' = (1,2)$, $X_2' = (5,3)$, and $X_3' = (3,8)$. From (2.5), let us solve the simultaneous system

$$\theta_1 X_1 + \theta_2 X_2 + \theta_3 X_3 = X$$
$$\theta_1 + \theta_2 + \theta_3 = 1$$

or

$$\theta_1 + 5\theta_2 + 3\theta_3 = 4$$

$$2\theta_1 + 3\theta_2 + 8\theta_3 = 4$$
$$\theta_1 + \theta_2 + \theta_3 = 1$$

to obtain $\theta_1 = 3/22$, $\theta_2 = 7/11$, and $\theta_3 = 5/22$. Note that the θ_j's are nonnegative as required. Next is the vector $\boldsymbol{X}' = (1, 1, 1)$, an element of the cone spanned by the vectors $\boldsymbol{a}_1' = \left(1, 0, \frac{1}{2}\right)$, $\boldsymbol{a}_2' = \left(\frac{1}{2}, 1, 4\right)$, and $\boldsymbol{a}_3' = \left(0, \frac{1}{4}, 0\right)$. Employing (2.6) and solving the system

$$\boldsymbol{A}\lambda = \boldsymbol{X} \quad \text{or} \quad \begin{aligned} \lambda_1 + \frac{1}{2}\lambda_2 \quad &= 1 \\ \lambda_2 + \frac{1}{4}\lambda_3 &= 1 \\ \frac{1}{2}\lambda_1 + 4\lambda_2 \quad &= 1 \end{aligned}$$

simultaneously we obtain $\lambda_1 = 14/15$, $\lambda_2 = 2/15$, and $\lambda_3 = 52/15$. Obviously $\boldsymbol{X} = \frac{14}{15}\boldsymbol{a}_1 + \frac{2}{15}\boldsymbol{a}_2 + \frac{52}{15}\boldsymbol{a}_3$, a nonnegative linear combination of the columns of $\boldsymbol{A}$, lies in the finite convex cone spanned by these vectors. Note that in this latter problem the λ_j coefficients are not required to sum to unity as in the initial problem. ■

A convex cone $\mathcal{C} \subset \mathcal{E}^n$ generated by a single vector $\boldsymbol{X}$ is termed a **ray** or **half-line** (denoted as $\mathcal{L}$), i.e. for $\boldsymbol{X} \in \mathcal{E}^n$, $\mathcal{L} = \{\boldsymbol{Y} | \boldsymbol{Y} = \lambda \boldsymbol{X}, \lambda \geq 0, \boldsymbol{Y} \in \mathcal{E}^n\}$ (see Figure 2.3 for $\mathcal{L} \subset \mathcal{E}^2$). There are two essential properties of cones (convex or finite):

a) If $\mathcal{C}_1, \mathcal{C}_2 \subset \mathcal{E}^n$ are convex cones (finite cones), their sum $\mathcal{C}_1 + \mathcal{C}_2 = \{\boldsymbol{X} | \boldsymbol{X} = \boldsymbol{X}_1 + \boldsymbol{X}_2, \boldsymbol{X}_1 \in \mathcal{C}_1, \boldsymbol{X}_2 \in \mathcal{C}_2\}$ is a convex cone (finite cone).
b) If $\mathcal{C}_1, \mathcal{C}_2 \subset \mathcal{E}^n$ are convex cones (finite cones), their intersection $\mathcal{C}_1 \cap \mathcal{C}_2$ is a convex cone (finite cone).

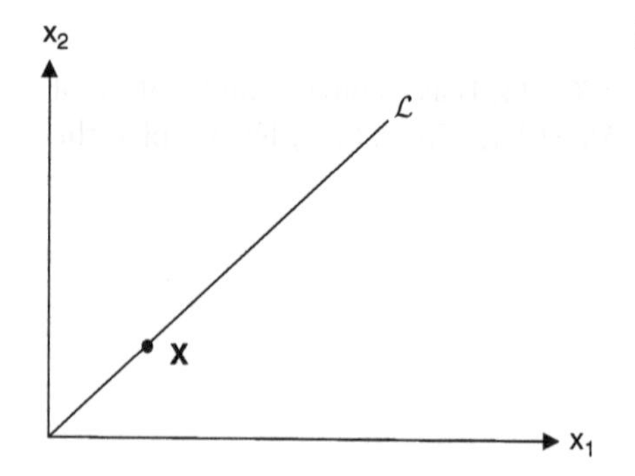

Figure 2.3 $\mathcal{L}$ is a ray or half-line.

In view of property (a), a convex cone $\mathcal{C} \subset \mathcal{E}^m$ is deemed a finite cone if there exists a finite number of half-lines $\mathcal{L}_j \subset \mathcal{E}^m, j = 1, \ldots, n$, such that $\mathcal{C} = \sum_{j=1}^{n} \mathcal{L}_j$, i.e. $\mathcal{C}$ is the sum of the n cones $\mathcal{L}_j$ or

$$\mathcal{C} = \left\{ Y | Y = \sum_{j=1}^{n} Y_j, Y_j \in \mathcal{L}_j, Y \in \mathcal{E}^m \right\}. \tag{2.7}$$

But since $\sum_{j=1}^{n} Y_j = \sum_{j=1}^{n} \lambda_j X_j = A\lambda, \lambda \geq O,$ (2.7) becomes

$$\mathcal{C} = \{Y | Y = A\lambda, \lambda \geq O, Y \in \mathcal{E}^m\}.$$

Clearly, this latter specification is equivalent to (2.6). In defining a finite cone, no mention was made of the linear independence of the generators $X_j, j = 1, \ldots, m$. However, for any linearly independent subset of the X_j's, the **dimension** of a finite cone $\mathcal{C}$, dim ($\mathcal{C}$), is the maximum number of linearly independent vectors in $\mathcal{C}$. So if $\mathcal{C} = \{X | X = A\lambda, \lambda \geq O, X \in \mathcal{E}^m\}$, dim ($\mathcal{C}$) = $\rho(A)$.

A point X is an **extreme point** or **vertex** of a convex set $\mathcal{S} \subset \mathcal{E}^n$ if and only if there "do not" exist points $X_1, X_2 \in \mathcal{S}$ ($X_1 \neq X_2$) such that $X = \theta X_2 + (1-\theta) X_1, 0 < \theta < 1$, i.e. X cannot be expressed as a convex combination of any two other distinct points of $\mathcal{S}$. In this regard, it is evident that X cannot lie between any other points of $\mathcal{S}$; thus X must be a boundary point of $\mathcal{S}$ and consequently must be contained in at least two distinct edges or boundary lines of $\mathcal{S}$ (see Figure 2.4a for $\mathcal{S} \subset \mathcal{E}^2$ wherein points X_1, X_2, X_3, and X_4 are all extreme points).

The **set of all convex combinations** of a finite number of points $X_j, j = 1, \ldots, m$, is the convex set formed by joining all these points with straight line segments; the resulting (shaded) polygon is the implied set (Figure 2.4b). Given a set that is not convex, the smallest convex set containing a given set is called the convex hull; it is the intersection of all convex sets that contain $\mathcal{S}$ (Figure 2.4c). The **convex hull** of a finite number of points $X_j, j = 1, \ldots, m$, is the set of all convex combinations of these points (Figure 2.4d) In this regard, the convex hull of a finite number of points is termed the **convex polyhedron** spanned by these points, i.e. if $\mathcal{S}$ consists of a finite number of points, the convex hull of $\mathcal{S}$ is a convex polyhedron. Some of the important properties of a convex polyhedron are:

a) It cannot have more than m extreme points since it is the set of all convex combinations of the m points $X_j, j = 1, \ldots, m$;
b) Any point in a convex polyhedron can be represented as a convex combination of the extreme points of the polyhedron.

On the basis of the preceding definitions we may now state an observation that will be of paramount importance when we discuss the set of admissible solutions to a linear programming problem in the next chapter. Specifically, any strictly bounded closed convex set with a finite number of extreme points is the convex hull of the extreme points, i.e. is a convex polyhedron.

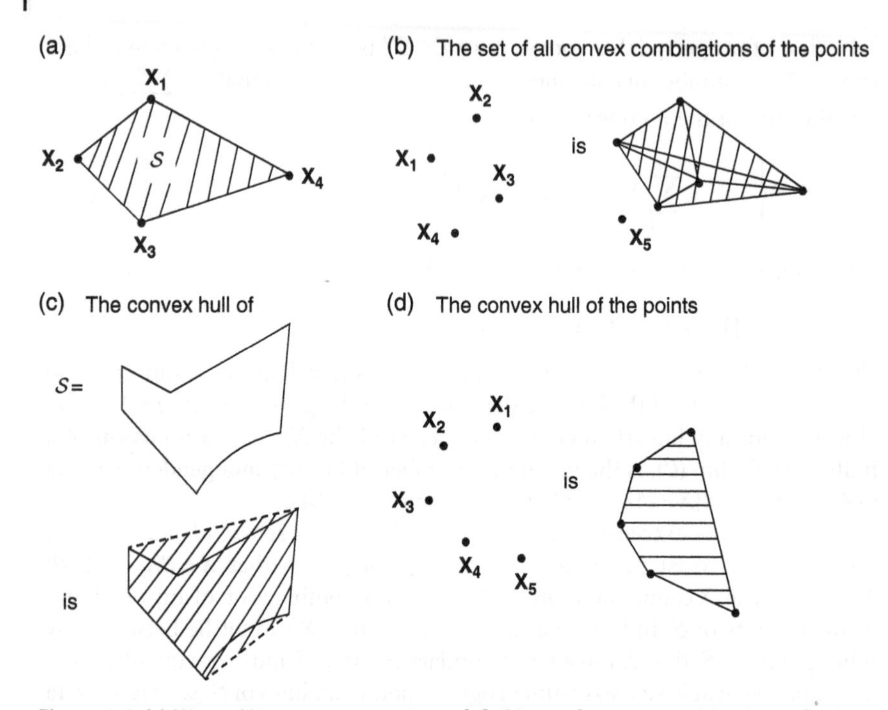

Figure 2.4 (a) X1, …, X4 are extreme points of $\mathcal{S}$; (b) set of convex combinations of points X1, …, X5; (c) convex hull of a set $\mathcal{S}$; and (d) convex hull of points X1, …, X5.

Finally, an n-dimensional convex polyhedron having $n + 1$ extreme points is called a **simplex,** so named because it refers to the "simplest" possible convex set that has exactly one more extreme point than the dimension of the space on which it is defined, e.g. a zero-dimensional simplex is a single point; a one-dimensional simplex is a line segment with two end points; a two-dimensional simplex is a triangle (it thus has three extreme points); and so on. Clearly, each of these figures is the smallest convex set containing its extreme points and thus represents the convex hull of the same.

3

Introduction to Linear Programming

3.1 Canonical and Standard Forms

We may depict the basic structure of a linear programming problem in what is termed **canonical form** as

$$\begin{aligned} &max f = c_1x_1 + c_2x_2 + \cdots + c_px_p \quad s.t. \\ &\quad a_{11}x_1 + a_{12}x_2 + \cdots + a_{1p}x_p \leq b_1 \\ &\quad a_{21}x_1 + a_{22}x_2 + \cdots + a_{2p}x_p \leq b_2 \\ &\quad \cdots\cdots\cdots\cdots\cdots\cdots\cdots\cdots\cdots \\ &\quad a_{m1}x_1 + a_{m2}x_2 + \cdots + a_{mp}x_p \leq b_m \\ &\quad x_1, x_2, \ldots, x_p \geq 0. \end{aligned} \tag{3.1}$$

Here f is the **objective function** (a hyperplane), the first m inequalities are the **structural constraints**, and the last p inequalities are simply called **nonnegativity conditions**. Taken altogether, the structural constraints and the nonnegativity conditions constitute a set of closed half-planes that will be called the **constraint system**. This system forms the p-dimensional **feasible region** $\mathcal{K}$, i.e. the area within which an admissible solution to (3.1) lies.

In matrix form, (3.1) may be rewritten as

$$\begin{aligned} &max f = \boldsymbol{C}'\boldsymbol{X} \quad s.t. \\ &\boldsymbol{AX} \leq \boldsymbol{b}, \boldsymbol{X} \geq \boldsymbol{O}, \end{aligned} \tag{3.2}$$

where

$$\underset{(px1)}{\boldsymbol{C}} = \begin{bmatrix} c_1 \\ c_2 \\ \vdots \\ c_p \end{bmatrix}, \underset{(px1)}{\boldsymbol{X}} = \begin{bmatrix} x_1 \\ x_2 \\ \vdots \\ x_p \end{bmatrix}, \underset{(mxp)}{\boldsymbol{A}} = \begin{bmatrix} a_{11} & a_{12} & \cdots & a_{1p} \\ a_{21} & a_{22} & \cdots & a_{2p} \\ \vdots & \vdots & \vdots & \vdots \\ a_{m1} & a_{m2} & \cdots & a_{mp} \end{bmatrix}, \underset{(mx1)}{\boldsymbol{b}} \begin{bmatrix} b_1 \\ b_2 \\ \vdots \\ b_m \end{bmatrix}.$$

Linear Programming and Resource Allocation Modeling, First Edition. Michael J. Panik.

Any $\boldsymbol{X}$ that satisfies $\boldsymbol{AX} \leq \boldsymbol{b}$ is called a **solution** to the linear programming problem while if such an $\boldsymbol{X}$ also satisfies $\boldsymbol{X} \geq \boldsymbol{O}$, i.e. $\boldsymbol{X} \varepsilon \mathcal{K} = \{X | \boldsymbol{AX} \leq \boldsymbol{b}, \boldsymbol{X} \geq \boldsymbol{O}\}$, then it is termed a **feasible solution**. Moreover, if X_0 is feasible and $f(X_0) \geq f(X)$, where $\boldsymbol{X}$ is any other feasible solution, then X_0 is an **optimal solution** to 3.1 (or (3.2)).

Rather than deal with the structural constraints as a set of inequalities, we may transform them into a set of equalities (called the **augmented structural constraint system**) by introducing m additional nonnegative **slack variables** x_{p+1}, x_{p+2}, …, x_{p+m} into (3.1). Once this is accomplished, we obtain the **augmented linear programing problem:**

$$
\begin{aligned}
&max f = c_1x_1 + c_2x_2 + \cdots + c_px_p + 0x_{p+1} + 0x_{p+2} + \cdots + 0x_{p+m} \ \ s.t.\\
&a_{11}x_1 + a_{12}x_2 + \cdots + a_{1p}x_p + x_{p+1} + 0x_{p+2} + \cdots + 0x_{p+m} = b_1\\
&a_{21}x_1 + a_{22}x_2 + \cdots + a_{2p}x_p + 0x_{p+1} + x_{p+2} + \cdots + 0x_{p+m} = b_2\\
&\cdots\cdots\cdots\cdots\cdots\cdots\cdots\cdots\cdots\cdots\cdots\cdots\cdots\cdots\cdots\cdots\cdots\cdots\cdots\cdots\\
&a_{m1}x_1 + a_{m2}x_2 + \cdots + a_{mp}x_p + 0x_{p+1} + 0x_{p+2} + \cdots + x_{p+m} = b_m\\
&x_1, x_2, \ldots, x_p, x_{p+1}, x_{p+2}, \ldots, x_{p+m} \geq 0.
\end{aligned} \tag{3.3}
$$

Any linear program written in this fashion is said to be in **standard form**.

The matrix equivalent of (3.3) is

$$
\begin{aligned}
&max\, f = \boldsymbol{C}'\boldsymbol{X} + \boldsymbol{O}'\boldsymbol{X}_s \ \ s.t.\\
&\boldsymbol{AX} + \boldsymbol{I}_m\boldsymbol{X}_s = \boldsymbol{b}, \boldsymbol{X} \geq \boldsymbol{O}, \boldsymbol{X}_s \geq \boldsymbol{O},
\end{aligned} \tag{3.4}
$$

where $\boldsymbol{I}_m$ is an mth order identity matrix and $\boldsymbol{X}_s$ is an (m × 1) vector of slack variables. If $p + m = n$ and

$$
\underset{(n\times 1)}{\bar{\boldsymbol{C}}} = \begin{bmatrix} \boldsymbol{C} \\ \cdots \\ \boldsymbol{O} \end{bmatrix}, \quad \underset{(n\times 1)}{\bar{\boldsymbol{X}}} = \begin{bmatrix} \boldsymbol{X} \\ \cdots \\ \boldsymbol{X}_s \end{bmatrix}, \quad \underset{(m\times n)}{\bar{\boldsymbol{A}}} = \left[\boldsymbol{A} \vdots \boldsymbol{I}_m\right],
$$

then (3.4) may be rewritten as

$$
\begin{aligned}
&max f = \bar{\boldsymbol{C}}'\bar{\boldsymbol{X}} \ \ s.t.\\
&\bar{\boldsymbol{A}}\bar{\boldsymbol{X}} = \boldsymbol{b}, \bar{\boldsymbol{X}} \geq \boldsymbol{O}.
\end{aligned} \tag{3.4.1}
$$

3.2 A Graphical Solution to the Linear Programming Problem

Let us solve

$$\begin{aligned}
\max f &= x_1 + 4x_2 \quad s.t.\\
x_1 & \leq 80\\
x_2 & \leq 35\\
\frac{1}{2}x_1 + x_2 & \leq 50\\
x_1, x_2 & \geq 0.
\end{aligned}$$

As Figure 3.1 reveals, the feasible region $\mathcal{K}$ is the intersection of the solution sets of the constraint system. It is apparent that if the objective function f is superimposed on $\mathcal{K}$ and we slide it upward over $\mathcal{K}$ parallel to itself so as to get it as high as possible while maintaining feasibility, then the optimal solutions occurs at point A. The coordinates of A may be obtained as the simultaneous solution of the **binding constraints** (i.e. those that hold as strict equalities) at that point. Hence solving the system

$$\begin{aligned}
x_2 &= 35\\
\frac{1}{2}x_1 + x_2 &= 50
\end{aligned}$$

simultaneously yields $x_1^0 = 30, x_2^0 = 35$. Then $f^0 = 30 + 4(35) = 170$. Clearly, the optimal solution occurs at a boundary point of $\mathcal{K}$ so that f is a supporting hyperplane of $\mathcal{K}$.

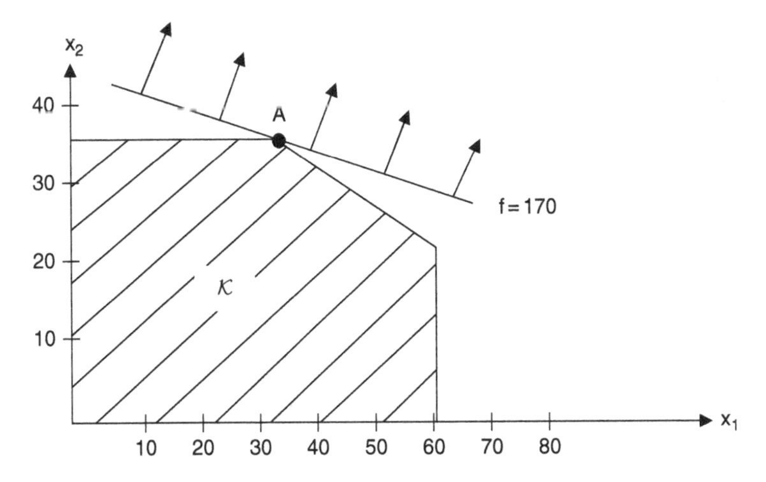

Figure 3.1 $f = 170$ is a supporting hyperplane of $\mathcal{K}$ at point A.

3.3 Properties of the Feasible Region

We noted earlier that $\mathcal{K}$ is formed as the intersection of a finite number of closed half-planes, each of which is a closed convex set. Hence, it follows that $\mathcal{K}$ itself is a closed convex set whose boundary consists of, at most, a finite number of edges and thus a finite number of extreme points. Since $\mathcal{K}$ will typically be bounded, we may legitimately assume in our subsequent analysis that $\mathcal{K}$ is a convex polyhedron, i.e. it is a strictly bounded closed convex set with a finite number of extreme points.

3.4 Existence and Location of Optimal Solutions

Given that f is continuous (since it is linear) and $\mathcal{K}$ is a convex polyhedron, we are assured that a finite maximum of f exists on $\mathcal{K}$. Moreover, this extremum will occur at a unique point of tangency (such as point A in Figure 3.2a) or along an entire edge of $\mathcal{K}$ (Figure 3.2b). More formally, these observations are summarized by Theorem 3.1.

Theorem 3.1 (Extreme Point Theorem). f assumes its maximum at an extreme point of $\mathcal{K}$. If it assumes its maximum at more than one extreme point, then it takes on the same value at every convex combination of those points.

The first part of this theorem is illustrated by Figure 3.2a, wherein a unique optimal solution occurs at point A. The second part is depicted in Figure 3.2b, the case of multiple optimal solutions. In this latter instance, there exist two optimal extreme point solutions (at points A and B) and an infinity of optimal

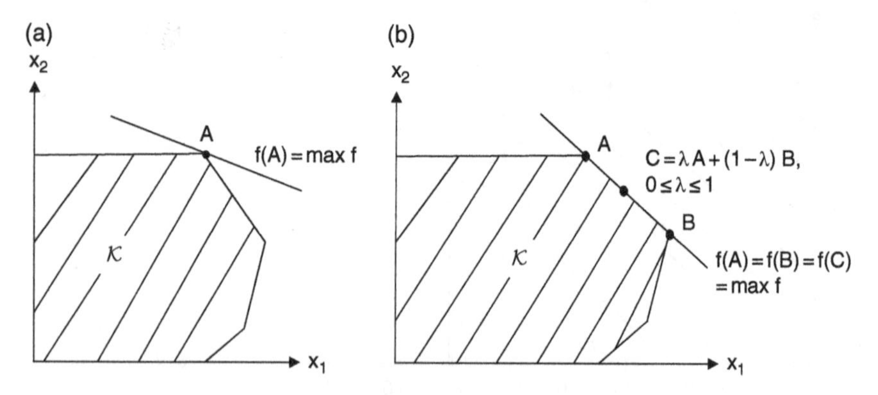

Figure 3.2 (a) Unique optimal solution at point A. (b) Multiple optimal solutions at points A, B, and C.

nonextreme point solutions (each a convex combination C of A and B, where $C = \lambda A + (1 - \lambda)B,\ 0 \le \lambda \le 1$).

How do we obtain extreme point solutions? The answer to this question is provided by the next section.

3.5 Basic Feasible and Extreme Point Solutions

In what follows, we shall demonstrate how to generate a basic solution to the augmented structural constraint system $\bar{A}\bar{X} = \boldsymbol{b}$. At the outset, let us impose the restriction that $\rho(\bar{A}) = \rho(\bar{A}, \boldsymbol{b})$.Under this assumption, $\bar{A}\bar{X} = \boldsymbol{b}$ is consistent and thus possesses at least one solution. In addition, if we further assume $\rho(\bar{A}) = m$ so that none of the augmented structural constraints is **redundant** (i.e. a particular structural constraint cannot be expressed as a linear combination of any others), then we may select m linearly independent columns from $\bar{A}$ and uniquely express the variables associated with these columns (which will be called **basic variables**) in terms of the remaining $n - m$ variables (which will be termed **nonbasic variables**). To accomplish this, let us rearrange the columns of $\bar{A}$ so that the first m is linearly independent and constitute the columns of the mth order **basis matrix** $\boldsymbol{B} = [\boldsymbol{b}_1, \ldots, \boldsymbol{b}_m]$. (This matrix is so named because its columns constitute a basis for $\mathcal{E}^m$.) The remaining $n - m$ columns of $\bar{A}$ then form the $(m \times n - m)$ matrix $\boldsymbol{R} = [\boldsymbol{r}_1, \ldots, \boldsymbol{r}_{n-m}]$. Hence $\bar{A}\bar{X} = \boldsymbol{b}$ may be expressed as

$$\bar{A}\bar{X} = \left[\boldsymbol{B} \vdots \boldsymbol{R}\right]\begin{bmatrix} X_B \\ \cdots \\ X_R \end{bmatrix} = \boldsymbol{B}X_B + \boldsymbol{R}X_R = \boldsymbol{b},$$

where $\bar{X}$ has been partitioned into an $(m \times 1)$ matrix X_B containing the basic variables $x_{B1}, \ldots, x_{Bm}$ and an $(n - m \times 1)$ matrix X_R containing the nonbasic variables $x_{R1}, \ldots, x_{R,n-m}$. Since $|\boldsymbol{B}| \neq 0$,

$$X_B = \boldsymbol{B}^{-1}\boldsymbol{b} - \boldsymbol{B}^{-1}\boldsymbol{R}X_R. \tag{3.5}$$

If $X_R = \boldsymbol{O}$, then $X_B = \boldsymbol{B}^{-1}\boldsymbol{b}$ and a **basic solution** to $\bar{A}\bar{X} = \boldsymbol{b}$ thus appears as $\bar{X}' = \left[\bar{X}_B' \vdots \boldsymbol{O}'\right]$. Furthermore, if $X_B \ge \boldsymbol{O}$ so that $\bar{X} \ge \boldsymbol{O}$, then $\bar{X}$ is termed a **basic feasible solution** to $\bar{A}\bar{X} = \boldsymbol{b}$. Finally, a basic solution to $\bar{A}\bar{X} = \boldsymbol{b}$ is **degenerated** if at least one of the basic variables is zero.

Example 3.1 Find a nondegenerate basic feasible solution to

$$x_1 - x_2 + 2x_3 + x_4 = 1$$
$$3x_1 + x_2 - x_3 + 2x_4 = 4.$$

Since

$$\begin{vmatrix} 1 & -1 \\ 3 & 1 \end{vmatrix} = 4 \neq 0,$$

$$\boldsymbol{B} = \begin{bmatrix} 1 & -1 \\ 3 & 1 \end{bmatrix}, \boldsymbol{B}^{-1} = \begin{bmatrix} 1/4 & 1/4 \\ -3/4 & 1/4 \end{bmatrix}, \boldsymbol{R} = \begin{bmatrix} 2 & -1 \\ -1 & 2 \end{bmatrix}, \boldsymbol{X_B} = \begin{bmatrix} x_1 \\ x_2 \end{bmatrix}, \boldsymbol{X_R} = \begin{bmatrix} x_3 \\ x_4 \end{bmatrix},$$

it follows, from (3.5), that

$$\begin{bmatrix} x_1 \\ x_2 \end{bmatrix} = \begin{bmatrix} 5/4 \\ 1/4 \end{bmatrix} - \begin{bmatrix} 1/4 & 3/4 \\ -7/4 & -1/4 \end{bmatrix} \begin{bmatrix} x_3 \\ x_4 \end{bmatrix}$$

or

$$x_1 = 5/4 - 1/4x_3 - 3/4x_4$$
$$x_2 = 1/4 + 7/4x_3 + 1/4x_4.$$

If $\boldsymbol{X_R} = \boldsymbol{O}, \boldsymbol{X}'_{\boldsymbol{B}} = (5/4, 1/4)$ and thus a nondegenerate basic feasible solution to $\bar{\boldsymbol{A}}\bar{\boldsymbol{X}} = \boldsymbol{b}$ is $\bar{\boldsymbol{X}}' = (5/4, 1/4, 0, 0)$. ■

What is the importance of a basic feasible solution to $\bar{\boldsymbol{A}}\bar{\boldsymbol{X}} = \boldsymbol{b}$? As we shall now see, we may undertake a search for an optimal solution to the linear programming problem by seeking only basic feasible solutions. Furthermore, we noted earlier that an optimal solution occurs at an extreme point or at some convex combination of extreme points of $\mathcal{K}$. So if it can be demonstrated that basic feasible solutions always correspond to extreme points of $\mathcal{K}$ and conversely, then we can locate extreme points of $\mathcal{K}$ and thus an optimal solution by examining only basic feasible solutions. In this regard, if we can find a set of m linearly independent vectors within $\bar{\boldsymbol{A}}$ and arrive at a basic feasible solution to $\bar{\boldsymbol{A}}\bar{\boldsymbol{X}} = \boldsymbol{b}$, it must yield an extreme point of $\mathcal{K}$, i.e. extreme point solutions are always generated by m linearly independent columns from $\bar{\boldsymbol{A}}$. The justification underlying this discussion is provided by Theorem 3.2.

Theorem 3.2 A vector $\bar{\boldsymbol{X}}$ is a basic feasible solution to $\bar{\boldsymbol{A}}\bar{\boldsymbol{X}} = \boldsymbol{b}$ if and only if $\bar{\boldsymbol{X}}$ is an extreme point of the region of feasible solutions.

3.6 Solutions and Requirement Spaces

Throughout this text we shall often refer to two types of Euclidean spaces that are intrinsic to any linear programming problem. We noted above that $\bar{X} \varepsilon \mathcal{E}^n$. The part of $\mathcal{E}^n$ that contains those $\bar{X}$ vectors satisfying the constrained equation system $\bar{A}\bar{X} = \boldsymbol{b}, \bar{X} \geq \boldsymbol{O}$, will be called **solutions space**. If a vector $\bar{X}$ satisfies $\bar{A}\bar{X} = \boldsymbol{b}$, then $\bar{X}$ lies within the intersection of a set of hyperplanes that comprise the rows of $[\bar{A} \vdots \boldsymbol{b}]$. If $\rho(\bar{A}) = m$, then the said intersection cannot be formed by fewer than m hyperplanes.

The second space in question will be referred to as **requirements space**. The elements of this space are the columns of $[\bar{A} \vdots \boldsymbol{b}]$, where $\boldsymbol{b}$ is termed the **requirement vector**. Since the columns of $[\bar{A} \vdots \boldsymbol{b}]$ represent a set of $(m \times 1)$ vectors, it is obvious that requirements space is a subset of $\mathcal{E}^m$. If $\bar{\boldsymbol{a}}_j$ denotes the jth column of $\bar{A}$, then

$$\bar{A}\bar{X} = \sum_{j=1}^{n} \boldsymbol{a}_j \bar{x}_j = \boldsymbol{b}.$$

Since we can reach a feasible solution to $\bar{A}\bar{X} = \boldsymbol{b}$ when $\boldsymbol{b}$ is expressible as a nonnegative linear combination of the columns of $\bar{A}$, $\boldsymbol{b}$ must be an element of the convex polyhedral cone generated by the columns of $\bar{A}$. Hence, the part of $\mathcal{E}^m$ that contains the convex polyhedral cone

$$\mathcal{C} = \left\{ Y | Y = \bar{A}\bar{X} = \sum_{j=1}^{n} \bar{\boldsymbol{a}}_j \bar{x}_j, \bar{X} \geq O \right\}$$

is requirements space. Now, if $\rho(\bar{A}) = m$, the columns of $\bar{A}$ yield a basis for $\mathcal{E}^m$ so that $\boldsymbol{b}$ cannot be expressible as a nonnegative linear combination of fewer than m of the $\bar{\boldsymbol{a}}_j$. Thus, the number of $\bar{x}_j$ s > 0 is never greater than m. In this regard, if the m linearly independent columns of $\bar{A}$ constitute the $(m \times m)$ matrix $\boldsymbol{B}$, then a basic feasible solution to $\bar{A}\bar{X} = \boldsymbol{b}$ results when $\boldsymbol{b}$ is expressible as a nonnegative linear combination of the columns of $\boldsymbol{B}$ so that $\boldsymbol{b}$ is an element of the convex polyhedral cone

$$\mathcal{C}_B = \left\{ Y | Y = [\boldsymbol{B} \vdots \boldsymbol{R}] \begin{bmatrix} X_B \\ \cdots \\ O \end{bmatrix} = \boldsymbol{B}X_B = \sum_{i=1}^{m} b_i x_{Bi}, X_B \geq \boldsymbol{O} \right\}.$$

3.6 Solutions and Requirement Spaces

Throughout this text we shall often refer to two types of Euclidean spaces that are intrinsic to any linear programming problem. We noted above that $X \in \mathcal{E}^n$. The part of $\mathcal{E}^n$ that contains those X vectors satisfying the constrained equation system $AX = b, X \geq O$, will be called **solutions space**. If a vector X satisfies $AX = b$, then X lies within the intersection of a set of hyperplanes that comprise the rows of $[A \vdots b]$. If $\rho(A) = m$, then the said intersection cannot be formed by fewer than m hyperplanes.

The second space in question will be referred to as **requirements space**. The elements of this space are the columns of $[A \vdots b]$, where b is termed the **requirement vector**. Since the columns of $[A \vdots b]$ represent a set of $(m \times 1)$ vectors, it is obvious that requirements space is a subset of $\mathcal{E}^m$. If a_j denotes the jth column of A, then

$$AX = \sum_{j=1}^{n} x_j a_j = b.$$

Since we can reach a feasible solution to $AX = b$ when b is expressible as a nonnegative linear combination of the columns of A, b must be an element of the convex polyhedral cone generated by the columns of A. Hence, the part of $\mathcal{E}^m$ that contains the convex polyhedral cone

$$C = \left\{ Y \mid Y = AX = \sum_{j=1}^{n} x_j a_j, X \geq O \right\}$$

is requirements space. Now, if $\rho(A) = m$, the columns of A yield a basis for $\mathcal{E}^m$ so that b cannot be expressible as a nonnegative linear combination of fewer than m of the a_j. Thus, the number of $x_j > 0$ is never greater than m. In this regard, if the m linearly independent columns of A constitute the $(m \times m)$ matrix B, then a basic feasible solution to $AX = b$ results when b is expressible as a nonnegative linear combination of the columns of B so that b is an element of the convex polyhedral cone

$$C_B = \left\{ Y \mid Y = [B \vdots R] \begin{bmatrix} X_B \\ \cdots \\ O \end{bmatrix} = BX_B = \sum_{i=1}^{m} x_{Bi} b_i, X_B \geq O \right\}$$

4

Computational Aspects of Linear Programming

4.1 The Simplex Method

We noted in the previous chapter that the region of feasible solutions has a finite number of extreme points. Since each such point has associated with it a basic feasible solution (unique or otherwise), it follows that there exists a finite number of basic feasible solutions. Hence, an optimum solution to the linear programming problem will be contained within the set of basic feasible solutions to $\bar{A}\bar{X} = \boldsymbol{b}$. How many elements does this set possess? Since a basic feasible solution has at most m of n variables different from zero, an upper bound to the number of basic feasible solutions is

$$\binom{n}{m} = \frac{n!}{m!(n-m!)},$$

i.e. we are interested in the total number of ways in which m basic variables can be selected (without regard to their order within the vector of basic variables $\boldsymbol{X}_B$) from a group of n variables. Clearly for large n and m it becomes an exceedingly tedious task to examine each and every basic feasible solution. What is needed is a computational scheme that examines, in a selective fashion, only some small subset of the set of basic feasible solutions. Such a scheme is the **simplex method** (Dantzig 1951). Starting from an initial basic feasible solution, this technique systematically proceeds to alternative basic feasible solutions and, in a finite number of steps or iterations, ultimately arrives at an optimal basic feasible solution. The path taken to the optimum is one for which the value of the objective function at any extreme point is at least as great as at an adjacent extreme point (two extreme points are said to be **adjacent** if they are joined by an edge of a convex polyhedron). For instance, if in Figure 4.1 the extreme point A represents our initial basic feasible solution, the first iteration slides f upwards parallel to itself over $\mathcal{K}$ until it passes through its adjacent extreme point B. The

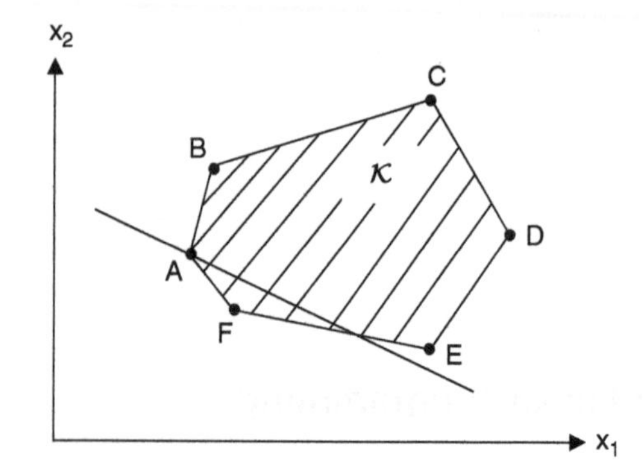

Figure 4.1 Movement to an optimal basic feasible solution.

next iteration advances f in a similar fashion to its optimal (maximal) basic feasible solution at C. So with f nondecreasing between successive basic feasible solutions, this search technique does not examine all basic feasible solutions but only those that yield a value of f at least as large as at the preceding basic feasible solution.

To see exactly how the simplex method works, let us maximize $f(\bar{X}) = \bar{C}'\bar{X}$ subject to $\bar{A}\bar{X} = \boldsymbol{b}$ and $\bar{X} \geq \boldsymbol{O}, \bar{X} \in \mathcal{E}^n$. To ensure that the simplex method terminates in a finite number of steps at an optimal basic feasible solution, let us invoke the following **nondegeneracy assumption**: every set of m columns from $\bar{A}$ is linearly independent. Under this restriction, we are guaranteed that: (i) every basic feasible solution has exactly m nonzero components and (ii) f increases in value between successive basic feasible solutions. With $\rho(\bar{A}) = m$, we may select from $\bar{A}$ a set of m linearly independent vectors that constitute a basis for $\mathcal{E}^m$ and thus the columns of the $(m \times m)$ nonsingular matrix $\boldsymbol{B}$. Since $|\boldsymbol{B}| \neq 0$, we have, from (3.5), $\boldsymbol{X_B} = \boldsymbol{B}^{-1}\boldsymbol{b} - \boldsymbol{B}^{-1}\boldsymbol{R}\boldsymbol{X_R}$.

Let us next partition the $(n \times 1)$ coefficient matrix $\bar{C}$ into the $(m \times 1)$ and $(n-m \times 1)$ matrices $\boldsymbol{C_B}$ and $\boldsymbol{C_R}$, respectively, where $\boldsymbol{C_B}$ contains the objective function coefficients corresponding to the basic variables in $\boldsymbol{X_B}$ and $\boldsymbol{C_R}$ contains the coefficients on the nonbasic variables in $\boldsymbol{X_R}$. Then

$$\begin{aligned} f &= \bar{C}'X \\ &= [C_B', C_R'] \begin{bmatrix} X_B \\ X_R \end{bmatrix} = C_B'X_B + C_R'X_R. \end{aligned} \tag{4.1}$$

Substituting $\boldsymbol{X_B}$ from above into (4.1) yields

$$f = C_B'B^{-1}b + (C_R' - C_B'B^{-1}R)X_R. \tag{4.2}$$

At a basic feasible solution to the linear programming problem, $\boldsymbol{X}_R = \boldsymbol{O}$ so that

$$\bar{X} = \begin{bmatrix} \boldsymbol{X}_B \\ \boldsymbol{O} \end{bmatrix} = \begin{bmatrix} \boldsymbol{B}^{-1}\boldsymbol{b} \\ \boldsymbol{O} \end{bmatrix} \geq \boldsymbol{O}, \qquad (4.3)$$

$$f = \boldsymbol{C}_B'\boldsymbol{B}^{-1}\boldsymbol{b}.$$

To see exactly how (4.3) is obtained in actual practice, let us solve

$$\begin{aligned} &max\, f = 2x_1 + 3x_2 + 4x_3 \quad \text{s.t.} \\ &\quad 2x_1 + x_2 + 4x_3 \leq 100 \\ &\quad x_1 + 3x_2 + x_3 \leq 80 \\ &\quad x_1, x_2, x_3 \geq 0. \end{aligned}$$

Since our starting point is the linear programming problem in standard form, we introduce two nonnegative slack variables x_4, x_5 and ultimately obtain

$$\begin{aligned} &max\, f = 2x_1 + 3x_2 + 4x_3 + 0x_4 + 0x_5 \quad \text{s.t.} \\ &\quad 2x_1 + x_2 + 4x_3 + x_4 + 0x_5 \leq 100 \\ &\quad x_1 + 3x_2 + x_3 + 0x_4 + x_5 \leq 80 \\ &\quad x_1, \dots, x_5 \geq 0. \end{aligned}$$

In terms of our previous notation,

$$\bar{\boldsymbol{A}} = \begin{bmatrix} 2 & 1 & 4 & 1 & 0 \\ 1 & 3 & 1 & 0 & 1 \end{bmatrix}, \bar{\boldsymbol{C}} = \begin{bmatrix} 2 \\ 3 \\ 4 \\ 0 \\ 0 \end{bmatrix}, \bar{\boldsymbol{X}} = \begin{bmatrix} x_1 \\ x_2 \\ x_3 \\ x_4 \\ x_5 \end{bmatrix}, \text{ and } \boldsymbol{b} = \begin{bmatrix} 100 \\ 80 \end{bmatrix}.$$

Since the determinant of the (2×2) submatrix formed from columns 4 and 5 of $\bar{\boldsymbol{A}}$ is different from zero, $\rho(\bar{\boldsymbol{A}}) = 2$. With these two columns linearly independent, we may consequently select x_4 and x_5 as basic variables with x_1, x_2, and x_3 nonbasic and proceed to generate our first basic feasible solution.[1] In this regard,

$$\boldsymbol{B} = \begin{bmatrix} 1 & 0 \\ 0 & 1 \end{bmatrix}, \boldsymbol{B}^{-1} = \begin{bmatrix} 1 & 0 \\ 0 & 1 \end{bmatrix}, \boldsymbol{R} = \begin{bmatrix} 2 & 1 & 4 \\ 1 & 3 & 1 \end{bmatrix}, \boldsymbol{X}_B = \begin{bmatrix} x_4 \\ x_5 \end{bmatrix},$$

1 Note that the columns of $\bar{\boldsymbol{A}}$ corresponding to the slack variables always provide us with an identity matrix and thus an initial basic feasible solution.

$$X_R = \begin{bmatrix} x_1 \\ x_2 \\ x_3 \end{bmatrix}, C_B = \begin{bmatrix} 0 \\ 0 \end{bmatrix}, \text{ and } C_R = \begin{bmatrix} 2 \\ 3 \\ 4 \end{bmatrix}.$$

From (3.5) we may solve for the basic variables in terms of the nonbasic variables as

$$X_B = B^{-1}b - B^{-1}RX_R$$

or

$$\begin{bmatrix} x_4 \\ x_5 \end{bmatrix} = \begin{bmatrix} 100 - 2x_1 - x_2 - 4x_3 \\ 80 - x_1 - 3x_2 - x_3 \end{bmatrix}. \quad (4.4)$$

If $X_R = O$, we obtain the nondegenerate basic feasible solution $\bar{X}' = (100,80,0,0,0)$.

What is the significance of the product $B^{-1}R$? Generally speaking, since the order of this matrix is (m × n – m) it is apparent that $B^{-1}R$ contains the same number of rows and columns as there are, respectively, basic and nonbasic variables. If we denote the jth column of $B^{-1}R$ by $Y_j = B^{-1}r_j, j = 1, \ldots, n - m$, then, for a fixed j, the ith component of Y_j, y_{ij}, $i = 1, \ldots, m$, represents the amount by which the ith basic variable would change in any basic solution given that the jth nonbasic variable increases from zero (its current nonbasic value) to one. For instance, in our example, y_{13} indicates that if x_3 increases from 0 to 1, x_4 changes by 4 units or $x_4 = 100–4(1) = 96$.

From (4.2),

$$f = C_B'B^{-1}b + (C_R' - C_B'B^{-1}R)X_R$$

$$= 0 + (2,3,4)\begin{bmatrix} x_{R1} \\ x_{R2} \\ x_{R3} \end{bmatrix} = 2x_1 + 3x_2 + 4x_3. \quad (4.5)$$

Upon setting $X_R = O$, it is easily seen that the value of the objective function corresponding to this basic feasible solution is zero. How may we interpret the elements within the $(1 \times n-m)$ matrix $C_R' - C_B'B^{-1}R$? Again considering the general case, if we denote the jth element in this $(1 \times n-m)$ matrix as $\bar{c}_j, j = 1, \ldots, n-m$, then this element is simply the coefficient on the jth nonbasic variable in (4.2). It indicates the *net change* in f given that x_{Rj} increases from zero (its present nonbasic value) to one. In what sense shall we interpret the notion of net change? Let us assume that, in the above example, x_3 increases from zero to one with the remaining nonbasic variables still equal to zero. The effect of this change is twofold. First, there is the direct influence of x_3 on f through its corresponding coefficient in C_R, i.e. since

$$C'_R X_R = (2,3,4)\begin{bmatrix} 0 \\ 0 \\ x_{R3} \end{bmatrix} = 4x_3,$$

it is easily seen that when $x_3 = 1, f = 4\ (1) = 4$. Hence f *directly* increases by four units. Next, x_3 affects f *indirectly* through its influence on the basic variables x_4 and x_5, which, in turn, precipitate a change in f. Moreover, the said effect is transmitted through the matrix $C'_B B^{-1} R$. That is to say, given

$$B^{-1} R X_R = \begin{bmatrix} 2 & 1 & 4 \\ 1 & 3 & 1 \end{bmatrix}\begin{bmatrix} 0 \\ 0 \\ x_{R3} \end{bmatrix} = \begin{bmatrix} 4x_3 \\ x_3 \end{bmatrix},$$

we see that if $x_3 = 1$, x_4 changes by four units and x_5 changes by one unit since, from (4.4),

$$\begin{bmatrix} x_4 \\ x_5 \end{bmatrix} = \begin{bmatrix} 100 - 4x_3 \\ 80 - x_3 \end{bmatrix}.$$

But the effect of a change in x_4 and x_5 upon f is issued through their corresponding coefficients in C_B, i.e.

$$C'_B B^{-1} R X_R = (0,0)\begin{bmatrix} 4x_3 \\ x_3 \end{bmatrix} = 0.$$

For $x_3 = 1$, the indirect effect of x_3 on f is thus $C'_B B^{-1} R X_R =$ zero units. Hence the net change in f, being the difference between the direct and indirect effects of x_3, is given by $C'_R - B^{-1} R = 4 - 0 = 4$ units, which is just the coefficient on x_3 in (4.5).

At this point in our discussion of the simplex method, an important question emerges: how do we know whether a particular basic feasible solution is optimal? To gain some insight into this problem let us examine (4.5). At the basic feasible solution $X' = (100, 80, 0, 0, 0)$ we found that $f = 0$. Is this the maximum value of f? As we shall now see, the answer is no. If $X_R \neq O$,

$$f = 2x_1 + 3x_2 + 4x_3$$

so that f can change if any one of the nonbasic variables x_1, x_2, or x_3 increases from its nonbasic value of zero to some positive level. Hence f increases if x_1 or x_2 or x_3 increases since the coefficients on these variables are all positive. So if all the elements $\bar{c}_j, j = 1,\ldots,n-m$, within the matrix $C'_R - C'_B B^{-1} R$ are nonpositive, it is obvious that f cannot increase in value if any nonbasic variable does since $\bar{c}_j x_R \leq 0, j = 1,\ldots,n-m$. Hence a particular basic feasible solution is maximal if $\bar{c}_j \leq 0, j = 1,\ldots,n-m$. In addition, if $\bar{c}_j \leq 0$ for all j and $\bar{c}_k = 0$ for some nonbasic variable x_{Rk}, then if x_{Rk} increases in value from zero by some positive amount, no change in f occurs. Hence the objective function assumes the same value at more than one extreme point so that the case of multiple optimal solutions emerges. We may formalize these observations in the following theorem.

Theorem 4.1 (Optimality Theorem). A basic feasible solution is optimal (maximal) if

$$C_R' - C_B' B^{-1} R \leq O', \text{i.e.}, \bar{c}_j \leq 0, j = 1, \ldots, n-m.$$

Moreover, an optimal basic feasible solution is unique if

$$C_R' - C_B' B^{-1} R < O', \text{i.e.}, \bar{c}_j < 0, j = 1, \ldots, n-m.$$

Hereafter, the terms $\bar{c}_j, j = 1, \ldots, n-m$, will be referred to as **optimality evaluators.**

We noted above that since $f = 2x_1 + 3x_2 + 4x_3$, $f = 0$ when $X_R = O$ is definitely not maximal since f can increase in value if any nonbasic variable does. Now at the moment each is equal to zero so that if, for instance, we increase the value of x_3 from the zero level by some positive amount, it is obvious that, with $\rho(B) = 2$, we must change the status of some variable, which is currently basic to nonbasic because there are at most two linearly independent vectors in a basis for $\mathcal{E}^2$ and thus at most two basic variables. Hence, we must determine which basic variable, x_4 or x_5, should now become nonbasic (i.e. decrease in value to zero). The answer to this question is the topic of the subsequent section wherein we consider the generation of a succession of extreme point solutions to a linear programming problem.

4.2 Improving a Basic Feasible Solution

To set the stage for the development of a computationally efficient version of the simplex procedure for solving a linear programming problem, let us form the following simultaneous system involving $m + 1$ equations and the $n + 1$ variables $\bar{X}, f$, namely

$$\begin{aligned} \bar{A}\bar{X} \quad &= b \\ -\bar{C}'\bar{X} + f &= 0 \end{aligned} \quad \text{or} \quad \begin{bmatrix} \bar{A} & O \\ -\bar{C}' & 1 \end{bmatrix} \begin{bmatrix} \bar{X} \\ f \end{bmatrix} = \begin{bmatrix} b \\ 0 \end{bmatrix}. \tag{4.6}$$

Here the objective function is treated "as if" it were an additional structural constraint. In this formulation of the linear programming problem we wish to obtain a solution to (4.6) wherein f is as large as possible (and unrestricted in sign) with $\bar{X} \geq O$. To consolidate our notation a bit, let

$$\hat{\boldsymbol{A}} = \begin{bmatrix} \bar{\boldsymbol{A}} & \boldsymbol{O} \\ -\bar{\boldsymbol{C}}' & 1 \end{bmatrix}, \hat{\boldsymbol{X}} = \begin{bmatrix} \bar{\boldsymbol{X}} \\ f \end{bmatrix}, \text{ and } \hat{\boldsymbol{b}} = \begin{bmatrix} \boldsymbol{b} \\ 0 \end{bmatrix},$$

where $\hat{\boldsymbol{A}}$ is of order $(m + 1 \times n + 1)$, $\hat{\boldsymbol{X}}$ is of order $(n + 1 \times 1)$, and $\hat{\boldsymbol{b}}$ is of order $(m + 1 \times 1)$. Then (4.6) can be written more succinctly as

$$\hat{\boldsymbol{A}}\hat{\boldsymbol{X}} = \hat{\boldsymbol{b}}. \tag{4.6.1}$$

Since the columns within $\hat{\boldsymbol{A}}$ contain $m + 1$ components, a basis for $\mathcal{E}^{m+1}$ may be formed by selecting $m + 1$ linearly independent vectors from $\hat{\boldsymbol{A}}$. Let the said collection of vectors constitute the columns of the $(m + 1)$ st order nonsingular basis matrix $\hat{\boldsymbol{B}} = \left[\hat{\boldsymbol{b}}_1, \ldots, \hat{\boldsymbol{b}}_{m+1}\right]$. Given that the objective function is to be maximized, the variable f must be a member of every basic solution. In this regard, since the last column of $\hat{\boldsymbol{A}}$ corresponds to the variable f, the unit column vector $\boldsymbol{e}_{m+1}$ will be a member of every basis matrix. If we adopt the convention of letting $\boldsymbol{e}_{m+1}$ always appear in the last column of the basis matrix, then $\hat{\boldsymbol{B}} = \left[\hat{\boldsymbol{b}}_1, \ldots, \hat{\boldsymbol{b}}_m, \boldsymbol{e}_{m+1}\right]$.

How does the task of obtaining a basic solution to $\hat{\boldsymbol{A}}\hat{\boldsymbol{X}} = \hat{\boldsymbol{b}}$ compare with our previous problem involving the determination of a basic solution to $\bar{\boldsymbol{A}}\bar{\boldsymbol{X}} = \bar{\boldsymbol{b}}$? To answer this question, let us see exactly how the vectors $\hat{\boldsymbol{b}}_i, i = 1, \ldots, m,$ are to be selected. Let the first n columns of $\hat{\boldsymbol{A}}$ be denoted as

$$\hat{\boldsymbol{a}}_j = \begin{bmatrix} \bar{\boldsymbol{a}}_j \\ -c_j \end{bmatrix}, j = 1, \ldots, n, \tag{4.7}$$

where $\bar{\boldsymbol{a}}_j$ is the jth column of $\bar{\boldsymbol{A}}$. If we can find m linearly independent vectors $\boldsymbol{b}_i$ within $\bar{\boldsymbol{A}}$, then, upon substituting them into (4.7), m of the $\hat{\boldsymbol{a}}_j$ can correspondingly be represented as

$$\hat{\boldsymbol{b}}_i = \begin{bmatrix} \boldsymbol{b}_i \\ -c_{Bi} \end{bmatrix}, i = 1, \ldots, m,$$

where, as before, $\boldsymbol{B} = [\boldsymbol{b}_1, \ldots, \boldsymbol{b}_m]$. Hence

$$\hat{\boldsymbol{B}} = \begin{bmatrix} \boldsymbol{B} & \boldsymbol{O} \\ -\boldsymbol{C}_B' & 1 \end{bmatrix},$$

i.e. since m linearly independent $\bar{\boldsymbol{a}}_j$ uniquely determine m linearly independent $\hat{\boldsymbol{a}}_j$, there exists a one-to-one correspondence between the basis matrix $\boldsymbol{B}$ for $\bar{\boldsymbol{A}}\bar{\boldsymbol{X}} = \boldsymbol{b}$ and the basis matrix $\hat{\boldsymbol{B}}$ for $\hat{\boldsymbol{A}}\hat{\boldsymbol{X}} = \hat{\boldsymbol{b}}$. If the variables associated with the $m + 1$ columns of $\hat{\boldsymbol{B}}$ are denoted as $x_{\hat{B}i}, i = 1, \ldots, m, f$, then a basic feasible solution to $\hat{\boldsymbol{A}}\hat{\boldsymbol{X}}=\hat{\boldsymbol{b}}$ will be defined as one for which f is as large as possible and $x_{\hat{B}i} \geq 0, i=1,\ldots,m$. In order to guarantee that the variant of the simplex process to be presented below converges to an optimal basic feasible solution in a finite number of iterations, let us again impose the nondegeneracy assumption set out at the beginning of the previous chapter, namely, that every set of m columns from $\bar{\boldsymbol{A}}$ is linearly independent.

To determine an initial basic feasible solution to $\hat{\boldsymbol{A}}\hat{\boldsymbol{X}}=\hat{\boldsymbol{b}}$ let us form the $(m+1 \times n+2)$ augmented matrix associated with this system as

$$
\left[\hat{A} \vdots \hat{b}\right]=\begin{bmatrix} \bar{A} & O & b \\ -\bar{C}' & 1 & 0 \end{bmatrix}
$$

$$
\begin{array}{c} \\ = \end{array}
\begin{array}{c}
\begin{array}{ccccccccccc} x_1 & x_2 & & x_p & x_{p+1} & x_{p+2} & & x_{p+m} & f & \end{array} \\
\begin{bmatrix}
a_{11} & a_{12} & \ldots & a_{1p} & 1 & 0 & \ldots & 0 & 0 & b_1 \\
a_{21} & a_{22} & \ldots & a_{2p} & 0 & 1 & \ldots & 0 & 0 & b_2 \\
\cdots & \cdots & \cdots & \cdots & \cdots & \cdots & \cdots & \cdots & \cdots & \cdots \\
a_{m1} & a_{m2} & \ldots & a_{mp} & 0 & 0 & \ldots & 1 & 0 & b_m \\
-c_1 & -c_2 & \ldots & -c_p & 0 & 0 & \ldots & 0 & 1 & 0
\end{bmatrix}
\end{array}. \tag{4.8}
$$

(For purposes of exposition, the variables corresponding to the first $m+1$ columns of (4.8) appear directly above this matrix.) Henceforth this augmented matrix will be referred to as the **simplex matrix**. For example, if we consider the problem presented at the outset of the previous chapter, i.e. we wish to

$$
\begin{aligned}
&max\, f=2x_1+3x_2+4x_3 \quad \text{s.t.}\\
&\quad 2x_1+x_2+4x_3 \leq 100\\
&\quad x_1+3x_2+x_3 \leq 80\\
&\quad x_1,x_2,x_3 \geq 0,
\end{aligned}
$$

then the simplex matrix becomes

$$
\begin{bmatrix}
2 & 1 & 4 & 1 & 0 & 0 & 100 \\
1 & 3 & 1 & 0 & 1 & 0 & 80 \\
-2 & -3 & -4 & 0 & 0 & 1 & 0
\end{bmatrix}. \tag{4.9}
$$

If the basic variables are taken to be those corresponding to the $m+1$ unit column vectors in (4.8), then we may express these $m+1$ variables $x_{p+1}, \ldots, x_{p+m}, f$ in terms of the p remaining or nonbasic variables $x_1, \ldots, x_p$ as

$$
\begin{aligned}
x_{p+1} &= b_1 - \sum_{j=1}^{p} a_{1j}x_j \\
x_{p+2} &= b_2 - \sum_{j=1}^{p} a_{2j}x_j \\
&\cdots\cdots\cdots\cdots \\
x_{p+m} &= b_m - \sum_{j=1}^{p} a_{mj}x_j \\
f &= \sum_{j=1}^{p} c_j x_j.
\end{aligned} \tag{4.10}
$$

Our initial basic feasible solution is then determined by setting the nonbasic variables in (4.10) equal to zero, to wit $x_{p+1} = b_1$, $x_{p+2} = b_2$, …, $x_{p+m} = b_m$, and $f = 0$. Have we obtained an optimal (maximal) basic feasible solution? To answer this question, let us examine the last equation in (4.10). It is obvious that f will increase in value if any nonbasic variable changes from its current zero level by some positive amount. In this regard, if any nonbasic variable has a positive coefficient in the $(m + 1)$ st equation in system (4.10), its associated coefficient in the $(m + 1)$ st row of (4.8) must be negative. Thus, if any negative values appear in the last or objective function row of the simplex matrix, the current basic feasible solution is not optimal. In terms of (4.9), if x_4 and x_5 are taken to be basic variables with x_1, x_2, and x_3 deemed nonbasic, then

$$
\begin{aligned}
x_4 &= 100 - 2x_1 - x_2 - 4x_3 \\
x_5 &= 80 - x_1 - 3x_2 - x_3 \\
f &= 2x_1 + 3x_2 + 4x_3
\end{aligned}
$$

and thus for $x_1 = x_2 = x_3 = 0$, $x_4 = 100$, $x_5 = 80$, and $f = 0$. Clearly, this initial basic feasible solution is not optimal since there are three negative entries in the last row of (4.9). In general, a particular basic feasible solution is optimal when all of the coefficients on the nonbasic variables within the last row of the simplex matrix are nonnegative or $-\bar{c}_j \geq 0, j = 1, \ldots, n - m$. Additionally, if the coefficient on some particular nonbasic variable within the last row of the simplex matrix is zero, then if that variable increases in value to some positive level, no change in f occurs. Hence the objective function assumes the same value for more than one basic feasible solution so that the case of multiple optimal solutions obtains. (As the reader has probably already recognized, the optimality criterion just stated is simply a tacit and informal application of Optimality Theorem 4.1 of the preceding section.)

Given that our initial basic feasible solution is not optimal, let us employ an iterative scheme that enables us to generate an optimal basic feasible solution using (4.8) as the starting point. To this end we shall first indicate some of the specific steps that will be executed in our search for an optimal basic feasible solution. Given that the simplex matrix has within it a set of $m + 1$ unit column vectors, let us consider a set of operations on this equation system that transforms it into an equivalent system whose associated simplex matrix also contains $m + 1$ unit column vectors. In the light of our discussion in Chapter 2 pertaining to the generation of equivalent systems of equations, the following two types of elementary row operations will transform (4.8) into a new simplex matrix which is row-equivalent to the latter:

TYPE 1. Multiply any equation E_h by a nonzero constant k and replace E_h by the new equation kE_h, $h = 1, \ldots, m$. (Note that since $h \neq m + 1$, this type of operation may not be performed on the objective function row of the simplex matrix.)

TYPE 2. Replace any equation E_i by the new equation $E_i + kE_h, i = 1, \ldots, m + 1$; $h = 1, \ldots, m$.

In actual practice (4.8) will be transformed into an equivalent system of equations yielding a new basic feasible solution, with an objective function value which is at least as great as that obtained from the previous one, by performing a particular sequence of elementary row operations called a **pivot operation**. Such an operation consists of an appropriate sequence of at most $m + 1$ elementary row operations, which transform (4.8) into an equivalent system in which a specified variable has a coefficient of unity in one equation and zeros elsewhere. The implementation of the process then proceeds as follows:

1) select a **pivotal term** $a_{rs} \neq 0$, $r \leq m$, $s \leq p + m = n$, within the simplex matrix (since $r \neq m + 1$, $s \neq n + 1$, $n + 2$, the elements within the last row and last two columns of this matrix may not serve as pivotal elements). Next,
2) replace E_r by $\dfrac{1}{a_{rs}}E_r$. Finally,
3) replace E_i by $E_i - \dfrac{a_{is}}{a_{rs}}E_r, i = 1, \ldots, m + 1; i \neq r$.

Given that each $-\bar{c}_j \geq 0, j = 1, \ldots, p$, in (4.8), our first step will be to determine which nonbasic variable is to become basic. We shall adopt the convention of choosing the nonbasic variable x_j, $j = 1, \ldots, p$, whose associated coefficient $-c_j$ within the objective function row of the simplex matrix is "most negative," i.e.

$$-c_k = \min_j\{-c_j < 0, j = 1, \ldots, p\}.$$

(Entry Criterion)

In terms of (4.10), such a choice produces the largest increment in f per unit change in x_j, $j = 1, \ldots, p$. If $-c_k$ happens to represent the most negative coefficient in the last row of (4.8), then x_k, the nonbasic variable corresponding to the kth column of (4.8), becomes basic. Since only one nonbasic variable will be introduced into the set of basic variables at a time, $x_1, \ldots, x_{k-1}, x_{k-1}, \ldots, x_p$ are held fixed at their current zero level. Hence, from (4.10),

$$\begin{aligned} x_{p+1} &= b_1 - a_{1k}x_k \\ x_{p+2} &= b_2 - a_{2k}x_k \\ &\ldots\ldots\ldots\ldots\ldots\ldots \\ x_{p+m} &= b_m - a_{mk}x_k \\ f &= c_k x_k. \end{aligned} \tag{4.10.1}$$

As x_k increases in value from zero, the new x_{p+i}, $i = 1, \ldots, m$, must still satisfy the nonnegativity conditions, i.e. we require that

$$x_{p+i} = b_i - a_{ik} \geq 0,, i = 1, \ldots, m, \tag{4.11}$$

Under what conditions will (4.11) be satisfied? If $a_{ik} \leq 0$, then $x_{p+i} > 0$ so that we need only concern ourselves with those $a_{ik} > 0$. In this regard, when at least one of the coefficients a_{ik} within the kth column of (4.8) is positive, there exists an upper bound to an increase in x_k that will not violate (4.11). Rearranging (4.11) yields $b_i/a_{ik} \geq x_k$ so that any increase in x_k that equals

$$\min_j \left\{ \frac{b_i}{a_{ik}}, a_{ik} > 0, i = 1, \ldots, m \right\}$$

preserves feasibility. Let us assume that for $i = r$,

$$\theta = \frac{b_r}{a_{rk}} = \underset{\text{(Exit Criterion)}}{\min_i \left\{ \frac{b_i}{a_{ik}}, a_{ik} > 0, i = 1, \ldots, m \right\}}, \tag{4.12}$$

i.e. in terms of (4.8), column k replaces column $p + r$ in the basis and thus x_k becomes a basic variable while x_{p+r} turns nonbasic. From (4.11), our new basic feasible solution becomes

$$\begin{aligned} x_k &= \theta \\ x_{p+i} &= b_i - a_{ik}\theta, i = 1, \ldots, m, i \neq r, \text{ and} \\ f &= c_k\theta. \end{aligned} \tag{4.11.1}$$

It is evident from this discussion that (4.12): (i) determines the vector to be removed from the current basis; (ii) determines the value of the variable corresponding to the incoming vector for the new basis; and (iii) ensures that the new basic solution to the augmented structural constraint system is feasible.

Turning to the simplex matrix, if x_k is to replace x_{p+r} in the set of basic variables, a pivot operation with a_{rk} as the (circled) pivotal element may be performed so as to express the new set of basic variables x_{p+i}, $i = 1, \ldots, m$, $i \neq r$,

x_k, and f in terms of the nonbasic variables $x_1, \ldots, x_{k-1}, x_{k+1}, \ldots, x_p$, and x_{p+r}. To this end, let us first divide E_r in (4.8) by a_{rk} to obtain

$$\begin{array}{cccccccccccccccccc}
x_1 & x_2 & & x_{k-1} & x_k & x_{k+1} & & x_p & x_{p+1} & x_{p+2} & & x_{p+r} & & x_{p+m} & f & \\
\end{array}$$
$$\left[\begin{array}{cccccccccccccccccc}
a_{11} & a_{12} & \cdots & a_{1,k-1} & a_{1k} & a_{1,k+1} & \cdots & a_{1p} & 1 & 0 & \cdots & 0 & \cdots & 0 & 0 & b_1 \\
\cdots & \cdots & \cdots & \cdots & \cdots & \cdots & \cdots & \cdots & \cdots & \cdots & \cdots & \cdots & \cdots & \cdots & \cdots & \cdots \\
\frac{a_{r1}}{a_{rk}} & \frac{a_{r2}}{a_{rk}} & \cdots & \frac{a_{r,k-1}}{a_{rk}} & \textcircled{1} & \frac{a_{r,k+1}}{a_{rk}} & \cdots & \frac{a_{rp}}{a_{rk}} & 0 & 0 & \cdots & \frac{1}{a_{rk}} & \cdots & 0 & 0 & \theta \\
\cdots & \cdots & \cdots & \cdots & \cdots & \cdots & \cdots & \cdots & \cdots & \cdots & \cdots & \cdots & \cdots & \cdots & \cdots & \cdots \\
a_{m1} & a_{m2} & \cdots & a_{m,k-1} & a_{mk} & a_{m,k+1} & \cdots & a_{mp} & 0 & 0 & \cdots & 0 & \cdots & 1 & 0 & b_m \\
-c_1 & -c_2 & \cdots & -c_{k-1} & -c_k & -c_{k+1} & \cdots & -c_p & 0 & 0 & \cdots & 0 & \cdots & 0 & 1 & 0
\end{array}\right]. \tag{4.8.1}$$

Next, to eliminate x_k from the remaining m equations, E_i, $i = 1, \ldots, m+1, i \neq r$, let us: (i) multiply E_r by $-a_{ik}$ and replace E_i by the new equation $E_i - a_{ik}E_r$, $i = 1, \ldots, m, i \neq r$; and (ii) multiply E_r by c_k and replace E_{m+1} by the new equation $E_{m+1} + c_kE_r$. Hence, (4.8.1) is ultimately transformed to

$$\begin{array}{cccccccccccccccccc}
x_1 & x_2 & & x_{k-1} & x_k & x_{k+1} & & x_p & x_{p+1} & x_{p+2} & & x_{p+r} & & x_{p+m} & f & \\
\end{array}$$
$$\left[\begin{array}{cccccccccccccccccc}
a'_{11} & a'_{12} & \cdots & a'_{1,k-1} & 0 & a'_{1,k+1} & \cdots & a'_{1p} & 1 & 0 & \cdots & a'_{1,p+r} & \cdots & 0 & 0 & b'_1 \\
\cdots & \cdots & \cdots & \cdots & \cdots & \cdots & \cdots & \cdots & \cdots & \cdots & \cdots & \cdots & \cdots & \cdots & \cdots & \cdots \\
\frac{a_{r1}}{a_{rk}} & \frac{a_{r2}}{a_{rk}} & \cdots & \frac{a_{r,k-1}}{a_{rk}} & \textcircled{1} & \frac{a_{r,k+1}}{a_{rk}} & \cdots & \frac{a_{rp}}{a_{rk}} & 0 & 0 & \cdots & \frac{1}{a_{rk}} & \cdots & 0 & 0 & \theta \\
\cdots & \cdots & \cdots & \cdots & \cdots & \cdots & \cdots & \cdots & \cdots & \cdots & \cdots & \cdots & \cdots & \cdots & \cdots & \cdots \\
a'_{m1} & a'_{m2} & \cdots & a'_{m,k-1} & 0 & a'_{m,k+1} & \cdots & a'_{mp} & 0 & 0 & \cdots & a'_{m,p+r} & \cdots & 1 & 0 & b'_m \\
c'_1 & c'_2 & \cdots & c'_{k-1} & 0 & c'_{k+1} & \cdots & c'_p & 0 & 0 & \cdots & c'_{p+r} & \cdots & 0 & 1 & c_k\theta
\end{array}\right], \tag{4.8.2}$$

where

$$a'_{ij} = a_{ij} - a_{ik}\frac{a_{rj}}{a_{rk}}, i = 1, \ldots, m, i \neq r; j = 1, \ldots, p+m$$

$$b'_1 = b_i - a_{ik}\theta, i = 1, \ldots, m, i \neq r$$

$$c'_j = -c_j + c_k\frac{a_{rj}}{a_{rk}}, j = 1, \ldots, p+m.$$

If the basic variables are again taken to be those corresponding to the $m + 1$ unit column vectors in the simplex matrix, then, from (4.8.2), these variables may be expressed in terms of the remaining or nonbasic variables as

$$x_k = \theta - \sum_{\substack{j=1 \\ j \neq k}}^{p} \frac{a_{rj}}{a_{rk}} x_j - \frac{1}{a_{rk}} x_{p+r}$$

$$x_{p+i} = b'_i - \sum_{\substack{j=1 \\ j \neq k}}^{p} a'_{ij} x_j - a_{i,p+r} x_{p+r}, i = 1, \dots, m, i \neq r \tag{4.13}$$

$$f = c_k \theta - \sum_{\substack{j=1 \\ j \neq k}}^{p} c'_j x_j - c'_{p+r} x_{p+r}.$$

Upon setting the nonbasic variables equal to zero, our new basic feasible solution appears as

$$\begin{aligned} x_k &= \theta \\ x_{p+i} &= b'_i, i = 1, \dots, m, i \neq r, \text{and} \\ f &= c_k \theta. \end{aligned} \tag{4.13.1}$$

(Note that this basic feasible solution may be obtained directly from the simplex matrix by ignoring all columns within (4.8.2) save those corresponding to the $m + 1$ unit column vectors and the (last) column of constants.) In addition, this solution is optimal if each $c'_j, j = 1, \dots, k-1, \dots, k+1, \dots, p, p+r$, within the objective function row of (4.8.2) is nonnegative. If negative coefficients still appear within the last row of (4.8.2), then the above pivotal process is repeated until an optimal basic feasible solution is attained.

We now return to the maximization problem started earlier. Let us duplicate the associated simplex matrix (4.9) as

$$\begin{bmatrix} 2 & 1 & \textcircled{4} & 1 & 0 & 0 & 100 \\ 1 & 3 & 1 & 0 & 1 & 0 & 80 \\ -2 & -3 & -4 & 0 & 0 & 1 & 0 \end{bmatrix} \begin{matrix} 100/4 = 25 \\ 80/1 = 80 \\ \end{matrix}. \tag{4.9}$$

Remembering that the basic variables are those corresponding to the $m + 1$ unit column vectors in this matrix, our initial basic feasible solution consists of $x_4 = 100$, $x_5 = 80$, and $f = 0$. Given that there are some negative coefficients within the last row of (4.9), the basic feasible solution just obtained is not optimal. To generate an improved solution, let us undertake Iteration 1. Since −4 is the most negative entry in this row, the nonbasic variable x_3 now becomes basic. What is the largest increment in x_3 that preserves feasibility? To determine this, let us take the positive coefficients on x_3 in the first two rows of (4.9), namely 4 and 1, and divide them, respectively, into the first two constants in the last column of this matrix (100 and 80). Carrying out the indicated division to the right of (4.9) yields $\theta = 25 = min\,\{25, 80\}$ as the maximum allowable increase in x_3. Hence, the (circled) pivotal element within simplex matrix (4.9) is 4. In the calculations to follow, column 3 of (4.9) will be replaced by a unit column vector with unity in the pivotal position and zeros elsewhere. Such a unit column vector lies in the fourth column of (4.9). Hence x_4 turns nonbasic (it loses its unit column vector to x_3). In order to achieve this transformation, the pivot operation on (4.9) proceeds as follows. If E_1 is replaced by $\frac{1}{4}E_1$, we obtain

$$\begin{bmatrix} \frac{1}{2} & \frac{1}{4} & \textcircled{1} & \frac{1}{4} & 0 & 0 & 25 \\ 1 & 3 & 1 & 0 & 1 & 0 & 80 \\ -2 & -3 & -4 & 0 & 0 & 1 & 0 \end{bmatrix}. \tag{4.9.1}$$

Next, replacing E_2, E_3 in (4.9.1) by $E_2 - E_1$, $E_3 + 4E_1$, respectively, produces

$$\begin{bmatrix} \frac{1}{2} & \frac{1}{4} & 1 & \frac{1}{4} & 0 & 0 & 25 \\ \frac{1}{2} & \textcircled{\frac{11}{4}} & 0 & -\frac{1}{4} & 1 & 0 & 55 \\ 0 & -2 & 0 & 1 & 0 & 1 & 100 \end{bmatrix} \begin{matrix} 25/(1/4) = 100 \\ 55/(11/4) = 20. \\ \end{matrix} \tag{4.9.2}$$

Our second basic feasible solution is thus $x_3 = 25$, $x_5 = 55$, and $f = 100$. Given that a negative value appears in the last row of (4.9.2), this basic feasible solution is still not optimal.

Iteration 2. To obtain a maximal solution, let us change the status of x_2 from nonbasic to basic. Since the largest increment in x_2 that does not violate the nonnegativity conditions is $\theta = 20 = min\,\{100, 20\}$, the pivotal element within (4.9.2) is 11/4. This time column 2 of (4.9.2) is to be replaced by the unit column vector appearing in column 5 of the same so that x_5 will be removed from the basis. To this end we now perform a pivot operation on (4.9.2). First, let us replace E_2 by $\frac{4}{11}E_2$ to obtain

$$\begin{bmatrix} \frac{1}{2} & \frac{1}{4} & 1 & \frac{1}{4} & 0 & 0 & 25 \\ \frac{2}{11} & \textcircled{1} & 0 & -\frac{1}{11} & \frac{4}{11} & 0 & 20 \\ 0 & -2 & 0 & 1 & 0 & 1 & 100 \end{bmatrix}. \tag{4.9.3}$$

Second, upon replacing E_1, E_3 in (4.9.3) by $E_1 - \frac{1}{4}E_2$, $E_3 + 2E_2$, respectively, we ultimately have

$$\begin{bmatrix} \frac{5}{11} & 0 & 1 & \frac{3}{11} & -\frac{1}{11} & 0 & 20 \\ \frac{2}{11} & \textcircled{1} & 0 & -\frac{1}{11} & \frac{4}{11} & 0 & 20 \\ \frac{4}{11} & 0 & 0 & \frac{9}{11} & \frac{8}{11} & 1 & 140 \end{bmatrix}. \tag{4.9.4}$$

This particular basic feasible solution consisting of $x_2 = 20, x_3 = 20$, and $f = 140$ is optimal since only positive entries appear in the objective function row of (4.9.4).

Example 4.1 Let us solve

$$\begin{aligned} &max\, f = 50x_1 + 40x_2 + 60x_3 \text{ s.t.} \\ &x_1 + 2x_2 + x_3 \le 70 \\ &2x_1 + x_2 + x_3 \le 40 \\ &3x_1 + x_2 + 3x_3 \le 90 \\ &x_1, x_2, x_3 \ge 0. \end{aligned}$$

To set up the simplex matrix, we may introduce three nonnegative slack variables, x_4, x_5, and x_6 and incorporate the objective function as an additional equation so as to obtain

$$\begin{aligned} x_1 + 2x_2 + x_3 + x_4 \qquad\qquad &= 70 \\ 2x_1 + x_2 + x_3 \quad + x_5 \qquad &= 40 \\ 3x_1 + x_2 + 3x_3 \qquad + x_6 &= 90 \\ -50x_1 - 40x_2 - 60x_3 \qquad + f &= 0. \end{aligned}$$

Then the associated simplex matrix

$$\begin{bmatrix} 1 & 2 & 1 & 1 & 0 & 0 & 0 & 70 \\ 2 & 1 & 1 & 0 & 1 & 0 & 0 & 40 \\ 3 & 1 & 3 & 0 & 0 & 1 & 0 & 90 \\ -50 & -40 & -60 & 0 & 0 & 0 & 1 & 0 \end{bmatrix}$$

yields the first basic feasible solution consisting of $x_4 = 70$, $x_5 = 40$, $x_6 = 90$, and $f = 0$. Using the initial simplex matrix as a starting point, let us proceed to a new basic feasible solution by undertaking the following:

Iteration 1.

$$\left[\begin{array}{cccccccc} 1 & 2 & 1 & 1 & 0 & 0 & 0 & 70 \\ 2 & 1 & 1 & 0 & 1 & 0 & 0 & 40 \\ 3 & 1 & \textcircled{3} & 0 & 0 & 1 & 0 & 90 \\ -50 & -40 & -60 & 0 & 0 & 0 & 1 & 0 \end{array}\right] \begin{array}{c} 70 \\ 40 \\ 30 \\ \end{array} \rightarrow \left[\begin{array}{cccccccc} 1 & 2 & 1 & 1 & 0 & 0 & 0 & 70 \\ 2 & 1 & 1 & 0 & 1 & 0 & 0 & 40 \\ 1 & \frac{1}{3} & \textcircled{1} & 0 & 0 & \frac{1}{3} & 0 & 30 \\ -50 & -40 & -60 & 0 & 0 & 0 & 1 & 0 \end{array}\right] \rightarrow$$

(Determining the pivot: x_3 replaces x_6 in the set of basic variables.) $\left(\frac{1}{3}E_3 \text{ replaces } E_3.\right)$

$$\left[\begin{array}{cccccccc} 0 & \frac{5}{3} & 0 & 1 & 0 & -\frac{1}{3} & 0 & 40 \\ 1 & \frac{2}{3} & 0 & 0 & 1 & -\frac{1}{3} & 0 & 10 \\ 1 & \frac{1}{3} & 1 & 0 & 0 & \frac{1}{3} & 0 & 30 \\ 10 & -20 & 0 & 0 & 0 & 20 & 1 & 1800 \end{array}\right].$$

($E_1 - E_3$, $E_2 - E_3$, $E_4 + 60E_3$ replace E_1, E_2, E_4, respectively.)

In this instance, the second basic feasible solution is $x_3 = 30$, $x_4 = 40$, $x_5 = 10$, and $f = 1800$. Since this solution is not optimal, we proceed to

Iteration 2.

$$\left[\begin{array}{cccccccc} 0 & \frac{5}{3} & 0 & 1 & 0 & -\frac{1}{3} & 0 & 40 \\ 1 & \textcircled{\frac{2}{3}} & 0 & 0 & 1 & -\frac{1}{3} & 0 & 10 \\ 1 & \frac{1}{3} & 1 & 0 & 0 & \frac{1}{3} & 0 & 30 \\ 10 & -20 & 0 & 0 & 0 & 20 & 1 & 1800 \end{array}\right] \begin{array}{c} 24 \\ 15 \\ 90 \\ \end{array} \rightarrow \left[\begin{array}{cccccccc} 0 & \frac{5}{3} & 0 & 1 & 0 & -\frac{1}{3} & 0 & 40 \\ \frac{3}{2} & \textcircled{1} & 0 & 0 & \frac{3}{2} & -\frac{1}{2} & 0 & 15 \\ 1 & \frac{1}{3} & 1 & 0 & 0 & \frac{1}{3} & 0 & 30 \\ 10 & -20 & 0 & 0 & 0 & 20 & 1 & 1800 \end{array}\right]$$

(Determining the pivot: x_2 replaces x_5 in the set of basic variables.) $\left(\frac{3}{2}E_2 \text{ replaces } E_2.\right)$

$$\rightarrow \left[\begin{array}{cccccccc} -\frac{5}{2} & 0 & 0 & 1 & -\frac{5}{2} & \frac{1}{2} & 0 & 15 \\ \frac{3}{2} & 1 & 0 & 0 & \frac{3}{2} & -\frac{1}{2} & 0 & 15 \\ \frac{1}{2} & 0 & 1 & 0 & \frac{1}{2} & \frac{1}{2} & 0 & 25 \\ 40 & 0 & 0 & 0 & 30 & 10 & 1 & 2100 \end{array}\right].$$

$\left(E_1 - \frac{5}{3}E_2,\ E_3 - \frac{1}{3}E_2,\ E_4 + 20E_2 \text{ replace } E_1, E_3, E_4, \text{ respectively.}\right)$

An examination of the last row of the final simplex matrix indicates that we have attained an optimal basic feasible solution composed of $x_2 = 15$, $x_3 = 25$, $x_4 = 15$, and $f = 2100$. ■

Example 4.2 Next, we shall solve

$$\begin{aligned} \max f &= 2x_1 + 3x_2 \text{ s.t.} \\ x_1 &\leq 10 \\ x_2 &\leq 8 \\ \frac{2}{3}x_1 + x_2 &\leq 12 \\ x_1, x_2 &\geq 0. \end{aligned}$$

For this problem, the simplex matrix will be derived from the following system involving four equations and six unknowns:

$$\begin{aligned} x_1 \quad\quad + x_3 \quad\quad &= 10 \\ x_2 \quad + x_4 \quad &= 8 \\ \frac{2}{3}x_1 + x_2 \quad\quad + x_5 &= 12 \\ -2x_1 - 3x_2 \quad\quad + f &= 0 \end{aligned}$$

where x_3, x_4, and x_5 are nonnegative slack variables. Hence, the implied matrix becomes

$$\begin{bmatrix} 1 & 0 & 1 & 0 & 0 & 0 & 10 \\ 0 & 1 & 0 & 1 & 0 & 0 & 8 \\ \frac{2}{3} & 1 & 0 & 0 & 1 & 0 & 12 \\ -2 & -3 & 0 & 0 & 0 & 1 & 0 \end{bmatrix}$$

and our initial basic feasible solution is $x_3 = 10$, $x_4 = 8$, $x_5 = 12$, and $f = 0$. Since this solution is not optimal, we undertake the following:

Iteration 1.

$$\begin{bmatrix} 1 & 0 & 1 & 0 & 0 & 0 & 10 \\ 0 & \textcircled{1} & 0 & 1 & 0 & 0 & 8 \\ \frac{2}{3} & 1 & 0 & 0 & 1 & 0 & 12 \\ -2 & -3 & 0 & 0 & 0 & 1 & 0 \end{bmatrix} \begin{matrix} \\ 8 \\ 12 \\ \\ \end{matrix} \rightarrow \begin{bmatrix} 1 & 0 & 1 & 0 & 0 & 0 & 10 \\ 0 & 1 & 0 & 1 & 0 & 0 & 8 \\ \frac{2}{3} & 0 & 0 & -1 & 1 & 0 & 4 \\ -2 & 0 & 0 & 3 & 0 & 1 & 24 \end{bmatrix}.$$

(Determining the pivot: x_2 replaces x_4 in the set of basic variables.)

($E_3 - E_2$, $E_4 + 3E_2$ replace E_3, E_4, respectively.)

Our second basic feasible solution is $x_2 = 8$, $x_3 = 10$, $x_3 = 4$, and $f = 24$. Since it, too, is not optimal, we look to

Iteration 2.

$$\begin{bmatrix} 1 & 0 & 1 & 0 & 0 & 0 & 10 \\ 0 & 1 & 0 & 1 & 0 & 0 & 8 \\ \textcircled{\tfrac{2}{3}} & 0 & 0 & -1 & 1 & 0 & 4 \\ -2 & 0 & 0 & 3 & 0 & 1 & 24 \end{bmatrix} \begin{matrix} 10 \\ \\ 6 \\ \\ \end{matrix} \rightarrow \begin{bmatrix} 1 & 0 & 1 & 0 & 0 & 0 & 10 \\ 0 & 1 & 0 & 1 & 0 & 0 & 8 \\ \textcircled{1} & 0 & 0 & -\frac{3}{2} & \frac{3}{2} & 0 & 6 \\ -2 & 0 & 0 & 3 & 0 & 1 & 24 \end{bmatrix} \rightarrow$$

(Determining the pivot: x_1 replaces x_5 in the set of basic variables.) $\left(\frac{3}{2}E_3 \text{ replaces } E_3.\right)$

$$\begin{bmatrix} 0 & 0 & 1 & \frac{3}{2} & -\frac{3}{2} & 0 & 4 \\ 0 & 1 & 0 & 1 & 0 & 0 & 8 \\ 1 & 0 & 0 & -\frac{3}{2} & \frac{3}{2} & 0 & 6 \\ 0 & 0 & 0 & 0 & 3 & 1 & 36 \end{bmatrix}.$$

$(E_1 - E_3, E_4 + 2E_3$ replace E_1, E_4, respectively.)

Given that all negative entries have been purged from the last row of our simplex matrix, we have obtained an optimal basic feasible solution wherein $x_1 = 6$, $x_2 = 8$, $x_3 = 4$, and $f = 36$. Further examination of the objective function row of this last simplex matrix indicates that since the coefficient on the nonbasic variable x_4 in this row is zero, we may increase x_4 by some positive amount without affecting the optimal value of the objective function, i.e. the status of x_4 can change from nonbasic to basic as we move from one extreme point to another and yet f will still equal 36. To see this let us introduce x_4 into the set of basic variables:

$$\begin{bmatrix} 0 & 0 & 1 & \textcircled{\tfrac{3}{2}} & -\frac{3}{2} & 0 & 4 \\ 0 & 1 & 0 & 1 & 0 & 0 & 8 \\ 1 & 0 & 0 & -\frac{3}{2} & \frac{3}{2} & 0 & 6 \\ -2 & 0 & 0 & 0 & 3 & 1 & 36 \end{bmatrix} \begin{matrix} \frac{8}{3} \\ 8 \\ \\ \\ \end{matrix} \rightarrow \begin{bmatrix} 0 & 0 & \frac{2}{3} & \textcircled{1} & -1 & 0 & \frac{8}{3} \\ 0 & 1 & 0 & 1 & 0 & 0 & 8 \\ 1 & 0 & 0 & -\frac{3}{2} & \frac{3}{2} & 0 & 6 \\ 0 & 0 & 0 & 0 & 3 & 1 & 36 \end{bmatrix} \rightarrow$$

(Determining the pivot: x_4 replaces x_3 in the set of basic variables.) $\left(\frac{2}{3}E_1 \text{ replaces } E_1.\right)$

$$\begin{bmatrix} 0 & 0 & \frac{2}{3} & 1 & -1 & 0 & \frac{8}{3} \\ 0 & 1 & -\frac{2}{3} & 0 & 1 & 0 & \frac{16}{3} \\ 1 & 0 & 1 & 0 & 0 & 0 & 10 \\ 0 & 0 & 0 & 0 & 3 & 1 & 36 \end{bmatrix}.$$

$\left(E_2 - E_1, E_3 + \frac{3}{2}E_1 \text{ replace } E_2, E_3, \text{ respectively.}\right)$

The second optimal basic feasible solution is $x_1 = 10, x_2 = \frac{16}{3}, x_4 = \frac{8}{3}$, and $f = 36$. Graphically, our first basic feasible solution corresponds to point A in Figure 4.2 while the second is located at point B. In addition, the set of all convex combinations of points A and B, namely $C = \lambda A + (1 - \lambda)B$, $0 \leq \lambda \leq 1$, gives rise to an infinity of optimal nonbasic feasible solutions.To summarize: the case of multiple optimal basic feasible solutions emerges if, in the objective function row of the optimal simplex matrix, a nonbasic variable has a zero coefficient. Is there another way of recognizing that the objective function will be tangent to the feasible region $\mathcal{K}$ at more than one extreme point? A glance at the objective function and the structural constraint inequalities indicates that the optimal objective function equation is three times the third structural constraint when the latter is treated as a strict equality. In this regard, when the objective function can be written as a multiple of some particular structural constraint, the case of multiple optimal solutions obtains. ■

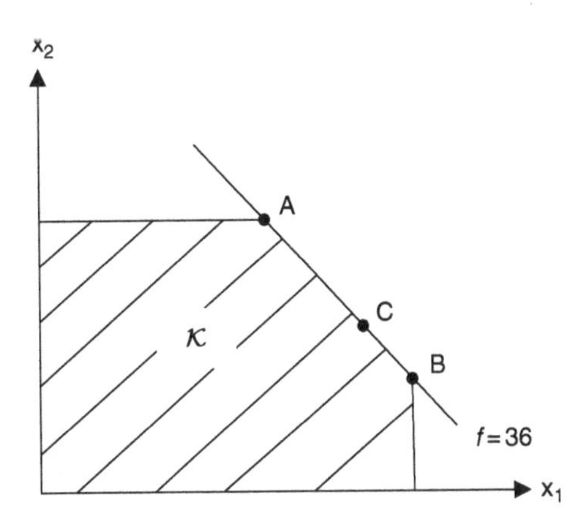

Figure 4.2 Multiple optimal solutions at A, B, and C.

Example 4.3 Solve

$$\begin{aligned} &max\, f = -x_1 + 2x_2 \text{ s.t.} \\ &-2x_1 + x_2 \le 3 \\ &-\frac{1}{2}x_1 + x_2 \le 6 \\ &\quad x_1, x_2 \ge 0. \end{aligned}$$

Given that x_3, x_4 are nonnegative slack variables, we form

$$\begin{aligned} -2x_1 + x_2 + x_3 \quad &= 3 \\ -\frac{1}{2}x_1 + x_2 \quad + x_4 \quad &= 6 \\ x_1 - 2x_2 \qquad + f &= 0. \end{aligned}$$

so as to obtain the simplex matrix

$$\begin{bmatrix} -2 & 1 & 1 & 0 & 0 & 3 \\ -\frac{1}{2} & 1 & 0 & 1 & 0 & 6 \\ 1 & -2 & 0 & 0 & 1 & 0 \end{bmatrix}.$$

Our initial basic feasible solution is $x_3 = 3$, $x_4 = 6$, and $f = 0$. Since it is not optimal, we undertake the following:

Iteration 1.

$$\begin{bmatrix} -2 & \textcircled{1} & 1 & 0 & 0 & 3 \\ -\frac{1}{2} & 1 & 0 & 1 & 0 & 6 \\ 1 & -2 & 0 & 0 & 1 & 0 \end{bmatrix} \begin{matrix} 3 \\ 6 \\ \end{matrix} \quad \rightarrow \quad \begin{bmatrix} -2 & 1 & 1 & 0 & 0 & 3 \\ \frac{3}{2} & 0 & -1 & 1 & 0 & 3 \\ -3 & 0 & 2 & 0 & 1 & 6 \end{bmatrix}.$$

(Determining the pivot : x_2 replaces x_3 in the set of basic variables.) ($E_2 - E_1$, $E_3 + 2E_1$ replace E_2, E_3, respectively.)

Here, the new basic feasible solution consists of $x_2 = 3$, $x_4 = 3$, and $f = 6$. Since this solution is still not optimal, we look to

Iteration 2.

$$\begin{bmatrix} -2 & 1 & 1 & 0 & 0 & 3 \\ \textcircled{\frac{3}{2}} & 0 & -1 & 1 & 0 & 3 \\ -3 & 0 & 2 & 0 & 1 & 6 \end{bmatrix} \begin{matrix} \\ 2 \\ \end{matrix} \rightarrow \begin{bmatrix} -2 & 1 & 1 & 0 & 0 & 3 \\ \textcircled{1} & 0 & -\frac{2}{3} & \frac{2}{3} & 0 & 2 \\ -3 & 0 & 2 & 0 & 1 & 6 \end{bmatrix} \rightarrow$$

(Determining the pivot : x_1 replaces x_4 in the set of basic variables.) $\left(\frac{2}{3}E_2 \text{ replaces } E_2.\right)$

$$\begin{bmatrix} 0 & 1 & -\frac{1}{3} & \frac{4}{3} & 0 & 7 \\ 1 & 0 & -\frac{2}{3} & \frac{2}{3} & 0 & 2 \\ 0 & 0 & 0 & 2 & 1 & 12 \end{bmatrix}.$$

($E_1 - 2E_2$, $E_3 + 3E_2$ replace E_1, E_3, respectively.)

The absence of any negative entries in the last row of the final simplex matrix indicates that this basic feasible solution consisting of $x_1 = 2$, $x_2 = 7$, and $f = 12$ is optimal. We noted earlier that if the objective function coefficient on a nonbasic variable in the optimal simplex matrix is zero, that variable (in this case x_3) can be increased from its current nonbasic value of zero to yield an alternative solution that has the same value of f. While this alternative solution is also optimal, it is, in this particular example, nonbasic, the reason being that the x_3 coefficients in the first two rows of the above matrix are negative. To see this, let us write the basic variables x_1,x_2 in terms of x_3 as

$$x_1 = 2 + \frac{2}{3}x_3$$

$$x_2 = 7 + \frac{1}{2}x_3.$$

In this instance, x_3, as well as x_1 and x_2, can be increased without bound so that an infinity of alternative optimal nonbasic feasible solutions obtains. For each of these alternative solutions we have three positive variables (excluding f) instead of only two, thus violating our requirement for an optimal basic feasible solution. As Figure 4.3 demonstrates, there exists but a single optimal basic feasible solution (point A) and an infinity of optimal nonbasic feasible solutions. ■

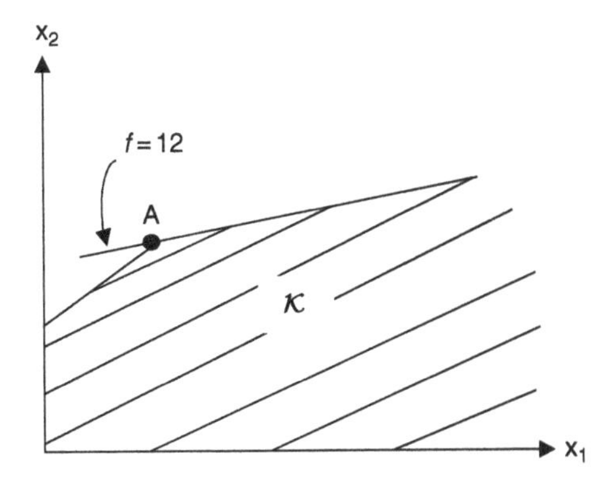

Figure 4.3 A single optimal basic feasible solution at A.

Example 4.4 Let us solve

$$\begin{aligned} &max\, f = -2x_1 + 3x_2 \text{ s.t.} \\ &-2x_1 + x_2 \le 3 \\ &-\frac{1}{2}x_1 + x_2 \le 6 \\ &\quad x_1, x_2 \ge 0. \end{aligned}$$

With x_3, x_4 taken to be nonnegative slack variables, we form

$$\begin{aligned} -2x_1 + x_2 + x_3 \qquad &= 3 \\ -\frac{1}{2}x_1 + x_2 \quad + x_4 \quad &= 6 \\ 2x_1 - 3x_2 \qquad + f \; &= 0 \end{aligned}$$

The associated simplex matrix

$$\begin{bmatrix} -2 & 1 & 1 & 0 & 0 & 3 \\ -\frac{1}{2} & 1 & 0 & 1 & 0 & 6 \\ -2 & -3 & 0 & 0 & 1 & 0 \end{bmatrix}$$

yields the nonoptimal initial basic feasible solution composed of $x_3 = 3$, $x_4 = 6$, and $f = 0$.

Iteration 1.

$$\begin{bmatrix} -2 & \textcircled{1} & 1 & 0 & 0 & 3 \\ -\frac{1}{2} & 1 & 0 & 1 & 0 & 6 \\ -2 & -3 & 0 & 0 & 1 & 0 \end{bmatrix} \begin{matrix} 3 \\ 6 \\ \end{matrix} \quad \rightarrow \quad \begin{bmatrix} -2 & 1 & 1 & 0 & 0 & 3 \\ \frac{3}{2} & 0 & -1 & 1 & 0 & 3 \\ -8 & 0 & 3 & 0 & 1 & 9 \end{bmatrix}$$

(Determining the pivot : x_2 replaces x_3 in the set of basic variables.) ($E_2 - E_1$, $E_3 + 3E_1$ replace E_2,E_3, respectively.)

Our new basic feasible solution consists of $x_2 = 3$, $x_4 = 3$, and $f = 9$. Since it is not optimal, we start

Iteration 2.

$$\begin{bmatrix} -2 & 1 & 1 & 0 & 0 & 3 \\ \textcircled{\frac{3}{2}} & 0 & -1 & 1 & 0 & 3 \\ -8 & 0 & 3 & 0 & 1 & 9 \end{bmatrix} \begin{matrix} \\ 2 \\ \end{matrix} \rightarrow \begin{bmatrix} -2 & 1 & 1 & 0 & 0 & 3 \\ \textcircled{1} & 0 & -\frac{2}{3} & \frac{2}{3} & 0 & 2 \\ -8 & 0 & 3 & 0 & 1 & 9 \end{bmatrix} \rightarrow$$

(Determining the pivot : x_1 replaces in the set of basic variables.) $\left(\frac{2}{3}E_2 \text{ replaces } E_2.\right)$

$$\begin{bmatrix} 0 & 1 & -\frac{1}{3} & \frac{4}{3} & 0 & 7 \\ 1 & 0 & -\frac{2}{3} & \frac{2}{3} & 0 & 2 \\ 0 & 0 & -\frac{7}{3} & \frac{16}{3} & 1 & 25 \end{bmatrix}.$$

($E_1 + 2E_2$, $E_3 + 8E_2$ replace E_1, E_3, respectively.)

Our third basic feasible solution is thus $x_1 = 2$, $x_2 = 7$, and $f = 25$. Even though this particular basic feasible solution is not optimal, we shall terminate the simplex process, the reason being that an examination of column 3 of the last simplex matrix indicates that the value of the objective function is unbounded. To see this, let us express the basic variables x_1, x_2 along with f in terms of x_3, the variable to be introduced into the basis. That is, since

$$x_2 = 7 + \frac{1}{3}x_3$$

$$x_1 = 2 + \frac{2}{3}x_3$$

$$f = 25 + \frac{7}{3}x_3,$$

x_3, as well as x_1, x_2, and f, may be increased without limit (Figure 4.4). Hence we obtain a feasible though nonbasic solution to this problem since the three variables x_1, x_2, and x_3 each assume positive values. In sum, if any column within the simplex matrix corresponding to a nonbasic variable contains all negative coefficients, the value of the objective function is unbounded. ■

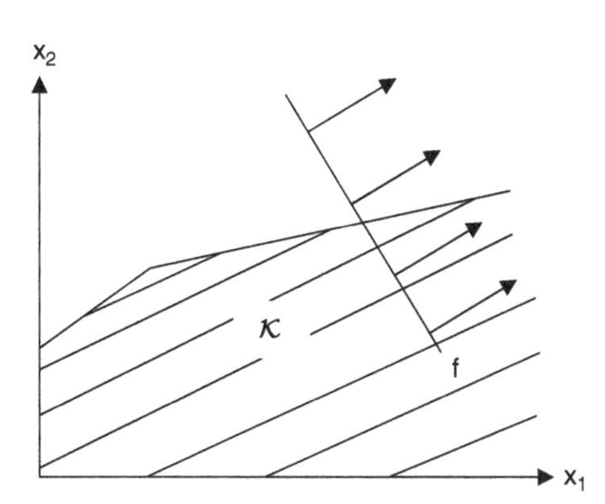

Figure 4.4 The objective function f is unbounded.

4.3 Degenerate Basic Feasible Solutions

Earlier in our discussion of the simplex method we instituted a nondegeneracy assumption, the purpose of which was to ensure that every basic feasible solution to the augmented structural constraint system $\bar{A}\bar{X} = \boldsymbol{b}$ was nondegenerate. In this section we shall examine the implications of the existence of degenerate basic feasible solutions to $\bar{A}\bar{X} = \boldsymbol{b}$. We noted in Chapter 3 that if $\rho(\bar{A}) = m$, we may select m linearly independent columns from $\bar{A}$ and form an $(m \times m)$ nonsingular matrix $\boldsymbol{B}$. We then obtain a basic solution to $\bar{A}\bar{X} = \boldsymbol{b}$ by solving $\boldsymbol{B}\boldsymbol{X}_{\boldsymbol{B}} = \boldsymbol{b}$. Hence such a solution occurs when $\boldsymbol{b}$ is expressible as a linear combination of the columns of $\boldsymbol{B}$. Moreover, the said solution is feasible if $\boldsymbol{b}$ can be represented as a nonnegative linear combination of the same, i.e. $\boldsymbol{b}$ is an element of the convex polyhedral cone $C_{\boldsymbol{B}} = \{Y | Y = \boldsymbol{B}\boldsymbol{X}_{\boldsymbol{B}}, \boldsymbol{X}_{\boldsymbol{B}} \geq \boldsymbol{0}\}$ generated by the columns of $\boldsymbol{B}$. Let us now get a bit more restrictive. If $\boldsymbol{b}$ is expressible with a *strictly positive* linear combination of all the basis vectors in $\boldsymbol{B}$, then the basic feasible solution to $\bar{A}\bar{X} = \boldsymbol{b}$ is **nondegenerate**. In this instance, none of the basic variables with $\boldsymbol{X}_{\boldsymbol{B}}$ vanishes. In sum, the point being made is that degeneracy occurs if $\boldsymbol{b}$ lies within a convex polyhedral cone spanned by only some subset of vectors of $\boldsymbol{B}$, i.e. $\boldsymbol{b}$ is a positive linear combination of fewer than m basis vectors. Hence $\boldsymbol{b}$ lies along an edge of $C_{\boldsymbol{B}}$ and not within $C_{\boldsymbol{B}}$ proper so that not all components of $\boldsymbol{X}_{\boldsymbol{B}}$ are positive.

To relate these observations to our discussion in the previous section pertaining to the generation of new or improved basic feasible solutions, let us address ourselves to a couple of remarks concerning the value of θ that constitutes the minimum of (4.12). As we shall now see, the presence or absence of degeneracy is reflected by the behavior of θ. Let us assume that θ is not unique, i.e. there is a tie for the minimum in (4.12). Hence, we may remove from the basis any one of the columns of $\boldsymbol{B}$ for which the same minimum value of θ is assumed and obtain a new (call it the kth) basic feasible solution with basis matrix $\bar{\boldsymbol{B}}$. In this instance, the kth basic feasible solution is degenerate since $x_{\bar{B}i} = 0$ for all i corresponding to those columns that have the same θ as the eliminated column. In this regard, if yet another iteration is performed and we obtain the $(k + 1)$st basic feasible solution, it will be degenerate also since now $\theta = 0$. Moreover, since f changes between successive basic feasible solutions by the amount $\bar{c}_j\theta$ (see (4.13.1)), it is clear that no improvement in the value of the objective function occurs, i.e. although we have actually obtained a new degenerate basic feasible solution from the old one, we are still at the same extreme point.

It must be mentioned, however, that a new basic feasible solution will not always be degenerate given that the previous one was. For example, if for some $b_s = 0$ we have $a_{sk} < 0$, then, since negative a_{sk}'s are not used in computing $\theta, {}^{b_s}/_{a_{sk}} = 0$ is not a candidate for the minimum in (4.12) so that $x_{\bar{B}i} = \theta \neq 0$ in the next iteration. Thus the new basic feasible solution will be nondegenerate.

To summarize, when a degenerate basic feasible solution occurs at some particular round of the simplex method, the value of the objective function is unchanged when the next iteration is carried out. In fact, it may remain constant for some large number of subsequent iterations. In addition, there also exists the possibility that all succeeding basic feasible solutions will be degenerate. In this latter instance, the simplex process is said to **cycle** forever in an endless loop, i.e. a given sequence of bases is repeated again and again so that convergence to an optimal solution never occurs. In the absence of degeneracy, the value of the objective function increases at each iteration so that we never return to a previous basis. And since there are a finite number of bases, we are guaranteed that the simplex method will terminate in a finite number of iterations.

When degeneracy presents itself, how may it be resolved? The answer is quite simple. Since in any ordinary linear program the threat of cycling is, for all practical purposes, virtually nonexistent, no anti-degeneracy procedure is really needed since the effect of degeneracy is simply to slow down the simplex routine along its path to the optimum rather than to preclude the optimum from ever being attained. All that one really needs to do is to make an arbitrary choice from among the tied ratios that constitute the minimum in (4.12).

Example 4.5 Solve

$$\begin{aligned} &max\, f = -2x_1 + 3x_2 \text{ s.t.} \\ &x_1 + x_2 \leq 60 \\ &x_1 + 2x_2 \leq 120 \\ &2x_1 + x_2 \leq 80 \\ &x_1, x_2 \geq 0. \end{aligned}$$

With x_3, x_4, and x_5 taken to be nonnegative slack variables, we form

$$\begin{aligned} x_1 + x_2 + x_3 \qquad\qquad &= 60 \\ x_1 + 2x_2 \quad + x_4 \qquad &= 120 \\ 2x_1 + x_2 \qquad + x_5 \quad &= 80 \\ -2x_1 - 3x_2 \qquad\quad + f &= 0 \end{aligned}$$

and thus

$$\begin{bmatrix} 1 & 1 & 1 & 0 & 0 & 0 & 60 \\ 1 & 2 & 0 & 1 & 0 & 0 & 120 \\ 2 & 1 & 0 & 0 & 1 & 0 & 80 \\ -2 & -3 & 0 & 0 & 0 & 1 & 0 \end{bmatrix}.$$

Associated with this simplex matrix is the nonoptimal initial basic feasible solution consisting of $x_3 = 60$, $x_4 = 120$, $x_5 = 80$, and $f = 0$.

Iteration 1.

$$\left[\begin{array}{cccccccc|c} 1 & 1 & 1 & 0 & 0 & 0 & 60 \\ 1 & \textcircled{2} & 0 & 1 & 0 & 0 & 120 \\ 2 & 1 & 0 & 0 & 1 & 0 & 80 \\ -2 & -3 & 0 & 0 & 0 & 1 & 0 \end{array}\right] \begin{array}{c} 60 \\ 80 \\ 80 \\ \\ \end{array}$$

Since there exists a tie for the minimum of the ratios to the right of this last matrix, we have the choice of replacing either x_3 or x_4 by x_2 in the set of basic variables. Let us choose x_4. Hence, the pivot is the second element in the second column of the above array. Then

$$\begin{bmatrix} 1 & 1 & 1 & 0 & 0 & 0 & 60 \\ \frac{1}{2} & \textcircled{1} & 0 & \frac{1}{2} & 0 & 0 & 60 \\ 2 & 1 & 0 & 0 & 0 & 0 & 80 \\ -2 & -3 & 0 & 0 & 1 & 1 & 0 \end{bmatrix} \rightarrow \begin{bmatrix} \frac{1}{2} & 0 & 1 & -\frac{1}{2} & 0 & 0 & 0 \\ \frac{1}{2} & 1 & 0 & \frac{1}{2} & 0 & 0 & 60 \\ \frac{3}{2} & 0 & 0 & -\frac{1}{2} & 1 & 0 & 20 \\ -\frac{1}{2} & 0 & 0 & \frac{3}{2} & 0 & 1 & 180 \end{bmatrix}.$$

$\left(\frac{1}{2}E_2 \text{ replaces } E_2.\right)$ ($E_1 - E_2$, $E_3 - E_2$, $E_4 + 3E_2$ replace E_1, E_3, E_4, respectively.)

This new basic feasible solution involving $x_2 = 60$, $x_3 = 0$, $x_5 = 20$, and $f = 180$, while being nonoptimal, is also degenerate since one of the basic variables vanishes. Moreover, by virtue of the operation of the simplex exit criterion, we may anticipate that the next round of the simplex process should also yield a degenerate basic feasible solution with the same value of f as at the previous degenerate basic feasible solution since now $\theta = 0$.

Iteration 2.

$$\begin{bmatrix} \textcircled{\frac{1}{2}} & 0 & 1 & -\frac{1}{2} & 0 & 0 & 0 \\ \frac{1}{2} & 1 & 0 & \frac{1}{2} & 0 & 0 & 60 \\ \frac{3}{2} & 0 & 0 & -\frac{1}{2} & 1 & 0 & 20 \\ -\frac{1}{2} & 0 & 0 & \frac{3}{2} & 0 & 1 & 180 \end{bmatrix} \begin{array}{c} 0 \\ 120 \\ 40/3 \\ \\ \end{array} \rightarrow \begin{bmatrix} \textcircled{1} & 0 & 2 & -1 & 0 & 0 & 0 \\ \frac{1}{2} & 1 & 0 & \frac{1}{2} & 0 & 0 & 60 \\ \frac{3}{2} & 0 & 0 & -\frac{1}{2} & 1 & 0 & 20 \\ -\frac{1}{2} & 0 & 0 & \frac{3}{2} & 0 & 1 & 180 \end{bmatrix} \rightarrow$$

(Determining the pivot : x_1 replaces x_3 in the set of basic variables.) ($2E_1$ replaces E_1.)

$$\begin{bmatrix} 1 & 0 & 2 & -1 & 0 & 0 & 0 \\ 0 & 1 & -1 & 1 & 0 & 0 & 60 \\ 0 & 0 & -3 & 1 & 1 & 0 & 20 \\ 0 & 0 & 1 & 1 & 0 & 1 & 180 \end{bmatrix}.$$

$$\left(E_2 - \frac{1}{2}E_1,\ E_3 - \frac{3}{2}E_1,\ E_4 + \frac{1}{2}\ E_1 \text{ replace } E_2,\ E_3,\ E_4, \text{ respectively.}\right)$$

As readily seen from this last simplex matrix, the indicated degenerate optimal basic feasible solution is $x_1 = 0$, $x_2 = 60$, $x_5 = 20$, and $f = 180$. ■

4.4 Summary of the Simplex Method

Let us now summarize the basic rudiments of the above iterative process. If we abstract from degeneracy, the principal steps leading to an optimal basic feasible solution may be stated as follows:

1) Form the simplex matrix $[\hat{A} \vdots \hat{b}]$ (4.8) from system (4.6).
2) To obtain an initial basic feasible solution to $\hat{A}\hat{X} = \hat{b}$, let the $m + 1$ basic variables be those corresponding to the $m + 1$ unit column vectors in the simplex matrix. The values of these variables, along with the objective function value, may be determined directly by examining the last column of (4.8).
3) Examine the entries in the last or objective function row of the simplex matrix:
   a) If they are all positive, the current basic feasible solution is optimal. Next, determine whether any coefficient corresponding to a nonbasic variable is zero. If all are positive, the optimal basic feasible solution is unique. If one happens to be zero, the case of multiple optimal basic feasible solutions obtains. Determine the alternative basic feasible solution.
   b) If the coefficient on some nonbasic variable is negative and all other components in the column associated with this variable are also negative, there exists a feasible solution in positive variables that is nonbasic and for which the value of the objective function is unbounded. In this instance, the simplex process should be terminated.
   c) If one or more of the coefficients on the nonbasic variables are negative and for each such variable there is at least one positive component in its associated column, proceed to the next step.
4) Locate the most negative coefficient in the last row of the simplex matrix in order to determine which nonbasic variable is to become basic.

5) If the kth nonbasic variable is to turn basic, take the positive components of the column associated with this variable and divide them into their corresponding components in the last column of the simplex matrix.
6) Compute $\theta = \min_i \left\{ \frac{b_i}{a_{ik}}, a_{ik} > 0 \right\}$ so as to determine which basic variable is to become nonbasic.
7) If the rth basic variable is to turn nonbasic, a_{rk}, the divisor that yields the smallest quotient in step (6) serves as the pivotal element and a pivot operation is performed.
8) Return to step (3).

By repeating the pivot process outlined in steps (4)–(7), each iteration takes us to a new basic feasible solution with the new value of the objective function being an improvement over the old one. The process is terminated when there are no negative coefficients remaining in the last row of the simplex matrix.

5

Variations of the Standard Simplex Routine

This chapter presents a variety of computational devices which can be employed to solve linear programming problems which depart from the basic structure depicted in Chapter 3 (3.1). In particular, we shall consider "≥" as well as "=" structural constraints, the minimization of the objective, and problems involving unrestricted variables.

5.1 The *M*-Penalty Method

Let us assume that we are faced with solving a linear program whose structural constraint system involves a mixture of structural constraints, e.g. the said problem with its **mixed structural constraint system** (involving "≤," "≥," and "=" structural constraints) might appear as

$$
\begin{aligned}
&max\, f = x_1 + x_2 \quad \text{s.t.}\\
&x_1 + 2x_2 \leq 1\\
&x_1 + \frac{1}{2}x_2 \geq 4\\
&\frac{1}{2}x_1 + 6x_2 = 2\\
&x_1, x_2 \geq 0.
\end{aligned}
$$

We noted in Chapter 3 that to convert a "≤" structural constraint to standard form (i.e. to an equality) we must add a nonnegative slack variable to its left-hand side. A similar process holds for "≥" structural constraints – this time we will subtract a nonnegative **surplus variable** from its left-hand side. So

Linear Programming and Resource Allocation Modeling, First Edition. Michael J. Panik.

for $x_3 \geq 0$ a slack variable and $x_4 \geq 0$ a surplus variable, the preceding problem, in $\hat{A}\hat{X} = \hat{b}$ form, becomes

$$\begin{aligned} x_1 + 2x_2 + x_3 \quad &= 1 \\ x_1 + \frac{1}{2}x_2 \quad - x_4 \quad &= 4 \\ \frac{1}{2}x_1 + 6x_2 \quad &= 2 \\ -x_1 - x_2 \quad + f \ &= 0. \end{aligned}$$

The associated simplex matrix thus appears as

$$\begin{bmatrix} 1 & 2 & 1 & 0 & 0 & 1 \\ 1 & \textcircled{\frac{1}{2}} & 0 & -1 & 0 & 4 \\ \frac{1}{2} & 6 & 0 & 0 & 0 & 2 \\ -1 & -1 & 0 & 0 & 1 & 0 \end{bmatrix}.$$

Since this matrix does not contain within it a fourth-order identity matrix, an initial basic feasible solution does not exist, i.e. we lack a starting point from which to initiate the various rounds of the simplex method because not all of the original structural constraints are of the "≤" variety. So in situations where some or all of the original structural constraints are of the "≥" or "=" form, specific steps must be taken to obtain an identity matrix and thus a starting basic feasible solution.

To this end, let us transform the augmented structural constraint system $\bar{A}\bar{X} = b$ into what is termed the **artificial augmented structural constraint system $A_*X_* = b$** by introducing an appropriate number of nonnegative **artificial variables** (one for each "≥" and "=" structural constraint), so named because these variables are meaningless in terms of the formulation of the original structural constraint system that contains only **legitimate variables.** If x_5, $x_6 \geq 0$ depict artificial variables (we will always introduce only enough artificial variables to obtain an initial identity matrix), then the $A_*X_* = b$ system appears as

$$\begin{aligned} x_1 + 2x_2 + x_3 \quad &= 1 \\ x_1 + \frac{1}{2}x_2 \quad - x_4 + x_5 \quad &= 4 \\ \frac{1}{2}x_1 + 6x_2 \quad + x_6 &= 2, \end{aligned}$$

where

$$A_* = \begin{bmatrix} 1 & 2 & 1 & 0 & 0 & 0 & 1 \\ 1 & \frac{1}{2} & 0 & -1 & 1 & 0 & 4 \\ \frac{1}{2} & 6 & 0 & 0 & 0 & 1 & 2 \end{bmatrix}, X_* = \begin{bmatrix} \bar{X} \\ \cdots \\ X_a \end{bmatrix} = \begin{bmatrix} x_1 \\ x_2 \\ x_3 \\ x_4 \\ \cdots \\ x_5 \\ x_6 \end{bmatrix},$$

and X_a is a vector of artificial variables. Here the vectors within A_* corresponding to artificial variables are termed **artificial vectors** and appear as a set of unit column vectors. Note that columns 3, 5, and 6 yield a third-order identity matrix and thus an initial basic feasible solution to $A_*X_* = b$ once the remaining variables are set equal to zero, namely $x_3 = 1$, $x_5 = 4$, and $x_6 = 2$.

While we have obtained an initial basic feasible solution to $A_*X_* = b$, it is not a feasible solution to the original structural constraint system $\bar{A}\bar{X} = b$, the reason being that $x_5, x_6 \neq 0$. Hence any basic feasible solution to the latter system must not admit any artificial variable at a positive level. In this regard, if we are to obtain a basic feasible solution to the original linear programming problem, we must utilize the simplex process to remove from the initial basis (i.e. from the identity matrix) all artificial vectors and substitute in their stead an alternative set of vectors chosen from those remaining in A_*. In doing so we ultimately obtain the original structural constraint system $\bar{A}\bar{X} = b$ itself. Hence all artificial variables are to become nonbasic or zero with the result that a basic feasible solution to $\bar{A}\bar{X} = b$ emerges wherein the basis vectors are chosen from the coefficient matrix of this system. To summarize: any basic feasible solution to $A_*X_* = b$ which is also a basic feasible solution to $\bar{A}\bar{X} = b$ has all artificial variables equal to zero.

The computational technique that we shall employ to remove the artificial vectors from the basis is the ***M*-penalty method** (Charnes et al. 1953). Here, extremely large unspecified negative coefficients are assigned to the artificial variables in the objective function in order to preclude the objective function from attaining a maximum as long as any artificial variable remains in the set of basic variables at a positive level. In this regard, for a sufficiently large $M > 0$, if the coefficient on each artificial variable in the objective function is $-M$, then the **artificial linear programming problem** may be framed as:

$$\begin{aligned} &max\, f = \bar{C}'\bar{X} - M\mathbf{1}'X_a \text{ s.t.} \\ &A_*X_* = b, X_* \geq O, \end{aligned} \tag{5.1}$$

wherein f is termed the **artificial objective function**. In terms of the preceding example problem, (5.1) becomes

$$\begin{aligned}
&max\, f = x_1 + x_2 - Mx_5 - Mx_6 \quad \text{s.t.} \\
&x_1 + 2x_2 + x_3 \qquad\qquad\qquad = 1 \\
&x_1 + \frac{1}{2}x_2 \quad - x_4 + x_5 \qquad = 4 \\
&\frac{1}{2}x_1 + 6x_2 \qquad\qquad + x_6 = 2 \\
&x_1, \ldots, x_6 \geq 0.
\end{aligned}$$

The simplex matrix associated with this problem is then

$$\left[\hat{A}_* \vdots \hat{b}\right] = \begin{bmatrix} 1 & 2 & 1 & 0 & 0 & 0 & 0 & 1 \\ 1 & \frac{1}{2} & 0 & -1 & 1 & 0 & 0 & 4 \\ \frac{1}{2} & 6 & 0 & 0 & 0 & 1 & 0 & 2 \\ -1 & -1 & 0 & 0 & M & M & 1 & 0 \end{bmatrix}. \tag{5.2}$$

To obtain an identity matrix within (5.2), let us multiply both the second and third rows of the same by $-M$ and form their sum as

$$\begin{array}{cccccccc}
-M & -\frac{1}{2}M & 0 & M & -M & 0 & 0 & -4M \\
-\frac{1}{2}M & -6M & 0 & 0 & 0 & -M & 0 & -2M \\
\hline
-\frac{3}{2}M & -\frac{13}{2}M & 0 & M & -M & -M & 0 & -6M
\end{array}.$$

If this row is then added to the objective function row of (5.2), the new objective function row is

$$-1 - \frac{3}{2}M \;\; -1 - \frac{13}{2}M \;\; 0 \;\; M \;\; 0 \;\; 0 \;\; 1 \;\; -6M.$$

(Note that the first two components of this composite row are each composed of an integer portion and a portion involving M.) To simplify our pivot operations, this transformed objective function row will be written as the sum of two rows. i.e. each objective function term will be separated into two parts, one that does not involve M and one appearing directly beneath it that does. Hence the preceding objective function row now appears as

$$\begin{array}{cccccccc}
-1 & -1 & 0 & 0 & 0 & 0 & 1 & 0 \\
-\frac{3}{2}M & -\frac{13}{2}M & 0 & M & 0 & 0 & 0 & -6M,
\end{array}$$

where this *double-row* representation actually depicts the *single* objective function row of the simplex matrix. In view of this convention, (5.2) becomes

$$\begin{bmatrix} 1 & 2 & 1 & 0 & 0 & 0 & 0 & 1 \\ 1 & \frac{1}{2} & 0 & -1 & 1 & 0 & 0 & 4 \\ \frac{1}{2} & 6 & 0 & 0 & 0 & 1 & 0 & 2 \\ -1 & -1 & 0 & 0 & 0 & 0 & 1 & 0 \\ -\frac{3}{2}M & -\frac{13}{2}M & 0 & M & 0 & 0 & 0 & -6M \end{bmatrix}$$

with columns 3, 5, 6, and 7 yielding a fourth-order identity matrix. Hence, the initial basic feasible solution to the artificial linear program is $x_3 = 1$, $x_5 = 4$, $x_6 = 2$, and $f = 6M$. And since f is expressed in terms of M, it is evident that the objective function cannot possibly attain a maximum with artificial vectors in the basis.

Before examining some additional example problems dealing with the implementation of this technique, a few summary comments pertaining to the salient features of the M-penalty method will be advanced. First, once an artificial vector leaves the basis, it never reenters the basis. Alternatively, once an artificial variable turns nonbasic, it remains so throughout the remaining rounds of the simplex process since, if an artificial variable increases in value from zero by a positive amount, a penalty of $-M$ is incurred and f concomitantly decreases. Since the simplex process is one for which f is nondecreasing between succeeding pivot rounds, this possibility is nonexistent. So as the various rounds of the simplex process are executed, we may delete from the simplex matrix those columns corresponding to artificial vectors that have turned nonbasic. Next, if the maximal solution to the artificial problem has all artificial variables equal to zero, then its $\bar{X}$ component is a maximal solution to the original problem with the values of the artificial and original objective functions being the same. Finally, if the maximal solution to the artificial problem contains at least one positive artificial variable, then the original problem possesses no feasible solution (we shall return to this last point later on when a detailed analysis pertaining to the questions on inconsistency and redundancy of the structural constraints is advanced).

Example 5.1 Using the M-penalty technique, solve

$$\begin{aligned} &max\, f = x_1 + x_2 \text{ s.t.} \\ &x_1 + 2x_2 \leq 12 \\ &x_1 + x_2 \geq 4 \\ &-x_1 + x_2 \geq 1 \\ &x_1, x_2 \geq 0. \end{aligned}$$

If we let x_3 denote a slack variable; x_4, x_5 depict surplus variables; and x_6, x_7 represent artificial variables, we obtain

$$
\begin{array}{llllllll}
x_1 + 2x_2 + x_3 & & & & & & = 12 \\
x_1 + \ x_2 & - x_4 & & + x_6 & & & = 4 \\
-x_1 + \ x_2 & & - x_5 & & + & x_7 & = 1 \\
-x_1 - \ x_2 & & & + Mx_6 & + Mx_7 + f & & = 0
\end{array}
$$

with associated simplex matrix

$$
\left[\hat{A}_{*} \vdots \hat{b}\right] = \begin{bmatrix} 1 & 2 & 1 & 0 & 0 & 0 & 0 & 12 \\ 1 & 1 & 0 & -1 & 1 & 0 & 0 & 4 \\ -1 & 1 & 0 & 0 & -1 & 1 & 0 & 1 \\ -1 & -1 & 0 & 0 & M & M & 1 & 0 \end{bmatrix}.
$$

To generate an initial basic feasible solution to this artificial linear programming problem, let us add

$$
\begin{array}{ccccccccc}
-M & -M & 0 & M & 0 & -M & 0 & 0 & -4M \\
M & -M & 0 & 0 & M & 0 & -M & 0 & -M \\
\hline
0 & -2M & 0 & M & M & -M & -M & 0 & -5M
\end{array}
$$

to the last row of the simplex matrix to obtain

$$
\begin{bmatrix} 1 & 2 & 1 & 0 & 0 & 0 & 0 & 0 & 12 \\ 1 & 1 & 0 & -1 & 0 & 1 & 0 & 0 & 4 \\ -1 & 1 & 0 & 0 & -1 & 0 & 1 & 0 & 1 \\ -1 & -1 & 0 & 0 & 0 & 0 & 0 & 1 & 0 \\ 0 & -2M & 0 & M & M & 0 & 0 & 0 & -5M \end{bmatrix}.
$$

The first basic feasible solution to the artificial problem thus becomes $x_3 = 12$, $x_6 = 4$, $x_7 = 1$, and $f = -5M$. Since this solution is clearly not optimal, let us undertake

Iteration 1.

$$
\left[\begin{array}{ccccccccc} 1 & 2 & 1 & 0 & 0 & 0 & 0 & 0 & 12 \\ 1 & 1 & 0 & -1 & 0 & 1 & 0 & 0 & 4 \\ -1 & \textcircled{1} & 0 & 0 & -1 & 0 & 1 & 0 & 1 \\ -1 & -1 & 0 & 0 & 0 & 0 & 0 & 1 & 0 \\ 0 & -2M & 0 & M & M & 0 & 0 & 0 & -5M \end{array}\right] \begin{array}{c} 6 \\ 4 \\ 1 \\ \\ \\ \end{array} \rightarrow \begin{bmatrix} 3 & 0 & 1 & 0 & 2 & 0 & -2 & 0 & 10 \\ 2 & 0 & 0 & -1 & 1 & 1 & -1 & 0 & 3 \\ -1 & 1 & 0 & 0 & -1 & 0 & 1 & 0 & 1 \\ -2 & 0 & 0 & 0 & -1 & 0 & 1 & 1 & 1 \\ -2M & 0 & 0 & M & -M & 0 & 2M & 0 & -3M \end{bmatrix}.
$$

(Determining the pivot: x_2 replaces x_7 in the set of basic variables.)

($E_1 - 2E_3, E_2 - E_3, E_4 + E_3, E_5 + 2ME_3$ replace E_1, E_2, E_4, E_5, respectively.)

Here the second basic feasible solution to the artificial problem $x_2 = 1, x_3 = 10$, $x_6 = 3$, and $f = 1 - 3M$. Since this solution is also not optimal, we proceed to

Iteration 2 (column 7 is deleted).

$$\left[\begin{array}{cccccccc} 3 & 0 & 1 & 0 & 2 & 0 & 0 & 10 \\ \textcircled{2} & 0 & 0 & -1 & 1 & 1 & 0 & 3 \\ -1 & 1 & 0 & 0 & -1 & 0 & 0 & 1 \\ -2 & 0 & 0 & 0 & -1 & 0 & 1 & 1 \\ -2M & 0 & 0 & M & -M & 0 & 0 & -3M \end{array}\right] \begin{array}{c} 10/3 \\ 3/2 \\ \\ \\ \\ \end{array} \rightarrow \left[\begin{array}{cccccccc} 3 & 0 & 1 & 0 & 2 & 0 & 0 & 10 \\ \textcircled{1} & 0 & 0 & -\frac{1}{2} & \frac{1}{2} & \frac{1}{2} & 0 & \frac{3}{2} \\ -1 & 1 & 0 & 0 & -1 & 0 & 0 & 1 \\ -2 & 0 & 0 & 0 & -1 & 0 & 1 & 1 \\ -2M & 0 & 0 & M & -M & 0 & 0 & -3M \end{array}\right] \rightarrow$$

(Determining the pivot: x_1 replaces x_6 in the set of basic variables.)

$\left(\frac{1}{2}E_2 \text{ replaces } E_2.\right)$

$$\left[\begin{array}{cccccccc} 0 & 0 & 1 & \frac{3}{2} & \frac{1}{2} & -\frac{3}{2} & 0 & \frac{11}{2} \\ 1 & 0 & 0 & -\frac{1}{2} & \frac{1}{2} & \frac{1}{2} & 0 & \frac{3}{2} \\ 0 & 1 & 0 & -\frac{1}{2} & -\frac{1}{2} & \frac{1}{2} & 0 & \frac{5}{2} \\ 0 & 0 & 0 & -1 & 0 & 1 & 1 & 4 \\ 0 & 0 & 0 & 0 & 0 & M & 0 & 0 \end{array}\right].$$

$(E_1 - 3E_2, E_3 + E_2, E_4 + 2E_3, E_5 + 2ME_2$ replace E_1, E_3, E_4, E_5, respectively.)

The third basic feasible solution to the artificial problem thus appears as $x_1 = \frac{3}{2}, x_2 = \frac{5}{2}, x_3 = \frac{11}{2}$, and $f = 4$. Since all artificial vectors have been removed from the basis, this particular basic feasible solution to the artificial problem represents an initial basic feasible solution to the original problem. In this regard, let us now undertake

Iteration 3 (column 6 and the "M-row" are deleted).

$$\left[\begin{array}{ccccccc} 0 & 0 & 1 & \textcircled{3/2} & \frac{1}{2} & 0 & \frac{11}{2} \\ 1 & 0 & 0 & -\frac{1}{2} & \frac{1}{2} & 0 & \frac{3}{2} \\ 0 & 1 & 0 & -\frac{1}{2} & -\frac{1}{2} & 0 & \frac{5}{2} \\ 0 & 0 & 0 & -1 & 0 & 1 & 4 \end{array}\right] \begin{array}{c} 11/3 \\ \\ \\ \\ \end{array} \rightarrow \left[\begin{array}{ccccccc} 0 & 0 & \frac{2}{3} & \textcircled{1} & \frac{1}{3} & 0 & \frac{11}{3} \\ 1 & 0 & 0 & -\frac{1}{2} & \frac{1}{2} & 0 & \frac{3}{2} \\ 0 & 1 & 0 & -\frac{1}{2} & -\frac{1}{2} & 0 & \frac{5}{2} \\ 0 & 0 & 0 & -1 & 0 & 1 & 4 \end{array}\right] \rightarrow$$

(Determining the pivot: x_4 replaces x_3 in the set of basic variables.)

$\left(\frac{2}{3}E_1 \text{ replaces } E_1.\right)$

$$\begin{bmatrix} 0 & 0 & \frac{2}{3} & 1 & \frac{1}{3} & 0 & \frac{11}{3} \\ 1 & 0 & \frac{1}{3} & 0 & \frac{2}{3} & 0 & \frac{10}{3} \\ 0 & 1 & \frac{1}{3} & 0 & -\frac{2}{3} & 0 & \frac{13}{3} \\ 0 & 0 & \frac{2}{3} & 0 & \frac{1}{3} & 1 & \frac{23}{3} \end{bmatrix}.$$

$$\left(E_2 + \frac{1}{2}E_1, E_3 + \frac{1}{2}E_1, E_4 + E_1 \text{ replace } E_2, E_3, E_4, \text{ respectively.}\right)$$

Since the optimality criterion is satisfied, the optimal basic feasible solution to the original linear programming problem is $x_1 = \frac{10}{3}, x_2 = \frac{13}{3}, x_4 = \frac{11}{3}$, and $f = \frac{23}{3}$. ■

5.2 Inconsistency and Redundancy

We now turn to a discussion of the circumstances under which all artificial vectors appearing in the initial basis of the artificial linear programming problem may be removed from the basis in order to proceed to an optimal basic feasible solution to the original problem. In addition, we shall also determine whether the original augmented structural constraint equalities $\bar{A}\bar{X} = \boldsymbol{b}$ are consistent, and whether any subset of them is redundant. Let us assume that from an initial basic feasible solution to the artificial problem we obtain a basic solution for which the optimality criterion is satisfied. Here one of three mutually exclusive and collectively exhaustive results will hold:

1) The basis contains no artificial vectors.
2) The basis contains one or more artificial vectors at the zero level.
3) The basis contains one or more artificial vectors at a positive level.

Let us discuss the implications of each case in turn.

First, if the basis contains no artificial vectors, then, given that the optimality criterion is satisfied, it is evident that we have actually obtained an optimal basic feasible solution to the original problem. In this instance the original structural constraints are consistent and none is redundant, i.e. if there exists at least one basic feasible solution to the original structural constraint system, the simplex process will remove all artificial vectors from the basis and ultimately reach an optimal solution to the same (a case in point was Example 5.1).

We next turn to the case where the basis contains one or more artificial vectors at the zero level. With all artificial variables equal to zero, we have a feasible solution to $\bar{A}\bar{X} = b$, i.e. the original structural constraint equations are consistent. However, there still exists the possibility of redundancy in the original structural constraint system. Upon addressing ourselves to this latter point, we find that two alternative situations merit our consideration. To set the stage for this discussion, let Y_j denote the jth legitimate nonbasic column of the optimal simplex matrix. In this regard, we first assume that y_{ij}, the ith component of Y_j, is different from zero for one or more j and for one or more $i \in \mathcal{A}$, where $\mathcal{A}$ denotes an index set consisting of all is corresponding to artificial basis vectors e_i. Now, if from some $j = j'$ we find that $y_{ij'} \neq 0$, it follows that the associated artificial basis vector e_i may be removed from the basis and replaced by $Y_{j'}$. Since the artificial vector currently appears in the basis at the zero level, $Y_{j'}$ also enters the basis at the zero level, with the result that the optimal value of the objective function is unchanged in the new basic feasible solution. If this process is repeated until all artificial vectors have been removed from the basis, we obtain a degenerate optimal basic feasible solution to the original problem involving only real or legitimate variables. In this instance none of the original structural constraints within $\bar{A}\bar{X} = b$ is redundant. Next, if this procedure does not remove all artificial vectors from the basis, we ultimately reach a state where $y_{ij} = 0$ for all Y_j and all remaining $i \in \mathcal{A}$. Under this circumstance we cannot replace any of the remaining artificial vectors by some Y_j and still maintain a basis. If we assume that there are k artificial vectors in the basis at the zero level, every column of the $(m \times m)$ matrix $\bar{A}$ may be written as a linear combination of the $m-k$ linearly independent columns of $\bar{A}$ appearing in the basis, i.e. the k artificial vectors are not needed to express any column of $\bar{A}$ in terms of basis vectors. Hence $\rho(\bar{A}) = m - k$ so that only $m-k$ rows of $\bar{A}$ are linearly independent, thus implying that k of the original structural constraints $\bar{A}\bar{X} = b$ are redundant. (As a practical matter, since inequality structural constraints can never be redundant – each is converted into an equality by the introduction of its "own" slack or surplus variable – our search for redundant structural constraints must be limited to only some subset of equations appearing in $\bar{A}\bar{X} = b$, namely those expressed initially as strict equalities.) To identify the specific structural constraints within $\bar{A}\bar{X} = b$ that are redundant, we note briefly that if the artificial basis vector e_h remains in the basis at the zero level and $y_{hk} = 0$ for all legitimate nonbasis vectors Y_j, then the hth structural constraint of $\bar{A}\bar{X} = b$ is **redundant**. In this regard, if at some stage of the simplex process we find that the artificial vector e_h appears in the basis at the zero level, with $y_{hj} = 0$ for all Y_j, then, before executing the next round of the simplex routine, we may delete the hth row of the simplex matrix containing the zero-valued artificial variable along with the associated artificial basis vector e_h. The implication here is that any redundant constraint may be omitted from the original structural constraint system without losing any information contained within the latter.

Example 5.2 To highlight the details of the preceding discussion, let us solve

$$\max f = x_1 + x_2 + x_3 \quad \text{s.t.}$$
$$x_1 + 4x_2 + 3x_3 = 6$$
$$3x_1 + 12x_2 + 9x_3 = 18$$
$$x_1, x_2, x_3 \geq 0.$$

Letting x_4, x_5 represent artificial variables, we obtain

$$\begin{aligned} x_1 + 4x_2 + 3x_3 + x_4 \qquad\qquad &= 6 \\ 3x_1 + 12x_2 + 9x_3 \qquad + x_5 \qquad &= 18 \\ -x_1 - x_2 \qquad + Mx_4 + Mx_5 + f &= 0 \end{aligned}$$

with associated simplex matrix

$$[\hat{A}_* \vdots \hat{b}] = \begin{bmatrix} 1 & 4 & 3 & 1 & 0 & 0 & 6 \\ 3 & 12 & 9 & 0 & 1 & 0 & 18 \\ -1 & -1 & -1 & M & M & 1 & 0 \end{bmatrix} \rightarrow \begin{bmatrix} 1 & 4 & 3 & 1 & 0 & 0 & 6 \\ 3 & 12 & 9 & 0 & 1 & 0 & 18 \\ -1 & -1 & -1 & 0 & 0 & 1 & 0 \\ -4M & -16M & -12M & 0 & 0 & 0 & -24M \end{bmatrix}.$$

Our initial basic feasible solution to this artificial problem is $x_4 = 6$, $x_5 = 18$, and $f = -24M$. Since the optimality criterion is not satisfied, we proceed to

Iteration 1.

$$\begin{bmatrix} 1 & \textcircled{4} & 3 & 1 & 0 & 0 & 6 \\ 3 & 12 & 9 & 0 & 1 & 0 & 18 \\ -1 & -1 & -1 & 0 & 0 & 1 & 0 \\ -4M & -16M & -12M & 0 & 0 & 0 & -24M \end{bmatrix} \rightarrow \begin{bmatrix} \frac{1}{4} & \textcircled{1} & \frac{3}{4} & \frac{1}{4} & 0 & 0 & \frac{3}{2} \\ 3 & 12 & 9 & 0 & 1 & 0 & 18 \\ -1 & -1 & -1 & 0 & 0 & 1 & 0 \\ -4M & -16M & -12M & 0 & 0 & 0 & -24M \end{bmatrix} \rightarrow$$

(Determining the pivot: x_2 replaces x_4 in the set of basic variables.) $\left(\frac{1}{2}E_1 \text{ replaces } E_1.\right)$

$$\begin{bmatrix} \frac{1}{4} & 1 & \frac{1}{4} & \frac{1}{4} & 0 & 0 & \frac{3}{2} \\ 0 & 0 & 0 & -3 & 1 & 0 & 0 \\ -\frac{3}{4} & 0 & -\frac{1}{4} & \frac{1}{4} & 0 & 1 & \frac{3}{2} \\ 0 & 0 & 0 & 4M & 0 & 0 & 0 \end{bmatrix}.$$

($E_2 - 12E_1, E_3 + E_1, E_4 + 16ME_4$ replace E_2, E_3, E_4, respectively.)

Then our second basic feasible solution to the artificial problem is $x_2 = \frac{3}{2}, x_5 = 0$, and $f = \frac{3}{2}$. Since the optimality criterion is still not satisfied, an additional round of the simplex routine is warranted. Before moving on to round 2, let us pause for a moment to inspect the preceding simplex matrix. Given that the artificial vector $\boldsymbol{e}_2$ enters the basis at the zero level and $y_{21} = y_{22} = 0$ for legitimate nonbasis vectors $\boldsymbol{Y}_1$, $\boldsymbol{Y}_2$, it is obvious that the second structural constraint is redundant so that we may, at this point, delete row 2 and column 5. Moreover, since all artificial vectors have been eliminated from the basis, round two starts with an initial basic feasible solution to the original problem. We now look to

Iteration 2 (rows 2 and 4 along with columns 4 and 5 are deleted).

$$\begin{bmatrix} \textcircled{\frac{1}{4}} & 1 & \frac{3}{4} & 0 & \frac{3}{2} & 6 \\ -\frac{3}{4} & 0 & -\frac{1}{4} & 1 & \frac{3}{2} \end{bmatrix} \rightarrow \begin{bmatrix} \textcircled{1} & 4 & 3 & 0 & 6 \\ -\frac{3}{4} & 0 & -\frac{1}{4} & 1 & \frac{3}{2} \end{bmatrix} \rightarrow$$

(Determining the pivot : x_1 replaces x_2 in the set of basic variables.)

($4E_1$ replaces E_1.)

$$\begin{bmatrix} 1 & 4 & 3 & 0 & 6 \\ 0 & 3 & 2 & 1 & 6 \end{bmatrix}.$$

$$\left(E_3 + \frac{3}{4}E_1 \text{ replaces } E_3.\right)$$

A glance at this last matrix indicates that we have obtained an optimal basic feasible solution to the original problem consisting of $x_1 = 6$ and $f = 6$. ■

We finally encounter the case where one or more artificial vectors appear in the basis at a positive level. In this instance, the basic solution generated is not meaningful, i.e. the original problem possesses **no feasible solution** since otherwise the artificial vectors would all enter the basis at a zero level. Here the original system either has no solution (the structural constraints are inconsistent) or has solutions that are not feasible. To distinguish between these two alternatives, we shall assume that the optimality criterion is satisfied and $y_{ij} \le 0$ for legitimate nonbasis vectors $\boldsymbol{Y}_j$ and for those $i \in \mathcal{A}$. If for some $j = j'$ we find that $y_{ij'} < 0$, we can insert $\boldsymbol{Y}_{j'}$ into the basis and remove the associated artificial basis vector $\boldsymbol{e}_i$ and still maintain a basis. However, with $y_{ij'} < 0$, the new basic solution will not be feasible since $\boldsymbol{Y}_{j'}$ enters the basis at a negative level. If this process is repeated until all artificial vectors have been driven from the basis, we obtain a basic though nonfeasible solution to the original problem involving only legitimate variables (see Example 5.3). Next, if this procedure does not remove all artificial vectors from the basis, we ultimately obtain a state where $y_{ij} = 0$ for all $\boldsymbol{Y}_j$ and for all i remaining in $\mathcal{A}$. Assuming that there are k artificial vectors in the basis, every column of $\bar{\boldsymbol{A}}$ may be written as a linear combination of the $m - k$ legitimate basis vector so that $\rho(\bar{\boldsymbol{A}}) = m-k$. But the fact that k artificial columns from $\bar{\boldsymbol{A}}$ appear in the basis at positive levels implies that $\boldsymbol{b}$ is expressible as a linear combination of more than $m - k$ basis vectors so that $\rho(\bar{\boldsymbol{A}}) \neq \rho[\bar{\boldsymbol{A}} \vdots \boldsymbol{b}]$. In this regard, the original structural constraint system $\bar{\boldsymbol{A}}\bar{\boldsymbol{X}} = \boldsymbol{b}$ is **inconsistent** and thus does not possess a solution (see Example 5.4).

Example 5.3 Let us generate a solution to

$$\begin{aligned} &max\, f = x_1 + x_2 \text{ s.t.} \\ &4x_1 - x_2 \le 1 \\ &x_1 - x_2 \ge 3 \\ &x_1, x_2 \ge 0. \end{aligned}$$

If x_3 denotes a slack variable, x_4 represents a surplus variable, and x_5 depicts an artificial variable, then we may form

$$\begin{aligned} 4x_1 - x_2 + x_3 \qquad\qquad\qquad &= 1 \\ x_1 \;\; - x_2 \qquad - x_4 + x_5 \qquad &= 3 \\ -x_1 - x_2 \qquad\qquad + Mx_5 + f &= 0 \end{aligned}$$

with associated simplex matrix

$$[\hat{A}_* \vdots \hat{b}] = \begin{bmatrix} 4 & -1 & 1 & 0 & 0 & 0 & 1 \\ 1 & -1 & 0 & -1 & 1 & 0 & 3 \\ -1 & -1 & 0 & 0 & M & 1 & 0 \end{bmatrix} \to \begin{bmatrix} 4 & -1 & 1 & 0 & 0 & 0 & 1 \\ 1 & -1 & 0 & -1 & 1 & 0 & 3 \\ -1 & -1 & 0 & 0 & 0 & 1 & 0 \\ -M & M & 0 & M & 0 & 0 & -3M \end{bmatrix}.$$

Since the initial basic feasible solution consisting of $x_3 = 1$, $x_5 = 3$, and $f = -3M$ is obviously not optimal, we proceed to

Iteration 1.

$$\begin{bmatrix} ④ & -1 & 1 & 0 & 0 & 0 & 1 \\ 1 & -1 & 0 & -1 & 1 & 0 & 3 \\ -1 & -1 & 0 & 0 & 0 & 1 & 0 \\ -M & M & 0 & M & 0 & 0 & -3M \end{bmatrix} \rightarrow \begin{bmatrix} ① & -\frac{1}{4} & \frac{1}{4} & 0 & 0 & 0 & \frac{1}{4} \\ 1 & -1 & 0 & -1 & 1 & 0 & 3 \\ -1 & -1 & 0 & 0 & 0 & 1 & 0 \\ -M & M & 0 & M & 0 & 0 & -3M \end{bmatrix} \rightarrow$$

(Determining the pivot: x_1 replaces x_4 in the set of basic variables.) $\left(\frac{1}{4}E_1 \text{ replaces } E_1.\right)$

$$\begin{bmatrix} 1 & -\frac{1}{4} & \frac{1}{4} & 0 & 0 & 0 & \frac{1}{4} \\ 0 & -\frac{3}{4} & -\frac{1}{4} & -1 & 1 & 0 & \frac{11}{4} \\ 0 & -\frac{5}{4} & \frac{1}{4} & 0 & 0 & 1 & \frac{1}{4} \\ 0 & \frac{3}{4}M & \frac{1}{4}M & M & 0 & 0 & -\frac{11}{4}M \end{bmatrix}.$$

$(E_2 - E_1, E_3 + E_1, E_4 + ME_1$ replace E_2, E_3, E_4, respectively.)

The second basic feasible solution to this artificial problem thus appears as $x_1 = \frac{1}{4}, x_5 = \frac{11}{4}$, and $f = \frac{1}{4} - \frac{11M}{4}$. Can the artificial basis vector $\boldsymbol{e}_2$ be eliminated from the basis? As we shall now see, the answer is yes. From the second row of the preceding simplex matrix, we find that

$$x_5 = \frac{11}{4} + \frac{3}{4}x_2 + \frac{1}{4}x_3 + x_4.$$

Hence, x_5 may be driven to zero only if x_2 or x_3 or x_4 assumes a negative value. More formally, since the artificial basis vector $\boldsymbol{e}_2$ remains in the basis at a positive level and $y_{2j} < 0$ for all legitimate nonbasis vectors $\boldsymbol{Y}_j$, $j = 1, 2, 3$, the original problem has no feasible solution (as Figure 5.1 indicates). ■

Example 5.4 Finally, let us solve

$$\begin{aligned} &max\, f = x_1 - x_2 \text{ s.t.} \\ &x_1 + x_2 = 6 \\ &x_1 + x_2 = 4 \\ &x_1, x_2 \geq 0. \end{aligned}$$

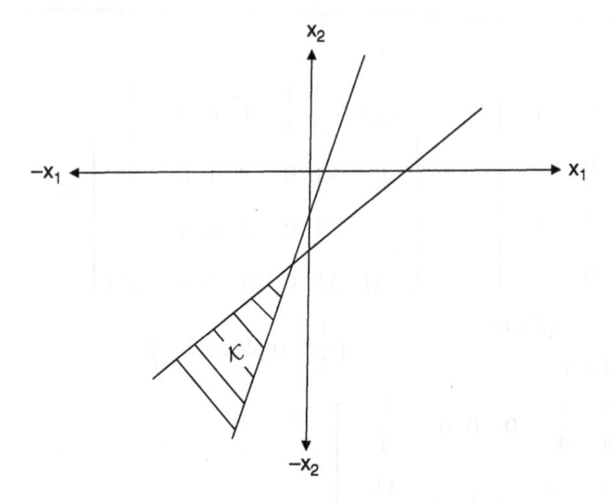

Figure 5.1 No feasible solution.

If x_3, x_4 represent artificial variables, we form

$$
\begin{aligned}
x_1 + x_2 \quad + x_3 \qquad\qquad &= 6 \\
x_1 + x_2 \qquad\quad + x_4 \quad &= 4 \\
-x_1 + \ x_2 + Mx_3 + Mx_4 + f &= 0
\end{aligned}
$$

from which we obtain

$$
\left[\hat{A}_{*} \vdots \hat{b}\right] = \begin{bmatrix} 1 & 1 & 1 & 0 & 0 & 6 \\ 1 & 1 & 0 & 1 & 0 & 4 \\ -1 & 1 & M & M & 1 & 0 \end{bmatrix} \to \begin{bmatrix} 1 & 1 & 1 & 0 & 0 & 6 \\ 1 & 1 & 0 & 1 & 0 & 4 \\ -1 & 1 & 0 & 0 & 1 & 0 \\ -2M & -2M & 0 & 0 & 0 & -10M \end{bmatrix}
$$

and thus an initial basic feasible solution consisting of $x_3 = 6$, $x_4 = 4$, and $f = -10M$. Since the optimality criterion is not satisfied, we look to

Iteration 1.

$$
\begin{bmatrix} 1 & 1 & 1 & 0 & 0 & 6 \\ \textcircled{1} & 1 & 0 & 1 & 0 & 4 \\ -1 & 1 & 0 & 0 & 1 & 0 \\ -2M & -2M & 0 & 0 & 0 & -10M \end{bmatrix} \to \begin{bmatrix} 0 & 0 & 1 & -1 & 0 & 2 \\ 1 & 1 & 0 & 1 & 0 & 4 \\ 0 & 2 & 0 & 1 & 1 & 4 \\ 0 & 0 & 0 & 2M & 0 & -2M \end{bmatrix}.
$$

(Determining the pivot: x_1 replaces x_4 in the set of basic variables.) ($E_1 - E_2, E_3 + E_2, E_4 + 2ME_2$ replace E_1, E_3, E_4, respectively.)

Figure 5.2 Inconsistent constraints.

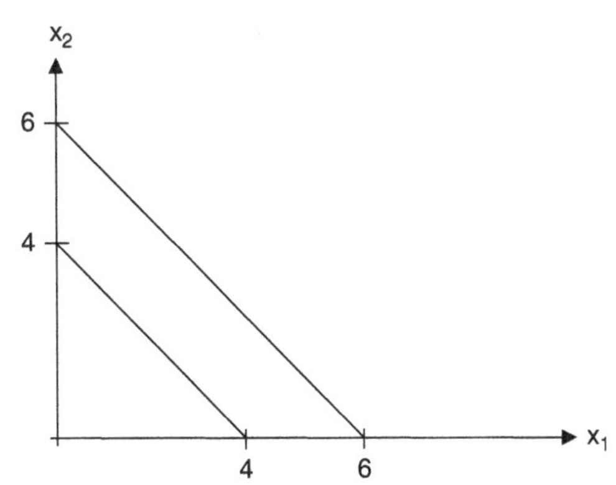

Hence, the optimal basic feasible solution to the artificial problem is $x_1 = 4$, $x_3 = 2$, and $f = 4 - 2M$. Since $y_{11} = 0$, $\boldsymbol{Y}_1$ cannot replace $\boldsymbol{e}_1$ in the basis. Therefore, the original structural constraints $\bar{\boldsymbol{A}}\bar{\boldsymbol{X}} = \boldsymbol{b}$ are inconsistent so that no feasible solution to the original problem exists (Figure 5.2). Generally speaking, when an artificial vector $\boldsymbol{e}_k$ remains in the basis at a positive level and $y_{kj} = 0$ for all legitimate nonbasis vectors $\boldsymbol{Y}_j$, then the original structural constraints are inconsistent so that the original problem exhibits no feasible solution. ■

A consolidation of the results obtained in this section now follows. Given that the optimality criterion is satisfied: (i) if no artificial vectors appear in the basis, then the solution is an optimal basic feasible solution to the original linear programming problem. In this instance, the original structural constraints $\bar{\boldsymbol{A}}\bar{\boldsymbol{X}} = \boldsymbol{b}$ are consistent and none is redundant; (ii) if a least one artificial vector appears in the basis at a zero level, then $\bar{\boldsymbol{A}}\bar{\boldsymbol{X}} = \boldsymbol{b}$ is consistent and we either obtain a degenerate optimal basic feasible solution to the original problem or at least one of the original structural constraints is redundant; and (iii) if at least one artificial vector appears in the basis at a positive level, the original problem has no feasible solution since either $\bar{\boldsymbol{A}}\bar{\boldsymbol{X}} = \boldsymbol{b}$ represents an inconsistent system or there are solutions but none is feasible.

5.3 Minimization of the Objective Function

Let us assume that we desire to

$$\begin{aligned} &min\, f = \boldsymbol{C}'\boldsymbol{X} \text{ s.t.} \\ &\quad \boldsymbol{AX} \geq \boldsymbol{b}, \boldsymbol{X} \geq \boldsymbol{O}. \end{aligned}$$

As we shall now see, to solve this minimization problem, simply transform it to a maximization problem, i.e. we shall employ the relationship $min\, f = -\, max\, \{-f\}$. To this end, we state Theorem 5.1:

Theorem 5.1 The minimum of $f = C'X$ occurs at the same point X_0 as the maximum of

$$-f = -C'X \text{ or } min\, f = -max\,\{-f\}. \tag{5.3}$$

Example 5.5 Using (5.3), solve

$$min\, f = x_1 + x_2 \text{ s.t.}$$
$$2x_1 + 4x_2 \geq 28$$
$$5x_1 + x_2 \geq 25$$
$$x_1, x_2 \geq 0.$$

To this end, let us maximize $g = -f = -x_1 - x_2$ subject to the above structural constraints and nonnegativity conditions. For x_3, x_4 surplus variables and x_5, x_6 artificial variables (each nonnegative), the initial and optimal simplex matrices are, respectively,

$$\begin{bmatrix} 2 & 4 & -1 & 0 & 1 & 0 & 0 & 28 \\ 5 & 1 & 0 & -1 & 0 & 1 & 0 & 25 \\ 1 & 1 & 0 & 0 & 0 & 0 & 1 & 0 \\ -7M & -5M & M & M & 0 & 0 & 0 & -53M \end{bmatrix} \rightarrow \begin{bmatrix} 0 & 1 & -\frac{5}{18} & \frac{1}{9} & 0 & 5 \\ 1 & 0 & \frac{1}{18} & -\frac{2}{9} & 0 & 4 \\ 0 & 0 & \frac{2}{9} & \frac{1}{9} & 1 & -9 \end{bmatrix}.$$

Thus, the optimal basic feasible solution consists of $x_1 = 4$, $x_2 = 5$, and $max\, g = max\,\{-f\} = -9$ or $min\, f = -max\,\{-f\} = -(-9) = 9$. ■

5.4 Unrestricted Variables

Another variation of the basic linear programming problem is that (one or more of) the components of X may be unrestricted in sign, i.e. may be positive, negative, or zero. In this instance, we seek to

$$max\, f = C'X \text{ s.t.}$$
$$AX \leq b, X \text{ unrestricted.}$$

To handle situations of this sort, let the variable x_k be expressed as the difference between two nonnegative variables as

$$x_k = x'_k - x''_k, \text{with } x'_k, x''_k \geq 0. \tag{5.4}$$

So by virtue of this transformation, the above problem may be rewritten, for $X = X' - X''$, as

$$max\, f(X', X'') = C'X' - C'X'' \text{ s.t.}$$
$$AX' - AX'' \leq b, \text{with } X', X'' \geq O.$$

Then the standard simplex technique is applicable since all variables are now nonnegative. Upon examining the various subcases that emerge, namely:

a) if $x'_k > x''_k$, then $x_k > 0$;
b) if $x'_k < x''_k$, then $x_k < 0$; and
c) if $x'_k = x''_k$, then $x_k = 0$.

we see that x_k, depending on the relative magnitude of x'_k and x''_k, is truly unrestricted in sign.

One important observation relative to using (5.4) to handle unrestricted variables is that any basic feasible solution in nonnegative variables cannot have both x'_k and x''_k in the set of basic variables since, for the kth column of $\bar{\boldsymbol{A}}, \bar{\boldsymbol{a}}_k, \left(x'_k - x''_k\right)\bar{\boldsymbol{a}}_k = x'_k\bar{\boldsymbol{a}}_k + x''_k(-\bar{\boldsymbol{a}}_k)$ so that $\bar{\boldsymbol{a}}_k$ and $-\bar{\boldsymbol{a}}_k$ are not linearly independent and thus cannot both constitute columns of the basis matrix.

Example 5.6 Using the preceding transformation, solve

$$\begin{aligned}
&max\, f = 3x_1 + x_2 + x_3 \text{ s.t.}\\
&2x_1 + 5x_2 - x_3 \le 10\\
&x_1 + 4x_2 + x_3 \le 12\\
&x_1, x_3 \text{ unrestricted}, x_2 \ge 0.
\end{aligned}$$

Setting $x_1 = x'_1 - x''_1$ and $x_3 = x'_3 - x''_3$, we wish to

$$\begin{aligned}
&max\, f = 3x'_1 - \text{etc s.t.}\\
&2x'_1 + \ldots + x''_3 \le 10\\
&x'_1 - x''_1 + 4x_2 + x'_3 - x''_3 \le 12\\
&x'_1, x''_1, x_2, x'_3, x''_3 \ge 0.
\end{aligned}$$

The standard simplex routine in all nonnegative variables now applies, with

$$x_j^0 = \left(x'_j\right)^0 - \left(x''_j\right)^0, j = 1, 3. \qquad \blacksquare$$

5.5 The Two-Phase Method

In this section, we shall develop an alternative procedure for solving a linear programming problem involving artificial variables. The approach will be to frame the simplex method in terms of two successive phases, Phase I (PI) and Phase II (PII) (Dantzig et al. 1955). In PI, we seek to drive all artificial variables to zero by maximizing a **surrogate objective function** involving only artificial variables. If the optimal basis for the PI problem contains no artificial vectors, or contains one or more artificial vectors at the zero level, then PII is initiated. However, if the optimal PI basis contains at least one artificial vector at a positive level, the process is terminated. If PII is warranted, we proceed by maximizing the original objective function using the optimal PI basis as a starting point, i.e. the optimal PI basis provides an initial though possibly nonbasic feasible solution to the original linear program.

PHASE I. For the PI objective function, the legitimate variables will be assigned zero coefficients while each artificial variable is given a coefficient of −1. Then the PI **surrogate linear programming problem** may be formulated as: maximize the **infeasibility form**

$$\begin{aligned} & g(\boldsymbol{X}_*) = \boldsymbol{O}'\boldsymbol{X} - \boldsymbol{1}'\boldsymbol{X}_a \quad \text{s.t.} \\ & \boldsymbol{A}_*\boldsymbol{X}_* = \boldsymbol{b}, \boldsymbol{X}_* \geq \boldsymbol{O}. \end{aligned} \tag{5.5}$$

Note that with $\boldsymbol{X}_a \geq \boldsymbol{O}$, it must be the case that $\max\{g\} \leq 0$. Additionally, the largest possible value that g can assume is zero, and this occurs only if the optimal PI basis contains no artificial vectors, or if any artificial vectors remaining in the said basis do so at the zero level. So if $\max\{g\} < 0$, at least one artificial vector remains in the optimal PI basis at a positive level.

Looking to some of the salient features of the PI routine: (i) as with the M-penalty method, an artificial vector never reenters the basis once it has been removed (here too the columns of the simplex matrix corresponding to artificial vectors that have turned nonbasis may be deleted as pivoting progresses); (ii) during PI, the sequence of vectors that enters and leaves the basis is the same as in the M-penalty method; and (iii) if g becomes zero before the optimality criterion is satisfied, we may terminate PI and proceed directly to PII.

Once the PI optimality criterion is satisfied, the optimal basis will exhibit one of three mutually exclusive and collectively exhaustive characteristics:

1) $\max\{g\} = 0$ and no artificial vectors remain in the basis;
2) $\max\{g\} = 0$ and one or more artificial vectors remain in the basis at the zero level; and
3) $\max\{g\} < 0$ and one or more artificial vectors remain in the basis at a positive level.

If case one obtains, the optimal PI basis provides a basic feasible solution to the original (primal) problem; the original structural constraints are consistent and none is redundant. For case two, we obtain a feasible solution to the original problem; the original structural constraints are consistent, though possibly redundant. And for case three, the surrogate problem has no feasible solution with $\boldsymbol{X}_a = \boldsymbol{O}$ and thus the original problem has no feasible solution as well. So, if either case one or case two emerges at the end of PI, there exists at least one feasible solution to the original problem. In this regard, we advance to PII.

PHASE II. Since we ultimately want to maximize $f(X)$, the optimal PI simplex matrix is transformed into the initial PII simplex matrix by respecifying the objective function coefficients. That is, during PII, the coefficients on the legitimate variables are the same as those appearing in the original objective function while the coefficients on the artificial variables are zero; the PII objective function is the original objective function $f(X)$ itself. Hence, the sequence of operations underlying the phase-two method may be expressed as:

$$\begin{aligned} &(\text{PI}) \max g(X_*) = O'X - 1'X_a \ \text{ s.t.} \\ &(\text{PII}) \max f(X_*) = C'X + O'X_a \ \text{ s.t.} \\ &\qquad A_*X_* = b, X_* \geq O. \end{aligned} \tag{5.6}$$

We now examine the implications for PII of the aforementioned characteristics of the optimal PI simplex matrix. For case one above, the optimal PI basis provides an initial basic feasible solution to the original problem. Once the PII objective function is introduced, pivoting proceeds until an optimal basic feasible solution to the original linear program is attained. Turning to case two, while we have obtained a feasible nonbasic solution to the original problem, there still remains the question of redundancy in the original structural constraints $AX = b$. Also, we need to develop a safeguard against the possibility that one or more artificial vectors currently in the initial PII basis at the zero level will appear in some future basis at a positive level as the PII pivot operations are executed.

Our discussion of redundancy here is similar to that advanced for the M-penalty method. Let us assume that within the optimal PI simplex matrix $y_{ij} = 0$ for all legitimate nonbasis vectors $\hat{Y}_j$ and for all $i \in \mathcal{A}$. In this instance, none of the artificial basis vectors e_i can be replaced by the $\hat{Y}_j$ so that there is redundancy in the original structural constraint equalities. In this regard, if there are k artificial vectors in the basis at the zero level, then, at the end of PI, we may simply delete the k rows of the simplex matrix containing the zero-valued artificial variables along with the associated artificial basis vectors e_i. Then PII begins with a basis of reduced size and pivoting is carried out until an optimal basic feasible solution to the original problem is attained.

We next examine a procedure that serves to guard against the possibility that an artificial vector appearing in the initial PII basis at the zero level manifests itself in some future basis at a positive level. Let us assume that for some $j = j'$ the vector $\hat{Y}_{j'}$ is to enter the basis and $y_{ij'} \leq 0$ with $y_{ij'} < 0$ for at least one $i \in \mathcal{A}$. In this circumstance, the artificial vectors for which $y_{ij'} < 0$ will not be considered as candidates for removal from the basis. Rather, some legitimate vector will be removed with the result that the artificial basis vectors for which $y_{ij'} < 0$ appear in the new basis at a positive level. To avoid this difficulty, instead of removing a legitimate vector from the basis, let us adopt the policy of arbitrarily removing any one of the artificial basis vectors e_i with $y_{ij'} < 0$. The result will be a feasible solution to the original problem wherein $\hat{Y}_{j'}$ enters the new basis at the zero level (since the artificial vector that it replaced was formerly at the zero level). With the value of the incoming variable equal to zero, the values of the other basic variables for the new solution are the same as in the preceding solution. Hence the value of the objective function for the new solution must be the same as the previous one (see Example 5.8).

Example 5.7 Let us solve the following problem using the two-phase method:

$$\begin{aligned} \max f &= x_1 + x_2 \quad \text{s.t.} \\ x_1 + 2x_2 &\leq 12 \\ x_1 + x_2 &\geq 4 \\ -x_1 + x_2 &\geq 1 \\ x_1, x_2 &\geq 0. \end{aligned}$$

Forming the PI infeasibility form $g = -x_6 - x_7$ we readily obtain the simplex matrix

$$\left[\hat{A}_{*} \vdots b_o\right] = \begin{bmatrix} 1 & 2 & 1 & 0 & 0 & 0 & 0 & 0 & 12 \\ 1 & 1 & 0 & -1 & 0 & 1 & 0 & 0 & 4 \\ -1 & 1 & 0 & 0 & -1 & 0 & 1 & 0 & 1 \\ 0 & 0 & 0 & 0 & 0 & 1 & 1 & 1 & 0 \end{bmatrix},$$

where $b'_o = [b', 0]$. To generate an initial basic feasible solution to this PI surrogate problem, let us add

$$\begin{array}{ccccccccc} -1 & -1 & 0 & 1 & 0 & -1 & 0 & 0 & -4 \\ 1 & -1 & 0 & 0 & 1 & 0 & -1 & 0 & -1 \\ \hline 0 & -2 & 0 & 1 & 1 & -1 & -1 & 0 & -5 \end{array}$$

to the last row of the above matrix to obtain

$$\begin{bmatrix} 1 & 2 & 1 & 0 & 0 & 0 & 0 & 0 & 12 \\ 1 & 1 & 0 & -1 & 0 & 1 & 0 & 0 & 4 \\ -1 & 1 & 0 & 0 & -1 & 0 & 1 & 0 & 1 \\ 0 & -2 & 0 & 1 & 1 & 0 & 0 & 1 & -5 \end{bmatrix}.$$

The first basic feasible solution to the surrogate problem thus becomes $x_3 = 12$, $x_6 = 4$, $x_7 = 1$, and $g = -5$. Since this solution is obviously not optimal, we undertake

PI, ROUND 1.

$$\left[\begin{array}{ccccccccc|c} 1 & 2 & 1 & 0 & 0 & 0 & 0 & 0 & 12 & 6 \\ 1 & 1 & 0 & -1 & 0 & 1 & 0 & 0 & 4 & 4 \\ -1 & \textcircled{1} & 0 & 0 & -1 & 0 & 1 & 0 & 1 & 1 \\ 0 & -2 & 0 & 1 & 1 & 0 & 0 & 1 & -5 & \end{array}\right] \rightarrow \begin{bmatrix} 3 & 0 & 1 & 0 & 2 & 0 & -2 & 0 & 10 \\ 2 & 0 & 0 & -1 & 1 & 1 & -1 & 0 & 3 \\ -1 & 1 & 0 & 0 & -1 & 0 & 1 & 0 & 1 \\ -2 & 0 & 0 & 1 & -1 & 0 & 2 & 1 & -3 \end{bmatrix}.$$

Here the second basic feasible solution to the surrogate problem is $x_1 = 1$, $x_3 = 10$, $x_6 = 3$, and $g = -3$. Since this solution is also not optimal, we turn to **PI, ROUND 2** (column 7 is deleted).

$$\left[\begin{array}{cccccccc|c} 3 & 0 & 1 & 0 & 2 & 0 & 0 & 10 & 10/3 \\ \textcircled{2} & 0 & 0 & -1 & 1 & 1 & 0 & 3 & 3/2 \\ -1 & 1 & 0 & 0 & -1 & 0 & 0 & 1 & \\ -2 & 0 & 0 & 1 & -1 & 0 & 1 & -3 & \end{array}\right] \rightarrow \begin{bmatrix} 0 & 0 & 1 & \frac{3}{2} & \frac{1}{2} & -\frac{3}{2} & 0 & \frac{11}{2} \\ 1 & 0 & 0 & -\frac{1}{2} & \frac{1}{2} & \frac{1}{2} & 0 & \frac{3}{2} \\ 0 & 1 & 0 & -\frac{1}{2} & -\frac{1}{2} & \frac{1}{2} & 0 & \frac{5}{2} \\ 0 & 0 & 0 & 0 & 0 & 1 & 1 & 0 \end{bmatrix}.$$

The third basic feasible solution to the surrogate problem thus appears as $x_1 = 3/2$, $x_2 = 5/2$, $x_3 = 11/2$, and $g = 0$. Since it is clearly optimal, PI ends. Moreover, since all artificial vectors have been removed from the basis, this particular basic feasible solution to the surrogate problem represents an initial basic feasible solution to the original problem. In this regard PII begins by deleting column 6 of the final PI simplex matrix and replacing the last row of the same by the original objective function $f = x_1 + x_2$. Thus,

$$\begin{bmatrix} 0 & 0 & 1 & \frac{3}{2} & \frac{1}{2} & 0 & \frac{11}{2} \\ 1 & 0 & 0 & -\frac{1}{2} & -\frac{1}{2} & 0 & \frac{3}{2} \\ 0 & 1 & 0 & -\frac{1}{2} & -\frac{1}{2} & 0 & \frac{5}{2} \\ -1 & -1 & 0 & 0 & 0 & 1 & 0 \end{bmatrix}$$

or, upon adding

$$\begin{array}{ccccccc} 1 & 0 & 0 & -\frac{1}{2} & \frac{1}{2} & 0 & \frac{3}{2} \\ 0 & 1 & 0 & -\frac{1}{2} & -\frac{1}{2} & 0 & \frac{5}{2} \\ \hline 1 & 1 & 0 & -1 & 0 & 1 & 4 \end{array}$$

to the last row of the preceding simplex matrix,

$$\begin{bmatrix} 0 & 0 & 1 & 3 & \frac{1}{2} & 0 & \frac{11}{2} \\ 1 & 0 & 0 & -\frac{1}{2} & \frac{1}{2} & 0 & \frac{3}{2} \\ 0 & 1 & 0 & -\frac{1}{2} & -\frac{1}{2} & 0 & \frac{5}{2} \\ 0 & 0 & 0 & -1 & 0 & 1 & 4 \end{bmatrix}.$$

Thus, the initial basic feasible solution to the PII original problem appears as $x_1 = 3/2$, $x_2 = 5/2$, $x_3 = 11/2$, and $f = 4$. Since the optimality criterion is not satisfied, we proceed to

PII, ROUND 1.

$$\begin{bmatrix} 0 & 0 & 1 & \circled{\frac{3}{2}} & \frac{1}{2} & 0 & \frac{11}{2} \\ 1 & 0 & 0 & -\frac{1}{2} & \frac{1}{2} & 0 & \frac{3}{2} \\ 0 & 1 & 0 & -\frac{1}{2} & -\frac{1}{2} & 0 & \frac{5}{2} \\ 0 & 0 & 0 & -1 & 0 & 1 & 4 \end{bmatrix} \begin{matrix} 11/3 \\ - \\ - \\ \end{matrix} \rightarrow \begin{bmatrix} 0 & 0 & \frac{2}{3} & 1 & \frac{1}{3} & 0 & \frac{11}{3} \\ 1 & 0 & \frac{1}{3} & 0 & \frac{2}{3} & 0 & \frac{10}{3} \\ 0 & 1 & \frac{1}{3} & 0 & -\frac{1}{3} & 0 & \frac{13}{3} \\ 0 & 0 & \frac{2}{3} & 0 & \frac{1}{3} & 1 & \frac{23}{3} \end{bmatrix}.$$

An inspection of the last row of this simplex matrix indicates that we have obtained an optimal basic feasible solution to the original problem consisting of $x_1 = 10/3$, $x_2 = 13/3$, $x_4 = 11/3$, and $f = 23/3$. ■

Example 5.8 Let us

$$\begin{aligned} \max f = 5x_1 + 4x_2 + \ x_3 + x_4 \ \text{s.t.} \\ x_1 + \ x_2 + \ x_3 \qquad\quad = 2 \\ x_2 + \ x_3 \qquad\quad = 2 \\ x_1 + 2x_2 + 3x_3 + 6x_4 = 4 \\ x_1, \ldots, x_4 \geq 0. \end{aligned}$$

Constructing the infeasibility form $g = -x_5 - x_6 - x_7$ we obtain

$$\left[\hat{A}_{*} \vdots b_o\right] = \begin{bmatrix} 1 & 1 & 1 & 0 & 1 & 0 & 0 & 0 & 2 \\ 0 & 1 & 1 & 0 & 0 & 1 & 0 & 0 & 2 \\ 1 & 2 & 3 & 6 & 0 & 0 & 1 & 0 & 4 \\ 0 & 0 & 0 & 0 & 1 & 1 & 1 & 1 & 0 \end{bmatrix} \rightarrow \begin{bmatrix} 1 & 1 & 1 & 0 & 1 & 0 & 0 & 0 & 2 \\ 0 & 1 & 1 & 0 & 0 & 1 & 0 & 0 & 2 \\ 1 & 2 & 3 & 6 & 0 & 0 & 1 & 0 & 4 \\ -2 & -4 & -5 & -6 & 0 & 0 & 0 & 1 & -8 \end{bmatrix},$$

where x_5, x_6 and x_6 are artificial variables. Here the initial basic feasible solution to this surrogate problem is $x_5 = 2$, $x_6 = 2$, $x_7 = 4$, and $g = -8$. To maximize g, we turn to

PI, ROUND 1.

$$\left[\begin{array}{ccccccccc} 1 & 1 & 1 & 0 & 1 & 0 & 0 & 0 & 2 \\ 0 & 1 & 1 & 0 & 0 & 1 & 0 & 0 & 2 \\ 1 & 2 & 3 & \textcircled{6} & 0 & 0 & 1 & 0 & 4 \\ -2 & -4 & -5 & -6 & 0 & 0 & 0 & 1 & -8 \end{array}\right] \begin{array}{c} - \\ - \\ 2/3 \\ \\ \end{array} \rightarrow \left[\begin{array}{ccccccccc} 1 & 1 & 1 & 0 & 1 & 0 & 0 & 0 & 2 \\ 0 & 1 & 1 & 0 & 0 & 1 & 0 & 0 & 2 \\ \frac{1}{6} & \frac{1}{3} & \frac{1}{2} & 1 & 0 & 0 & \frac{1}{6} & 0 & \frac{2}{3} \\ -1 & -2 & -2 & 0 & 0 & 0 & 1 & 1 & -4 \end{array}\right].$$

The second basic feasible solution to the surrogate problem consists of $x_4 = 2/3$, $x_5 = 2$, $x_6 = 2$, and $g = -4$. Given that the optimality criterion is still not satisfied, we proceed to

PI, ROUND 2 (column 7 is deleted).

$$\left[\begin{array}{cccccccc} 1 & 1 & 1 & 0 & 1 & 0 & 0 & 2 \\ 0 & 1 & 1 & 0 & 0 & 1 & 0 & 2 \\ \frac{1}{6} & \frac{1}{3} & \frac{1}{2} & 1 & 0 & 0 & 0 & \frac{2}{3} \\ -1 & -2 & -2 & 0 & 0 & 0 & 1 & -4 \end{array}\right] \rightarrow \left[\begin{array}{cccccccc} 1 & 1 & 1 & 0 & 1 & 0 & 0 & 2 \\ -1 & 0 & 0 & 0 & -1 & 1 & 0 & 0 \\ -\frac{1}{6} & 0 & \frac{1}{6} & 1 & -\frac{1}{3} & 0 & 0 & 0 \\ 1 & 0 & 0 & 0 & 2 & 0 & 1 & 0 \end{array}\right].$$

With the optimality criterion satisfied, the optimal basic feasible solution to the PI surrogate problem is $x_2 = 2$, $x_4 = x_6 = 0$, and $g = 0$. Turning now to PII, let us delete column 5 of the final PI simplex matrix and replace the last row of the same by the original objective function $f = 5x_1 + 4x_2 + x_3 + x_4$. To this end, we obtain

$$\left[\begin{array}{ccccccc} 1 & 1 & 1 & 0 & 0 & 0 & 2 \\ -1 & 0 & 0 & 0 & 1 & 0 & 0 \\ -\frac{1}{6} & 0 & \frac{1}{6} & 1 & 0 & 0 & 0 \\ -5 & -4 & -1 & -1 & 0 & 1 & 0 \end{array}\right] \rightarrow \overset{\hat{Y}_1}{\left[\begin{array}{ccccccc} 1 & 1 & 1 & 0 & 0 & 0 & 2 \\ -1 & 0 & 0 & 0 & 1 & 0 & 0 \\ -\frac{1}{6} & 0 & \frac{1}{6} & 1 & 0 & 0 & 0 \\ -\frac{7}{6} & 0 & \frac{19}{6} & 0 & 0 & 1 & 8 \end{array}\right]}.$$

Since the initial PII basic feasible solution to the original problem consisting of $x_2 = 2$, $x_4 = x_6 = 0$, and $f = 8$ is not optimal, round one of PII is warranted. However, before attempting to generate an improved solution to the PII objective function, we note that since $y_{21} < 0$, the artificial vector $\boldsymbol{e}_2$ will not be considered for removal from the basis. Rather, simplex pivoting dictates that the vector $\boldsymbol{e}_1$ will be deleted from the basis, with the result that the current artificial basis vector $\boldsymbol{e}_2$ will appear in the new basis at a positive level. To preclude the latter from occurring, instead of removing $\boldsymbol{e}_1$, let us remove the artificial basis vector $\boldsymbol{e}_2$. To this end we proceed to

PII, ROUND 1.

$$\begin{bmatrix} 1 & 1 & 1 & 0 & 0 & 0 & 2 \\ \textcircled{-1} & 0 & 0 & 0 & 1 & 0 & 0 \\ -\frac{1}{6} & 0 & \frac{1}{6} & 1 & 0 & 0 & 0 \\ -\frac{7}{6} & 0 & \frac{19}{6} & 0 & 0 & 1 & 8 \end{bmatrix} \rightarrow \begin{bmatrix} 0 & 1 & 1 & 0 & 1 & 0 & 2 \\ 1 & 0 & 0 & 0 & -1 & 0 & 0 \\ 0 & 0 & \frac{1}{6} & 1 & -\frac{1}{6} & 0 & 0 \\ 0 & 0 & \frac{19}{6} & 0 & -\frac{7}{6} & 1 & 8 \end{bmatrix}.$$

Thus, the optimal (degenerate) basic feasible solution to the PII original problem consists of $x_1 = 0$, $x_2 = 2$, $x_4 = 0$, and $f = 8$. ■

6

Duality Theory

6.1 The Symmetric Dual

Given a particular linear programming problem, there is associated with it another (unique) linear program called its dual. In general, duality theory addresses itself to the study of the connection between two related linear programming problems, where one of them, the **primal**, is a maximization (minimization) problem and the other, the **dual**, is a minimization (maximization) problem. Moreover, this association is such that the existence of an optimal solution to any one of these two problems guarantees an optimal solution to the other, with the result that their extreme values are equal. To gain some additional insight into the role of duality in linear programming, we note briefly that the solution to any given linear programming problem may be obtained by applying the simplex method to either its primal or dual formulation, the reason being that the simplex process generates an optimal solution to the primal-dual pair of problems simultaneously. So with the optimal solution to one of these problems obtainable from the optimal solution of its dual, there is no reason to solve both problems separately. Hence the simplex technique can be applied to whichever problem requires the least computational effort.

We now turn to the expression of the primal problem in canonical form and then look to the construction of its associated dual. Specifically, if the primal problem appears as

$$
\begin{aligned}
&max\, f = c_1x_1 + \cdots + c_px_p \quad \text{s.t.} \\
&\quad a_{11}x_1 + \cdots + a_{1p}x_p \leq b_1 \\
&\quad a_{21}x_1 + \cdots + a_{2p}x_p \leq b_2 \\
&\quad \cdots\cdots\cdots\cdots\cdots\cdots\cdots\cdots \\
&\quad a_{m1}x_1 + \cdots + a_{mp}x_p \leq b_m \\
&\quad x_1, \ldots, x_p \geq 0,
\end{aligned}
$$

Linear Programming and Resource Allocation Modeling, First Edition. Michael J. Panik.

then its dual is

$$min\ g = b_1u_1 + \cdots + b_mu_m \quad \text{s.t.}$$

$$a_{11}u_1 + \cdots + a_{m1}u_m \geq c_1$$

$$a_{12}u_1 + \cdots + a_{m2}u_m \geq c_2$$

....

$$a_{1p}u_1 + \cdots + a_{mp}u_m \geq c_p$$

$$u_1, \ldots, u_m \geq 0.$$

Henceforth, this form of the dual will be termed the **symmetric dual**, i.e. in this instance, the primal and dual problems are said to be *mutually symmetric*. In matrix form, the symmetric primal-dual pair of problems appears as

Primal	**Symmetric Dual**
$max\ f = \mathbf{C}'\mathbf{X}$ s.t.	$min\ g = \mathbf{b}'\mathbf{U}$ s.t.
$\mathbf{AX} \leq \mathbf{b}, \mathbf{X} \geq \mathbf{O}$	$\mathbf{A}'\mathbf{U} \geq \mathbf{C}, \mathbf{U} \geq \mathbf{O}$

where $\mathbf{X}$ and $\mathbf{C}$ are of order $(p \times 1)$, $\mathbf{U}$ and $\mathbf{b}$ are of order $(m \times 1)$, and $\mathbf{A}$ is of order $(m \times p)$. In standard form, the primal and its symmetric dual may be written as

Primal	**Symmetric Dual**
$max\ f = \mathbf{C}'\mathbf{X} + \mathbf{O}'\mathbf{X}_s$ s.t.	$min\ g = \mathbf{b}'\mathbf{U} + \mathbf{O}'\mathbf{U}_s$ s.t.
$\mathbf{AX} + \mathbf{I}_m\mathbf{X}_s = \mathbf{b}$	$\mathbf{A}'\mathbf{U} - \mathbf{I}_p\mathbf{U}_s \geq \mathbf{C}$
$\mathbf{X}, \mathbf{X}_s \geq \mathbf{O}$	$\mathbf{U}, \mathbf{U}_s \geq \mathbf{O}$

where $\mathbf{X}_s$, $\mathbf{U}_s$ denote $(m \times 1)$ and $(p \times 1)$ vectors of primal slack and dual surplus variables, respectively.

An examination of the structure of the primal and symmetric dual problems reveals the following salient features:

a) If the primal problem involves maximization, then the dual problem involves minimization, and conversely. A new set of variables appears in the dual.
b) In this regard, if there are p variables and m structural constraints in the primal problem, there will be m variables and p structural constraints in the dual problem, and conversely. In particular, there exists a one-to-one correspondence between the jth primal variable and the jth dual structural constraint, and between the ith dual variable and the ith primal structural constraint.
c) The primal objective function coefficients become the right-hand sides of the dual structural constraints and, conversely. The coefficient matrix of the dual structural constraint system is the transpose of the coefficient matrix of the primal structural constraint system, and vice-versa. Thus, the coefficients on x_j in the primal problem become the coefficients of the jth dual structural constraint.
d) The inequality signs in the structural constraints of the primal problem are reversed in the dual problem, and vice-versa.

Example 6.1 Given that the primal problem appears as:

$$\begin{aligned}
&max\, f = 2x_1 + 3x_2 + 5x_3 \quad \text{s.t.}\\
&x_1 + x_2 + 3x_3 \le 5\\
&2x_1 - x_2 + x_3 \le 6\\
&x_1 + 3x_2 - 4x_3 \le 8\\
&x_2 + x_3 \le 7\\
&x_1, x_2, x_3 \ge 0,
\end{aligned}$$

find its dual. In the light of the preceding discussion, it is obvious that the symmetric dual problem is:

$$\begin{aligned}
&min\, g = 5u_1 + 6u_2 + 8u_3 + 7u_4 \quad \text{s.t.}\\
&u_1 + 2u_2 + u_3 \ge 2\\
&u_1 - u_2 + 3u_3 + u_4 \ge 3\\
&3u_1 + u_2 - 4u_3 + u_4 \ge 5\\
&u_1, \dots, u_4 \ge 0.
\end{aligned}$$

■

6.2 Unsymmetric Duals

This section will develop a variety of duality relationships and forms that result when the primal maximum problem involves structural constraints other than of the "≤" variety and variables that are unrestricted in sign. In what follows, our approach will be to initially write the given primal problem in canonical form, and then, from it, obtain the corresponding symmetric dual problem. To this end, let us first assume that the primal problem appears as

$$\begin{aligned}
&max\, f = \boldsymbol{C}'\boldsymbol{X} \quad \text{s.t.}\\
&\boldsymbol{AX} = \boldsymbol{b}, \boldsymbol{X} \ge \boldsymbol{O}.
\end{aligned}$$

In this instance two sets of inequality structural constraints are implied by $\boldsymbol{AX} = \boldsymbol{b}$, i.e. $\boldsymbol{AX} \le \boldsymbol{b}$, $-\boldsymbol{AX} \le -\boldsymbol{b}$ must hold simultaneously. Then the above primal problem has the canonical form

$$\begin{aligned}
&max\, f = \boldsymbol{C}'\boldsymbol{X} \quad \text{s.t.}\\
&\begin{bmatrix} \boldsymbol{A} \\ \cdots \\ -\boldsymbol{A} \end{bmatrix} \boldsymbol{X} \le \begin{bmatrix} \boldsymbol{b} \\ \cdots \\ -\boldsymbol{b} \end{bmatrix}, \boldsymbol{X} \ge \boldsymbol{O}.
\end{aligned}$$

The symmetric dual associated with this problem becomes

$$min\ g(\boldsymbol{W},\boldsymbol{V}) = \boldsymbol{b}'\boldsymbol{W} - \boldsymbol{b}'\boldsymbol{V} \quad \text{s.t.}$$

$$\left[\boldsymbol{A}' \vdots -\boldsymbol{A}'\right]\begin{bmatrix}\boldsymbol{W}\\ \cdots \\ \boldsymbol{V}\end{bmatrix} \geq \boldsymbol{C} \text{ or } \boldsymbol{A}'\boldsymbol{W} - \boldsymbol{A}'\boldsymbol{V} \geq \boldsymbol{C}, \boldsymbol{W}, \boldsymbol{V} \geq \boldsymbol{O},$$

where the dual variables $\boldsymbol{W}$, $\boldsymbol{V}$ are associated with the primal structural constraints $\boldsymbol{AX} \leq \boldsymbol{b}$, $-\boldsymbol{AX} \leq -\boldsymbol{b}$, respectively.

If $\boldsymbol{U} = \boldsymbol{W} - \boldsymbol{V}$, then we ultimately obtain

$$\begin{aligned} &min\ g = \boldsymbol{b}'\boldsymbol{U} \quad \text{s.t.} \\ &\boldsymbol{A}'\boldsymbol{U} \geq \boldsymbol{C}, \boldsymbol{U} \text{ unrestricted,} \end{aligned}$$

since if both $\boldsymbol{W}, \boldsymbol{V} \geq \boldsymbol{O}$, their difference may be greater than or equal to, less than or equal to, or exactly equal to, the null vector.

Next, suppose the primal problem appears as

$$\begin{aligned} &max\ f = \boldsymbol{C}'\boldsymbol{X} \quad \text{s.t.} \\ &\boldsymbol{AX} \geq \boldsymbol{b}, \boldsymbol{X} \geq \boldsymbol{O}. \end{aligned}$$

In canonical form, we seek to

$$\begin{aligned} &max\ f = \boldsymbol{C}'\boldsymbol{X} \quad \text{s.t.} \\ &-\boldsymbol{AX} \leq -\boldsymbol{b}, \boldsymbol{X} \geq \boldsymbol{O}. \end{aligned}$$

The symmetric dual associated with this problem is

$$\begin{aligned} &min\ g = -\boldsymbol{b}'\boldsymbol{V} \quad \text{s.t.} \\ &-\boldsymbol{A}'\boldsymbol{V} \geq \boldsymbol{C}, \boldsymbol{V} \geq \boldsymbol{O}. \end{aligned}$$

If $\boldsymbol{U} = -\boldsymbol{V}$, then we finally

$$\begin{aligned} &min\ g = \boldsymbol{b}'\boldsymbol{U} \quad \text{s.t.} \\ &\boldsymbol{A}'\boldsymbol{U} \geq \boldsymbol{C}, \boldsymbol{U} \leq \boldsymbol{O}. \end{aligned}$$

Lastly, if the primal problem is of the form

$$\begin{aligned} &max\ f = \boldsymbol{C}'\boldsymbol{X} \quad \text{s.t.} \\ &\boldsymbol{AX} \leq \boldsymbol{b}, \boldsymbol{X} \text{ unrestricted,} \end{aligned}$$

then, with X expressible as the difference between the two vectors X_1, X_2 whose components are all nonnegative, the primal problem becomes

$$max\ f = \left[C' \vdots -C'\right] \begin{bmatrix} X_1 \\ \cdots \\ X_2 \end{bmatrix} \quad \text{s.t.}$$

$$\left[A \vdots -A\right] \begin{bmatrix} X_1 \\ \cdots \\ X_2 \end{bmatrix} \leq b, X_1, X_2 \geq O.$$

The symmetric dual of this latter problem is

$$min\ g = b'U \quad \text{s.t.}$$

$$\begin{bmatrix} A' \\ \cdots \\ -A' \end{bmatrix} U \geq \begin{bmatrix} C \\ \cdots \\ -C \end{bmatrix} \text{ or } A'U = C, U \geq O.$$

To summarize, let us assume that the primal maximum problem contains a mixture of structural constraints. Then:

1) If the primal variables are all nonnegative, the dual minimum problem has the following properties:
   a) The form of the dual structural constraint system is independent of the form of the primal structural constraint system, i.e. the former always appears as $A'U \geq C$. Moreover, only the dual variables are influenced by the primal structural constraints. Specifically:
   b) The dual variable u_r corresponding to the rth "≤" primal structural constraint is nonnegative.
   c) The dual variable u_s corresponding to the sth "≥" primal structural constraint is nonpositive.
   d) The dual variable u_t corresponding to the tth "=" primal structural constraint is unrestricted in sign.
2) If some variable x_j in the primal problem is unrestricted in sign, then the jth dual structural constraint will hold as an equality.

Example 6.2 The dual of the primal problem

$$\begin{aligned} max\ f &= 3x_1 - x_2 + x_3 - x_4 \quad \text{s.t.} \\ x_1 + x_2 + 2x_3 + 3x_4 &\leq 5 \\ x_3 - x_4 &\geq 4 \\ x_1 - x_2 &= -1 \\ x_1, x_2, x_4 \geq 0, x_3 &\text{ unrestricted} \end{aligned}$$

is

$$\begin{aligned} &\min g = 5u_1 + 4u_2 - u_3 \quad \text{s.t.} \\ &u_1 + u_3 \geq 3 \\ &u_1 - u_3 \geq -1 \\ &2u_1 + u_2 = 1 \\ &3u_1 - u_2 \geq -1 \\ &u_1 \geq 0, u_2 \leq 0, u_3 \text{ unrestricted.} \end{aligned}$$

■

6.3 Duality Theorems

To further enhance our understanding of the relationship between the primal and dual linear programming problems, let us consider the following sequence of duality theorems (Theorems 6.1–6.3).

Theorem 6.1 The dual of the dual is the primal problem itself.

The implication of this theorem is that it is immaterial which problem is called the dual.

The next two theorems address the relationship between the objective function values of the primal-dual pair of problems. First,

Theorem 6.2 If $\boldsymbol{X}_0$ is a feasible solution to the primal maximum problem and $\boldsymbol{U}_0$ is a feasible solution to the dual minimum problem, then $\boldsymbol{C}'\boldsymbol{X}_0 \leq \boldsymbol{b}'\boldsymbol{U}_0$.

In this regard, the dual objective function value provides an upper bound to the value of the primal objective function. Second,

Theorem 6.3 If $\boldsymbol{X}_0$, $\boldsymbol{U}_0$ are feasible solutions to the primal maximum and dual minimum problems, respectively, and $\boldsymbol{C}'\boldsymbol{X}_0 = \boldsymbol{b}'\boldsymbol{U}_0$, then both $\boldsymbol{X}_0$, $\boldsymbol{U}_0$ are optimal solutions.

In addition, either both the primal maximum and dual minimum problems have optimal vectors $\boldsymbol{X}_0$, $\boldsymbol{U}_0$, respectively, with $\boldsymbol{C}'\boldsymbol{X}_0 = \boldsymbol{b}'\boldsymbol{U}_0$, or neither has an optimal vector.

We noted in Chapter 4 that it is possible for the primal objective function to be unbounded over the feasible region $\mathcal{K}$. What is the implication of this outcome as far as the behavior of the dual objective is concerned? As will be seen in Example 6.4, if the primal maximum (dual minimum) problem possesses a feasible solution and $C'X(b'U)$ is unbounded from above (below), then the dual minimum (primal maximum) problem is infeasible. Looked at in another fashion, one (and thus both) of the primal maximum and dual minimum problems has an optimal solution if and only if either $C'X$ or $b'U$ is bounded on its associated nonempty feasible region.

We now turn to what may be called the **fundamental theorems of linear programming**. In particular, we shall examine the existence, duality, and complementary slackness theorems. We first state Theorem 6.4.

Theorem 6.4 Existence Theorem (Goldman and Tucker 1956). The primal maximum and dual minimum problems possess optimal solutions if and only if both have feasible solutions.

We next have Theorem 6.5.

Theorem 6.5 Duality Theorem (Goldman and Tucker 1956). A feasible solution $\hat{X}$ to the primal maximum problem is optimal if and only if the dual minimum problem has a feasible solution $\hat{U}$ with $b'\hat{U} = C'\hat{X}$. (Likewise, a feasible solution $\hat{U}$ is optimal if and only if a feasible $\hat{X}$ exists for which $C'\hat{X} = b'\hat{U}$.)

Example 6.3 The optimal solution to the primal problem

$$\begin{aligned} & max\, f = 24x_1 + 25x_2 \qquad \text{s.t.} \\ & x_1 + 5x_2 \le 1 \\ & 4x_1 + x_2 \le 3 \\ & x_1, x_2 \ge 0 \end{aligned}$$

is $X_0' = (14/19, 1/19)$ and $f^0 = 19$ (Figure 6.1a) while the same for the dual problem

$$\begin{aligned} & min\, g = u_1 + 3u_2 \qquad \text{s.t.} \\ & u_1 + 4u_2 \ge 24 \\ & 5u_1 + u_2 \ge 25 \\ & u_1, u_2 \ge 0 \end{aligned}$$

is $U_0' = (4,5)$ and $g^0 = 19$ (Figure 6.1b). Here both the primal and dual problems possess optimal vectors X_0, U_0, respectively, with $f^0 = g^0$, as required by the duality theorem. ■

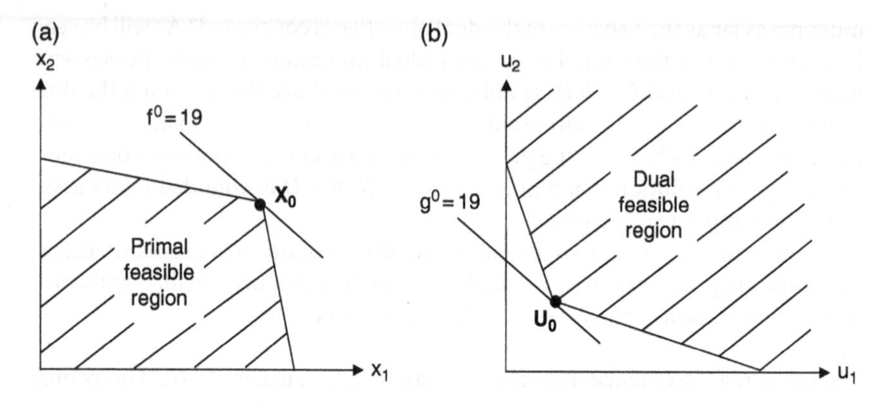

Figure 6.1 (a) Unique optimal primal solution; (b) Unique optimal dual solution.

Example 6.4 From Example 4.4 we determine that the primal problem

$$\begin{aligned} & max\, f = 2x_1 + 3x_2 \qquad \text{s.t.} \\ & -2x_1 + x_2 \le 3 \\ & -\frac{1}{2}x_1 + x_2 \le 6 \\ & x_1, x_2 \ge 0 \end{aligned}$$

had an unbounded objective function. From our preceding discussion we should anticipate that its dual,

$$\begin{aligned} & min\, g = 3u_1 + 6u_2 \qquad \text{s.t.} \\ & -2u_1 - \frac{1}{2}u_2 \ge 2 \\ & u_1 + u_2 \ge 3 \\ & u_1, u_2 \ge 0, \end{aligned}$$

does not possess a feasible solution, as Figure 6.2 clearly indicates. ■

We now turn to the third fundamental theorem of linear programming, namely a statement of the complementary slackness conditions. Actually, two such theorems will be presented – the strong and the weak cases. We first examine the weak case provided by Theorem 6.6.

Theorem 6.6 Weak Complementary Slackness Theorem. The solutions $(X_0', (X_s)_0'), (U_0', (U_s)_0')$ to the primal maximum and dual minimum problems,

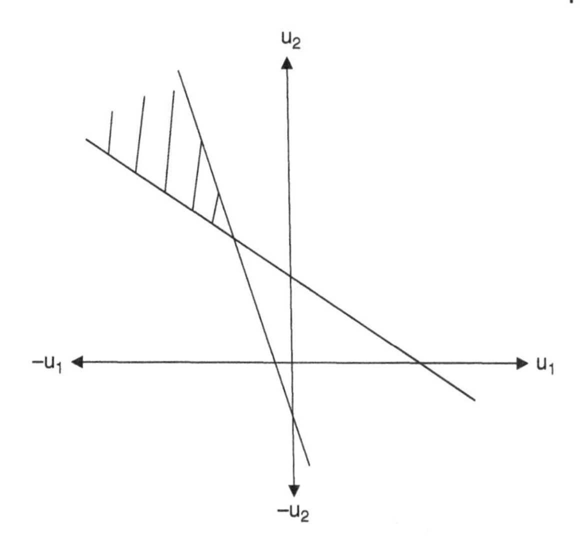

Figure 6.2 When the primal objective is unbounded, the dual problem is infeasible.

respectively, are optimal if and only if they satisfy the complementary slackness conditions

$$(\boldsymbol{U}_s)_0'\boldsymbol{X}_0 + \boldsymbol{U}_0'(\boldsymbol{X}_S)_0 = 0 \text{ or } \begin{cases} u_i^0 x_{p+i}^0 = 0, i = 1,\ldots,m; \\ x_j^0 u_{m+j}^0 = 0, j = 1,\ldots,p. \end{cases} \tag{6.1}$$

That is, for optimal solutions $(\boldsymbol{X}_0', (\boldsymbol{X}_s)_0'), (\boldsymbol{U}_0', (\boldsymbol{U}_s)_0')$ to the primal maximum and dual minimum problems, respectively:

a) If a primal slack variable x_{p+i}^0 is positive, then the ith dual variable $u_i^0, i = 1,\ldots,m,$ must equal zero. Conversely, if u_i^0 is positive, then $x_{p+i}^0, i = 1,\ldots,m,$ equals zero;
b) If the jth primal variable x_j^0 is positive, then the dual surplus variable $u_{m+j}^0, j = 1,\ldots,p,$ equals zero. Conversely, if u_{m+j}^0 is positive, then $x_j^0, j = 1,\ldots,p,$ must equal zero.

How may we interpret (6.1)? Specifically, $u_i^0 x_{p+i} = 0$ implies that either $x_{p+i}^0 > 0$ and $u_i^0 = 0$ or $u_i^0 > 0$ and $x_{p+i}^0 = 0$ or both $u_i^0, x_{p+i}^0 = 0, i = 1,\ldots,m$. Similarly, $x_j^0 u_{m+j}^0 = 0$ implies that either $x_j^0 > 0$ and $u_{m+j}^0 = 0$ or $u_{m+j}^0 > 0$ and $x_j^0 = 0$ or both $x_j^0, u_{m+j}^0 = 0, j = 1,\ldots,p$. Alternatively, if the constraint system for the primal-dual pair of problems is written as

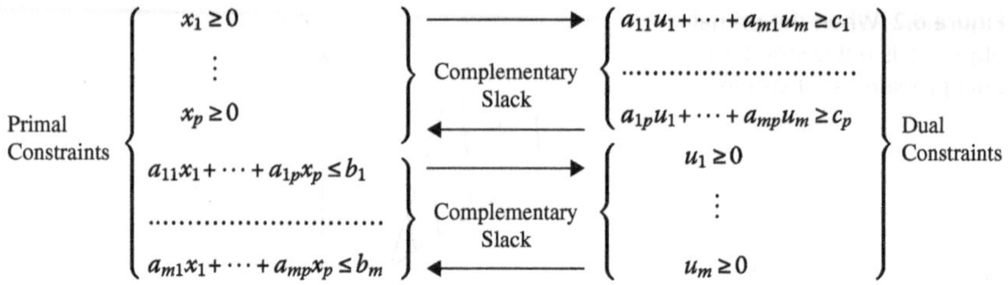

Primal Constraints:

$$\left.\begin{array}{c} x_1 \geq 0 \\ \vdots \\ x_p \geq 0 \end{array}\right\} \xrightarrow[\longleftarrow]{\text{Complementary Slack}} \left\{\begin{array}{c} a_{11}u_1 + \cdots + a_{m1}u_m \geq c_1 \\ \cdots\cdots\cdots \\ a_{1p}u_1 + \cdots + a_{mp}u_m \geq c_p \end{array}\right.$$

$$\left.\begin{array}{c} a_{11}x_1 + \cdots + a_{1p}x_p \leq b_1 \\ \cdots\cdots\cdots \\ a_{m1}x_1 + \cdots + a_{mp}x_p \leq b_m \end{array}\right\} \xrightarrow[\longleftarrow]{\text{Complementary Slack}} \left\{\begin{array}{c} u_1 \geq 0 \\ \vdots \\ u_m \geq 0 \end{array}\right.$$

Dual Constraints

and (6.1) is rewritten as

$$\begin{aligned} &u_i^0\left(b_i - \sum_{j=1}^{p} a_{ij}x_j^0\right) = 0, i = 1,\ldots,m, \\ &x_j^0\left(\sum_{i=1}^{m} a_{ij}u_i^0 - c_j\right) = 0, j = 1,\ldots,p, \end{aligned} \tag{6.1.1}$$

it is easily seen that if the kth dual variable is positive, then its complementary primal structural constraint is binding (i.e. holds as an equality). Conversely, if the kth primal structural constraint holds with strict inequality, then its complementary dual variable is zero. Likewise, if the lth primal variable is positive, then its complementary dual structural constraint is binding. Conversely, if the lth dual structural constraint holds as a strict inequality, its complementary primal variable is zero. In short, $u_i^0 \geq 0$ and $b_i - \sum_{j=1}^{p} a_{ij}x_j^0 \geq 0$ as well as $x_j^0 \geq 0$ and $\sum_{i=1}^{m} a_{ij}u_i^0 - c_j \geq 0$ are *complementary slack* since within each pair of inequalities at least one of them must hold as an equality.

As a final observation, the geometric interpretation of the weak complementary slackness conditions is that at an optimal solution to the primal-dual pair of problems, the vector of primal variables $\boldsymbol{X}_0$ is orthogonal to the vector of dual surplus variables $(\boldsymbol{U}_s)_0$ while the vector of dual variables $\boldsymbol{U}_0$ is orthogonal to the vector of primal slack variables $(\boldsymbol{X}_s)_0$.

It is important to note that the weak complementary slackness conditions (6.1), (6.1.1) must hold for every pair of optimal solutions to the primal and dual problems. However, it may happen that, say, for some particular value of j (i), both $x_j^0\left(u_i^0\right)$ and $u_{m+j}^0\left(x_{n+i}^0\right)$ vanish at the said solutions. In this instance, can we be sure that there exists at least one pair of optimal solutions to the primal and dual problems for which this cannot occur? The answer is in the affirmative, as Theorem 6.7 demonstrates.

Theorem 6.7 Strong Complementary Slackness Theorem. For optimal solutions $\left(\boldsymbol{X}_0', (\boldsymbol{X}_s)_0'\right), \left(\boldsymbol{U}_0', (\boldsymbol{U}_s)_0'\right)$ to the primal maximum and dual minimum problems, respectively:

a) If the primal slack variable x^0_{p+i} equals zero, then the ith dual variable $u^0_i, i = 1,\ldots,m,$ is positive. Conversely, if u^0_i equals zero, then $x^0_{p+i}, i = 1,\ldots,m,$ is positive.
b) If the jth primal variable x^0_j equals zero, then the dual surplus variable $u^0_{m+j}, j = 1,\ldots,p,$ is positive. Conversely, if u^0_{m+j} equals zero, then $x^0_j, j = 1,\ldots,p,$ is positive. Parts (a), (b) taken together may be summarized as:

$$\left\{\begin{array}{l}(\boldsymbol{X}_s)_0 + \boldsymbol{U}_0 > \boldsymbol{O} \\ (\boldsymbol{U}_s)_0 + \boldsymbol{X}_0 > \boldsymbol{O}\end{array}\right\} \text{ or } \left\{\begin{array}{l}u^0_i + x^0_{p+i} > 0, i = 1,\ldots,m, \\ x^0_j + u^0_{m+j} > 0, j = 1,\ldots,p.\end{array}\right. \tag{6.2}$$

By way of an interpretation: $u^0_i + x^0_{p+i} > 0, i = 1,\ldots,m,$ implies that either $u^0_i = 0$ and $x^0_{p+i} > 0$ or $x^0_{p+i} = 0$ and $u^0_i > 0,$ but not both $u^0_i, x^0_{p+i} = 0$; while $x^0_j + u^0_{m+j} > 0, j = 1,\ldots,p,$ implies that either $x^0_j > 0$ and $u^0_{m+j} = 0$ or $u^0_{m+j} > 0$ and $x^0_j = 0,$ but not both $x^0_j, u^0_{m+j} = 0.$

Let us now establish the connection between the weak and strong complementary slackness conditions by first rewriting (6.2) as

$$\begin{aligned} &u^0_i + \left(b_i - \sum_{j=1}^{p} a_{ij}x^0_j\right) > 0, i = 1,\ldots,m, \\ &x^0_j + \left(\sum_{i=1}^{m} a_{ij}u^0_i - c_j\right) > 0, j = 1,\ldots,p. \end{aligned} \tag{6.2.1}$$

Then, for the strong case, if the kth dual variable is zero, then its complementary primal structural constraint holds as a strict inequality. Conversely, if the kth primal structural constraint is binding, then its complementary dual variable is positive. Note that this situation cannot prevail under the weak case since if $u^0_i = 0\left(b_i - \sum_{j=1}^{p} a_{ij}x^0_j = 0\right)$, it does not follow that $b_i - \sum_{j=1}^{p} a_{ij}x^0_j > 0\left(u^0_i > 0\right)$, i.e. under the weak conditions $u^0_i = 0\left(b_i - \sum_{j=1}^{p} a_{ij}x^0_j = 0\right)$ implies that $b_i - \sum_{j=1}^{p} a_{ij}x^0_j \geq 0\left(u^0_i > 0\right)$. Similarly (again considering the strong case), if the lth primal variable is zero, it follows that its complementary dual structural constraint holds with strict inequality. Conversely, if the lth dual structural constraint is binding, it follows that its complementary primal variable is positive. Here, too, a comparison of the strong and weak cases indicates that, for the latter, if $x^0_j = 0\left(\sum_{i=1}^{m} a_{ij}u^0_i - c_j = 0\right)$, we may conclude that $\sum_{i=1}^{m} a_{ij}u^0_i - c_j \geq 0\left(x^0_j \geq 0\right)$ but not that $\sum_{i=1}^{m} a_{ij}u^0_i - c_j > 0\left(x^0_j > 0\right)$ as in the former.

To summarize:

Weak Complementary Slackness

$u_i^0 > 0$ implies $b_i - \sum_{j=1}^{p} a_{ij}x_j^0 = 0$ and

$b_i - \sum_{j=1}^{p} a_{ij}x_j^0 > 0$ implies $u_i^0 = 0, i = 1, \ldots, m;$

while $x_j^0 > 0$ implies $\sum_{i=1}^{m} a_{ij}u_i^0 - c_j = 0$ and

$\sum_{i=1}^{m} a_{ij}u_i^0 - c_j > 0$ implies $x_j^0 = 0, j = 1, \ldots, p.$

Strong Complementary Slackness

$u_i^0 = 0$ implies $b_i - \sum_{j=1}^{p} a_{ij}x_j^0 > 0$ and

$b_i - \sum_{j=1}^{p} a_{ij}x_j^0 = 0$ implies $u_i^0 > 0, i = 1, \ldots, m;$

while $x_j^0 = 0$ implies $\sum_{i=1}^{m} a_{ij}u_i^0 - c_j > 0$ and

$\sum_{i=1}^{m} a_{ij}u_i^0 - c_j = 0$ implies $x_j^0 > 0, j = 1, \ldots, p.$

6.4 Constructing the Dual Solution

It was mentioned in the previous section that either both of the primal-dual pair of problems have optimal solutions, with the values of their objective functions being the same, or neither has an optimal solution. Hence, we can be sure that if one of these problems possesses an optimal solution, then the other does likewise. In this light, let us now see exactly how we may obtain the optimal dual solution from the optimal primal solution. To this end, let us assume that the primal maximum problem

$$\begin{aligned} &max\, f = \mathbf{C}'\mathbf{X} + \mathbf{O}'\mathbf{X}_s \quad \text{s.t.} \\ &\mathbf{A}\mathbf{X} + \mathbf{I}_m\mathbf{X}_s = \mathbf{b} \\ &\mathbf{X}, \mathbf{X}_s \geq \mathbf{O} \end{aligned}$$

has an optimal basic feasible solution consisting of

$$\begin{aligned} &\mathbf{X}_B = \mathbf{B}^{-1}\mathbf{b}, \\ &f = \mathbf{C}_B'\mathbf{X}_B = \mathbf{C}_B'\mathbf{B}^{-1}\mathbf{b}. \end{aligned}$$

As an alternative to expressing the optimal value of the primal objective function in terms of the optimal values of the primal basic variables x_{Bi}, $i = 1, \ldots, m$, let us respecify the optimal value of the primal objective function in terms of the optimal dual variables by defining the latter as

$$\begin{aligned} \mathbf{C}_B'\mathbf{B}^{-1} &= (c_{B1}, \ldots, c_{Bm}) \begin{bmatrix} \beta_{11} \cdots \beta_{1m} \\ \vdots \quad \vdots \\ \beta_{m1} \cdots \beta_{mm} \end{bmatrix} \\ &= \left(\sum_{i=1}^{m} c_{Bi}\beta_{i1}, \ldots, \sum_{i=1}^{m} c_{Bi}\beta_{im} \right) = (u_1, \ldots, u_m) = \mathbf{U}' \text{ or} \\ \mathbf{U} &= (\mathbf{B}^{-1})'\mathbf{C}_B, \end{aligned} \tag{6.3}$$

where c_{Bi}, $i = 1, \ldots, m$, is the ith component of C_B and β_{ij}, $i, j = 1, \ldots, m$, is the element in the ith row and jth column of B^{-1}. Then $f = U'b$ so that the value of f corresponding to an optimal basic feasible solution may be expressed in terms of the optimal values of either the primal or dual variables. Moreover, if the primal optimality criterion is satisfied, the U is a feasible solution to the dual structural constraints. And since $g = b'U = U'b = C_B'B^{-1}b = C_B'X_B = f$, U is an optimal solution to the dual minimum given that X represents the optimal basic feasible solution to the primal maximum problem.

By way of an interpretation of these dual variables we have Theorem 6.8.

Theorem 6.8 The optimal value of the ith dual variable u_i^0 is (generally) a measure of the amount by which the optimal value of the primal objective function changes given a small (incremental) change in the amount of the ith requirement b_i, with all other requirements constant, i.e. $u_i^0 = \partial f / \partial b_i, i = 1, \ldots, m$.

However, there may exist values of some requirement, say b_k, for which $\partial f^0 / \partial b_k$ is not defined, i.e. points where $\partial f^0 / \partial b_k$ is discontinuous. Such points are extreme points of the feasible region $\mathcal{K}$. In view of this characterization of the dual variables u_i, it is evident that the components of U may be treated as **Lagrange multipliers** in the **Lagrangean expression** $L(X, U) = C'X + U'(b - AX)$, $X \geq O$, $U \geq O$ associated with the primal problem

$$\max f = C'X \quad \text{s.t.}$$
$$AX \leq b, X \geq O.$$

In fact, this interpretation of the u_i will be employed once we construct a set of resource allocation models in the next chapter.

From a computational viewpoint, how are the optimal values of the dual variables obtained from the optimal primal solution? The simultaneous augmented form of the above primal maximum problem is

$$\begin{aligned} AX + I_m X_s \quad &= b \\ -C'X + O'X_s + f &= 0 \end{aligned}$$

with simplex matrix

$$\begin{bmatrix} A & I_m & O & b \\ -C' & O' & 1 & 0 \end{bmatrix}.$$

For this system the $(m + 1 \times m + 1)$ basis matrix and its inverse appear, respectively, as

$$\hat{B} = \begin{bmatrix} B & O \\ -C_B' & 1 \end{bmatrix}, \hat{B}^{-1} \begin{bmatrix} B^{-1} & O \\ C_B'B^{-1} & 1 \end{bmatrix},$$

where $\boldsymbol{B}$ represents the $(m \times m)$ basis matrix associated with the primal structural constraint system. Then

$$\begin{bmatrix} \boldsymbol{B}^{-1} & \boldsymbol{O} \\ \boldsymbol{C}_B'\boldsymbol{B}^{-1} & 1 \end{bmatrix} \begin{bmatrix} \boldsymbol{A} & \boldsymbol{I}_m & \boldsymbol{O} & \boldsymbol{b} \\ -\boldsymbol{C}' & \boldsymbol{O}' & 1 & 0 \end{bmatrix} = \begin{bmatrix} \boldsymbol{B}^{-1}\boldsymbol{A} & \boldsymbol{B}^{-1} & \boldsymbol{O} & \boldsymbol{B}^{-1}\boldsymbol{b} \\ \boldsymbol{C}_B'\boldsymbol{B}^{-1}\boldsymbol{A} - \boldsymbol{C}' & \boldsymbol{C}_B'\boldsymbol{B}^{-1} & 1 & \boldsymbol{C}_B'\boldsymbol{B}^{-1}\boldsymbol{b} \end{bmatrix} \tag{6.4}$$

yields a basic feasible solution to the primal maximum problem. Assuming that the optimality criterion is satisfied, the optimal values of the dual variables appear as the circled array of elements in the last row of (6.4), i.e. $\boldsymbol{U}' = \boldsymbol{C}_B'\boldsymbol{B}^{-1} \geq \boldsymbol{O}'$. Moreover, from the augmented dual structural constraint system $\boldsymbol{A}'\boldsymbol{U} - \boldsymbol{I}_m\boldsymbol{U}_s = \boldsymbol{C}$, we may solve for the dual surplus variables as $\boldsymbol{U}_s' = \boldsymbol{U}'\boldsymbol{A} - \boldsymbol{C}'$. But this expression is just the double-circled array of elements $\boldsymbol{U}_s' = \boldsymbol{C}_B'\boldsymbol{B}^{-1}\boldsymbol{A} - \boldsymbol{C}' \geq \boldsymbol{O}'$ in the last row of (6.4). In sum, our rules for determining the optimal values of the dual variables (both original and surplus) may be stated as follows:

a) To find the values of the dual variables $\boldsymbol{U}$ that correspond to an optimal basic feasible solution of the primal maximum problem, look in the last row of the final primal simplex matrix under those columns that were, in the original simplex matrix, the columns of the identity matrix. The ith dual variable u_i, $i = 1, \ldots, m$, will appear under the column that was initially the ith column of the identity matrix.
b) To find the values of the dual surplus variables $\boldsymbol{U}_s$ that correspond to an optimal basic feasible solution of the primal maximum problem, look in the last row of the final primal simplex matrix under those columns that were, in the original simplex matrix, the columns of the matrix $\boldsymbol{A}$. The jth dual surplus variable u_{m+j}, $j = 1, \ldots, p$, will appear under the column that was initially the jth column of $\boldsymbol{A}$.

Example 6.5 Let us solve the primal program

$$\begin{aligned} \max f &= x_1 + x_2 \quad \text{s.t.} \\ -2x_1 + x_2 &\leq 20 \\ x_1 &\leq 8 \\ x_2 &\leq 6 \\ x_1, x_2 &\geq 0. \end{aligned}$$

Given that x_3, x_4, and x_5 depict nonnegative slack variables, we obtain the simultaneous augmented form

$$\begin{array}{llllllll} 2x_1 & +x_2 & +x_3 & & & & = 20 \\ x_1 & & & +x_4 & & & = 8 \\ & x_2 & & & +x_5 & & = 6 \\ -x_1 & -x_2 & & & & +f & = 0 \end{array}$$

with simplex matrix

$$\begin{bmatrix} 2 & 1 & 1 & 0 & 0 & 0 & 20 \\ 1 & 0 & 0 & 1 & 0 & 0 & 8 \\ 0 & 1 & 0 & 0 & 1 & 0 & 6 \\ -1 & -1 & 0 & 0 & 0 & 1 & 0 \end{bmatrix}.$$

It is easily demonstrated that the optimal simplex matrix is

$$\begin{bmatrix} 0 & 1 & 0 & 0 & 1 & 0 & 6 \\ 1 & 0 & \frac{1}{2} & 0 & -\frac{1}{2} & 0 & 7 \\ 0 & 0 & -\frac{1}{2} & 1 & \frac{1}{2} & 0 & 1 \\ 0 & 0 & \frac{1}{2} & 0 & \frac{1}{2} & 1 & 13 \end{bmatrix}$$

with $X_0' = (7,6)$ and $f^0 = 13$ (see point A of Figure 6.3). In addition, $(X_s)_0' = (0,1,0)$ and, according to our previous discussion concerning the construction of the dual solution, $U_0' = \left(\frac{1}{2}, 0, \frac{1}{2}\right), (U_s)_0' = (0,0)$. ■

Turning now to the complementary slackness conditions we have, for the weak and strong cases, respectively,

Weak	*Strong*
$u_1^0 x_3^0 = \frac{1}{2} \cdot 0 = 0$	$u_1^0 + x_3^0 = \frac{1}{2} + 0 = \frac{1}{2}$
$u_2^0 x_4^0 = 0 \cdot 1 = 0$	$u_2^0 + x_4^0 = 0 + 1 = 1$
$u_3^0 x_5^0 = \frac{1}{2} \cdot 0 = 0$	$u_3^0 + x_5^0 = \frac{1}{2} + 0 = \frac{1}{2}$
and	and
$x_1^0 u_4^0 = 7 \cdot 0 = 0$	$x_1^0 + u_4^0 = 7 + 0 = 7$
$x_2^0 u_5^0 = 6 \cdot 0 = 0$	$x_2^0 + u_5^0 = 6 + 0 = 6.$

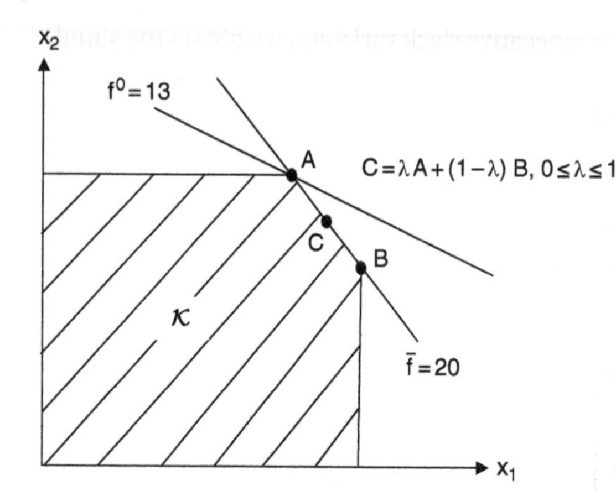

Figure 6.3 Optimal primal solution at point A.

Suppose we replace the original objective function by $\bar{f} = 2x_1 + x_2$ while leaving the constraint system unchanged. One can easily verify that the new optimal simplex matrix appears as

$$\begin{bmatrix} 0 & 1 & 1 & -2 & 0 & 0 & 4 \\ 1 & 0 & 0 & 1 & 0 & 0 & 8 \\ 0 & 0 & -1 & 2 & 1 & 0 & 2 \\ 0 & 0 & 1 & 0 & 0 & 1 & 20 \end{bmatrix}$$

with $\bar{X}' = (8,4)$ and $\bar{f} = 20$ (see point B of Figure 6.3). Moreover, $\bar{X}_s' = (0,0,2)$ with $\bar{U}' = (1,0,0)$ and $\bar{U}_s' = (0,0)$. Now the weak and strong complementary slackness conditions become

Weak	*Strong*
$\bar{u}_1\bar{x}_3 = 1{\cdot}0 = 0$	$\bar{u}_1 + \bar{x}_3 = 1 + 0 = 1$
$\bar{u}_2\bar{x}_4 = 0{\cdot}0 = 0$	$\bar{u}_2 + \bar{x}_4 = 0 + 0 = 0$
$\bar{u}_3\bar{x}_5 = 0{\cdot}2 = 0$	$\bar{u}_3 + \bar{x}_5 = 0 + 2 = 2$
and	and
$\bar{x}_1\bar{u}_4 = 8{\cdot}0 = 0$	$\bar{x}_1 + \bar{u}_4 = 8 + 0 = 8$
$\bar{x}_2\bar{u}_5 = 4{\cdot}0 = 0$	$\bar{x}_2 + \bar{u}_5 = 4 + 0 = 4.$

In this instance only the weak complementary slackness conditions hold since an additional zero-valued dual variable appears in the optimal set of the same, i.e. now $\bar{u}_3 = 0$. To offset the effect of this extra zero-valued dual variable on the strong complementary slackness conditions, let us arbitrarily set $\lambda = \frac{1}{2}$ and choose point C (Figure 6.3) as our optimal primal solution. (Here C has coordinates (15/2, 5) and $\bar{f}(C) = 20$.) Then the complementary slackness conditions become

Weak	*Strong*
$u_1^c x_3^c = 1 \cdot 0 = 0$	$u_1^c + x_3^c = 1 + 0 = 1$
$u_2^c x_4^c = 0 \cdot \frac{1}{2} = 0$	$u_2^c + x_4^c = 0 + \frac{1}{2} = \frac{1}{2}$
$u_3^c x_5^c = 0 \cdot 1 = 0$	$u_3^c + x_5^c = 0 + 1 = 1$
and	and
$x_1^c u_4^c = (15/2) \cdot 0 = 0$	$x_1^c + u_4^c = \frac{15}{2} + 0 = \frac{15}{2}$
$x_2^c u_5^c = 5 \cdot 0 = 0$	$x_2^c + u_5^c = 5 + 0 = 5.$

Notice that now both weak and strong complementary slackness hold. ■

We noted earlier that the simplex method can be applied to either the primal or dual formulation of a linear programming problem, with the choice of the version actually solved determined by considerations of computational ease. In this regard, if the dual problem is selected, how may we determine the optimal values of the primal variables from the optimal dual solution? Given that the dual problem appears as

$$
\begin{aligned}
&\min g = \boldsymbol{b}'\boldsymbol{U} \quad \text{s.t.}\\
&\boldsymbol{A}'\boldsymbol{U} \geq \boldsymbol{C}, \boldsymbol{U} \geq \boldsymbol{O}
\end{aligned}
$$

or

$$
\begin{aligned}
&\max h = -g = -\boldsymbol{b}'\boldsymbol{U} \quad \text{s.t.}\\
&\boldsymbol{A}'\boldsymbol{U} \geq \boldsymbol{C}, \boldsymbol{U} \geq \boldsymbol{O},
\end{aligned}
$$

we may utilize the M-penalty method to derive the artificial simultaneous augmented form

$$
\begin{aligned}
\boldsymbol{A}'\boldsymbol{U} - \boldsymbol{I}_p\boldsymbol{U}_s + \boldsymbol{I}_p\boldsymbol{U}_a \quad &= \boldsymbol{C}\\
\boldsymbol{b}'\boldsymbol{U} \quad + M\boldsymbol{1}'\boldsymbol{U}_a + h &= 0,
\end{aligned}
$$

where $\boldsymbol{U}_a$ denotes an $(m \times 1)$ vector of nonnegative artificial variables and the associated simplex matrix is

$$
\begin{bmatrix} \boldsymbol{A}' & -\boldsymbol{I}_p & \boldsymbol{I}_p & \boldsymbol{O} & \boldsymbol{C} \\ \boldsymbol{b}' & \boldsymbol{O}' & M\boldsymbol{1}' & 1 & 0 \end{bmatrix} \rightarrow \begin{bmatrix} \boldsymbol{A}' & -\boldsymbol{I}_p & \boldsymbol{I}_p & \boldsymbol{O} & \boldsymbol{C} \\ \boldsymbol{b}' - M\boldsymbol{1}'\boldsymbol{A} & M\boldsymbol{1}' & \boldsymbol{O}' & 1 & -M\boldsymbol{1}'\boldsymbol{C} \end{bmatrix}.
$$

Relative to this system the $(p+1 \times p+1)$ basis matrix and its inverse may be written, respectively, as

$$
\hat{\boldsymbol{D}} = \begin{bmatrix} \boldsymbol{D} & \boldsymbol{O} \\ \boldsymbol{b}_D' - M\boldsymbol{1}'\boldsymbol{D} & 1 \end{bmatrix}, \hat{\boldsymbol{D}}^{-1} = \begin{bmatrix} \boldsymbol{D}^{-1} & \boldsymbol{O} \\ -\boldsymbol{b}_D'\boldsymbol{D}^{-1} + M\boldsymbol{1}' & 1 \end{bmatrix},
$$

where $\boldsymbol{D}$ depicts the $(m \times m)$ basis matrix associated with the dual artificial structural constraint system. Then

$$\begin{bmatrix} \boldsymbol{D}^{-1} & \boldsymbol{O} \\ -\boldsymbol{b}_D'\boldsymbol{D}^{-1} + M\boldsymbol{1}' & 1 \end{bmatrix} \begin{bmatrix} \boldsymbol{A}' & -\boldsymbol{I}_p & \boldsymbol{I}_p & \boldsymbol{O} & \boldsymbol{C} \\ \boldsymbol{b}' - M\boldsymbol{1}'\boldsymbol{A} & M\boldsymbol{1}' & \boldsymbol{O}' & 1 & -M\boldsymbol{1}'\boldsymbol{C} \end{bmatrix} = \begin{bmatrix} \boldsymbol{D}^{-1}\boldsymbol{A}' & -\boldsymbol{D}^{-1} & \boldsymbol{O} & \boldsymbol{D}^{-1}\boldsymbol{C} \\ -\boldsymbol{b}_D'\boldsymbol{D}^{-1}\boldsymbol{A}' + \boldsymbol{b}' & \boldsymbol{b}_D'\boldsymbol{D}^{-1} & 1 & -\boldsymbol{b}_D'\boldsymbol{D}^{-1}\boldsymbol{C} \end{bmatrix} \tag{6.5}$$

yields a basic feasible solution to the dual minimum problem, where the vectors associated with the artificial variables, i.e. those corresponding to the third partitioned column of (6.5), have been deleted since the artificial vectors have been driven from the basis. If we assume that the optimality criterion is satisfied, then the optimal values of the primal variables appear as the circled array of elements in the objective function row of (6.5), i.e. $\boldsymbol{X}' = \boldsymbol{b}_D\boldsymbol{D}^{-1} \geq \boldsymbol{O}'$. In addition, from the augmented primal structural constraint system $\boldsymbol{AX} + \boldsymbol{I}_m\boldsymbol{X}_s = \boldsymbol{b}$, we may isolate the primal slack variables as $\boldsymbol{X}_s' = \boldsymbol{b}' - \boldsymbol{X}'\boldsymbol{A}'$. But this expression corresponds to the double-circled array of elements $\boldsymbol{X}_s' = \boldsymbol{b}_D'\boldsymbol{D}^{-1}\boldsymbol{A}' + \boldsymbol{b}' \geq \boldsymbol{O}'$ in the objective function row of (6.5). To summarize, our rules for isolating the optimal values of the primal variables (original and slack) may be determined as follows:

a) To find the values of the primal variables $\boldsymbol{X}$ that correspond to an optimal basic feasible solution of the dual minimum problem, look in the last row of the final dual simplex matrix under those columns that were, in the original simplex matrix, the columns of the negative of the identity matrix. The jth primal variable x_j, $j = 1, \ldots, p$, will appear under the column that was initially the jth column of the negative of the identity matrix.
b) To find the values of the primal slack variables $\boldsymbol{X}_s$ that correspond to an optimal basic feasible solution of the dual minimum problem, look in the last row of the final dual simplex matrix under those columns that were, in the original simplex matrix, the columns of the matrix $\boldsymbol{A}'$. The ith primal slack variable x_{p+i}, $i = 1, \ldots, m$, will appear under the column that was initially the ith column of $\boldsymbol{A}'$.

In this section we have demonstrated how the dual (primal) variables may be obtained from the optimal primal (dual) simplex matrix. Hence, no matter which of the primal-dual pair of problems is solved, all of the pertinent information generated by an optimal basic feasible solution is readily available, i.e. from either the final primal or dual simplex matrix we can isolate both the primal and dual original variables as well as the primal slack and dual surplus variables.

Example 6.6 The simultaneous artificial augmented form of the dual minimum problem presented in Example 6.3 is

$$\begin{aligned} u_1 + 4u_2 - u_3 \quad\quad + u_5 \quad\quad &= 24 \\ 5u_1 + u_2 \quad - u_4 \quad\quad + u_6 &= 25 \\ u_1 + 3u_2 \quad\quad + Mu_5 + Mu_6 + h &= 0, \end{aligned}$$

where $h = -g$; u_3, u_4 represent surplus variables and u_5, u_6 depict artificial variables. In addition, the associated simplex matrix is

$$\begin{bmatrix} 1 & 4 & -1 & 0 & 1 & 0 & 0 & 24 \\ 5 & 1 & 0 & -1 & 0 & 1 & 0 & 25 \\ 1 & 3 & 0 & 0 & M & M & 1 & 0 \end{bmatrix} \rightarrow \begin{bmatrix} 1 & 4 & -1 & 0 & 1 & 0 & 0 & 24 \\ 5 & 1 & 0 & -1 & 0 & 1 & 0 & 25 \\ 1 & 3 & 0 & 0 & 0 & 0 & 1 & 0 \\ -6M & -5M & M & M & 0 & 0 & 0 & -49M \end{bmatrix}.$$

It is easily shown that the optimal simplex matrix is

$$\begin{bmatrix} 0 & 1 & -\frac{5}{19} & \frac{1}{19} & 0 & 5 \\ 1 & 0 & \frac{1}{19} & -\frac{4}{19} & 0 & 4 \\ 0 & 0 & \frac{14}{19} & \frac{1}{19} & 1 & -19 \end{bmatrix},$$

where the columns associated with the artificial variables have been deleted. Upon examining the objective function row of this matrix we see that $X_0' = \left(14/19, \frac{1}{19}\right)$ and $X_s' = (0,0)$ as expected. ■

6.5 Dual Simplex Method (Lemke 1954)

In attempting to generate an optimal basic feasible solution to a linear programming problem, our procedure, as outlined in Chapters 4 and 5, was to employ the simplex method or, as we shall now call it, the **primal simplex method**. To briefly review this process, we start from an initial basic feasible solution for which the optimality criterion is not satisfied, i.e. not all $\bar{c}_j \leq 0, j = 1, \ldots, n-m$ (the dual problem is infeasible). We then make changes in the basis, one vector at a time, maintaining nonnegativity of the basic variables at each iteration until we obtain a basic feasible solution for which all $\bar{c}_j \leq 0, j = 1, \ldots, n-m$ (so that dual feasibility now holds). Hence the primal simplex method is a systematic technique that preserves primal feasibility while striving for primal optimality, thereby reducing, at each round, dual infeasibility, until primal optimality, and thus dual feasibility, is ultimately achieved. This last observation precipitates a very interesting question. Specifically, can we solve a linear programming problem by a procedure that works in a fashion opposite to that of the primal simplex algorithm? That is to say, by a method which relaxes primal feasibility while maintaining primal optimality. As we shall now see, the answer is yes. In this regard, we now look to an alternative simplex routine called the **dual simplex method**, so named because it involves an application of the simplex method to the dual problem, yet is constructed so as to have its pivotal operations carried out within the standard primal simplex matrix. Here the starting point is an initial basic but nonfeasible solution to the primal problem for which optimality (and thus dual feasibility) holds. As is the usual case, we move from this solution to an optimal basic feasible solution through a sequence of pivot operations which involve changing the status of a single basis vector at a time. Hence the dual simplex method preserves primal optimality (dual feasibility) while reducing, at each iteration, primal infeasibility, so that an optimal basic feasible solution to both the primal and dual problems is simultaneously achieved. Moreover, while the primal objective starts out at a suboptimal level and monotonically increases to its optimal level, the dual objective starts out at a superoptimal level and monotonically decreases to its optimal feasible value.

6.6 Computational Aspects of the Dual Simplex Method

One important class of problems that lends itself to the application of the dual simplex method is when the primal problem is such that a surplus variable is added to each "≥" structural constraint. As we shall see below, the dual simplex method enables us to circumvent the procedure of adding artificial variables to the primal structural constraints. For instance, if the primal problem appears as:

$$\left.\begin{array}{l} \min f = \mathbf{C}'\mathbf{X} \quad \text{s.t.} \\ \mathbf{AX} \geq \mathbf{b} \text{ with } \mathbf{C},\mathbf{X} \geq \mathbf{O} \end{array}\right\} or \left\{\begin{array}{l} \max h = -f = -\mathbf{C}'\mathbf{X} \quad \text{s.t.} \\ -\mathbf{AX} \leq -\mathbf{b} \text{ with } \mathbf{C},\mathbf{X} \geq \mathbf{O}, \end{array}\right.$$

then the (symmetric) dual problem may be formed as

$$\left.\begin{array}{l} \min l = -\mathbf{b}'\mathbf{U} \quad \text{s.t.} \\ -\mathbf{A}'\mathbf{U} \geq -\mathbf{C}, \mathbf{U} \geq \mathbf{O} \end{array}\right\} or \left\{\begin{array}{l} \max g = -\mathbf{b}'\mathbf{U} \quad \text{s.t.} \\ \mathbf{A}'\mathbf{U} \leq \mathbf{C}, \mathbf{U} \geq \mathbf{O}. \end{array}\right.$$

To determine an initial basic solution to the primal problem, let us examine its associated simplex matrix:

$$\begin{array}{cccccccccccccccc} x_1 & x_2 & & x_k & & x_p & x_{p+1} & x_{p+2} & & x_{p+r} & & x_{p+m} & h & \\ \end{array}$$

$$\left[\begin{array}{ccccccccccccccc} -a_{11} & -a_{12} & \dots & -a_{1k} & \dots & -a_{1p} & 1 & 0 & \dots & 0 & \dots & 0 & 0 & -b_1 \\ -a_{21} & -a_{22} & \dots & -a_{2k} & \dots & -a_{2p} & 0 & 1 & \dots & 0 & \dots & 0 & 0 & -b_2 \\ \dots & \dots & \dots & \dots & \dots & \dots & \dots & \dots & \dots & \dots & \dots & \dots & \dots & \dots \\ -a_{r1} & -a_{r2} & \dots & -a_{rk} & \dots & -a_{rp} & 0 & 0 & \dots & 1 & \dots & 0 & 0 & -b_r \\ \dots & \dots & \dots & \dots & \dots & \dots & \dots & \dots & \dots & \dots & \dots & \dots & \dots & \dots \\ -a_{m1} & -a_{m2} & \dots & -a_{mk} & \dots & -a_{mp} & 0 & 0 & \dots & 0 & \dots & 1 & 0 & -b_m \\ c_1 & c_2 & \dots & c_k & \dots & c_p & 0 & 0 & \dots & 0 & \dots & 0 & 1 & 0 \end{array}\right] \tag{6.6}$$

where $x_{p+1}, \ldots, x_{p+m}$ are nonnegative slack variables. If the basic variables are taken to be those corresponding to the $m+1$ unit column vectors in (6.6), then

$$\begin{aligned} x_{p+1} &= -b_1 + \sum_{j=1}^{p} a_{1j}x_j \\ x_{p+2} &= -b_2 + \sum_{j=1}^{p} a_{2j}x_j \\ &\dots\dots\dots\dots \\ x_{p+r} &= -b_r + \sum_{j=1}^{p} a_{rj}x_j \\ &\dots\dots\dots\dots \\ x_{p+m} &= -b_m + \sum_{j=1}^{p} a_{mj}x_j \\ h &= \sum_{j=1}^{p} c_jx_j \end{aligned} \tag{6.7}$$

and thus our initial primal basic solution amounts to $x_{p+i} = -b_i$, $i = 1, \ldots, m$, and $h = 0$. Since this current solution is dual feasible (primal optimal) but not primal feasible, let us reduce primal infeasibility by choosing the vector to be removed from the basis according to

$$x_{p+r} = \min_i \{-b_i, -b_i < 0\}.$$
$$\text{(Exit Criterion)}$$

Hence, column $p + r$ is deleted from the current basis so that x_{p+r} decreases in value to zero. By virtue of the weak complementary slackness conditions

$$u_i^0 x_{p+i}^0 = 0, i = 1, \ldots, m,$$
$$x_j^0 u_{m+j}^0 = 0, j = 1, \ldots, p,$$

it is evident that the dual variable u_r can increase in value from zero to a positive level. To determine the largest allowable increase in u_r that preserves dual feasibility, let us construct the simplex matrix associated with the dual problem as

$$\begin{array}{cccccccccccccc} u_1 & u_2 & & u_r & & u_m & u_{m+1} & u_{m+2} & & u_{m+k} & & u_{m+p} & g & \\ \end{array}$$
$$\left[\begin{array}{cccccccccccccc} a_{11} & a_{12} & \ldots & a_{r1} & \ldots & a_{m1} & 1 & 0 & \ldots & 0 & \ldots & 0 & 0 & c_1 \\ a_{21} & a_{22} & \ldots & a_{r2} & \ldots & a_{m2} & 0 & 1 & \ldots & 0 & \ldots & 0 & 0 & c_2 \\ \cdots & & & & & & & & & & & & & \\ a_{1k} & a_{2k} & \ldots & a_{rk} & \ldots & a_{mk} & 0 & 0 & \ldots & 1 & \ldots & 0 & 0 & c_k \\ \cdots & & & & & & & & & & & & & \\ a_{1p} & a_{2p} & \ldots & a_{rp} & \ldots & a_{mp} & 0 & 0 & \ldots & 0 & \ldots & 1 & 0 & c_p \\ -b_1 & -b_2 & \ldots & -b_r & \ldots & -b_m & 0 & 0 & \ldots & 0 & \ldots & 0 & 1 & 0 \end{array}\right] \tag{6.8}$$

wherein $u_{m+1}, \ldots, u_{m+p}$ are nonnegative slack variables. From the $m + 1$ unit column vectors in (6.8) we have

$$\begin{aligned} u_{m+1} &= c_1 - \sum_{i=1}^{m} a_{i1} u_i \\ u_{m+2} &= c_2 - \sum_{i=1}^{m} a_{i2} u_i \\ &\cdots\cdots \\ u_{m+k} &= c_k - \sum_{i=1}^{m} a_{ik} u_i \\ &\cdots\cdots \\ u_{m+p} &= c_p - \sum_{i=1}^{m} a_{ip} u_i \\ g &= \sum_{i=1}^{m} b_i u_i. \end{aligned} \tag{6.9}$$

Thus, an initial dual basic feasible solution consists of $u_{m+j} = c_j$, $j = 1, \ldots, p$, and $g = 0$. If u_r is to enter the set of dual basic variables, (6.9) becomes

$$\begin{aligned}
u_{m+1} &= c_1 - a_{r1}u_r \\
u_{m+2} &= c_2 - a_{r2}u_r \\
&\cdots\cdots\cdots\cdots\cdots\cdots \\
u_{m+k} &= c_k - a_{rk}u_r \\
&\cdots\cdots\cdots\cdots\cdots\cdots \\
u_{m+p} &= c_p - a_{rp}u_r \\
g &= b_r u_r.
\end{aligned} \tag{6.9.1}$$

As u_r increases in value from zero, the new values of u_{m+j}, $j = 1, \ldots, p$, must remain feasible, i.e. we require that

$$u_{m+j} = c_j - a_{rj}u_r \geq 0, j = 1, \ldots, p. \tag{6.9.2}$$

Under what conditions will this set of inequalities be satisfied? If $a_{rj} \leq 0$, then $u_{m+j} > 0$ for any positive level of u_r, whence (6.9.1) indicates that g is unbounded from above and thus the primal problem is infeasible. Hence, we need only concern ourselves with those $a_{rj} > 0$. In this regard, when at least one of the coefficients a_{rj} within the rth row of (6.8) is positive, there exists an upper bound to an increase in u_r that will not violate (6.9.2). Rearranging (6.9.2) yields $u_r \leq c_j/a_{rj} = \bar{c}_j/a_{rj}$ so that any increase in u_r that equals

$$\min_j \left\{ \frac{\bar{c}_j}{a_{rj}}, a_{rj} > 0 \right\}$$

preserves dual feasibility. Let us assume that, for $j = k$,

$$\hat{\theta} = \frac{\bar{c}_k}{a_{rk}} = \min_j \left\{ \frac{\bar{c}_j}{a_{rj}}, a_{rj} > 0 \right\}, \tag{6.10}$$

i.e. the dual basic variable u_{m+k} decreases in value to zero. But in terms of the preceding complementary slackness conditions, if $u_{m+k} = 0$, then x_k can increase to a positive level. Hence, from (6.6), column k replaces column $p + r$ in the primal basis and thus x_k becomes a basic variable while x_{p+r} turns nonbasic. From (6.9.1), our new dual basic feasible solution becomes:

$$\begin{aligned}
u_r &= \hat{\theta} \\
u_{m+j} &= c_j - a_{rj}\hat{\theta}, j = 1, \ldots, p, j \neq k, \text{ and} \\
g &= b_r\hat{\theta}.
\end{aligned}$$

In terms of the primal problem, since

$$\hat{\theta} = \frac{\bar{c}_k}{-a_{rk}} = \max_j \left\{ \frac{\bar{c}_j}{-a_{rj}}, -a_{rj} < 0 \right\}$$

(Entry Criterion)

is equivalent to (6.10), x_k enters the set of primal basic variables and thus (6.7) simplifies to

$$
\begin{aligned}
x_{p+1} &= -b_1 + a_{1k}x_k \\
x_{p+2} &= -b_2 + a_{2k}x_k \\
&\cdots\cdots\cdots\cdots\cdots\cdots \\
x_{p+r} &= -b_r + a_{rk}x_k = 0 \\
&\cdots\cdots\cdots\cdots\cdots\cdots \\
x_{p+m} &= -b_m + a_{mk}x_k \\
h &= -c_k x_k
\end{aligned}
$$

or

$$
\begin{aligned}
& x_k = b_r/a_{rk} \\
& x_{p+i} = -b_i + a_{ik}(b_r/a_{rk}), i = 1, \ldots, m, i \neq r, \text{ and} \\
& h = \hat{\theta} b_r.
\end{aligned}
$$

As noted earlier, the dual simplex method is constructed so as to have its pivotal operations carried out within the standard primal simplex matrix. So from (6.6), if x_k is to replace x_{p+r} in the set of primal basic variables, a pivot operation with $-a_{rk}$ as the pivotal element may be performed to express the new set of basic variables x_{p+i}, $i = 1, \ldots, m, i \neq r, x_k$, and h in terms of the nonbasic variables $x_1, \ldots, x_{k-1}, x_{k+1}, \ldots, x_p$, and x_{p+r}. To this end, (6.6) is transformed to

$$
\begin{array}{ccccccccccccccccc}
x_1 & x_2 & & x_{k-1} & x_k & x_{k+1} & & x_p & x_{p+1} & x_{p+2} & & x_{p+r} & & x_{p+m} & h & \\
\end{array}
$$

$$
\left[\begin{array}{ccccccccccccccccc}
-a_{11} & -a_{12} & \cdots & -a_{1,k-1} & -a_{1k} & -a_{1,k+1} & \cdots & -a_{1p} & 1 & 0 & \cdots & 0 & \cdots & 0 & 0 & -b_1 \\
\cdots & & & & & & & & & & & & & & & \cdots \\
\frac{a_{r1}}{a_{rk}} & \frac{a_{r2}}{a_{rk}} & \cdots & \frac{a_{r,k-1}}{a_{rk}} & \textcircled{1} & \frac{a_{r,k+1}}{a_{rk}} & \cdots & \frac{a_{rp}}{a_{rk}} & 0 & 0 & \cdots & \frac{1}{-a_{rk}} & \cdots & 0 & 0 & \frac{b_r}{a_{rk}} \\
\cdots & & & & & & & & & & & & & & & \cdots \\
-a_{m1} & -a_{m2} & \cdots & -a_{m,k-1} & -a_{mk} & -a_{m,k+1} & \cdots & -a_{mp} & 0 & 0 & \cdots & 0 & \cdots & 1 & 0 & -b_m \\
c_1 & c_2 & \cdots & c_{k-1} & c_k & c_{k+1} & \cdots & c_p & 0 & 0 & \cdots & 0 & \cdots & 0 & 1 & 0
\end{array}\right] \rightarrow
$$

$$
\left(\frac{1}{-a_{rk}} E_r \text{ replaces } E_r\right)
$$

$$
\left[\begin{array}{cccccccccccccccc}
\bar{a}_{11} & \bar{a}_{12} & \cdots & \bar{a}_{1,k-1} & 0 & \bar{a}_{1,k+1} & \cdots & \bar{a}_{1p} & 1 & 0 & \cdots & \bar{a}_{1,p+r} & \cdots & 0 & 0 & -b_1 \\
\cdots & & & & & & & & & & & & & & & \cdots \\
\frac{a_{r1}}{a_{rk}} & \frac{a_{r2}}{a_{rk}} & \cdots & \frac{a_{r,k-1}}{a_{rk}} & \textcircled{1} & \frac{a_{r,k+1}}{a_{rk}} & \cdots & \frac{a_{rp}}{a_{rk}} & 0 & 0 & \cdots & \frac{1}{-a_{rk}} & \cdots & 0 & 0 & \frac{b_r}{a_{rk}} \\
\cdots & & & & & & & & & & & & & & & \cdots \\
\bar{a}_{m1} & \bar{a}_{m2} & \cdots & \bar{a}_{m,k-1} & 0 & \bar{a}_{m,k+1} & \cdots & \bar{a}_{mp} & 0 & 0 & \cdots & \bar{a}_{m,p+r} & \cdots & 1 & 0 & b_m \\
c'_1 & c'_2 & \cdots & c'_{k-1} & 0 & c'_{k+1} & \cdots & c'_p & 0 & 0 & \cdots & c'_{p+r} & \cdots & 0 & 1 & \hat{\theta}
\end{array}\right]
$$

$$
(E_r \text{ is replaced by } E_i + a_{ik}E_r, i = 1, \ldots, m, i \neq r, \\
\text{while } E_{m+1} \text{ is replaced by } E_{m+1} - c_k E_r) \tag{6.11}
$$

where

$$\bar{a}_{ij} = -a_{ij} + a_{ik}\left(a_{rj}/a_{rk}\right), i = 1,\ldots,m, i \neq r; j = 1,\ldots,p+m$$
$$\bar{b}_i = b_i + a_{ik}\left(b_r/a_{rk}\right), i = 1,\ldots,m, i \neq r$$
$$c_j' = c_j + \hat{\theta} a_{rj}, j = 1,\ldots,p+m.$$

Hence our new primal basic solution is readily obtainable from (6.11) as

$$x_k = b_r/a_{rk}$$
$$x_{p+i} = \bar{b}_i, i = 1,\ldots,m, i \neq r, \text{and}$$
$$h = \hat{\theta} b_r \text{ or } f = -\hat{\theta} b_r.$$

This pivotal process is continued until we obtain a basic feasible solution to the primal problem, in which case we have derived optimal basic feasible solutions to both the primal and dual problems.

Example 6.7 Using the dual simplex routine, solve

$$\begin{aligned} &\min f = 2x_1 + 4x_2 \quad \text{s.t.} \\ &x_1 + 3x_2 \geq 20 \\ &x_1 + x_2 \geq 15 \\ &2x_1 + x_2 \geq 25 \\ &x_1, x_2 \geq 0. \end{aligned}$$

Upon converting this problem to a maximization problem and subtracting nonnegative surplus variables x_3, x_4, and x_5 from the left-hand sides of the structural constraints we obtain the simplex matrix

$$\begin{bmatrix} 1 & 3 & -1 & 0 & 0 & 0 & 20 \\ 1 & 1 & 0 & -1 & 0 & 0 & 15 \\ 2 & 1 & 0 & 0 & -1 & 0 & 25 \\ 2 & 4 & 0 & 0 & 0 & 1 & 0 \end{bmatrix} \rightarrow \begin{bmatrix} -1 & -3 & 1 & 0 & 0 & 0 & -20 \\ -1 & -1 & 0 & 1 & 0 & 0 & -15 \\ \textcircled{-2} & -1 & 0 & 0 & 1 & 0 & -25 \\ 2 & 4 & 0 & 0 & 0 & 1 & 0 \end{bmatrix}$$

and an initial primal optimal (dual feasible) basic solution $x_3 = -20$, $x_4 = -15$, $x_5 = -25$ and $h = 0$, where $h = -f$. Since the current solution is not primal feasible, let us undertake the first round of the dual simplex method by determining

$$-25 = \min_i \{x_{Bi}, x_{Bi} < 0\} = \min\{-20, -15, -25\},$$

$$\hat{\theta} = -1 = \max_j \left\{\frac{\bar{c}_j}{-a_{rj}}, -a_{rj} < 0\right\} = \max\{2/(-2) = -1, 4/(-1) = -4\}.$$

Here x_1 is to replace x_5 in the set of primal basic variables (alternatively, $\boldsymbol{r}_1$ replaces $\boldsymbol{b}_3$ in the primal basis) so that the (circled) pivotal element is –2. An appropriate pivot operation yields

$$\begin{bmatrix} 0 & \boxed{-\frac{5}{2}} & 1 & 0 & -\frac{1}{2} & 0 & -\frac{15}{2} \\ 0 & -\frac{1}{2} & 0 & 1 & -\frac{1}{2} & 0 & -\frac{5}{2} \\ 1 & \frac{1}{2} & 0 & 0 & -\frac{1}{2} & 0 & \frac{25}{2} \\ 0 & 3 & 0 & 0 & 1 & 1 & -25 \end{bmatrix}$$

and a second primal basic solution consisting of $x_1 = \frac{25}{2}, x_3 = -\frac{15}{2}, x_4 = -\frac{5}{2}$, and $h = -25$. Again primal feasibility does not hold so that we next compute:

$$-\frac{15}{2} = min\left\{-\frac{15}{2}, -\frac{5}{2}\right\},$$
$$\hat{\theta} = -\frac{6}{5} = max\left\{3\Big/\left(-\frac{5}{2}\right) = -\frac{6}{5}, 1\Big/\left(-\frac{1}{2}\right) = -2\right\}.$$

If we now undertake an additional pivotal operation with $-\frac{5}{2}$ as the pivot, we obtain

$$\begin{bmatrix} 0 & 1 & -\frac{2}{5} & 0 & \frac{1}{5} & 0 & 3 \\ 0 & 0 & -\frac{1}{5} & 1 & \boxed{-\frac{2}{5}} & 0 & -1 \\ 1 & 0 & \frac{1}{5} & 0 & -\frac{3}{5} & 0 & 11 \\ 0 & 0 & \frac{6}{5} & 0 & \frac{2}{5} & 1 & -34 \end{bmatrix}$$

and thus a third primal basic solution wherein $x_1 = 11$, $x_2 = 3$, $x_4 = -1$, and $h = -34$. Since primal infeasibility is still evident, we again look to our respective exit and entry criteria

$$-1 = min\{-1\},$$
$$\hat{\theta} = -1 = max\{(6/5)/(-1/5) = -6, (2/5)/(-2/5) = -1\}$$

to obtain a pivot of $\frac{-2}{5}$. A final iteration yields

$$\begin{bmatrix} 0 & 1 & -\frac{1}{2} & \frac{1}{2} & 0 & 0 & \frac{5}{2} \\ 0 & 0 & \frac{1}{2} & -\frac{5}{2} & 1 & 0 & \frac{5}{2} \\ 1 & 0 & \frac{1}{2} & -\frac{3}{2} & 0 & 0 & \frac{25}{2} \\ 0 & 0 & 1 & 1 & 0 & 1 & -35 \end{bmatrix}$$

and an optimal primal basic feasible solution with $x_1 = \frac{25}{2}, x_2 = \frac{5}{2}, x_5 = \frac{5}{2}$, and $f = 35$. It is important to note that if at some stage of the dual-simplex pivot operations we encounter a simplex matrix wherein for some row, say, the kth, $\bar{b}_k < 0$ and $y_{kj} > 0$ for all nonbasic vectors $\boldsymbol{r}_j$, then the primal problem is infeasible in the sense that the basic variable x_{Bk} can be driven to zero only if some current nonbasic variable enters the set of basic variables at a negative level. In this instance, the dual objective value is **unbounded**.

Additionally, it may be the case that at some point in the dual simplex routine it turns out that $\hat{\theta}$ is not unique. Hence we may introduce into the primal basis any one of the columns of $\boldsymbol{R}$ for which the same maximum value of $\hat{\theta}$ is assumed and obtain a new (call it the kth) primal basic solution. In this instance, $\bar{c}_j = 0$ for all j corresponding to those columns that have the same $\bar{c}_j/(-a_{rj}), -a_{rj} < 0$, as the incoming basis vector and thus the $(k+1)$st iteration is executed with $\hat{\theta} = 0$, i.e. the case of **dual degeneracy** emerges. Moreover, for the $(k+1)$st primal basic solution, the primal objective value is unchanged. (Note that the dual objective value must also remain invariant.) In fact, the primal objective value may remain unchanged for several successive iterations. However, in the absence of cycling, the dual simplex algorithm will eventually converge to an optimal basic feasible solution to both the primal and dual problems simultaneously. ■

6.7 Summary of the Dual Simplex Method

Let us now consolidate our discussion of the theory and major computational aspects of the dual simplex method. If we abstract from the possibility of degeneracy, then a summary of the principal steps leading to an optimal basic feasible solution to both the primal and dual problems simultaneously is:

1) Obtain an initial dual feasible (primal optimal) solution with $\bar{c}_j \leq 0$, $j = 1, \ldots, n - m$.

2) If primal feasibility also holds, then the current solution is an optimal basic feasible solution to both the primal and dual problems. If not all $x_{Bi} \geq 0$, $i = 1, \ldots, m$ (primal feasibility is relaxed), then proceed with the next step.
3) Find $x_{Br} = min_i\{x_{Bi}, x_{Bi} < 0\}$. The vector $\boldsymbol{b}_r$ corresponding to the minimum is to be removed from the basis.
4) Determine whether $-a_{rj} \geq 0$, $j = 1, \ldots, m$. If the answer is *yes*, the process is terminated since, in this instance, the primal problem is infeasible and thus the dual objective function is unbounded. If the answer is *no*, step 5 is warranted.
5) Compute $\hat{\theta} = max_j\left\{\frac{\bar{c}_j}{-a_{rj}}, -a_{rj} < 0\right\}$. If $\frac{\bar{c}_k}{-a_{rk}} = \hat{\theta}$, then $\boldsymbol{r}_k$ replaces $\boldsymbol{b}_r$ in the basis.

   At this point, let us contrast the pivot operation involved in the dual simplex method with the one encountered previously in the primal simplex method. First, only negative elements can serve as pivotal elements in the dual simplex method whereas only positive pivotal elements are admitted in the primal simplex routine. Second, the dual simplex method first decides which vector is to leave the basis and next determines the vector to enter the basis. However, just the opposite order of operations for determining those vectors entering and leaving the basis holds for the primal simplex method.
6) Determine the new dual feasible (primal optimal) solution by pivoting on $-a_{rk}$. Return to step 2.

By continuing this process of replacing a single vector in the basis at a time, we move to a new dual feasible solution with the value of the dual objective function diminished and primal infeasibility reduced. The process is terminated when we attain our first primal feasible solution, in which case we attain an optimal basic feasible solution to both the primal and dual problems.

7

Linear Programming and the Theory of the Firm[1]

7.1 The Technology of the Firm

To gain some insight into how linear programming may be utilized to solve a variety of resource allocation or general economic decision problems facing the firm, we shall first examine the structure of what may be called the firm's **linear technology**. To this end, let us define **a production process** or **activity** as a particular method for performing an economic task, e.g. a physical operation that transforms a flow of scarce inputs into a flow of output in a prescribed or technologically determined fashion. Moreover, the factor input proportions are assumed to be fixed and **returns to scale** (the output response to a proportionate increase of all inputs) are constant as the inputs are varied. Hence, increasing the output of a particular process by a fixed amount can only occur if we increase "each" factor input by the same amount so that factor proportions are invariant.

More specifically, if we have m scarce inputs, each of which is used by p separate activities in fixed amounts, then the jth **simple activity** may be denoted as the $(m \times 1)$ vector

$$\boldsymbol{a}_j = \begin{bmatrix} a_{ij} \\ \vdots \\ a_{mj} \end{bmatrix}, j = 1, \ldots, p,$$

where the (constant) element a_{ij} is interpreted as "the minimum amount of factor i, $i = 1, \ldots, m$, required to operate activity $j, j = 1, \ldots, p$, at the unit level." Thus, to produce one unit of output using activity j, we must utilize a_{1j} units of factor 1, a_{2j} units of factor 2, and so on. If we momentarily concentrate on a typical activity $\boldsymbol{a}_k$, then we may express the relationship between the inputs and the input-output coefficient a_{ik} as $a_{ik}x_k = v_{ik}$, $i = 1, \ldots, m$, or

1 Dano (1966); Kogiku (1971); Vandermullen (1971); Hadar (1971); Naylor and Vernon (1969); Allen (1960); Ferguson (1971); Baumol (1977); Thompson (1971); Dorfman et al. (1958); Panik (1993, 1996).

Linear Programming and Resource Allocation Modeling, First Edition. Michael J. Panik.

$$\boldsymbol{a}_k x_k = \begin{bmatrix} a_{1k} \\ \vdots \\ a_{mk} \end{bmatrix} x_k = \begin{bmatrix} v_{1k} \\ \vdots \\ v_{mk} \end{bmatrix} = \boldsymbol{v}_k, \; x_k \geq 0, \tag{7.1}$$

where x_k denotes the **level of operation** of (i.e. output produced by) activity k and v_{ik}, the ith component of the $(m \times 1)$ **resource requirements vector** $\boldsymbol{v}_k$, is the amount of the ith input needed to produce x_k units of output by the kth activity. It is assumed that x_k and v_{ik} vary continuously so that the factor inputs as well as the output produced are **perfectly divisible.** As (7.1) indicates, there exists strict proportionality between output x_k and the inputs v_{ik}, where the constants of proportionality are the a_{ik} values. And as alluded to above, if the inputs v_{ik} are each increased by, say, $100\lambda\%$, the output of activity k increases by $100\lambda\%$ also, i.e. for $v'_{ik} = (1+\lambda)v_{ik}, v'_{ik} = a_{ik}x'_k$, where $x'_k = (1+\lambda)x_k$ is the new level of operation of activity k. Thus, *returns to scale are constant.* In terms of resource utilization, the total factor used for the operation of activity k at the x'_k level is $(1+\lambda)\boldsymbol{a}_k$.

It is evident from the structure of the **unit level of activity** k, $\boldsymbol{a}_k$, that all inputs are **limitational.** That is to say, this fixed coefficients technology does not admit any factor substitution. Using more of one input and less of another is not technologically feasible. If the amounts are to be changed, they must all be changed together so that there exists **perfect complementarity** between the inputs. We may use more of each or less of each so long as the inputs are combined in fixed proportions for all activity levels, where the absolute amount of each input used depends on the level of the activity alone. In this regard, from $v_{ik} = a_{ik}x_k$, $i = 1, \ldots, m$, we require that $v_{ik}/v_{mk} = a_{ik}/a_{mk}$ (a constant), $i = 1, \ldots, m-1$. So once the level of one of the inputs, say, the mth, is given, the levels of the remaining $m-1$ inputs are uniquely determined. Moreover, increasing each input by $100\lambda\%$ does not change the proportions in which the factors are utilized since $(1+\lambda)v_{ik}/(1+\lambda)v_{mk} = v_{ik}/v_{mk}$ for all $i = 1, \ldots, m-1$.

For example, if $\boldsymbol{a}'_k = (a_{1k}, a_{2k}) = (3,2)$, then the kth process or activity may be depicted as the locus of all input combinations involving unchanged input proportions, i.e. it corresponds to a **process ray** or **half-line** OR through the origin (Figure 7.1) with slope $a_{2k}/a_{1k} = 2/3$. Thus, process k uses 2/3 units of input 2 per unit of input 1.

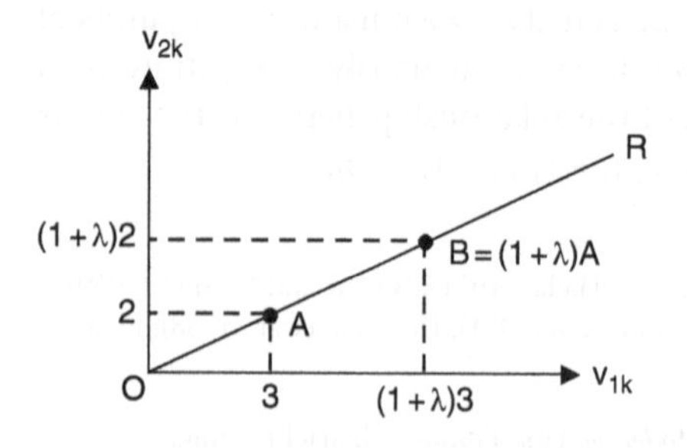

Figure 7.1 Process ray OR.

If point A depicts one unit of output produced by (the unit level of) process k, then point B denotes an increase in the output produced by process k by $100\lambda\%$. Note that, as mentioned above: (i) returns to scale are constant since, from $\boldsymbol{a}_k x_k = \boldsymbol{v}_k$, input requirements at A are

$$\begin{bmatrix} 3 \\ 2 \end{bmatrix}(1) = \begin{bmatrix} 3 \\ 2 \end{bmatrix}$$

while at B, total factor use required to support activity k at the $(1 + \lambda)$ level is

$$\begin{bmatrix} 3 \\ 2 \end{bmatrix}(1 + \lambda) = \begin{bmatrix} (1+\lambda)3 \\ (1+\lambda)2 \end{bmatrix};$$

and (ii) factor proportions are the same at both outputs A, B since $2/3 = (1 + \lambda)2/(1 + \lambda)3$.

7.2 The Single-Process Production Function

For the simple activity $\boldsymbol{a}'_k = (a_{1k}, a_{2k})$, the associated **single-process production function** (to be derived shortly) is homogeneous of degree one[2] (the constant returns to scale assumption) and has right-angled **isoquants** or **constant product curves** (Figure 7.2). To actually construct this technological relationship, let us define the input variables as the quantities of the first and second resources *actually available* (denoted as b_{1k}, b_{2k}, respectively) per production period. Then $\boldsymbol{a}_k x_k = \boldsymbol{v}_k$ is replaced by

$$\boldsymbol{a}_k x_k = \begin{bmatrix} a_{1k} \\ a_{2k} \end{bmatrix} x_k \le \begin{bmatrix} b_1 \\ b_2 \end{bmatrix} = \boldsymbol{b}, x_k \ge 0.$$

(Note that the components of $\boldsymbol{v}_k$ represent the amounts of the resources *consumed* to support activity k at the level x_k and thus correspond to points on ray OR whereas the components of $\boldsymbol{b}$ depict input upper limits that, in effect, provide an upper bound to the level of operation of activity k. In this regard, $v_{1k} \le b_1$, $v_{2k} \le b_2$.) Then a single-valued production function may be obtained by defining x_k as the maximum amount of output obtainable given the fixed resource levels indicated by the components of $\boldsymbol{b}$. Since $a_{1k} x_k \le b_1$, $a_{2k} x_k \le b_2$ may be transformed to

$$x_k \le b_1/a_{1k}, x_k \le b_2/a_{2k} \tag{7.2}$$

2 A function f is **homogeneous of degree t** in the variables $x_1, \ldots, x_n$ if, when each variable is multiplied by a scalar λ, the function itself is multiplied by λ^t, i.e., $f(\lambda x_1, \ldots, \lambda x_n) = \lambda^t f(x_1, \ldots, x_n)$.

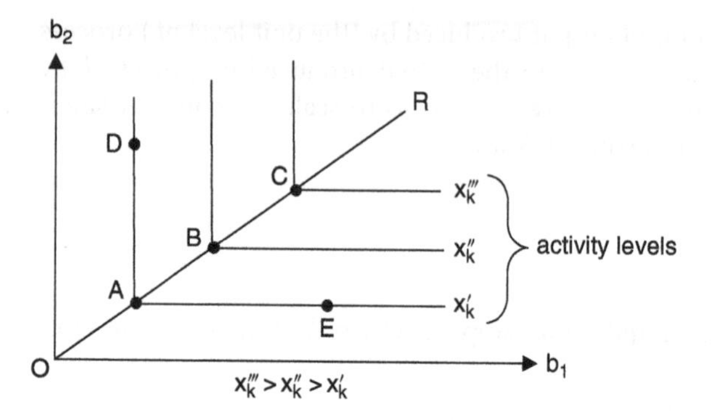

Figure 7.2 Single-process production isoquants.

respectively, it is evident that the upper limit to x_k will correspond to the lowest of the ratios given in (7.2) or

$$x_k = min\left\{\frac{b_1}{a_{1k}}, \frac{b_2}{a_{2k}}\right\}.^3 \tag{7.3}$$

An examination of the above isoquant map indicates that any input combination on isoquant DAE supports the same activity level x'_k. At point A the inputs *consumed* are exactly $v'_{1k} = a_{1k}x'_k, v'_{2k} = a_{2k}x'_k$. At point D (E) input 2 (1) is **redundant** in that the excess amount DA (AE) is wasted, i.e. the same output level may be produced with less of input 2 (1) and the same amount of input 1 (2). Thus, point A is the only **technologically efficient** point on the isoquant since it minimizes the amount of inputs needed to produce x'_k, i.e. no input is redundant there. And if positive prices must be paid for the inputs, then point A is also **economically efficient** since it minimizes the cost of producing the x'_k level of output. If we vary the level of activity k from x'_k to x''_k and so on, then each point on ray OR may be viewed as an economically efficient activity level. Hence, OR may be termed the firm's **expansion path** – the locus of economically efficient input combinations corresponding to successively higher levels of activity k (and thus successively higher levels of total factor cost) along which inputs 1, 2 are combined in the ratio a_{2k}/a_{1k}. Since along OR we require that $b_1/a_{1k} = b_2/a_{2k}$, the equation of the expansion path is $b_2 = b_1a_{2k}/a_{1k}$. Thus along ray OR the inputs are indeed **mutually limitational**. Input 1 is termed **limitative** at point D (as well as at any other point on the vertical portion of isoquant DAE) in the sense that if more of input 1 were available, the output of activity k

3 Since $min\{\lambda b_1/a_{1k}, \lambda b_2/a_{2k}\} = \lambda\, min\{b_1/a_{1k}, b_2/a_{2k}\} = \lambda x_k$, clearly (7.3) is homogeneous of degree one.

would increase; input 2 is limitative at point E (and at every point on the horizontal portion of the DAE isoquant) in a similar sense.

To expand our discussion of the single-process production function a bit, let us examine the **total product function** of input 1 given a fixed level of input 2 along with its associated marginal and average productivity functions. In panel (a) of Figure 7.3, let the level of input 2 be held fixed at $\bar{b}_2$. Then (7.3) may be rewritten as

$$x_k = \begin{cases} \dfrac{b_1}{a_{1k}}, b_1 \leq \dfrac{\bar{b}_2 a_{1k}}{a_{2k}}, \\ \dfrac{\bar{b}_2}{a_{2k}}, b_1 > \dfrac{\bar{b}_2 a_{1k}}{a_{2k}}. \end{cases} \tag{7.4}$$

Since for $b_2 = \bar{b}_2$, the maximum amount of b_1 available before mutual limitationality occurs is $b_1 = \bar{b}_2 a_{1k}/a_{2k}$ (between the origin and this b_1 value input 1 is limitative while input 2 is redundant). Then as we move along the line $\bar{b}_2$, the output of activity k increases until we come to point A, the point of intersection between $\bar{b}_2$ and the expansion path. At this point input 1 ceases to be limitative so that the level of activity k given $b_2 = \bar{b}_2$ ceases to increase since, at $b_1 = \bar{b}_2 a_{1k}/a_{2k}$ the inputs become mutually limitational (panel (b) of Figure 7.3). From (7.4) we may determine the **marginal productivity function** for input 1 as

$$MP_1 = \frac{dx_k}{db_1} = \begin{cases} \dfrac{1}{a_{1k}}, b_1 \leq \dfrac{\bar{b}_2 a_{1k}}{a_{2k}}, \\ 0, b_1 > \dfrac{\bar{b}_2 a_{1k}}{a_{2k}} \end{cases} \tag{7.5}$$

(panel (c) of Figure 7.3). In addition, the **average productivity function** for input 1 is

$$AP_1 = \frac{x_k}{b_1} = \begin{cases} \dfrac{1}{a_{1k}}, b_1 \leq \dfrac{\bar{b}_2 a_{1k}}{a_{2k}}, \\ \dfrac{\bar{b}_2}{a_{2k} b_1}, b_1 > \dfrac{\bar{b}_2 a_{1k}}{a_{2k}}. \end{cases} \tag{7.6}$$

For both $\boldsymbol{a}_k, \boldsymbol{b} \in \mathcal{E}^m$, (7.3) may be generalized to

$$x_k = min\left\{\frac{b_1}{a_{1k}}, \frac{b_2}{a_{2k}}, \ldots, \frac{b_m}{a_{mk}}\right\}. \tag{7.3.1}$$

Since fixed proportions must hold among all factor inputs, $b_1/a_{1k} = b_2/a_{2k} = \cdots = b_m/a_{mk}$ at any technologically efficient point, i.e. no input is redundant so that the m inputs are mutually limitational. At any such point the cost of producing the

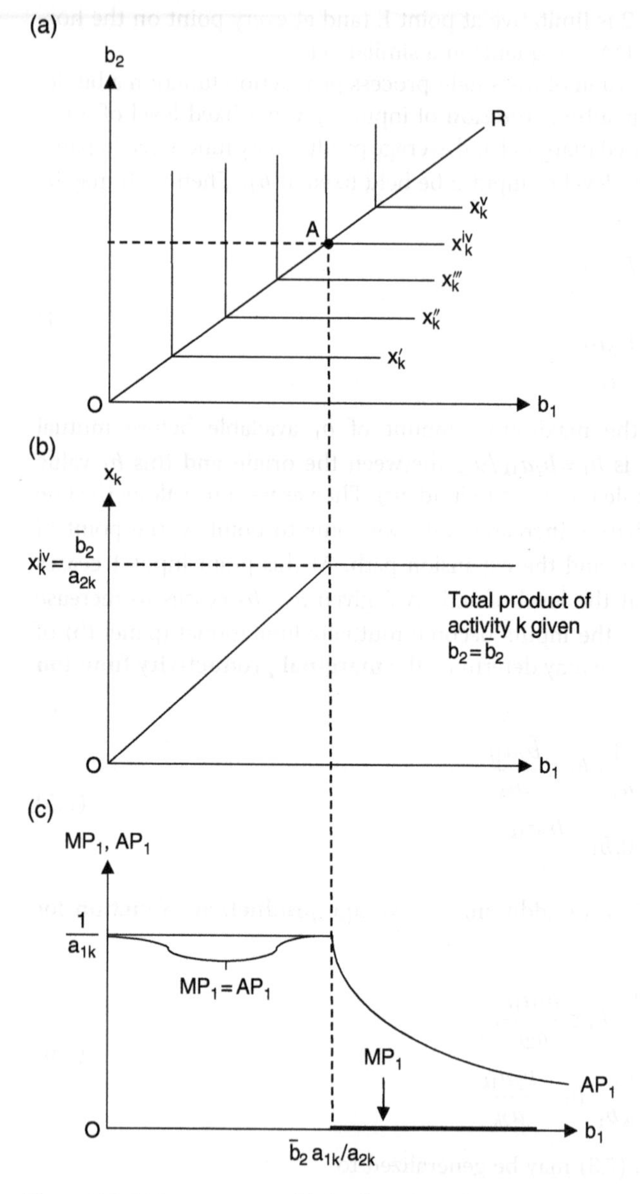

Figure 7.3 (a) Isoquant map; (b) Total product of input 1; (c) Marginal and average products of input 1.

corresponding output level x_k must be at a minimum so that the expansion path in this instance is given by $b_i/a_{ik} = b_m/a_{im}$, $i = 1, \ldots, m-1$. If there is some **limiting subset** containing the smallest possible number of inputs such that (7.3.1) is satisfied, e.g. $b_1/a_{1k} = b_2/a_{2k} = \cdots = b_h/a_{hk} < b_{lk}/a_{lk}$, $l = h+1, \ldots, m$, then the first h inputs are mutually limitational and the remaining $m-h$ inputs are redundant. For any input within the limiting subset, expressions similar to (7.5), (7.6) may be derived, provided that all inputs within this subset are varied in the same proportion so that the equalities $b_1/a_{1k} = \cdots = b_h/a_{hk}$ are preserved. Moreover, as is indicated in Figure 7.3c, at the point of mutual limitationality, the factor's marginal productivity function is discontinuous and its average productivity function possesses a kink.

Example 7.1 Given the single process vector $\boldsymbol{a}_1' = (6, 8)$ and the requirements vector $\boldsymbol{b}' = (30, 40)$, derive the total, marginal, and average productivity functions for input 1. Then from (7.4), (7.5), and (7.6), respectively (with $b_2 = \bar{b}_2 = 40$):

$$x_1 = \begin{cases} b_1/6, & b_1 \le 30, \\ 5, & b_1 > 30; \end{cases}$$

$$MP_1 = \begin{cases} 1/6, & b_1 \le 30, \\ 0, & b_1 > 30; \end{cases}$$

$$AP_1 = \begin{cases} 1/6, & b_1 \le 30, \\ 5/b_1, & b_1 > 30. \end{cases}$$

■

7.3 The Multiactivity Production Function

To generalize the single-process production function to the case where a given output level may be produced by a single process or a linear combination of a finite number of processes (the simple activities are **additive**), let us assume that each process or any combination of processes uses the same inputs in fixed proportions and thus exhibits constant returns to scale. For the unit levels of activities $\boldsymbol{a}_j$, $j = 1, \ldots, p$, their nonnegative linear combination is, from (7.1), the **composite activity**

$$\begin{aligned} &\sum_{j=1}^{p} \boldsymbol{a}_j x_j = \sum_{j=1}^{p} \boldsymbol{v}_j \ge 0, j = 1, \ldots, p, \text{or} \\ &\boldsymbol{A}\boldsymbol{X} = \boldsymbol{V}, \boldsymbol{X} \ge \boldsymbol{O}, \end{aligned} \tag{7.7}$$

where the **technology matrix** of input-output coefficients $\boldsymbol{A} = [\boldsymbol{a}_1, \ldots, \boldsymbol{a}_p]$ is of order $(m \times p)$, the $(p \times 1)$ **activity vector** $\boldsymbol{X}$ has as its components the activity levels x_j, and the elements within the $(m \times 1)$ resource requirements vector

$$V = \begin{bmatrix} v_1 \\ \vdots \\ v_m \end{bmatrix} = \begin{bmatrix} \sum_{j=1}^{p} v_{1j} \\ \vdots \\ \sum_{j=1}^{p} v_{mj} \end{bmatrix}$$

depict the total amounts of each of the m resources required to support the j activities at their respective levels x_j. As far as these latter quantities are concerned, upon summing across the rows of (7.7),

$$v_i = \sum_{j=1}^{p} v_{ij} = \sum_{j=1}^{p} a_{ij}x_j, i = 1,\ldots,m. \tag{7.8}$$

The composite activity (7.7) thus produces a total output level of

$$f = \sum_{j=1}^{p} x_j = \mathbf{1}'X \tag{7.9}$$

and *consumes* the m resources in the amounts provided by the components of V. It is further assumed that the activities are **independent**, i.e. the input requirements of a particular activity do not depend on the levels at which the other activities are operated. In view of the preceding discussion, the **joint process linear production model** may be summarized as

$$\begin{aligned} f &= \sum_{j=1}^{p} x_j \\ v_i &= \sum_{j=1}^{p} a_{ij}x_j, i = 1,\ldots,m \\ x_j &\geq 0, j = 1,\ldots,p. \end{aligned} \tag{7.10}$$

As (7.10) reveals, the p activities can be used jointly, where the final quantities of the outputs produced and the inputs consumed by the p activities altogether are simply the arithmetic sum of the quantities that would result if the activities were used individually.

While factor substitution within any simple activity is not permissible since the m inputs are limitational, the possibility of factor substitution when more than one activity is present is admissible by virtue of **process substitution**, e.g. by taking nonnegative linear combinations of simple activities in a fashion such that if the level of operation of one activity is increased, then the level of operation of another must be decreased in order to produce the same fixed level of output. But this can occur only if more of one input is utilized at the expense of another. To see exactly how this process (and thus factor) substitution takes place, we must restrict our movements to the **joint process isoquant** $f = \bar{f} =$ constant. Then from (7.10), the **parametric representation of the isoquant** $f = \bar{f}$ is

$$\begin{aligned} \bar{f} &= \sum_{j=1}^{p} x_j \\ v_i &= \sum_{j=1}^{p} a_{ij} x_j, i = 1, \ldots, m \\ x_j &\geq 0, j = 1, \ldots, p, \end{aligned} \tag{7.10.1}$$

where the parameters are the p activity levels x_j. To simplify our analysis a bit, let $p = m = 2$. Then (7.10.1) becomes, for $\bar{f} = 1$, the parametric form of the **unit isoquant**

$$\begin{aligned} 1 &= x_1 + x_2 \\ v_1 &= a_{11}x_1 + a_{12}x_2 \\ v_2 &= a_{21}x_1 + a_{22}x_2 \\ x_1&, x_2 \geq 0. \end{aligned}$$

Here $\boldsymbol{a}_1' = (a_{11}, a_{21})$, $\boldsymbol{a}_2' = (a_{12}, a_{22})$. These activities are illustrated in Figure 7.4a, where OP, OP' are process rays (each representing the locus of input combinations involving fixed factor proportions) and $v_1 = v_{11} + v_{12}$, $v_2 = v_{21} + v_{22}$. The slope of OP is a_{22}/a_{12}, i.e. activity 1 uses a_{21}/a_{11}, units of v_2/unit of v_1 and activity

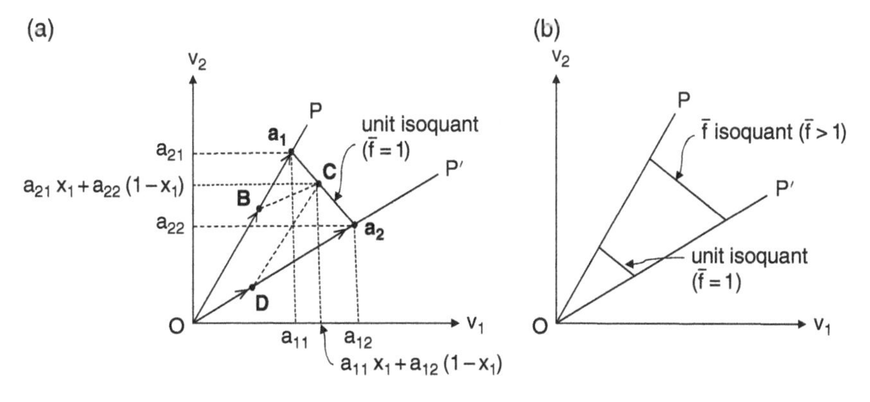

Figure 7.4 (a) Process rays OP, OP'; (b) A radial projection of the unit isoquant.

2 uses a_{22}/a_{12} units of v_2/unit of v_1. (Note that we have two different processes since $a_{21}/a_{11} \neq a_{22}/a_{12}$. If equality held between these two ratios, we still need not consider $\boldsymbol{a}_1, \boldsymbol{a}_2$ as the same process since one may be more costly to operate than the other.)

It is evident that one unit of output may be produced either by utilizing only the first activity at the $x_1 = 1$ level ($x_2 = 0$) with $v_1 = a_{11}$, $v_2 = a_{21}$ (see point $\boldsymbol{a}_1$) or by employing only the second activity at the $x_2 = 1$ level ($x_1 = 0$) with $v_1 = a_{12}$, $v_2 = a_{22}$(point $\boldsymbol{a}_2$) or by operating the activities jointly by forming a composite activity as the convex combination of $\boldsymbol{a}_1$, $\boldsymbol{a}_2$, namely $\boldsymbol{C} = x_1\boldsymbol{a}_1 + (1 - x_1)\boldsymbol{a}_2$, $0 \leq x_1 \leq 1$. Thus, the proportion of the unit level of output produced by activity 1 is x_1 while the proportion of the same produced by activity 2 is $1 - x_1$.[4] In this regard, any point on the unit isoquant may be produced by the input combination

$$\boldsymbol{C} = \begin{bmatrix} v_1 \\ v_2 \end{bmatrix} = \begin{bmatrix} a_{11}x_1 + a_{12}(1-x_1) \\ a_{21}x_1 + a_{22}(1-x_1) \end{bmatrix}.$$

(Note that $\boldsymbol{C}$ is obtained by applying the parallelogram law for vector addition, i.e. $\boldsymbol{C} = \boldsymbol{B} + \boldsymbol{D}$, where $\boldsymbol{B} = x_1\boldsymbol{a}_1$, $\boldsymbol{D} = (1 - x_1)\boldsymbol{a}_2$.)[5] Any point such as $\boldsymbol{C}$ on the unit isoquant is technologically efficient in that no other composite activity can be

---

4 To obtain the actual equation of the unit isoquant, let us solve the system

$$v_1 = a_{11}x_1 + a_{12}(1-x_1)$$
$$v_2 = a_{21}x_1 + a_{22}(1-x_1)$$

simultaneously to obtain

$$v_2 = \frac{a_{11}a_{22} - a_{12}a_{21}}{a_{11} - a_{12}} - \left(\frac{a_{22} - a_{21}}{a_{11} - a_{12}}\right)v_1. \tag{7.11}$$

If $x_1 + x_2 = \bar{f}$, then the associated isoquant is given by

$$v_2 = \bar{f}\left(\frac{a_{11}a_{22} - a_{12}a_{21}}{a_{11} - a_{12}}\right) - \left(\frac{a_{22} - a_{21}}{a_{11} - a_{12}}\right)v_1. \tag{7.12}$$

Here (7.12) is just the radial projection of (7.11) (Figure 7.4b), i.e. the isoquant in (7.11) is shifted outward parallel to itself (because of constant returns to scale) until the $\bar{f}$ level (depicted by (7.12)) is attained.

5 It is important to note that the closer the composite activity $\boldsymbol{C}$ is to one of the simple activities $\boldsymbol{a}_1$ or $\boldsymbol{a}_2$, the greater the level of use of that activity in the production of the unit output level. To see this, we need only note that the fraction of the unit level of output produced at $\boldsymbol{B}$ (the x_1 level of activity 1) is, from Figure 7.4, $x_1 = \|\boldsymbol{B}\|/\|\boldsymbol{a}_1\| = \|\boldsymbol{C} - \boldsymbol{a}_2\|/\|\boldsymbol{a}_1 - \boldsymbol{a}_2\|$, while the fraction of the unit output level produced at $\boldsymbol{D}$ (the $1 - x_1$ level of activity 2) is $\|\boldsymbol{D}\|/\|\boldsymbol{a}_2\| = \|\boldsymbol{a}_1 - \boldsymbol{C}\|/\|\boldsymbol{a}_1 - \boldsymbol{a}_2\| = \|\boldsymbol{a}_1 - \boldsymbol{a}_2\| - \|\boldsymbol{C} - \boldsymbol{a}_2\|/\|\boldsymbol{a}_1 - \boldsymbol{a}_2\| = 1 - x_1$. Thus points $\boldsymbol{B}, \boldsymbol{D}$ taken together represent the output level $x_1 + (1 - x_1) = 1$, only if $\boldsymbol{C}$ is midway between $\boldsymbol{a}_1$, $\boldsymbol{a}_2$ are the activities operated at the same level. Moreover, the total amounts of resources consumed by activities $\boldsymbol{B}, \boldsymbol{D}$ together correspond to exactly the input requirements of the composite activity $\boldsymbol{C}$. That is, since $\boldsymbol{B}' = (x_1a_{11}, x_1a_{21})$, $\boldsymbol{D}' = ((1 - x_1)a_{12}, (1 - x_1)a_{22})$, the total amount of v_1 used by both of these process levels is $x_1a_{11} + (1 - x_1)a_{12}$ while the total amount of v_2 utilized by the same is $x_1a_{21} + (1 - x_1)a_{22}$. However, these latter values are just the components of $\boldsymbol{C}$.

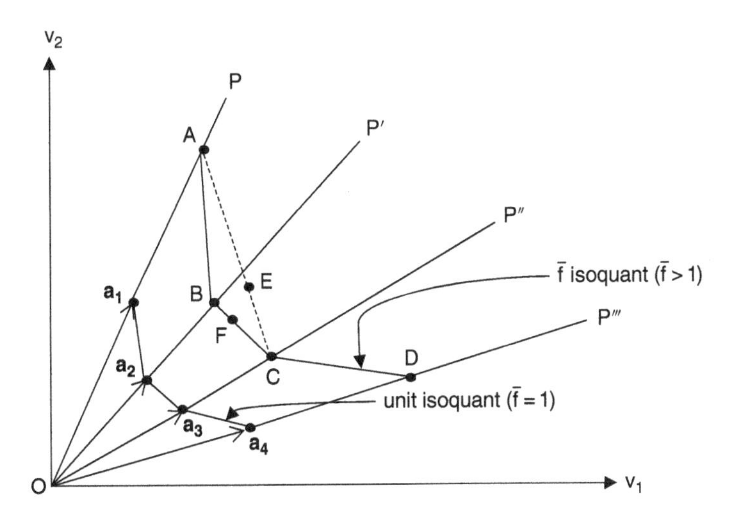

Figure 7.5 Isoquants for multiple processes.

found which produces the same level of output by using less of one of the inputs and no more of the other. Geometrically, any composite activity lies within the area bounded by *POP′*, the latter being a convex polyhedral cone in requirements space representing the set of a nonnegative linear combinations of the vectors.

When more than two processes are under consideration, the isoquants, defined by (7.10.1), are convex polygons (Figure 7.5 considers the case where four activities are present). As noted above, any isoquant (such as ABCD) lying above the unit isoquant defined by the activities $\boldsymbol{a}_1, \ldots, \boldsymbol{a}_4$ is just a radial projection of the unit isoquant with $\bar{f} > 1$. That this latter statement is true in general is easily verified by considering the operation of p activities at the unit level, where each activity uses fixed amounts of the same m inputs. If the present activity levels are x_j, $j = 1, \ldots, p$, with a total of one unit of output produced, then, if the activities are to be jointly operated at the $\bar{f}$ level, the system

$$\left.\begin{aligned}&\sum_{j=1}^{p} x_j = 1\\&\sum_{j=1}^{p} a_{ij}x_j = v_i,\ i = 1,\ldots,m\\&x_j \geq 0,\ j = 1,\ldots,p\end{aligned}\right\}\text{becomes}\left\{\begin{aligned}&\sum_{j=1}^{p}\left(\bar{f}x_j\right) = \bar{f}(1) = \bar{f}\\&\sum_{j=1}^{p} a_{ij}\left(\bar{f}x_j\right) = \bar{f}v_i,\ i = 1,\ldots,m\\&x_j \geq 0,\ j = 1,\ldots,p,\end{aligned}\right.$$

where the *j*th activity is now operated at the $\bar{f}x_j$ level. But this is just a formal demonstration of our previous assumption that the production function is homogeneous of degree one or that constant returns to scale prevails.

To further determine the properties of the $\bar{f}$ isoquant ABCD, let us note that any point such as E on the line segment AC also yields $\bar{f}$ units of output but will never be considered as a candidate for an optimal input combination since it represents an inefficient or wasteful allocation of resources, e.g. it supports the production of the same level of output as point B yet uses more of each input. Thus, only input combinations on the ABCD isoquant are technologically efficient. In this regard, an efficient use of processes is confined to either a single simple activity or to a nonnegative linear combination of two adjacent activities, thus eliminating any point such as E that can be expressed as a nonnegative linear combination of the two nonadjacent processes P, P''.

If point F represents an optimal input combination for the production of $\bar{f}$ units of output, then the information content of this solution is: (i) total output or $\bar{f}$; (ii) the optimal levels of the process P', P'', namely $\bar{f}x_2 = \bar{f}FC/BC$, $\bar{f}x_3 = \bar{f}(1-x_2) = \bar{f}BF/BC$, respectively; (iii) the factor inputs required by each activity, i.e. activity 2 uses $a_{12}\bar{f}x_2$ units of v_1 and $a_{22}\bar{f}x_2$ units of v_2 while activity 3 uses $a_{13}\bar{f}(1-x_2)$ units of v_1 and $a_{23}\bar{f}(1-x_2)$ units of v_2; and (iv) the aggregate levels of the factor inputs employed are, for $v_1, \bar{f}(a_{12}x_2 + a_{13}(1-x_2))$, and for $v_2, \bar{f}(a_{22}x_2 + a_{13}(1-x_2))$.

At present, the quantities v_i, $i = 1, \ldots, m$, denote the amounts of the m resources actually consumed in support of the p activities at their respective levels x_j, $j = 1, \ldots, p$. If the v_i are replaced by b_i, the amounts of the m resources available per production period, then $v_i \leq b_i$, $i = 1, \ldots, m$, and thus the f value in (7.10) now becomes the maximum amount of output obtainable given the fixed resource level b_1. Hence (7.10) is replaced by

$$\left.\begin{array}{l} \max f = \sum_{j=1}^{p} x_j \text{ s.t.} \\ \sum_{j=1}^{p} a_{ij}x_j \leq b_1,\ i = 1, \ldots, m \\ x_j \geq 0,\ j = 1, \ldots, p \end{array}\right\} \text{ or } \left\{\begin{array}{l} \max f = \mathbf{1}'\mathbf{X} \text{ s.t.} \\ \mathbf{AX} \leq \mathbf{b} \\ \mathbf{X} \geq \mathbf{O}, \end{array}\right. \tag{7.13}$$

where $\mathbf{b}' = (b_1, \ldots, b_m)$. Here the maximum of f defines a single-valued production function which is dependent on the fixed resource levels b_i, $i = 1, \ldots, m$. In this instance, an isoquant corresponding to an attainable value of f has the shape depicted in Figure 7.6. Note that the isoquant possesses both vertical and horizontal segments, indicating that any input combination on the same is redundant or technologically inefficient, i.e. it is possible to product the same output level with less of one input and the same amount of the other. For movements along the isoquant given in Figure 7.6, the **technical rate of substitution** of b_1 for b_2 (or b_2 for b_1), written $TRS_{1,2} = \left|\frac{db_2}{db_1}\right|_{f=\bar{f}}$, depicts the rate at which b_1 must

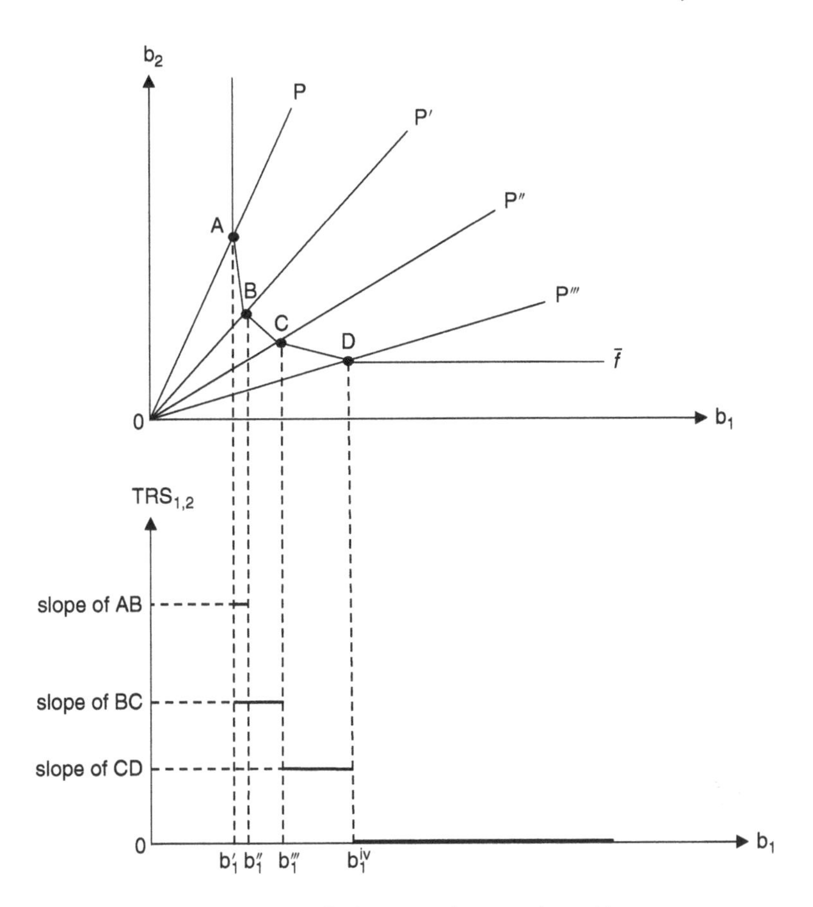

Figure 7.6 The technical rate of substitution between b_1 and b_2.

be substituted for b_2 (or conversely) to maintain the same fixed level of output. So for this joint process linear production model, the technical rate of substitution of one factor for another is discontinuous and diminishing (or at least nonincreasing) along any "well-behaved" isoquant, as Figure 7.6 indicates. If input b_2 is held fixed at the $\bar{b}_2$ level, then the total product function of input b_1 given $b_2 = \bar{b}_2$, along with the associated marginal and average productivity functions for b_1, determined, respectively, as $MP_1 = df/db_1$ and $AP_1 = f/b_1$, are illustrated in Figure 7.7. Note that MP_1 is discontinuous (indicating the abrupt transition from one activity to the next as b_1 increases) and diminishing, with this latter property indicating that the model exhibits **diminishing returns**, i.e. the marginal output response to increased use of one input declines given that the remaining input is held constant.

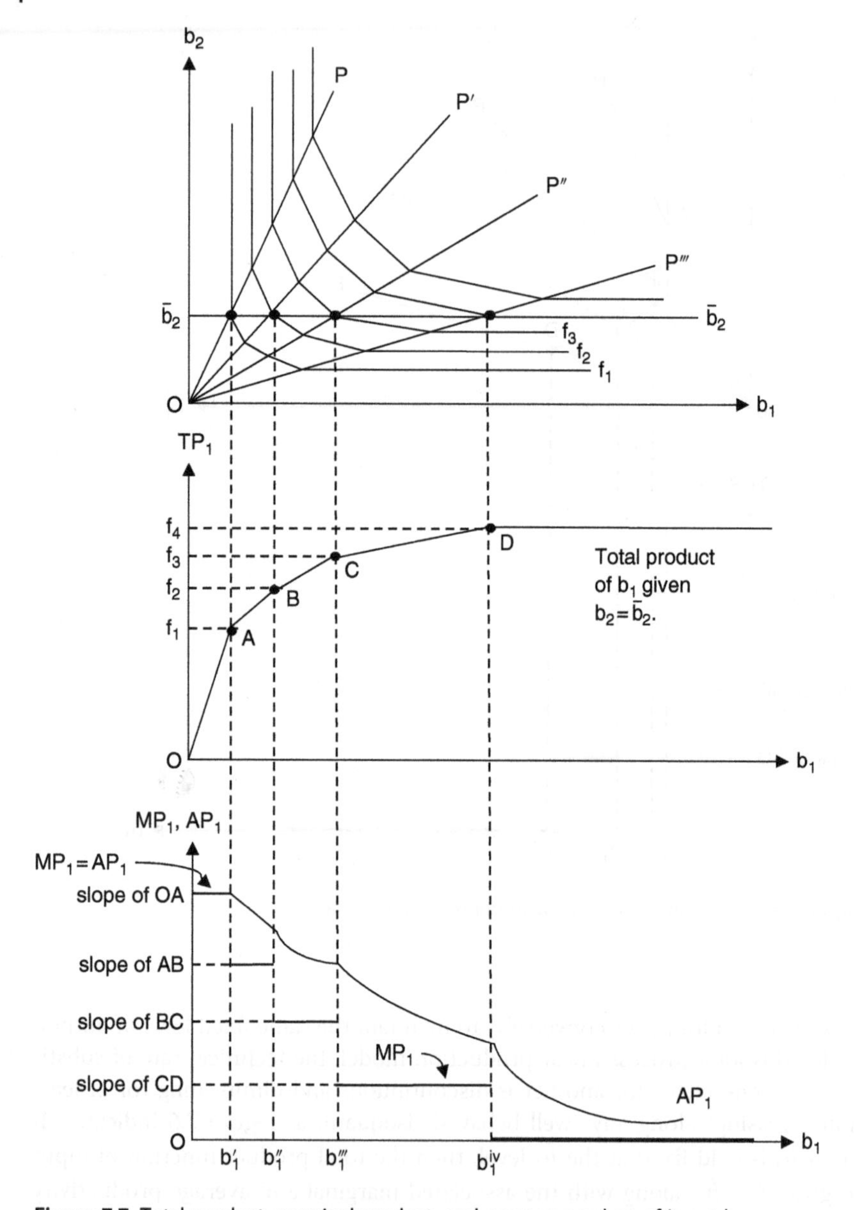

Figure 7.7 Total product, marginal product, and average product of input b_1.

Example 7.2 Given the activity vectors $\boldsymbol{a}_1' = (2,8)$, $\boldsymbol{a}_2' = (4,6)$, and $\boldsymbol{a}_3' = (7,5)$, let us determine the total, marginal, and average productivity functions for b_1 when $\bar{b}_2 = 20$. As b_1 increases from the zero level, the first activity to be utilized is $\boldsymbol{a}_1$. Thus the system

$$\begin{aligned} & f = x_1 + x_2 + x_3 \\ & x_1 \begin{bmatrix} 2 \\ 8 \end{bmatrix} + x_2 \begin{bmatrix} 4 \\ 6 \end{bmatrix} + x_3 \begin{bmatrix} 7 \\ 5 \end{bmatrix} = \begin{bmatrix} 16 \\ 20 \end{bmatrix} \\ & x_j \geq 0, j = 1,2,3 \end{aligned} \tag{7.14}$$

becomes

$$\left. \begin{aligned} & f = x_1 \\ & 2x_1 = b_1 \\ & 8x_1 = 20 \end{aligned} \right\} \text{or } f = b_1/2,\ 0 \leq b_1 \leq 5. \tag{7.14.1}$$

Next, if activity $\boldsymbol{a}_2$ is brought on line and $\boldsymbol{a}_1$, $\boldsymbol{a}_2$ are run in combination, (7.14) appears as

$$\left. \begin{aligned} & f = x_1 + x_2 \\ & 2x_1 + 4x_2 = b_1 \\ & 8x_1 + 6x_2 = 20 \end{aligned} \right\} \text{or } f = 2 + \frac{1}{10} b_1, 5 \leq b_1 \leq 40/3. \tag{7.14.2}$$

If activity $\boldsymbol{a}_3$ replaces activity $\boldsymbol{a}_1$ so that now $\boldsymbol{a}_2$, $\boldsymbol{a}_3$ operate jointly, (7.14) yields

$$\left. \begin{aligned} & f = x_2 + x_3 \\ & 4x_2 + 7x_3 = b_1 \\ & 6x_2 + 5x_3 = 20 \end{aligned} \right\} \text{or } f = \frac{30}{11} + \frac{1}{22} b_1, 40/3 \leq b_1 \leq 28. \tag{7.14.3}$$

Finally, if $\boldsymbol{a}_3$ is run alone, (7.14) reduces to

$$\left. \begin{aligned} & f = x_3 \\ & 7x_3 = b_1 \\ & 8x_3 = 20 \end{aligned} \right\} \text{or } f = 4,\ b_1 \geq 28. \tag{7.14.4}$$

Upon combining (7.14.1)–(7.14.4), the total productivity function for input b_1 given $\bar{b}_2 = 20$ is

$$f = \begin{cases} b_1/2, 0 \leq b_1 \leq 5, \\ 2 + \dfrac{1}{10} b_1, 5 \leq b_1 \leq 40/3, \\ \dfrac{30}{11} + \dfrac{1}{22} b_1, 40/3 \leq b_1 \leq 28, \\ 4,\ b_1 \geq 28 \end{cases} \tag{7.15}$$

(see Figure 7.8).

From (7.15), the marginal and average productivity functions for b_1 given $\bar{b}_2 = 20$ are, respectively,

$$MP_1 = \frac{df}{db_1} = \begin{cases} 1/2, & 0 \le b_1 < 5, \\ 1/10, & 5 \le b_1 < 40/3, \\ 1/22, & 40/3 \le b_1 < 28, \\ 0, & b_1 \ge 28; \end{cases} \tag{7.16}$$

$$AP_1 = \frac{f}{b_1} = \begin{cases} 1/2, & 0 \le b_1 \le 5, \\ \dfrac{2}{b_1} + \dfrac{1}{10}, & 5 \le b_1 < 40/3, \\ \dfrac{30}{11b_1} + \dfrac{1}{22}, & 40/3 \le b_1 < 28, \\ 4/b_1, & b_1 \ge 28 \end{cases} \tag{7.17}$$

(Figure 7.9). ■

In subsequent sections/chapters we shall employ a numerical solution technique (the simplex method) for solving a problem such as the one given in (7.13). For the time being, however, a graphical solution for $\boldsymbol{b} \varepsilon \mathcal{E}^2$ will suffice. If in Figure 7.10 $\bar{b}_1$, $\bar{b}_2$ represent upper limits to the availability of inputs b_1, b_2, respectively, then the maximum output level is f^o, the value of f corresponding to point A, where the region of feasible solutions $O\bar{b}_1 A\bar{b}_2$ has a point in common with the highest possible isoquant attainable. Here the optimal levels of the

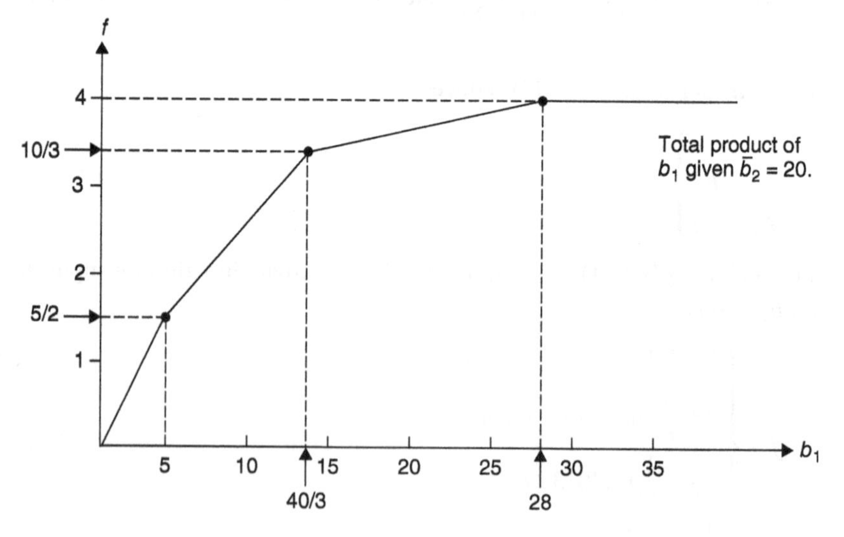

Figure 7.8 Total product of b_1 for $\bar{b}_2 = 20$.

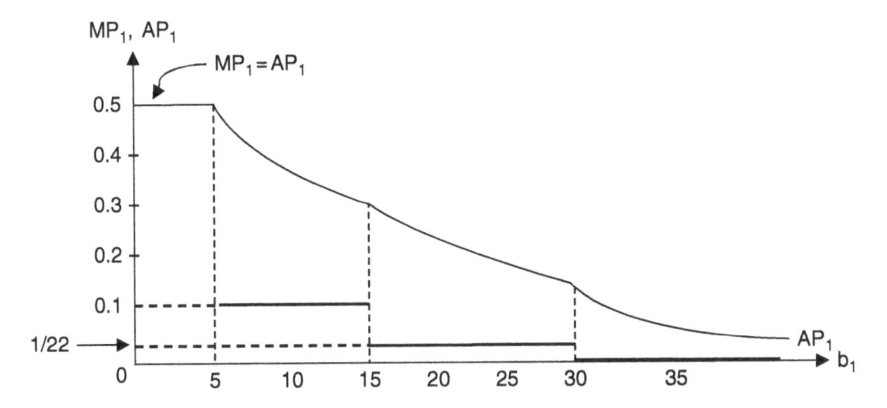

Figure 7.9 Marginal and average product functions for b_1.

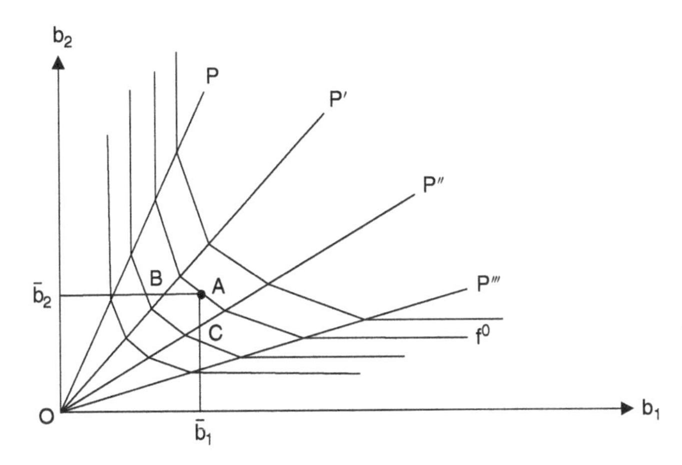

Figure 7.10 Output maximized at point A.

second and third activities are, respectively, $f^o x_2 = f^o AC/BC$ and $f^o x_3 = f^o(1 - x_2) = f^o BA/BC$. In addition, all of the available quantities of both b_1, b_2 are fully utilized in the determination of the f^o output level.

7.4 The Single-Activity Profit Maximization Model

(Dano 1966; Hadar 1971; and Ferguson 1971)

Given that the flow of output of some product x_k results from the flow of m inputs and that the production process employs the fixed coefficients linear technological relation $\boldsymbol{a}_k x_k = \boldsymbol{v}_k$, let the first l factors be fixed in quantity (the firm operates in the short-run) so that this equality may be rewritten as

$$a_{ik}x_k \leq b_{ik},\ i = 1, \ldots, l,$$

$$a_{ik}x_k = v_{ik},\ i = l + 1, \ldots, m.$$

Let us further assume that a state of perfect competition exists in both the product and factor markets. Then with the price (p) of a unit of x_k constant, total revenue is $TR = px_k$. And with the prices (q_i, $i = l + 1, \ldots, m$) of the $m - l$ variable factors also taken as constants, total variable cost is

$$TVC = \sum_{i=l+1}^{m} q_i v_{ik} = \sum_{i=l+1}^{m} q_i a_{ik} x_k,$$

where $\sum_{i=l+1}^{m} q_i a_{ik}$ represents unit variable cost.

For the l fixed inputs, $q_i = 0$. Thus, the contribution of these fixed inputs to total factor cost amounts to a fixed charge assigned or imputed to their very presence as an integral part of the production process (whether any or all of the costs of these fixed inputs are fully "written off" or not) since production cannot take place without them. In this regard, since the available amounts of the fixed factors set an upper limit to total profit, their services are valued on an opportunity cost basis, i.e. u_i, $i = 1, \ldots, m$, the **shadow price** (or **marginal imputed value**) of the services of the ith fixed factor, measures the amount of profit forgone for lack of an additional unit of the ith fixed factor. So if u_i depicts the incremental profit response due to an extra unit in the available capacity of the ith fixed input, it may be alternatively viewed as the maximum amount the firm would be willing to pay for an additional unit of the services of b_{ik}. It must be emphasized that these shadow prices are not market prices but only internal prices for the scarce resources, i.e. its imputation involves an internal pricing mechanism which is based on alternative fixed factor uses within the firm. Once calculated, these fictitious prices serve as accounting prices that may be used to determine the total imputed value of the firm's fixed resources.

In view of the previous discussion, our **short-run fixed-coefficients profit-maximization model** under perfect competition appears as

$$
\begin{aligned}
max f(x_k) &= TR - TVC = px_k - \sum_{i=l+1}^{m} q_i a_{ik} x_k \\
&= \left(p - \sum_{i=l+1}^{m} q_i a_{ik}\right) x_k \quad \text{s.t.} \qquad (7.18) \\
a_{ik}x_k &\leq b_{ik},\ i = 1, \ldots, l, \\
x_k &\geq 0
\end{aligned}
$$

The symmetric dual associated with this problem is

$$\begin{aligned} &\min g(u_i, \ldots, u_l) = \sum_{i=l}^{l} b_{ik} u_i \quad \text{s.t.} \\ &\sum_{i=l}^{l} a_{ik} u_i \geq p - \sum_{i=l+1}^{m} q_i a_{ik} \\ &u_i \geq 0, \; i = 1, \ldots, l. \end{aligned} \tag{7.19}$$

This dual problem requires that the total imputed value of the fixed resources be minimized subject to the condition that the imputed cost of the fixed factors' services per unit of activity k be at least as great as the gross profit margin. For an optimal solution in which activity k is operated at a positive level ($x_k > 0$), the dual structural constraint must hold as a strict equality in (7.19), i.e. gross profit per unit of output equals the imputed cost of the fixed factors of production per unit of x_k produced. If, in an optimal solution, the dual structural constraint holds as an inequality, gross profit per unit of output falls short of the imputed cost of the fixed factors per unit of x_k so that $x_k = 0$, i.e. activity k is unprofitable to operate. In this latter instance, the dual surplus variable associated with the dual structural constraint in (7.19), which may be interpreted as an **accounting loss** figure, is positive. In the former case where $x_k > 0$, the accounting loss associated with activity k is zero. Moreover, since at an optimal solution to the primal-dual pair of problems we have $f(x_k) = g(u_1, \ldots, u_l)$ or

$$\left(p - \sum_{i=l+1}^{m} q_i a_{ik} \right) x_k = \sum_{i=1}^{l} b_{ik} u_i, \text{[6]} \tag{7.20}$$

it follows from the structure of this very simple model and the preceding equality that the output of activity k should be expanded to the point where the gross profit margin equals the imputed value of the fixed resources per unit of activity k, or

$$p - \sum_{i=l+1}^{m} q_i a_{ik} = \sum_{i=1}^{l} b_{ik} u_i / x_k.$$

Thus, the optimal level of activity k is

$$x_k = \sum_{i=1}^{l} b_{ik} u_i \Big/ \left(p - \sum_{i=l+1}^{m} q_i a_{ik} \right). \tag{7.21}$$

---

6 This equality is equivalent to the condition that net profit = gross profit − Imputed value of the scarce resources = TR − TVC $-\sum_{i=1}^{l} b_{ik} u_i = 0$ at the optimal level of activity k when $x_k > 0$. if $x_k = 0$, net profit is negative.

Looked at from another perspective, since the $m - l$ fixed inputs are limitational, the optimal level of activity k is alternatively given by

$$x_k = b_{rk}/a_{rk} = \min_i \left\{ \frac{b_{1k}}{a_{1k}}, \ldots, \frac{b_{lk}}{a_{lk}} \right\}. \tag{7.22}$$

That (7.22) is indeed equivalent to (7.21) follows from the fact that for $x_k > 0$, equality holds in the structural constraint in (7.19) so that, from (7.21), $x_k = \sum_{i=1}^{l} b_{ik}u_i / \sum_{i=l}^{l} a_{ik}u_i$. For all fixed factors for which the minimum in (7.22) is not attained, we must have $u_i = 0$, $i \neq r$, since these resources are in excess supply. In view of this requirement, this last expression for x_k simplifies to $x_k = b_{rk}/a_{rk}$.

For $p > \sum_{i=l+1}^{m} q_i a_{ik}$, the optimal level of activity k may be indicated graphically in Figure 7.11a, where, as determined above, $x_k = b_{rk}/a_{rk}$. Moreover, the determination of the optimum level of activity k may also be depicted in terms of the familiar **marginal revenue** and **marginal cost relationships**, as exhibited in Figure 7.11b. That is to say, for marginal revenue $(MR = dTR/dx_k) = p$ and marginal cost $(MC) = dTVC/dx_k = \sum_{i=l+1}^{m} q_i a_{ik}$, the familiar $p = MR = MC$ equilibrium condition under perfect competition results when $x_k = b_{rk}/a_{rk}$. Since **average variable cost** $(AVC) = \sum_{i=l+1}^{m} q_i a_{ik} (= MC)$, the firm's **short-run supply curve** is that (vertical) portion of the marginal cost curve above minimum average variable cost.

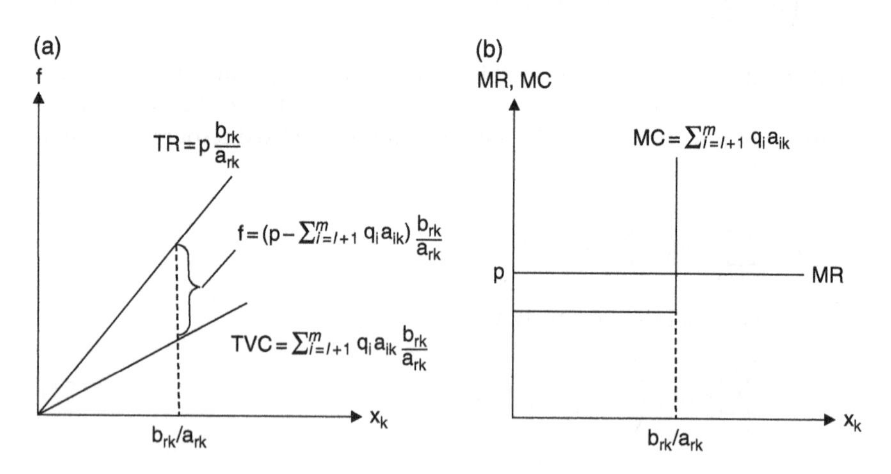

Figure 7.11 (a) Determining the optimal level of activity k; (b) Marginal revenue and marginal cost functions for activity k.

7.5 The Multiactivity Profit Maximization Model

(Dano 1966; Hadar 1971; and Ferguson 1971)

This section presents a generalization of the preceding single-activity profit maximization model to the case where a given output level may be produced by a single activity or by a nonnegative linear combination of a finite number of activities. In this regard, the production process employs a fixed-coefficients linear technology and exhibits constant returns to scale. If the unit level of the jth activity is $\boldsymbol{a}_j, j = 1, \ldots, p$, then a nonnegative linear combination of activities is depicted as

$$\sum_{j=1}^{p} \boldsymbol{a}_j x_j = \sum_{j=1}^{p} \boldsymbol{v}_j,\ x_j \geq 0 \text{ or } \boldsymbol{AX} = \boldsymbol{V}, \boldsymbol{X} \geq \boldsymbol{O},$$

where the components within the resource requirements vector

$$\boldsymbol{V} = \begin{bmatrix} v_1 \\ \vdots \\ v_m \end{bmatrix} = \begin{bmatrix} \sum_{j=1}^{p} v_{1j} \\ \vdots \\ \sum_{j=1}^{p} v_{mj} \end{bmatrix}$$

depict the total amount of each of the m resources required to support the p activities at their respective levels x_j. If the first l factors are fixed in quantity (again it is assumed that the firm operates in the short-run), then the preceding composite activity may be rewritten as

$$\sum_{j=1}^{p} a_{ij} x_j \leq b_i,\ i = 1, \ldots, l,$$

$$\sum_{j=1}^{p} a_{ij} x_j \leq v_i,\ i = l + 1, \ldots, m.$$

Thus, the firm is constrained in its ability to select those activity levels that yield maximum profit by its given endowment of the l fixed factors.

If we again invoke the assumption of perfect competition in both the product and factor markets, the price per unit of output produced by the jth simple activity is $p_j = c =$ constant for all $j = 1, \ldots, p$ so that the revenue generated by operating activity j at the level x_j is cx_j. Then the total revenue from all p activities is $c\sum_{j=1}^{p} x_j = c\boldsymbol{1}'\boldsymbol{X}$. With the prices q_i of the $m - l$ variable inputs also taken to be constants, the cost of input i incurred by operating activity j at the x_j level is $q_i a_{ij} x_j$, $i = l + 1, \ldots, m$. Thus, the total variable input cost of operating activity

j at the x_j level is, upon summing over all variable inputs, $TVC_j = \sum_{i=l+1}^{m} q_i a_{ij} x_j = Q' a_j^* x_j$, where the ith component of the $(m - l \times 1)$ vector Q is q_i and the components of the $(m - l \times 1)$ vector a_j^* are the last $m - l$ components of the activity vector a_j, $j = 1, \ldots, p$. If we next sum over all p activities, we obtain total variable input cost $TVC = \sum_{j=1}^{p} \left(\sum_{i=l+1}^{m} q_i a_{ij}\right) x_j = Q' A^* X$, where $A^* = [a_1^*, \ldots, a_p^*]$ is of order $(m - l \times p)$ and the unit variable cost of the jth activity is $\sum_{i=l+1}^{m} q_i a_{ij}$. As was our approach in the previous section, $q_i = 0$ for the l fixed inputs. Here, too, the contribution of the fixed resources to total factor cost is determined on an opportunity cost basis by assigning shadow or accounting prices to these factors. Once this imputation is executed, the resulting fixed charge is simply the total imputed value of the firm's fixed resources.

In view of this discussion, the **short-run linear technology profit-maximization model** under perfect competition appears as

$$\max f(x_1, \ldots, x_p) = TR - TVC = c \sum_{j=1}^{p} x_j - \sum_{j=1}^{p} \left(\sum_{i=l+1}^{m} q_i a_{ij} \right) x_j$$

$$= \sum_{j=1}^{p} \left(c - \sum_{i=l+1}^{m} q_i a_{ij} \right) x_j \quad \text{s.t.} \tag{7.23}$$

$$\sum_{j=1}^{p} a_{ij} x_j \leq b_i, \; i = 1, \ldots, l,$$

$$x_j \geq 0, \; j = 1, \ldots, p,$$

where $c - \sum_{i=l+1}^{m} q_i a_{ij}$ depicts the gross profit margin per unit of activity j. Alternatively, (7.23) may be rewritten in matrix form as

$$\begin{aligned} &\max f(X) = c1'X - Q'A^*X = \left(c1 - (A^*)'Q\right)'X \quad \text{s.t.} \\ &\bar{A}X \leq b, X \geq O, \end{aligned} \tag{7.23.1}$$

where $\bar{A} = [\bar{a}_1, \ldots, \bar{a}_p]$ is of order $(l \times p)$, the components of the $(l \times 1)$ vector $\bar{a}_j$ are the first l components of the activity vector a_j, $j = 1, \ldots, p$, and $b' = (b_1, \ldots, b_l)$.

The symmetric dual to this problem is

$$\begin{aligned} &\min g(U) = b'U \quad \text{s.t.} \\ &\bar{A}U \geq c1 - (A^*)'Q, U \geq O \end{aligned} \tag{7.24}$$

or, in terms of the individual linear expressions subsumed therein,

$$\min g(u_1, \ldots, u_l) = \sum_{i=1}^{l} b_1 u_i \quad \text{s.t.}$$

$$\bar{a}_j' U \geq c - \left(a_j^*\right)' Q \text{ or} \sum_{i=1}^{l} a_{ij} u_i \geq c - \sum_{i=l+1}^{m} q_i a_{ij}, \; j = 1, \ldots, p, \tag{7.24.1}$$

$$u_i \geq 0, \; i = 1, \ldots, l.$$

As these two formulations of the dual indicate, the firm seeks to minimize the total imputed value of its fixed resources subject to the requirement that the imputed cost of operating activity j at the unit level is at least as great as the per-unit contribution to gross profit of activity j, $j = 1, \ldots, p$. That is to say, the firm desires to determine the smallest imputed value of the stock of fixed resources which fully accounts for all of the gross profits generated by the p activities.[7] If at an optimal solution to the dual problem a dual structural constraint holds as an equality, the imputed value of the inputs going into the operation of activity j at the unit level just matches the gross profit which the firm makes by operating activity j at that level. In this instance, activity j is run at a positive level (its accounting loss is zero) and thus $x_j > 0$. If, however, strict inequality prevails in a dual structural constraint at the optimum, activity j is unprofitable and will be terminated at ($x_j = 0$) since it costs more per unit of output to run activity j than the unit profit it yields (i.e. the accounting loss figure for activity j is positive). These observations may be reinforced by examining the complementary slackness conditions that are in force at the optimal solution to the primal and dual problems. In this regard, at an optimal solution to the said pair of problems, if $(\boldsymbol{U}_s)_0$, $(\boldsymbol{X}_s)_0$, respectively, depict the $(p \times 1)$ and $(m - l \times 1)$ vectors of the dual surplus and primal slack variables obtained at the optimum, then

$$(\boldsymbol{U}_s)_0' \boldsymbol{X}_0 + \boldsymbol{U}_0'(\boldsymbol{X}_s)_0 = 0 \text{ or } \sum_{j=1}^{p} x_j^0 u_{m+j}^0 + \sum_{i=l+1}^{m} u_i^0 x_{p+i}^0 = 0$$

so that the following conditions exist:

a) If $x_j^0 > 0$, then $u_{m+j}^0 = 0$ – if an activity is undertaken at a positive level in the optimal primal solution, its corresponding accounting loss figure must be zero.
b) If $u_{m+j}^0 > 0$, then $x_j^0 = 0$ – whenever a surplus variable appears in the optimal dual solution, the activity to which it refers is not undertaken.
c) If $u_i^0 > 0$, then $x_{p+i}^0 = 0$ – if u_i^0 appears in the optimal dual solution, input i is used to capacity so that the firm would be willing to pay up to $\$u_i$ to have the ith constraint relaxed by one unit.
d) If $x_{p+i}^0 > 0$, then $u_i^0 = 0$ – if any resource is not fully utilized in the optimal primal solution, its corresponding shadow price is zero, i.e. the firm is not

7 In attempting to determine the smallest imputed value of its stock of fixed resources which completely explains all of the gross profits resulting from the operation of the p activities, the firm desires to allocate its gross profit (on an accounting basis) among its l fixed inputs since it is their employment in the various processes which creates the profit. But what share of profit should be allocated to each fixed resource? Moreover, how is the allocation to be carried out? To answer these questions. We need only note that the firm allocates the profit from each activity to "the inputs used in that activity." Thus, the accounting values assigned must be such that "the cost of each activity should completely exhaust the profit obtained from each activity."

willing to pay a positive price for an extra unit of a resource that is presently in excess supply.

7.6 Profit Indifference Curves (Baumol 1977)

To transform a given production indifference curve (represented by, say, ABCD of Figure 7.12 and corresponding to $\bar{f}$ units of output) into a profit indifference curve, we must answer the following question. Specifically, what points yield $\$\bar{f}$ in total profit? For any activity x_j, unit profit is $c-\sum_{i=l+1}^{m} q_i a_{ij}, j=1,\ldots,p$. So for the four activities depicted in Figure 7.12, the unit profit levels are, respectively,

$$c_1 = c - \sum_{i=l+1}^{m} q_i a_{i1}, \qquad c_3 = c - \sum_{i=l+1}^{m} q_i a_{i3},$$

$$c_2 = c - \sum_{i=l+1}^{m} q_i a_{i2}, \qquad c_4 = c - \sum_{i=l+1}^{m} q_i a_{i4}.$$

We note first that if $c_1 = c_2 = c_3 = c_4 = \$1$, then each of the points, A, B, C, and D, represent both $\bar{f}$ units of output and $\$\bar{f}$ in profit (e.g. at A, $\bar{f}\,(\$1) = \$\bar{f}$). Next, if on the one hand $c_1 < \$1$, then point A involves less than $\$\bar{f}$ in profit. That is, since every unit of output by activity P_1 yields $\$c_1$, point A represents a total profit of $\$\bar{f}c_1 < \$\bar{f}$. So to earn $\$\bar{f}$ using activity P_1, a total output level of $f' = \bar{f}/c_1 > \bar{f}$ must be produced at point A′ of Figure 7.12 (note that $f' \cdot c_1 = \bar{f}$) so that output is increased by the multiple $\frac{1-c_1}{c_1}$, i.e. $f' - \bar{f} = \bar{f}\left(\frac{1-c_1}{c_1}\right)$. On

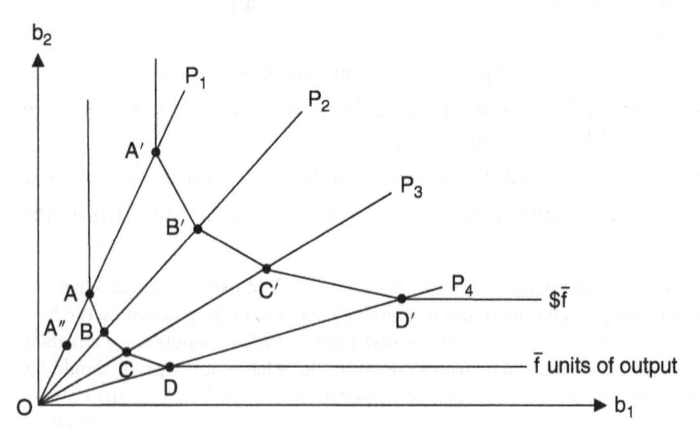

Figure 7.12 Profit indifference curve $\$\bar{f}$.

the other hand, if $c_1 > \$1$, point A now involves more than $\$\bar{f}$ profit. Here every unit of output by activity P_1 generates $\$c_1$ so that point A depicts a total profit level of $\$\bar{f}c_1 > \$\bar{f}$. Thus, to earn $\$\bar{f}$ by process P_1, a total output level of $f'' = \bar{f}/c_1 < \bar{f}$ must be produced (see point A″ of Figure 7.12). In this instance, output decreases by the factor $\frac{c_1 - 1}{c_1}$ since $\bar{f} - f'' = \bar{f} - \bar{f}/c_1 = \bar{f}\left(1 - \frac{1}{c_1}\right) = \bar{f}\left(\frac{c_1 - 1}{c_1}\right)$. If a similar discussion is engaged in relative to activities P_2, P_3, and P_4, then a hypothetical profit indifference curve for $\$\bar{f}$ total profit is A′ B′ C′ D′, which is obtained by connecting the points A′, B′, C′, and D′, by straight-line segments (note that we are tacitly assuming that $\$1 > c_1 > c_2 > c_3 > c_4$).

As was the case with production indifference curves, successive profit indifference curves are radial projections of the one corresponding to a profit level of $\$\bar{f}$ (or, more fundamentally, the unit profit or \$1 indifference curve) and exhibit constant (profit) returns to scale, i.e. they reflect the linear homogeneity property of the fixed coefficients linear technology. In this regard, if all inputs are increased by $100\lambda\%$, total profit also increases $100\lambda\%$. Moreover, the less profitable a particular activity is, the further out will be the shift along the activity ray of a point representing $\bar{f}$ units of output to a point representing $\$\bar{f}$ of profit. For instance, since in Figure 7.12 it was assumed that $c_1 > c_2 > c_3 > c_4$, it follows that distance AA′ < distance BB′ < distance CC′ < distance DD′ or that $\frac{1 - c_1}{c_1} < \frac{1 - c_2}{c_2} < \frac{1 - c_3}{c_3} < \frac{1 - c_4}{c_4}$.

If from a set of output indifference curves we derive a set of profit indifference curves, then the graphical solution to a problem such as (7.23) is indicated, for $\mathbf{b} \in \mathcal{E}^2$, in Figure 7.13. Here the maximum quantities of b_1, b_2 available are $\bar{b}_1$, $\bar{b}_2$,

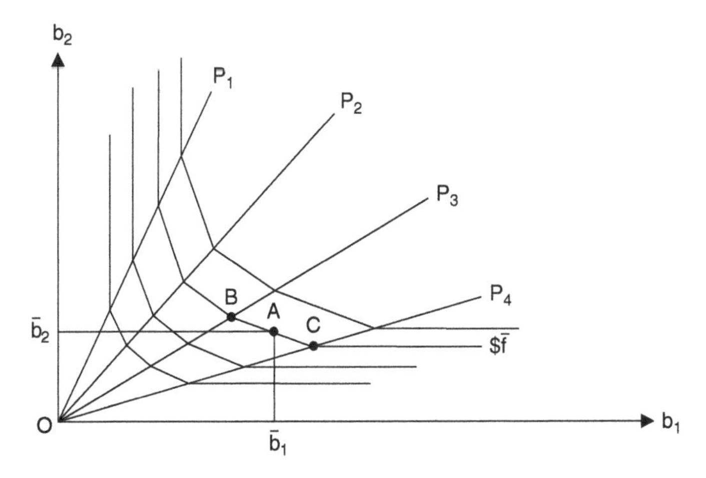

Figure 7.13 Determining the optimum profit level.

respectively, so that the region of feasible solutions is $O\bar{b}_1 A\bar{b}_2$. Clearly, the optimum profit level ($\$\bar{f}$) occurs at point A where the $\$\bar{f}$ profit indifference curve is the highest one attainable that also preserves feasibility. At point A, activity 3 contributes $(\$\bar{f})x_3 = (\$\bar{f})AC/BC$ to a total profit value of $\$\bar{f}$ while activity 4 contributes $(\$\bar{f})x_4 = (\$\bar{f})(1-x_3) = (\$\bar{f})BA/BC$ to the same. In addition, all of the available capacities of both b_1, b_2 are fully utilized in the determination of the $\$\bar{f}$ profit level.

7.7 Activity Levels Interpreted as Individual Product Levels

If the firm does not produce a single product but actually produces p different products, one by each individual activity, then each activity has associated with it a separate constant price per unit of output p_j, $j = 1, \ldots, p$. Letting the price of the jth activity correspond to the jth component of the $(p \times 1)$ price vector $\boldsymbol{P}$, the total revenue generated from all p activities is $TR = \sum_{j=1}^{p} p_j x_j = \boldsymbol{P}'\boldsymbol{X}$. In view of this modification, (7.23), (7.23.1) may be rewritten as

$$\begin{aligned} &max\, f(x_1,\ldots,x_p) = \sum_{j=1}^{p}\left(p_j - \sum_{i=l+1}^{m} q_i a_{ij}\right)x_j \quad \text{s.t.} \\ &\sum_{j=1}^{p} a_{ij}x_j \le b_i,\ i = 1,\ldots,l, \\ &x_j \ge 0,\ j = 1,\ldots,p, \end{aligned} \tag{7.25}$$

or

$$\begin{aligned} &max\, f(\boldsymbol{X}) = (\boldsymbol{P} - (\boldsymbol{A}^*)'\boldsymbol{Q})'\boldsymbol{X} \quad \text{s.t.} \\ &\bar{\boldsymbol{A}}\boldsymbol{X} \le \boldsymbol{b}, \boldsymbol{X} \ge \boldsymbol{O}. \end{aligned} \tag{7.25.1}$$

The symmetric dual thus becomes

$$\begin{aligned} &min\, g(\boldsymbol{U}) = \boldsymbol{b}'\boldsymbol{U} \quad \text{s.t.} \\ &\bar{\boldsymbol{A}}'\boldsymbol{U} \ge \boldsymbol{P} - (\boldsymbol{A}^*)'\boldsymbol{Q}, \boldsymbol{U} \ge \boldsymbol{O}, \end{aligned}$$

or

$$\begin{aligned} &min\, g(u_1,\ldots,u_l) = \sum_{i=1}^{l} b_i u_i \quad \text{s.t.} \\ &\bar{\boldsymbol{a}}_j'\boldsymbol{U} \ge p_j - \left(\boldsymbol{a}_j^*\right)'\boldsymbol{Q} \text{ or } \sum_{i=1}^{l} a_{ij}u_i \ge p_j - \sum_{i=l+1}^{m} q_i a_{ij},\ j = 1,\ldots,p, \\ &u_i \ge 0,\ i = 1,\ldots,l. \end{aligned} \tag{7.26.1}$$

For an economic interpretation of this dual problem, see the discussion underlying (7.24).

Example 7.3 Given that a profit-maximizing firm produces a single product using the activities

$$\boldsymbol{a}_1 = \begin{bmatrix} 10 \\ 2 \\ 3 \\ 1 \end{bmatrix}, \boldsymbol{a}_2 = \begin{bmatrix} 8 \\ 3 \\ 1 \\ 5 \end{bmatrix}, \boldsymbol{a}_3 = \begin{bmatrix} 6 \\ 5 \\ 4 \\ 1 \end{bmatrix}, \text{and } \boldsymbol{a}_4 = \begin{bmatrix} 4 \\ 8 \\ 3 \\ 3 \end{bmatrix},$$

let us assume that the first two inputs are fixed at the levels $b_1 = 40$, $b_2 = 30$, respectively, while the third and fourth factors are freely variable in quantity. In addition, the price per unit of factor three is $q_3 = \$3$ while that of factor four is $q_4 = \$2$. If the price per unit of output is $c = \$25$, then $TR = 25\sum_{j=1}^{4} x_j$. With

$$TVC_1 = \sum_{i=3}^{4} q_i a_{i1} = 3{\cdot}3 + 2{\cdot}1 = 11,$$

$$TVC_2 = \sum_{i=3}^{4} q_i a_{i2} = 3{\cdot}1 + 2{\cdot}5 = 13,$$

$$TVC_3 = \sum_{i=3}^{4} q_i a_{i3} = 3{\cdot}4 + 2{\cdot}1 = 14,$$

$$TVC_4 = \sum_{i=3}^{4} q_i a_{i4} = 3{\cdot}3 + 2{\cdot}3 = 15,$$

it follows that $TVC = \sum_{j=1}^{4} (TVC_j) x_j = 11x_1 + 13x_2 + 14x_3 + 15x_4$ and thus total profit appears as $f = (25 - 11)x_1 + (25 - 13)x_2 + (25 - 14)x_3 + (25 - 15)x_4$. Then the final form of this profit maximization problem is

$$\begin{aligned} max\, f = 14x_1 + 12x_2 + 11x_3 + 10x_4 \quad \text{s.t.} \\ 10x_1 + 8x_2 + 6x_3 + 4x_4 \le 40 \\ 2x_1 + 3x_2 + 5x_3 + 8x_4 \le 30 \\ x_1, \ldots, x_4 \ge 0. \end{aligned}$$

If x_5, x_6 represent nonnegative slack variables, the optimal simplex matrix corresponding to this problem is

$$\begin{bmatrix} 1 & \frac{11}{19} & 0 & -\frac{14}{19} & \frac{5}{38} & -\frac{3}{19} & 0 & \frac{10}{19} \\ 0 & \frac{7}{19} & 1 & \frac{36}{19} & -\frac{1}{19} & \frac{5}{19} & 0 & \frac{110}{19} \\ 0 & \frac{3}{19} & 0 & \frac{10}{19} & \frac{24}{19} & \frac{13}{19} & 1 & \frac{1350}{19} \end{bmatrix}$$

with $\frac{10}{19}$ units of output produced by activity 1 and $\frac{110}{19}$ units of output produced by activity 3, thus yielding a total of $\frac{120}{19}$ units of output produce for a (gross) profit level of \$1350/19. From this solution we obtain four separate data sets that are of importance to the firm: (i) the activity mix $\left(x_1 = \frac{10}{19}, x_2 = 0, x_3 = \frac{110}{19}, x_4 = 0\right)$; (ii) resource utilization levels (the slack variables $x_5, x_6 = 0$, indicating that there is no excess or unused capacity with respect to the fixed inputs)[8]; (iii) shadow prices (the firm would be willing to pay up to $u_1 = \$24/19$ to have the first structural constraint relaxed by one unit and up to $u_2 = \$13/19$ to have the second structural constraint relaxed by one unit – or – an extra unit of $b_1(b_2)$ employed efficiently adds \$24/19 (\$13/19) to total profit); and (iv) accounting loss figures (the dual surplus variables $u_{s1} = u_{s3} = 0$, indicating that activities 1 and 3 just break even (net profit is zero) or are run at an optimal level while $u_{s2} = \$13/19(u_{s4} = \$10/19)$ reveals that the imputed cost of the fixed resources used to produce a unit of output using activity 2 (4) exceeds the unit profit obtained from this activity by \$13/19 (\$10/19) so that activity 2 (4) is unprofitable and thus idle). This information is conveniently summarized by directly employing the complementary slackness conditions:

Activity mix vs. Accounting loss values	**Resource utilization vs. Shadow prices**
$x_1^o u_{s1}^0 = \frac{10}{19} \cdot 0 = 0$	$u_1^o x_5^o = \frac{24}{19} \cdot 0 = 0$
$x_2^o u_{s2}^0 = 0 \cdot \frac{3}{19} = 0$	$u_2^o x_6^o = \frac{14}{19} \cdot 0 = 0$
$x_3^o u_{s3}^0 = \frac{110}{19} \cdot 0 = 0$	
$x_4^o u_{s4}^0 = 0 \cdot \frac{10}{19} = 0$	

8 The amounts of the fixed factors b_1, b_2 needed to operate activities one and three at their optimal levels may be determined, respectively, as

$$x_1 \bar{a}_1 = \frac{10}{19}\begin{bmatrix}10\\2\end{bmatrix} = \begin{bmatrix}100/19\\20/19\end{bmatrix}$$

(activity one uses 110/19 units of b_1 and 20/19 units of b_2);

$$x_3 \bar{a}_3 = (1 - x_1)\bar{a}_3 = \frac{110}{19}\begin{bmatrix}6\\5\end{bmatrix} = \begin{bmatrix}660/19\\550/19\end{bmatrix}$$

(activity three utilizes 660/19 units of b_1 and 550/19 units of b_2).

Let us next determine the profit indifference curve corresponding to $f^o = \frac{120}{19}$ units of output by first finding the $f^o = \frac{120}{19}$ isoquant in input space (i.e. the $f^o = 120/19$ isoquant is to be transformed into the $\$f^o = \$120/19$ profit indifference curve). To do so let us look to the following question. Specifically, how much of each of the fixed inputs, b_1 and b_2, is needed to operate each individual activity at the 120/19 level? To solve this problem we need only employ the set of technological relationships $a_{ij}x_j = b_i$, with $x_j = 120/19$, $i = 1, 2$; $j = 1, 2, 3, 4$. So from the first two components of activities 1–4, we obtain, respectively,

$$\begin{bmatrix} 10 \\ 2 \end{bmatrix} \frac{120}{19} = \begin{bmatrix} b_1 \\ b_2 \end{bmatrix} = \begin{bmatrix} 63.15 \\ 12.63 \end{bmatrix},$$
$$\begin{bmatrix} 8 \\ 3 \end{bmatrix} \frac{120}{19} = \begin{bmatrix} b_1 \\ b_2 \end{bmatrix} = \begin{bmatrix} 50.52 \\ 18.95 \end{bmatrix},$$
$$\begin{bmatrix} 6 \\ 5 \end{bmatrix} \frac{120}{19} = \begin{bmatrix} b_1 \\ b_2 \end{bmatrix} = \begin{bmatrix} 37.89 \\ 31.58 \end{bmatrix},$$
$$\begin{bmatrix} 4 \\ 8 \end{bmatrix} \frac{120}{19} = \begin{bmatrix} b_1 \\ b_2 \end{bmatrix} = \begin{bmatrix} 25.26 \\ 50.52 \end{bmatrix}.$$

These four sets of input combinations correspond to the points D, C, B, A, respectively, of Figure 7.14. The $f^o = 120/19$ isoquant is then constructed by joining these points with straight line segments.

If $c_1 = c_2 = c_3 = c_4 = \$1$, each point on ABCD (Figure 7.14) would represent both $f^o = 120/19$ units of output as well as $\$f^o = \$120/19$ in profit. With $c_1 = \$14 > 1$, point D involves more than $\$f^o = \$120/19$ profit. Hence the output of activity 1 must decrease by the factor $\frac{c_1 - 1}{c_1} = 13/14$ or a total output level of $f^o/c_1 = \frac{60}{133} = 0.451$ must be produced by activity 1. Similarly, for $c_2 = \$12 > 1$, point C also involves more than $\$f^o = \$120/19$ profit so that the output of activity 2 must be decreased by the factor $\frac{c_2 - 1}{c_2} = 11/12$ or the output level $f^o/c_2 = \frac{10}{19} = 0.526$ must be produced by the second activity. Finally, with $c_3 = \$11 > 1(c_4 = \$10 > 1)$, point B (A) yields more than $\$f^o = \$120/19$ profit so that the output of activity 3 (4) must be decreased by $\frac{c_3 - 1}{c_3} = \frac{10}{11}\left(\frac{c_4 - 1}{c_4} = \frac{9}{10}\right)$ or an output value of $f^o/c_3 = \frac{120}{209} = 0.574$ $\left(f^o/c_4 = \frac{12}{19} = 0.632\right)$ is warranted by activity 3 (4). We may plot the $\$f^o =$

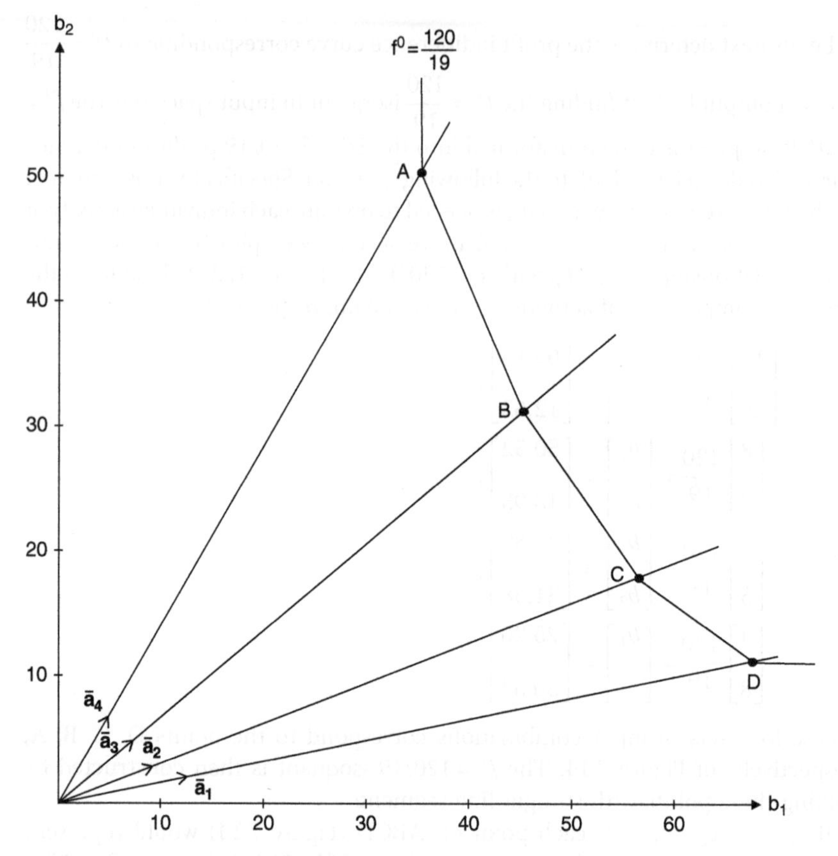

Figure 7.14 Constructing the f^o isoquant.

\$120/19 profit indifference curve in input space by performing a set of calculations similar to those undertaken to derive the $f^o = 120/19$ isoquant, i.e. again using the first two components of activities 1–4, we have, respectively,

$$\begin{bmatrix} 10 \\ 2 \end{bmatrix} 0.451 = \begin{bmatrix} b_1 \\ b_2 \end{bmatrix} = \begin{bmatrix} 4.510 \\ 0.902 \end{bmatrix},$$

$$\begin{bmatrix} 8 \\ 3 \end{bmatrix} 0.526 = \begin{bmatrix} b_1 \\ b_2 \end{bmatrix} = \begin{bmatrix} 4.208 \\ 1.578 \end{bmatrix},$$

$$\begin{bmatrix} 6 \\ 5 \end{bmatrix} 0.574 = \begin{bmatrix} b_1 \\ b_2 \end{bmatrix} = \begin{bmatrix} 3.444 \\ 2.870 \end{bmatrix},$$

$$\begin{bmatrix} 4 \\ 8 \end{bmatrix} 0.632 = \begin{bmatrix} b_1 \\ b_2 \end{bmatrix} = \begin{bmatrix} 2.528 \\ 5.056 \end{bmatrix}.$$

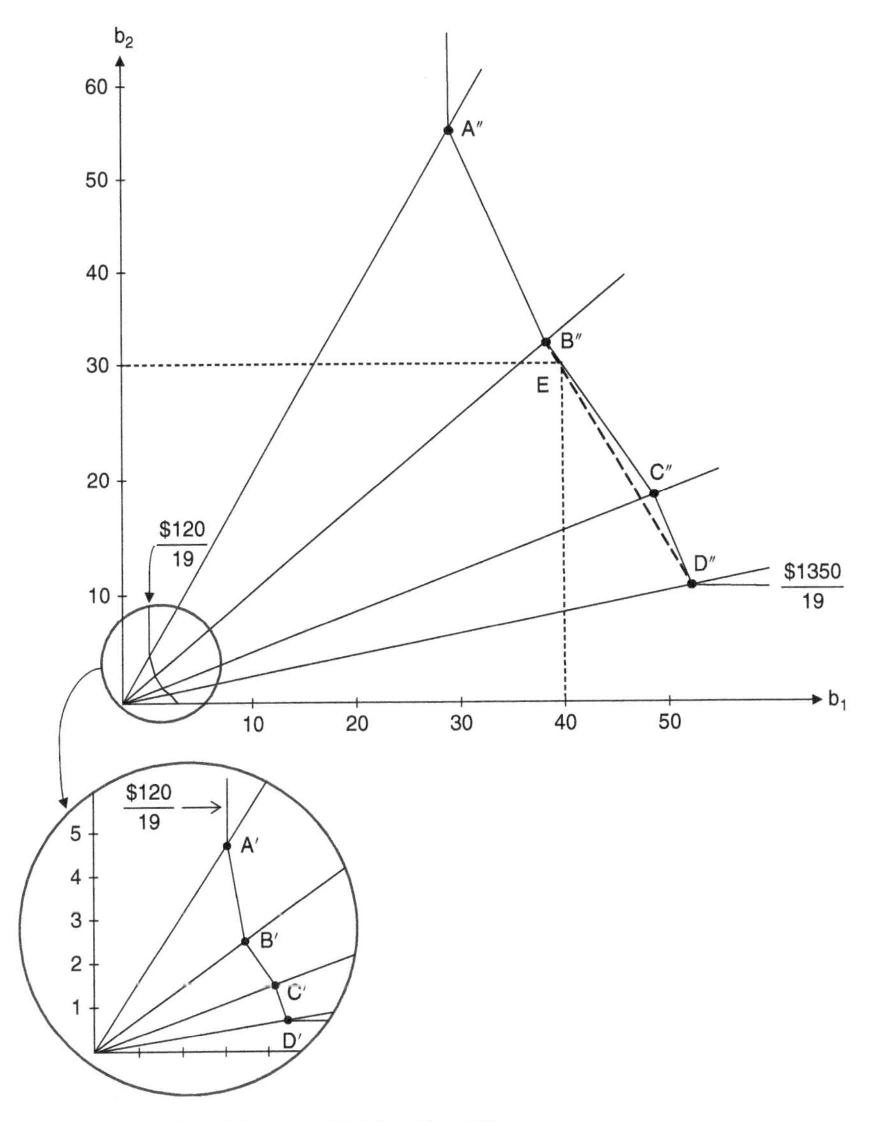

Figure 7.15 The \$120/19 and \$1350/19 profit indifference curves.

These latter sets of input combinations correspond to points D′, C′, B′, A′, respectively, of Figure 7.15. The $\$f^o = \$120/19$ profit indifference curve is then drawn by joining these points using straight line segments.

We may also find the optimal profit indifference curve corresponding to a $\$\frac{1350}{19}$ profit level by forming an appropriate radial projection of the $\$\frac{120}{19}$ profit indifference curve, i.e. since $\$\frac{1350}{19} = 11.25\left(\$\frac{120}{19}\right)$, we may increase each of the components of the points A′, B′, C′, D′ by the scalar multiple 11.25 to generate a new set of points A″, B″, C″, D″ found on the $\$\frac{1350}{19}$ profit indifference curve (Figure 7.15).[9] The coordinates of these points are, respectively,

$$11.25\begin{bmatrix}4.510\\0.902\end{bmatrix}=\begin{bmatrix}50.738\\10.148\end{bmatrix},$$

$$11.25\begin{bmatrix}4.208\\1.578\end{bmatrix}=\begin{bmatrix}47.340\\17.753\end{bmatrix},$$

$$11.25\begin{bmatrix}3.444\\2.870\end{bmatrix}=\begin{bmatrix}38.745\\32.288\end{bmatrix},$$

$$11.25\begin{bmatrix}2.528\\5.056\end{bmatrix}=\begin{bmatrix}28.440\\56.880\end{bmatrix}.$$

We noted earlier that if two separate activities are to be operated simultaneously (by forming their nonnegative linear combination) to maximize output, then these activities must be adjacent given that the isoquant is convex to the origin (the diminishing technical rate of substitution property) and the firm seeks to economize on its input utilization. Yet, if we examine Figure 7.15, activities 1 and 3 are operated simultaneously yet these activities are obviously not adjacent. How may we reconcile this apparent contradiction? The answer is that

9 If we were initially interested in only the $\$\frac{1350}{19}$ profit indifference curve, then we could have derived it in a more straightforward fashion by utilizing the structure of the initial problem itself, e.g. if from the objective function we set $f = \frac{1350}{19} = 14x_1$, then $x_1 = 5.075$ and thus, from the first two components of the $\boldsymbol{a}_1$ activity vector, we obtain

$$5.075\begin{bmatrix}10\\2\end{bmatrix}=\begin{bmatrix}b_1\\b_2\end{bmatrix}=\begin{bmatrix}50.750\\10.150\end{bmatrix}$$

as the coordinates of point D″ in Figure 7.15 (here the final b_1, b_2 values may differ slightly from the ones obtained previously because of rounding). Similarly, to get the coordinates of point C″, $f = \frac{1350}{19} = 12x_2$ yields $x_2 = 5.921$ so that

$$5.921\begin{bmatrix}8\\3\end{bmatrix}=\begin{bmatrix}b_1\\b_2\end{bmatrix}=\begin{bmatrix}47.368\\17.763\end{bmatrix}.$$

The coordinates of B″, A″ may also be obtained in like fashion.

we are now maximizing profit and not output. The relative magnitudes of the coefficients associated with the activities in the profit function play a key role in determining the shape of the profit indifference curves in that these coefficients reflect the relative profitability of the individual activities themselves. Clearly the $\$\frac{1350}{19}$ profit indifference curve in Figure 7.15 is not convex to the origin for all input combinations. In fact, the portion of this curve from B″ and D″ is concave to the origin, thus implying that activities 1 and 3 in combination are more efficient from a profitability standpoint than activity 2. That activity 2 is an inefficient profit earner when compared with a nonnegative linear combination of activities 1 and 3 follows from the fact that, for any point on B″ C″ D″, one can find a point on B″ED″ that yields the same profit yet utilizes less of each input. Thus, input combinations on B″EF″ are preferred to those on B″ C″ D″. In sum, the profit-maximizing solution is found on the $\$\frac{1350}{19}$ profit indifference curve at point E where activity 1 is operated at the $x_1 = \frac{10}{19}$ level and activity 3 is run at the $x_3 = (1 - x_1) = \frac{110}{19}$ level. ■

7.8 The Simplex Method as an Internal Resource Allocation Process

We noted earlier that the dual variables u_i, $i = 1, \ldots, l$, were interpreted as shadow prices, i.e. they amount to imputed or internal prices for the services of the fixed resources that serve to direct the allocation of these inputs within the firm itself. Hence these prices: (i) foster the efficient allocation of inputs which results in the most profitable output or activity mix; and (ii) signal the firm to expand (contract) its use of a fixed factor's services if its internal price per unit exceeds (falls short of) its unit market price. Thus, the worth to the firm of an extra unit of a particular fixed input is what that extra unit can ultimately add to total profit once it is optimally employed; it is the factor's internal worth (since ostensibly production cannot occur without it) that matters and not the factor's unit market value since the factor may be fully depreciated, though still usable, and thus command only a salvage price if it were sold outright. To repeat, the u_i values serve to allocate resources among competing activities within a given firm and not between firms as is the case with market prices.

How is this internal allocation process carried out? Let us assume that we have a basic feasible solution to a problem involving the determination of a profit-maximizing activity or product mix. Per our previous notation, if in the augmented primal structural constraint system $\boldsymbol{AX} = \boldsymbol{b}$ we partition the coefficient matrix as $A = [B \vdots R]$, where B is an $l + h$ order basis matrix, then we obtain, from the dual structural constraint system, $B'U = C_B$, i.e. for the l basic activities b_i we have $b_i'U = c_{Bi}$ or the ith basic activity just breaks even since the cost of operating activity i at the unit level exactly equals the gross profit per unit of activity $i, i = 1, \ldots, l$. Moreover, again from the dual structural constraints, let us form the difference $C_R' - U'R = C_R' - C_B'B^{-1}R$. In terms of the components of this vector, we have $n - l$ optimality evaluators $\bar{C}_j = C_{Rj} - C_B'B^{-1}r_j, j = 1, \ldots, n - l$, where r_j is the jth column of R. For those nonbasic activities for which $\bar{C}_j = C_{Rj} - C_B'B^{-1}r_j > 0$, the jth activity is a candidate for entry into the basis since, in this case, there exists an excess of gross profit per unit over imputed unit cost. Let us explore this point a bit further.

If we set $B^{-1}R = [Y_1, \ldots, Y_{n-l}]$, then $Y_j = B^{-1}r_j$ has as its ith component y_{ij}, the amount by which the ith basic variable changes in any basic solution given that the jth nonbasic variable x_{Rj} increases from zero to one $j = 1, \ldots, n - l$. In all probability, the operation of most, if not all, of the basic activities will have to be curtailed as x_{Rj} increases. Additionally, any fixed resource excess capacity, as indicated by a slack variable appearing in the basis, will typically diminish as x_{Rj} becomes basic. Hence, the y_{ij}'s measure the extent to which the output of basic activities and the quantities of unused resources are reduced if x_{Rj} turns basic. These reductions may be viewed as the real opportunity costs (in terms of outputs and resources forgone) of introducing a nonbasic activity into the activity mix at the unit level. Then the total monetary value of the real opportunity cost of introducing x_{Rj} into the set of basic variables at the unit level is $C_B'Y_j = \sum_{i=1}^{l} c_{Bi}y_{ij}, j = 1, \ldots, n - l$. Clearly, this figure represents the gross profit forgone from the most lucrative alternative production scheme, which is indicated by the current basic feasible solution. And since c_{Rj} depicts the gross profit per unit of the jth nonbasic activity, the difference $\bar{c}_j = c_{Rj} - C_B'Y_j$ or excess of gross profit per unit of activity j over the monetary value of the opportunity cost per unit of activity j, provides a test of the attractiveness, in terms of marginal profitability, of the activities not being operated, $j = 1, \ldots, n - l$. Since $\bar{c}_j$ represents the marginal gain from replacing one basic feasible solution by another, the firm thus chooses for entry into the basis the nonbasic activity with the largest $\bar{c}_j > 0$ coefficient. This selection procedure mirrors a quantity-adjustment process in the market for the firm's outputs. That is, we may think of $C_B'Y_j = U'r_j$ as the minimum profit level at which the firm is willing to operate the jth activity or to supply the jth product at the unit level. If the growth profit per unit of activity j, $\bar{c}_j$, exceeds this minimum, the firm expands the output of activity j until a limit to the capacity of at least one of its fixed resources is reached and the imputed cost per unit of the activity rises to a level coincident with the activities gross profit per unit. Hence $c_{Rj} - C_B'Y_j = c_{Rj} - U'r_j = 0$ once x_{Rj} turns basic.

7.9 The Dual Simplex Method as an Internalized Resource Allocation Process

A close examination of the dual problem and, in particular, the dual simplex routine reveals that it resembles a price-adjustment process for the internal allocation of the firm's fixed inputs. At an initial optimal basic but infeasible solution to the dual problem all dual surplus variables $u_{l+j}, j = 1, \ldots, p$, are basic, thus implying that each primal activity incurs an accounting loss (the imputed value of the inputs used to run each activity at the unit level exceeds its gross profit per unit) or, what amounts to the same thing, a subsidy must be paid to each activity in order for it to break even, i.e. in order for it to be operated at an optimal level.

In addition, since each shadow price $u_i, i = 1, \ldots, l$, is nonbasic and thus zero, there exists a positive excess demand for each scarce fixed input under the subsidy plan. (That a positive excess demand for a fixed input exists when its internal price is zero is easily understood if this price is interpreted as the rate at which management leases the services of the fixed input to various production units within, or activities undertaken by, the firm.)

To move toward optimality, we choose some basic variable to turn nonbasic. Once a dual simplex pivot operation is undertaken, the internal price of some fixed input, say u_r, turns positive (here supply equals demand for the input in question since $b_r = \sum_{j=1}^{p} a_{rj}x_j$) while demand exceeds supply for all other fixed inputs. Moreover, since some dual surplus variable or accounting loss figure turns nonbasic, its complementary activity now breaks even (the subsidy is withdrawn) while the remaining activities still require subsidies. As long as any excess demand exists for some of the firm's fixed inputs, the dual simplex process continues to raise the imputed prices of these inputs to a level which systematically eliminates these excess demands while, at the same time, adjusting the current values of the basic (dual) variables so that zero excess demand for their complementary set of fixed factors is maintained. Ultimately, a set of prices is determined that eliminates all excess demands for the fixed inputs.

7.10 A Generalized Multiactivity Profit-Maximization Model

As was the case in the model presented in Section 7.5, we shall assume that a given output level may be produced by a single activity or by a nonnegative linear combination of a finite number of activities. Thus, the fixed coefficients linear technology exhibits constant returns to scale. By taking a nonnegative linear combination of the simple activities $\boldsymbol{a}_j, j = 1, \ldots, p$, we may form the composite activity

$$\sum_{j=1}^{p} \boldsymbol{a}_j x_j = \sum_{j=1}^{p} \boldsymbol{v}_j,\ x_j \geq 0 \text{ or } \boldsymbol{AX} = \boldsymbol{V}, \boldsymbol{X} \geq \boldsymbol{O},$$

where the components of the resource requirements vector

$$\boldsymbol{V} = \begin{bmatrix} v_1 \\ \vdots \\ v_m \end{bmatrix} = \begin{bmatrix} \sum_{j=1}^{p} v_{1j} \\ \vdots \\ \sum_{j=1}^{p} v_{mj} \end{bmatrix}$$

represents the total amount of each of the m resources required to support the j activities at their respective levels x_j. Again, taking the first l factors as fixed in quantity (the short run still prevails), the preceding composite activity may be rewritten as

$$\sum_{j=1}^{p} a_{ij} x_j \leq b_i,\ i = 1, \ldots, l,$$
$$\sum_{j=1}^{p} a_{ij} x_j \leq v_i,\ i = l+1, \ldots, m.$$

Under the assumption of perfect competition in the product and factor markets, the price per unit of activity j is p_j = constant, $j = 1, \ldots, p$. Thus, the total revenue from all p activities is $TR = \sum_{j=1}^{p} p_j x_j = \boldsymbol{P}'\boldsymbol{X}$, where p_j is the j component of the $(p \times 1)$ vector $\boldsymbol{P}$. And with the prices q_i, $i = l+1, \ldots, m$, of the $m - l$ variable inputs also treated as constants, the cost of input i incurred by operating activity j at the x_j level is $q_i a_{ij} x_j$. Then the total variable cost of operating activity j at the x_j level is $TVC_j = \sum_{i=l+1}^{m} q_i a_{ij} x_j = \boldsymbol{Q}'\boldsymbol{a}_j^* x_j$, where the ith component of the $(m - l \times 1)$ vector $\boldsymbol{Q}$ is q_i and the components of the $(m - l \times 1)$ vectors $\boldsymbol{a}_j^*$ are the last $m - l$ components of $\boldsymbol{a}_j$, $j = 1, \ldots, p$. Upon summing over all p activities, the total variable input cost is $TVC = \sum_{j=1}^{p} \left(\sum_{i=l+1}^{m} q_i a_{ij} \right) x_j = \boldsymbol{Q}'\boldsymbol{A}^*\boldsymbol{X}$, where the matrix $\boldsymbol{A}^* = [\boldsymbol{a}_1^*, \ldots, \boldsymbol{a}_p^*]$ is of order $(m - l \times p)$. One additional cost dimension will now be admitted to our discussion (Pfouts 1961). This new cost element pertains to the cost of converting or switching one unit of the ith fixed factor for use in the jth activity. Specifically, this type of cost does not strictly belong to either the variable or fixed categories but changes only as the activity mix or product mix of the firm is altered. If r_{ij} = constant represents the cost of converting one unit of the ith fixed factor to the jth activity, then $r_{ij} a_{ij}$ depicts the cost of converting a_{ij} units of the ith fixed factor to the jth activity,

$i = l + 1, \ldots, m; j = 1, \ldots, p$. Hence the cost of converting the $m - l$ fixed factors to the operation of the jth activity at the unit level is $h_j = \sum_{i=l+1}^{m} r_{ij} a_{ij} = \boldsymbol{r}_j' \bar{\boldsymbol{a}}_j$, where r_{ij} is the ith component of the $(m - l \times 1)$ vector $\boldsymbol{r}_j$ and the components of the $(l \times 1)$ vector $\bar{\boldsymbol{a}}_j$ are the first l components of the jth activity $\boldsymbol{a}_j$. Then the **total conversion cost for activity *j*** is $TCC_j = h_j x_j$. Upon summing over all p activities, the **total conversion cost** is thus $TCC = \sum_{j=1}^{p} h_j x_j = \boldsymbol{H}'\boldsymbol{X}$, where the jth component of the $(p \times 1)$ vector $\boldsymbol{H}$ is h_j.

Again borrowing from the model presented in (7.5), $q_i = 0$ for all fixed inputs since the contribution of the fixed resources to total factor cost is based on the notion of an implicit opportunity cost calculation (i.e. shadow or accounting prices are assigned to the l fixed inputs). Hence, the resulting fixed charge amounts to the total imputed value of the firm's fixed resources.

Based on the previous discussion, the **generalized short-run fixed-coefficients profit-maximization model** under perfect competition assumes the form

$$\begin{aligned} max f(x_1,\ldots,x_p) &= TR - TVC - TCC = \sum_{j=1}^{p} p_j x_j - \sum_{j=1}^{p}\left(\sum_{i=l+1}^{m} q_i a_{ij}\right) x_j - \sum_{j=1}^{p} h_j x_j \\ &= \sum_{j=1}^{p}\left(p_j - \sum_{i=l+1}^{m} q_i a_{ij} - h_j\right) x_j \quad \text{s.t.} \\ &\sum_{j=1}^{p} a_{ij} x_j \le b_i, i = 1, \ldots, l, \\ &x_j \ge 0, j = 1, \ldots, p, \end{aligned} \tag{7.27}$$

where the gross profit margin per unit of activity j is $p_j - \sum_{i=l+1}^{m} q_i a_{ij} - h_j$. Alternatively, (7.27) may be rewritten in matrix form as

$$\begin{aligned} max f(\boldsymbol{X}) &= \boldsymbol{P}'\boldsymbol{X} - \boldsymbol{Q}'\boldsymbol{A}^*\boldsymbol{X} - \boldsymbol{H}'\boldsymbol{X} = \left(\boldsymbol{P} - (\boldsymbol{A}^*)'\boldsymbol{Q} - \boldsymbol{H}\right)' \boldsymbol{X} \text{ s.t.} \\ &\bar{\boldsymbol{A}}\boldsymbol{X} \le \boldsymbol{b}, \boldsymbol{X} \ge \boldsymbol{O}, \end{aligned} \tag{7.27.1}$$

where the matrix $\bar{\boldsymbol{A}} = [\bar{\boldsymbol{a}}_1, \ldots, \bar{\boldsymbol{a}}_p]$ is of order $(l \times p)$.

To generate a solution to this primal problem, let us form the primal Lagrangian of f as

$$\begin{aligned} L(\boldsymbol{X}, \boldsymbol{S}_1, \boldsymbol{S}_2, \boldsymbol{U}_1, \boldsymbol{U}_2) &= \left(\boldsymbol{P} - (\boldsymbol{A}^*)'\boldsymbol{Q} - \boldsymbol{H}\right)'\boldsymbol{X} \\ &\quad + \boldsymbol{U}_1'\left(\boldsymbol{b} - \bar{\boldsymbol{A}}\boldsymbol{X} - \boldsymbol{S}_1\right) + \boldsymbol{U}_2'(\boldsymbol{X} - \boldsymbol{S}_2), \end{aligned}$$

where

$$S_1 = \begin{bmatrix} x_{p+1}^2 \\ \vdots \\ x_{p+l}^2 \end{bmatrix}, S_2 = \begin{bmatrix} x_{p+1+1}^2 \\ \vdots \\ x_{2p+l}^2 \end{bmatrix}$$

are, respectively, $(l \times 1)$ and $(p \times 1)$ vectors of squares of slack variables (to ensure their nonnegativity) and

$$\boldsymbol{U}_1 = \begin{bmatrix} u_{11} \\ \vdots \\ u_{l1} \end{bmatrix}, \boldsymbol{U}_2 = \begin{bmatrix} u_{12} \\ \vdots \\ u_{p2} \end{bmatrix}$$

are, respectively, $(l \times 1)$ and $(p \times 1)$ vectors of Lagrange multipliers. Then the Kuhn-Tucker-Lagrange necessary and sufficient conditions (Panik 1976, pp. 241–252) for the points $\boldsymbol{X}_o \varepsilon \mathcal{E}^p$, $\boldsymbol{U}_1^o \varepsilon \mathcal{E}^l$ to solve (7.27.1) are

$$\begin{aligned}
&\text{(a) } (\boldsymbol{P} - (\boldsymbol{A}^*)'\boldsymbol{Q} - \boldsymbol{H}) - \bar{\boldsymbol{A}}'\boldsymbol{U}_1^o \leq \boldsymbol{O} \\
&\text{(b) } \boldsymbol{X}_o' \left[(\boldsymbol{P} - (\boldsymbol{A}^*)'\boldsymbol{Q} - \boldsymbol{H}) - \bar{\boldsymbol{A}}'\boldsymbol{U}_1^o \right] = 0 \\
&\text{(c) } (\boldsymbol{U}_1^o)'(\boldsymbol{b} - \bar{\boldsymbol{A}}\boldsymbol{X}_o) = 0 \\
&\text{(d) } \boldsymbol{b} - \bar{\boldsymbol{A}}\boldsymbol{X}_o \geq \boldsymbol{O} \\
&\text{(e) } \boldsymbol{U}_1^o \geq \boldsymbol{O} \\
&\text{(f) } \boldsymbol{X}_o \geq \boldsymbol{O}.
\end{aligned} \tag{7.28}$$

In terms of the individual components of the matrices presented in (7.28), an alternative representation of this necessary and sufficient condition is

$$\begin{aligned}
&\text{(a) } \left(p_j - \sum_{i=l+1}^{m} q_i a_{ij} - h_j \right) - \sum_{i=1}^{l} u_{i1}^o a_{ij} \leq 0, j = 1, \ldots, p \\
&\text{(b) } \sum_{j=1}^{p} x_j^o \left[\left(p_j - \sum_{i=l+1}^{m} q_i a_{ij} - h_j \right) - \sum_{i=1}^{l} u_{i1}^o a_{ij} \right] = 0 \\
&\text{(c) } \sum_{i=1}^{l} u_{i1}^o \left(b_i - \sum_{j=1}^{p} a_{ij} x_j^o \right) = 0 \\
&\text{(d) } b_i - \sum_{j=1}^{p} a_{ij} x_j^o \geq 0, i = 1, \ldots, l \\
&\text{(e) } u_{i1}^o \geq 0, i = 1, \ldots, l \\
&\text{(f) } x_j^o \geq 0, j = 1, \ldots, p.
\end{aligned} \tag{7.28.1}$$

By way of an interpretation of these optimality conditions presented in (7.28.1), (a) indicates that positive profits are not permitted for any of the p activities, i.e. the gross profit margin for the jth activity cannot exceed the imputed cost per unit of the jth activity, $j = 1, \ldots, p$. From (b), if the jth activity is unprofitable in terms of the shadow price $u_{i1}^0, i = 1, \ldots, l$ (the associated accounting loss figure for activity j is positive), then this activity is not operated so that we must have $x_j^0 = 0, j = 1, \ldots, p$. However, if the jth activity breaks even in terms of the shadow prices u_{i1}^0, $i = 1, \ldots, l$ (the accounting loss for this activity is zero), then activity j may be run at a positive level so that $x_j^0 > 0, j = 1, \ldots, p$. Looking to (c), if equality holds in the ith structural constraint (the associated slack variable is zero), then the shadow price u_{i1}^0 can be positive, i.e. the firm would be willing to pay up to \$$u_{i1}^0$ to have the ith structural constraint relaxed by one unit, $i = 1, \ldots, l$. But if the ith structural constraint admits excess capacity (the corresponding slack variable is positive so that the structural constraint holds as an equality), then u_{i1}^0 must be zero, $i = 1, \ldots, l$. (As the reader may have already noted, conditions (b), (c) simply represent the usual complementary slackness conditions.)

Turning to the symmetric dual of (7.27.1) we have

$$
\begin{aligned}
&\min g(\boldsymbol{U}_1) = \boldsymbol{b}'\boldsymbol{U}_1 \quad \text{s.t.} \\
&\bar{\boldsymbol{A}}'\boldsymbol{U}_1 \geq \boldsymbol{P} - (\boldsymbol{A}^*)'\boldsymbol{Q} - \boldsymbol{H} \\
&\boldsymbol{U}_1 \geq \boldsymbol{O}
\end{aligned}
$$

or, in terms of the individual components of these matrices,

$$
\begin{aligned}
&\min g(u_{11}, \ldots, u_{l1}) = \sum_{i=1}^{l} b_i u_{i1} \quad \text{s.t.} \\
&\sum_{i=1}^{l} u_{i1} a_{ij} \geq p_j - \sum_{i=l+1}^{m} q_i a_{ij} - h_j, \; j = 1, \ldots, p \\
&u_{i1} \geq 0, \; i = 1, \ldots, l.
\end{aligned}
$$

The standard economic interpretation of this dual problem holds, i.e. the firm attempts to find imputed values or shadow prices u_{i1}^0, $i = 1, \ldots, l$, for its fixed resources that minimizes their total (imputed) value given the requirement that the imputed cost of the fixed factors' services per unit of activity j be at least as great as the gross profit margin for activity j, $j = 1, \ldots, p$.

7.11 Factor Learning and the Optimum Product-Mix Model

In the preceding sections, we examined a variety of optimum product-mix models constructed under the assumption of constant returns to scale in production, i.e. for each activity $\boldsymbol{a}_j$ the set of a_{ij} input/output coefficients is assumed constant so that increasing each of the factors by the same proportion θ results in an increase in the output produced by that activity by the same proportion. However, an abundance of statistical evidence (Nanda and Alder 1982) has indicated that the aforementioned strict proportionality assumption may, in reality, be inappropriate – especially in instances where special projects are undertaken that involve newly designed products and somewhat unfamiliar production techniques. In particular, the existence of **learning economies** in production is well documented, e.g. the amount of input required per unit of output decreases by a constant percentage each time total output is doubled. Before we embellish the standard linear programming optimum product-mix model to take account of this phenomenon, let us examine the properties of one of the usual learning or progress functions typically used in empirical work.

Let us assume that the **learning rate** is Ø%, i.e. the cumulative average number of units of a resource needed to "double" the number of units produced is Ø% of the amount required to produce the previous increment (doubling) of output. For instance, if it took K units of some input to produce the first unit of output and the learning rate is Ø%, then the cumulative average number of units needed to double the current output from one to two units is $K \times Ø\%$. (Since this is the average number of units of some input needed to produce a unit of output, the total amount of input used is $K \times Ø\% \times 2$.) To again "double" production, this time from two to four units, the cumulative average number of input units needed decreases to Ø% of the previous average or $(K \times Ø\%) \times Ø\% = \mathrm{K}\,(Ø\%)^2$. (Here the total amount of the input used is $K \times Ø\% \times Ø\% \times 4$.) As an example, let us assume that the first unit of output required $K = 100$ units of an input and that the learning rate is 80%. Table 7.1 then summarizes the preceding discussion. It is evident from the second column of Table 7.1 that the values presented therein represent K times the learning rate (Ø%) to a power, the latter being the number of times output is doubled. Moreover, from Figure 7.16, the

Table 7.1 Output vs. Input in the presence of learning.

Number of units of output produced (N)	Cumulative average number of units of input needed (Y)	Total amount of input used ($N \times Y$)
1	$K = 100$	100
2	$K \times Ø\% = 100\ (0.80) = 80$	$K \times Ø\% \times 2 = 160$
4	$K \times (Ø\%)^2 = 100\ (0.64) = 64$	$K \times (Ø\%)^2 \times 4 = 256$
8	$K \times (Ø\%)^3 = 100\ (0.512) = 51.2$	$K \times (Ø\%)^3 \times 8 = 409.6$
⋮	⋮	⋮

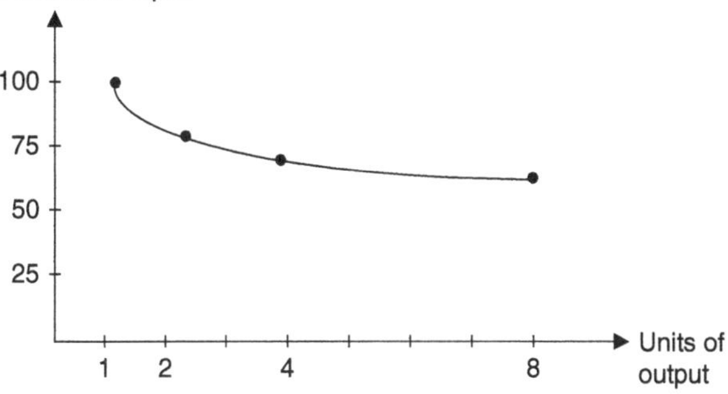

Figure 7.16 Negative exponential learning curve.

functional relationship connecting the indicated variables is the **negative exponential**, i.e.

$$Y = KN^b, N \geq 1, -\frac{1}{2} < b \leq 0, \tag{7.29}$$

where b, the **learning index** or **progress elasticity**, is the exponent associated with the learning rate (Ø%). To determine b, let Y_i, Y_{i+1} represent two successive ordinate values in Figure 7.16. Then $(Y_i/Y_{i+1}) = (N_i/N_{i+1})^b$. Since $N_{i+1} = 2N_i$, $(Y_i/Y_{i+1}) = 2^{-b}$ and thus

$$b = \frac{\log(Y_i/Y_{i+1})}{\log 2} = -\frac{\log Ø}{\log 2}. \tag{7.30}$$

For instance, from Table 7.1, if $Y_i = 80$, $Y_{i+1} = 64$, then $b = -\log 1.25/\log 2 = -0.322$ and thus

$$Y = 100N^{-0.322}.$$

Looked at in another fashion, if a learning rate of Ø% is given (say, Ø% = 0.90), then b can be obtained from the expression $2^b = 0.90$ and thus $b = \log 0.90/\log 2 = (-0.0458/0.3010) = -0.1522$. And if b is determined empirically (say $b = -0.2566$) from (7.29) using

$$\log Y = \log K + b \log N, \tag{7.29.1}$$

then the learning rate can be found from the percentage slope of the logarithmic learning curve (7.29.1) as $2^b = 2^{-0.2566} = 0.8371$, i.e. we have an 83.71% = Ø% learning rate. In addition, we may determine the *total amount of input used to produce any cumulative level of output* N_i as $N_iY_i = KN_i^{1+b}$ (column 3 of Table 7.1) while the *(expected) amount of input needed to produce an individual unit of output* is $Y_i = K\left[N_i^{1+b} - (N_1 - 1)^{1+b}\right]$. For example, the amount of input needed to produce the third unit of output is $Y_3 = 100(3^{0.678} - 2^{0.678}) = 50.61$ units. Moreover, it is easily shown that, say $Y_1 + Y_2 + Y_3 + Y_4 = 100 + 60 + 50.61 + 45.39 = 256$, the total amount of input used to produce four units of output (see the last column of Table 7.1).

Let us now look to the incorporation of learning into the standard linear programming optimum product mix model (Liao 1979; Reeves 1980; Reeves and Sweigart 1981). As mentioned above, we shall modify the assumption of strict proportionality between inputs and output (to reflect the possibility that the amount of input required per unit of output decreases as total output increases) as well as the assumption of a constant gross profit contribution for each product. For the linear program at hand, let us start with

$$\begin{aligned} &max\, f = \sum_{j=1}^{p} c_j x_j \quad \text{s.t.} \\ &\sum_{j=1}^{p} a_{ij} x_j \le b_i,\ i = 1, \ldots, l, \\ &x_j \ge 0,\ j = 1, \ldots, p \end{aligned} \tag{7.31}$$

(see (7.25)) and assume that: (i) each successive unit of a given product requires a decreasing number of units of those fixed resources subject to learning (i.e. if the first r of the l fixed inputs are subject to learning, then the constant a_{ij} coefficient is replaced by

$$a_{ij}^* = a_{ij} x_j^{b_{ij}},\ i = 1, \ldots, r \le l; j = 1, \ldots, p, \tag{7.32}$$

where a_{ij}^* is the cumulative average number of units of resource i required to produce one unit of product j when the total production of the product is x_j units, $a_{ij} > 0$ is the number of units of resource i required to produce the first unit of product j, and b_{ij}, $-\frac{1}{2} < b_{ij} \le 0$, is the progress elasticity of resource i in the production of product j); and (ii) the gross profit contribution of product j is an increasing function of the level of production of that product (i.e. if the first s of the $m - l$ variable inputs are subject to learning, then the objective function coefficient $c_j = p_j - TVC_j = p_j - \sum_{i=l+1}^{m} q_i a_{ij}$ is replaced by

$$c_j^* = p_j - TVC_j^o - \sum_{i=l+1}^{l+s} q_i a_{ij}^*,\ j = 1, \ldots, p, \tag{7.33}$$

where $TVC_j^o = \sum_{i=l+s+1}^{m} q_i a_{ij}$ is that portion of unit variable cost attributed to those resources not subject to learning). In view of these adjustments, (7.31) becomes

$$\begin{aligned}
max\, f^* &= \sum_{j=1}^{p} c_j^* x_j = \sum_{j=1}^{p} \left[p_j - TVC_j^0 - \sum_{i=l+1}^{l+s} q_i a_{ij}^* \right] x_j \\
&= \sum_{j=1}^{p} \left[\left(p_j - TVC_j^o \right) x_j - \sum_{i=l+1}^{l+s} q_i a_{ij} x_j^{1+b_{ij}} \right] \quad \text{s.t.} \\
&\sum_{j=1}^{p} a_{ij}^* x_j = \sum_{j=1}^{p} a_{ij} x_j^{1+b_{ij}} \le b_i,\ i = 1, \ldots, r, \\
&\sum_{j=1}^{p} a_{ij} x_j \le b_i,\ i = r+1, \ldots, l, \\
&x_j \ge 0,\ j = 1, \ldots, p.
\end{aligned} \tag{7.34}$$

How does the incorporation of learning in (7.31) affect the structure of the optimum product-mix model? First, it is evident that (7.34) is not a linear program as is (7.31), i.e. f^* is a convex function of the x_js and the first r structural constraints are concave in the same arguments (for details on a numerical technique for solving (7.34) see Reeves and Sweigart 1981). Second, the feasible region corresponding to (7.34) is larger than that associated with (7.31) (with learning, more of each product can be produced with the same resource base). Next, although the feasible region for (7.31) is a convex set, the set of admissible solutions in (7.34) is nonconvex. Finally, while a local solution to (7.31) is also global in character, such is not the case for (7.34); in fact, there may exist multiple optimal local solutions to (7.34).

7.12 Joint Production Processes

In previous sections of this chapter, a multiactivity profit maximization model was presented wherein each individual product was produced by a single activity, i.e. the various activities were assumed to be **technologically independent**. However, if we now assume that the production activities are **technologically interdependent** in that each product can be produced by employing more than one process (here we have a separate set of processes available for the production of each product), then the firm's objective is to choose those processes that will fulfil the various production quotas at minimum resource cost.

If we have a single scarce input that is used in fixed amounts by j separate processes to produce k different products, then the jth $(k \times l)$ vector of output (-input) coefficients or **output activity vector** (defined as a particular way of using a given input) may be expressed as

$$\boldsymbol{b}_j = \begin{bmatrix} b_{ij} \\ \vdots \\ b_{kj} \end{bmatrix}, j = 1, \ldots, p, \tag{7.35}$$

where b_{ij}, $i = 1, \ldots, k$, represents the maximum amount of the ith product produced by the jth process for each unit of input employed. So if one unit of some input is employed by activity j, b_{1j} units of product 1 are produced, b_{2j} units of product 2 are produced, and so on. For a typical output activity $\mathbf{b}_r$ we may express the relationship between the outputs and the output coefficients b_{ir} as $b_{ir}x_r = g_{ir}$, $i = 1, \ldots, k$, or

$$\mathbf{b}_r x_r = \begin{bmatrix} b_{ir} \\ \vdots \\ b_{kr} \end{bmatrix} x_r = \begin{bmatrix} g_{ir} \\ \vdots \\ g_{kr} \end{bmatrix} = \mathbf{g}_r, \tag{7.36}$$

where x_r is the **level of resource utilization** (i.e. the amount of input used) by the rth process and g_{ir}, the ith component of the $(k \times 1)$ *output vector* $\mathbf{g}_r$, is the amount of the ith product produced by the rth process when x_r units of the input are employed. It is assumed that x_r and g_{ir} vary continuously so that the scarce input as well as the various outputs are **perfectly divisible**. Moreover, from (7.35), there exists strict proportionality between the input level x_r and the outputs g_{ir}, where the constants of proportionality are the b_{ir} values. In this regard, it is evident that if the outputs g_{ir} are increased by a fixed percentage, then the input level required by the process must increase by the same percentage.

From the structure of the process $\mathbf{b}_r$ it is clear that the various outputs are **limitational** in nature. That is to say, no output substitution is admitted by this fixed-coefficients technology so that producing more of one output and less of another is not technically feasible. If the output levels are to be changed, they must all change together so that the outputs are **perfectly complementary**. Thus, from (7.36), for $g_{ir} = b_{ir}x_r$, $i = 1, \ldots, k$, we must have $g_{ir}/g_{kr} = b_{ir}/b_{kr} =$ constant, $i = 1, \ldots, k-1$, i.e. once the level of the kth output is given, the levels of the remaining $k-1$ products are uniquely determined.

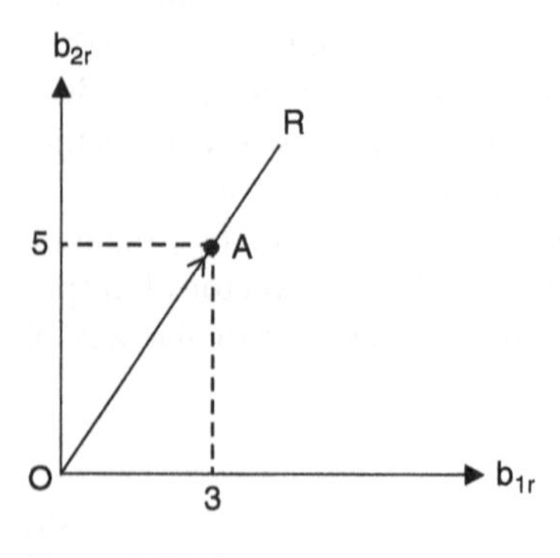

Figure 7.17 Output process ray.

For example, if $\mathbf{b}'_r = (b_{1r}, b_{2r}) = (3,5)$, then the rth output activity vector is the locus of all output combinations involving unchanged output proportions, i.e. it corresponds to the **output process ray** OR through the origin (Figure 7.17) with slope $b_{2r}/b_{1r} = 5/3$. Thus, activity r produces 5/3 unit of output 2 per unit of output 1 produced. Here point A depicts one unit of the single input used by (i.e. it is the unit level of) process r.

7.13 The Single-Process Product Transformation Function

Given the output activity $\boldsymbol{b}_r' = (b_{1r}, b_{2r})$, if we define the associated output quantities on the right-hand side of (7.36) as those which are minimally acceptable (i.e. they are *output quotas* for the first and second products) per production period, then $\boldsymbol{b}_r x_r = \boldsymbol{g}_r$ is replaced by

$$\boldsymbol{b}_r x_r = \begin{bmatrix} b_{1r} \\ b_{2r} \end{bmatrix} x_r \geq \begin{bmatrix} q_1 \\ q_2 \end{bmatrix} = \boldsymbol{q}.$$

(Note that the components within $\boldsymbol{g}_r$ represent the output quantities produced when the rth production activity is operated at the x_r level and thus correspond to points on the ray OR while the components of $\boldsymbol{q}$ depict output lower limits, which, in effect, provide a lower bound on the level of operation of the rth activity. Clearly $q_1 \leq g_{1r}, q_2 \leq g_{2r}$.) In this regard, the single factor **iso-input** or **transformation curve** (depicting the locus of all possible output combinations resulting from the employment of a "fixed" amount of the input) may be obtained by defining x_r as the minimum amount of the input needed to meet the production quotas q_1, q_2. Since $b_{1r}x_r \geq q_1$, $b_{2r}x_r \geq q_2$ may be converted to

$$x_r \geq q_1/b_{1r}, \; x_r \geq q_2/b_{2r}, \tag{7.37}$$

respectively, the lower limit on x_r will correspond to the largest of the ratios given in (7.37) or

$$x_r = \max\left\{\frac{q_1}{b_{1r}}, \frac{q_2}{b_{2r}}\right\}. \tag{7.38}$$

An examination of the transformation curve BAC in Figure 7.18 reveals that any output combination on it may be produced using the input activity level x_r'. At point A, the outputs produced are exactly $g_{1r} = b_{1r}x_r', g_{2r} = b_{2r}x_r'$. At point D (E), output 1 (2) will have to be disposed of or destroyed costlessly. Hence, point A is the only **technologically efficient** point on the transformation curve in that it maximizes the quantities of the outputs obtained when the input is used at the x_r' level, i.e. no output can be increased without increasing the level of resource utilization while the latter cannot be decreased without decreasing at least one output. Moreover, point A is also **economically efficient** since it minimizes the resource cost of obtaining this output combination.

If we vary the level of resource utilization from x_r' to x_r'' and so on, then each point on OR may be interpreted as an economically efficient output combination. Thus, OR may be viewed as the firm's **joint output expansion path** – the locus of economically efficient output combinations corresponding to successively higher levels of resource utilization (and thus to successively higher levels

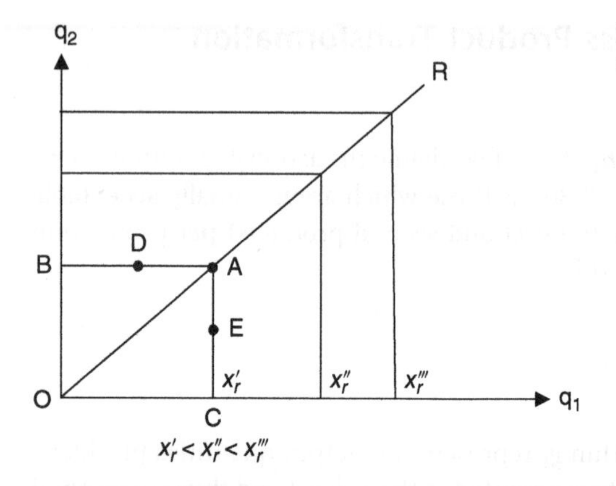

Figure 7.18 Output transformation curves.

of resource cost). Along ray OR, the outputs are produced in the ratio b_{2r}/b_{1r} and since $b_{2r}/b_{1r} = q_2/q_1$, the equation of the expansion path is $q_2 = q_1\, b_{2r}/b_{1r}$.

Let us hold the level of output 2 fixed at $\bar{q}_2$(Figure 7.19a). Then (7.38) may be written as

$$x_r = \begin{cases} \dfrac{\bar{q}_2}{b_{2r}}, q_1 < \dfrac{\bar{q}_2 b_{1r}}{b_{2r}}, \\ \dfrac{q_1}{b_{1r}}, q_1 \geq \dfrac{\bar{q}_2 b_{1r}}{b_{2r}} \end{cases} \tag{7.39}$$

since, for $q_2 = \bar{q}_2$, the maximum amount of q_1 produced before mutual limitationality between the outputs occurs is $q_1 = \bar{q}_2 b_{1r}/b_{2r}$. Here (7.39) represents the **total (real) resource cost function** of output 1 given a fixed level of output 2 (Figure 7.19b). Notice that as we move along the line $\bar{q}_2$, the input requirement remains constant at x_r'' until we come to point A; beyond A we encounter successively higher transformation curves, and thus the input requirement increases at a fixed rate.

From (7.39) we may determine the **marginal (real) resource cost function** for output 1 as

$$MC_1 = \frac{dx_r}{dq_1} = \begin{cases} 0, q_1 < \dfrac{\bar{q}_2 b_{1r}}{b_{2r}}, \\ \dfrac{1}{b_{1r}}, q_1 \geq \dfrac{\bar{q}_2 b_{1r}}{b_{2r}} \end{cases} \tag{7.40}$$

(Figure 7.19c) while the **average (real) resource cost function** for output 1 (Figure 7.19c) is

$$AC_1 = \frac{x_r}{q_1} = \begin{cases} \dfrac{\bar{q}_2}{q_1 b_{2r}}, q_1 < \dfrac{\bar{q}_2 b_{1r}}{b_{2r}}, \\ \dfrac{1}{b_{1r}}, q_1 \geq \dfrac{\bar{q}_2 b_{1r}}{b_{2r}}. \end{cases} \tag{7.41}$$

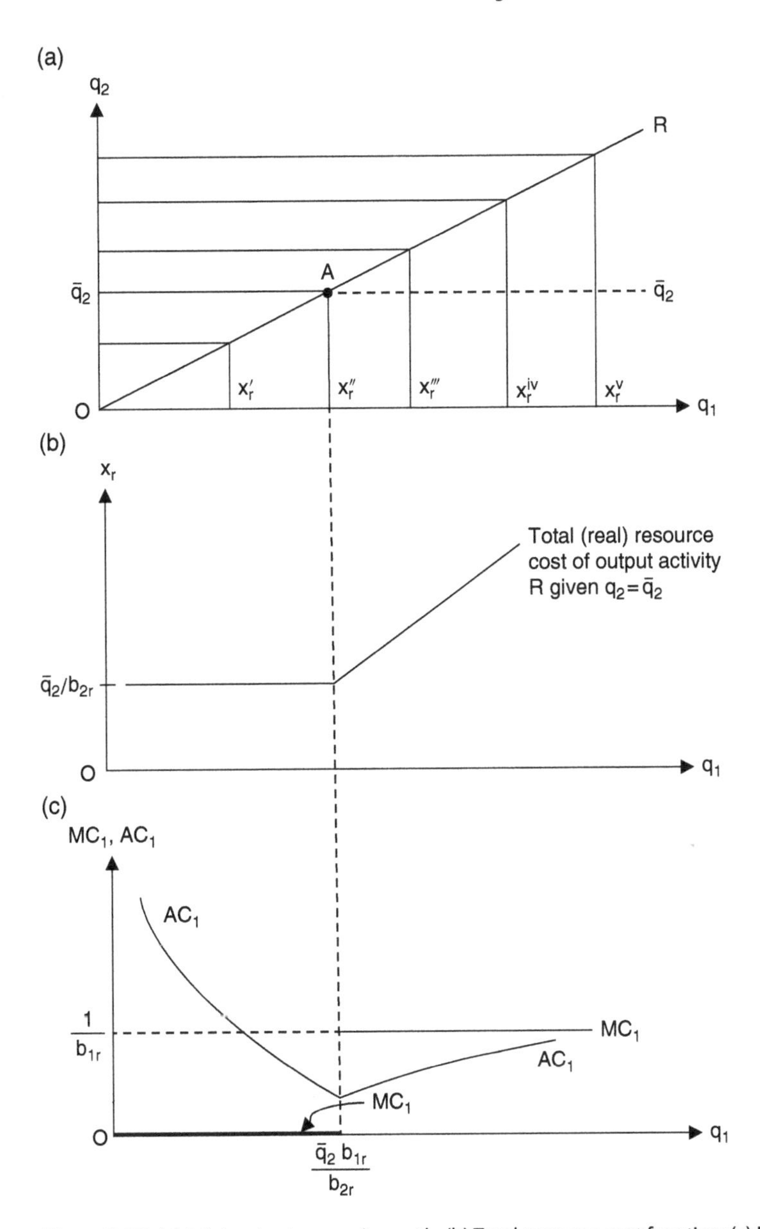

Figure 7.19 (a) Joint output expansion path; (b) Total resource cost function; (c) Marginal and average resource cost functions.

It is also of interest to note that the **total product function** of the input in the production of good 1 is

$$TP_1 = q_1 = \begin{cases} \text{indeterminate}, x_r = \bar{q}_2/b_{2r}, \\ x_r b_{1r}, x_r > \bar{q}_2/b_{2r} \end{cases} \tag{7.42}$$

(Figure 7.20a), from which we may obtain the **marginal product function** of the input in the production of output 1 as

$$MP_1 = \frac{dq_1}{dx_r} = \frac{1}{MC_1} = \begin{cases} \text{undefined}, x_r = \bar{q}_2/b_{2r}, \\ b_{1r}, x_r > \bar{q}_2/b_{2r} \end{cases} \tag{7.43}$$

(Figure 7.20b).

Example 7.4 Given the single output activity vector $\boldsymbol{b}_1' = (5,3)$ and the output quota vector $\boldsymbol{q}' = (15, 20)$, let us derive the total, marginal, and average (real) resource cost functions for product 1. Using (7.39), (7.40), and (7.41), respectively (with $q_2 = \bar{q}_2 = 20$):

$$x_1 = \begin{cases} 20/3, & q_1 < 100/3, \\ 3, & q_1 \geq 100/3; \end{cases}$$

$$MC_1 = \begin{cases} 0, & q_1 < 100/3, \\ 1/5 & q_1 \geq 100/3; \text{ and} \end{cases}$$

$$AC_1 = \begin{cases} 4/15, & q_1 < 100/3, \\ 1/5 & q_1 \geq 100/3. \end{cases}$$

■

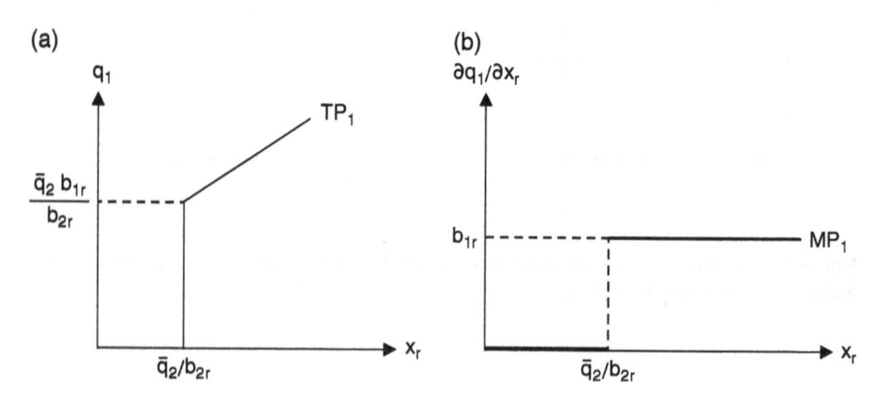

Figure 7.20 (a) Total product function; (b) Marginal product function.

7.14 The Multiactivity Joint-Production Model

In this section we shall generalize the above iso-input or single-process output transformation function to the case where the outputs may be produced by a single process or by a linear combination of a finite number of output processes. If each process or combination of processes used produces the same outputs in fixed proportions using a single variable input, then, for the unit levels of output activities $\boldsymbol{b}_j$, $j = 1, \ldots, p$, their nonnegative linear combination is, using (7.35), the **composite output activity**

$$\begin{aligned} &\sum_{j=1}^{p} \boldsymbol{b}_j x_j = \sum_{j=1}^{p} \boldsymbol{g}_j,\ x_j \geq 0, \text{ or} \\ &\boldsymbol{BX} = \boldsymbol{G}, \boldsymbol{X} \geq \boldsymbol{O}, \end{aligned} \tag{7.44}$$

where the **output technology matrix** of output–input coefficients $\boldsymbol{B} = [\boldsymbol{b}_1, \ldots, \boldsymbol{b}_p]$ is of order $(k \times p)$, the $(p \times 1)$ *activity vector* $\boldsymbol{X}$ has as its components the resource utilization levels x_j, and the elements within the $(k \times 1)$ output vector

$$\boldsymbol{G} = \begin{bmatrix} g_1 \\ \vdots \\ g_k \end{bmatrix} = \begin{bmatrix} \sum_{j=1}^{p} g_{ij} \\ \vdots \\ \sum_{j=1}^{p} g_{kj} \end{bmatrix}.$$

represent the total amounts of each of the k products produced by the j output activities when the latter are operated at the resource utilization levels x_j. In view of the structure of (7.44), it is clear that

$$g_i = \sum_{j=1}^{p} g_{ij} = \sum_{j=1}^{p} b_{ij} x_j,\ i = 1, \ldots, k. \tag{7.45}$$

On the basis of (7.44), the composite output activity utilizes a total amount of the input corresponding to

$$f = \sum_{j=1}^{p} x_j = \boldsymbol{1}'\boldsymbol{X} \tag{7.46}$$

(x_j is the amount of the input used by process j) and generates the k outputs in the amounts provided by the components of $\boldsymbol{G}$. Furthermore, the output activities are assumed to be **independent** in that the outputs produced by a given activity do not depend on the levels of resource utilization employed by the other activities. Upon consolidating the above discussion, the **joint output linear production model** may be summarized as

$$f = \sum_{j=1}^{p} x_j$$
$$g_i = \sum_{j=1}^{p} b_{ij} x_j, \; i = 1, \ldots, k, \tag{7.47}$$
$$x_j \geq 0, \; j = 1, \ldots p.$$

While output substitution within any given output activity vector is inadmissible, since the k outputs are limitational, the possibility of output substitution when more than one output activity is present is certainly possible, i.e. **process substitution** may occur in the sense that, for a given nonnegative linear combination of output activities, if the level of one output activity is increased, then the level of operation of another must be decreased in order to preserve the same level of utilization of the input. But this can occur only if more of one output is produced at the expense of another. To see exactly how this process of output substitution works, let us restrict our movements to the **joint process transformation curve** $f = \bar{f}$(a constant). Then from (7.47), the **parametric representation of the transformation curve** $f = \bar{f}$ is

$$\bar{f} = \sum_{j=1}^{p} x_j$$
$$g_i = \sum_{j=1}^{p} b_{ij} x_j, \; i = 1, \ldots, k, \tag{7.47.1}$$
$$x_j \geq 0, \; j = 1, \ldots, p,$$

where the parameters are the resource utilization levels x_j.

To make this discussion a bit more transparent, let us assume that $p = k = 2$. Then (7.47.1) becomes, for $\bar{f} = 1$, the parametric form of the **unit transformation curve** (representing the various combinations of the outputs obtainable by employing one unit of the input)

$$1 = x_1 + x_2$$
$$g_1 = b_{11} x_1 + b_{12} x_2$$
$$g_2 = b_{21} x_1 + b_{22} x_2$$
$$x_1, x_2 \geq 0.$$

Here $b_1' = (b_{11}, b_{21}), b_{12}' = (b_{12}, b_{22})$ are the output activities depicted by the process rays OR, OR', respectively (each representing the locus of output combinations involving fixed output proportions), and $g_1 = g_{11} + g_{12}$, $g_2 = g_{21} + g_{22}$. The slope of OR is b_{21}/b_{11} while the slope of OR' is b_{22}/b_{12}, i.e. output activity 1

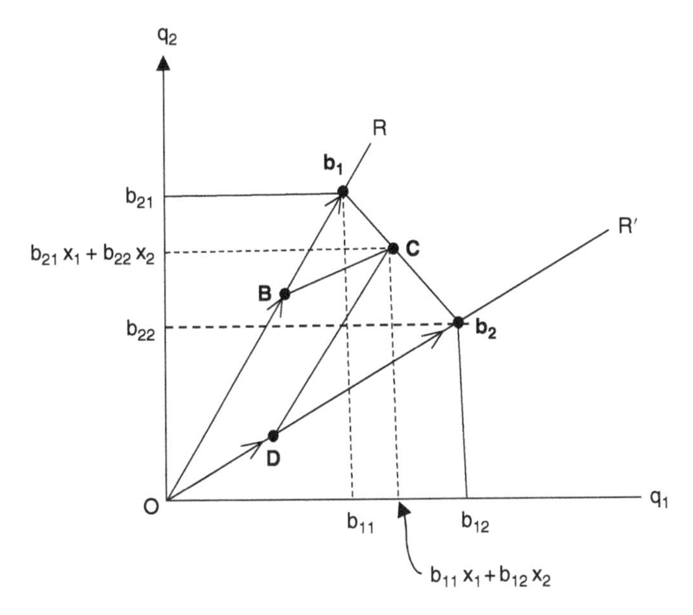

Figure 7.21 Unit transformation curve.

produces b_{21}/b_{11} units of g_2 per unit of g_1; and activity 2 produces b_{22}/b_{12} units of g_2 per unit of g_1 (Figure 7.21). It is apparent that the single unit of the input may be used at the x_1 level by the output activity $\boldsymbol{b}_1$ alone ($x_2 = 0$) with $g_1 = b_{11}$, $g_2 = b_{21}$(point $\boldsymbol{b}_1$) or at the x_2 level by $\boldsymbol{b}_2$ alone ($x_1 = 0$) with $g_1 = b_{21}$, $g_2 = b_{22}$(point $\boldsymbol{b}_2$) or by operating these output activities jointly by forming a composite activity as the convex combination of $\boldsymbol{b}_1$, $\boldsymbol{b}_2$, namely $\boldsymbol{C} = x_1\boldsymbol{b}_1 + (1 - x_1)\boldsymbol{b}_2$, $0 \leq x_1 \leq 1$.Thus, the proportion of the single unit of the scarce input used by $\boldsymbol{b}_1$ is x_1 while the proportion of the same used by $\boldsymbol{b}_2$ is $1 - x_1$.[10] In this

10 The equation of the unit transformation curve may be obtained by solving the system

$$g_1 = b_{11}x_1 + b_{12}(1-x_1)$$
$$g_2 = b_{12}x_1 + b_{22}(1-x_1)$$

simultaneously. Upon doing so we have

$$g_2 = \frac{b_{11}b_{22} - b_{12}b_{21}}{b_{11} - b_{12}} - \left(\frac{b_{22} - b_{21}}{b_{11} - b_{12}}\right)g_1.$$

For $x_1 + x_2 = \bar{f} > 1$, the associated output transformation curve is given by

$$g_2 = \bar{f}\left(\frac{b_{11}b_{22} - b_{12}b_{21}}{b_{11} - b_{12}}\right) - \left(\frac{b_{22} - b_{21}}{b_{11} - b_{12}}\right)g_1.$$

regard, any output combination on that portion of the unit transformation curve between $\boldsymbol{b}_1$, $\boldsymbol{b}_2$ can be written as

$$\boldsymbol{C} = \begin{bmatrix} g_1 \\ g_2 \end{bmatrix} = \begin{bmatrix} b_{11}x_1 + b_{12}(1-x_1) \\ b_{21}x_1 + b_{22}(1-x_1) \end{bmatrix},$$

Moreover, any such point is technologically efficient in that no other composite activity can be found that utilizes the same amount of the input and produces more of one of the outputs without producing less of the other. It is important to note that the closer the composite activity $\boldsymbol{C}$ is to one of the output activities $\boldsymbol{b}_1$ or $\boldsymbol{b}_2$, the greater the proportion of the input used by that activity relative to the other. Geometrically, each composite activity lies within the convex polyhedral cone ROR' in output space representing the set of all nonnegative linear combinations of the vectors $\boldsymbol{b}_1$, $\boldsymbol{b}_2$.

When more than two output processes are considered, the iso-input or joint product transformation curves, defined by (7.47.1), are convex polygons (Figure 7.22 represents the instance where four output activity vectors are present). Moreover, any transformation curve (say, ABCD) lying above the unit transformation curve specified by the output activities $\boldsymbol{b}_1, \ldots, \boldsymbol{b}_4$ is just a radial projection of the unit transformation curve with $\bar{f} > 1$.

If we now focus our attention on the $\bar{f}$ output transformation curve, ABCD, we may note that any point such as E on the line segment AC utilizes $\bar{f}$ units of the input but will not be considered as a candidate for an optimal output combination since it represents an inefficient output bundle, e.g. it uses the same amount of the input as point B does yet produces less of each output. Hence only output combinations on the transformation curve ABD are technologically efficient. In this regard, an efficient use of processes is confined to either a single

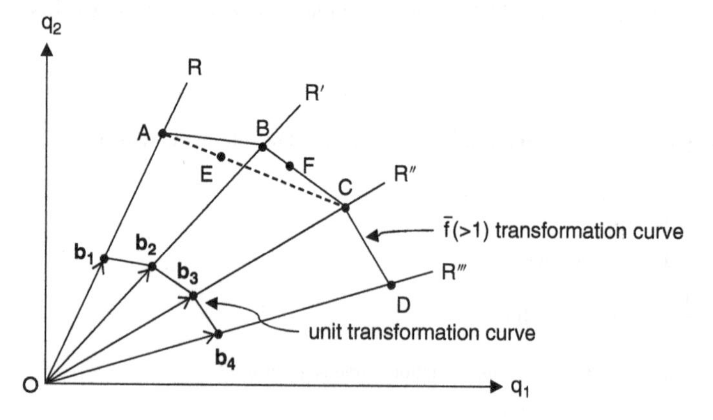

Figure 7.22 Joint product transformation curves.

output activity or to a nonnegative linear combination of two "adjacent" activities. Thus, point E is omitted from the set of technically efficient points.

Assuming that point F depicts an optimum combination when $\bar{f}$ units of the input are employed, the information conveyed by a solution point such as this is: (i) total input utilization or $\bar{f}$; (ii) the optimal levels of the output processes R', R'' namely $\bar{f}x_2 = \bar{f}\,FC/BC, \bar{f}x_3 = \bar{f}(1-x_2) = \bar{f}BF/BC$, respectively; (iii) the output levels produced by each activity, i.e. activity 2 produces $b_{12}\bar{f}x_2$ units of g_1 and $b_{22}\bar{f}x_2$ units of g_2 while activity 3 produces $b_{13}\bar{f}(1-x_2)$ units of g_1 and $b_{23}\bar{f}(1-x_2)$ units of g_2; and (iv) the aggregate levels of production of the two outputs are, for g_1, $\bar{f}(b_{12}x_2 + b_{13}(1-x_2))$; and for g_2, $\bar{f}(b_{22}x_2 + b_{23}(1-x_2))$.

At present, the quantities g_i, $i = 1, \ldots, k$, denote the amounts of the products actually produced when the activities b_j, $j = 1, \ldots, p$, are operated at their respective levels x_j. If the g_1 are replaced by their output quotas, q_i, $i = 1, \ldots. k$, then $q_i \leq g_i$ and thus the f value in (7.47) now becomes the minimum amount of the input required to achieve the q_i output levels. Hence (7.47) is replaced by

$$\left.\begin{array}{l} min\, f = \sum_{j=1}^{p} x_j \text{ s.t.} \\ \sum_{j=1}^{p} b_{ij}x_j \geq q_i,\ i = 1, \ldots, k \\ x_j \geq 0,\ j = 1, \ldots, p \end{array}\right\} \text{or} \left\{\begin{array}{l} min f = \mathbf{1}'\mathbf{X} \text{ s.t.} \\ \mathbf{BX} \geq \mathbf{q} \\ \mathbf{X} \geq \mathbf{O}, \end{array}\right. \tag{7.48}$$

where $\mathbf{q}' = (q_1, \ldots, q_k)$. Here, the minimum of f defines a single-valued input function that depends on the fixed-output quotas q_i, $i = 1, \ldots, k$. In this instance, an iso-input contour or output transformation curve corresponding to an attainable value of f has the shape depicted in Figure 7.23. Note that the transformation curve exhibits both vertical and horizontal segments, indicating that any output combination on the same curve is technologically inefficient, i.e. requires overproduction and costless disposal.

For movements along the output transformation curve $f = \bar{f}$ given in Figure 7.23, the **rate of product transformation** of q_1 for q_2 (or q_2 for q_1), denoted $RPT_{1,2} = |dq_2/dq_1|_{f=\bar{f}}$, depicts the rate at which q_2 must be sacrificed to obtain more of q_1 (or conversely) given a fixed level of the scarce input.[11]

11 If MC_1 denotes the marginal cost of product 1 "in terms of the scarce input" (it is the real resource cost of producing an extra unit of product 1) and MC_2 denotes the marginal cost of product 2 "in terms of the scarce input," then at a point on the output transformation curve, $RPT_{1,2} = MC_1/MC_2$. alternatively, if MP_1 denotes the marginal product of the scarce resource in the production of product 1 (it is the increase in the output of product 1 resulting from the employment of an additional unit of the single input) and MP_2 represents the marginal product of the scarce input in the production of product 2, then, since $MC_1 = 1/MP_1$ and $MC_2 = 1/MP_2$, it follows that $RPT_{1,2} = MP_1/MP_2$.

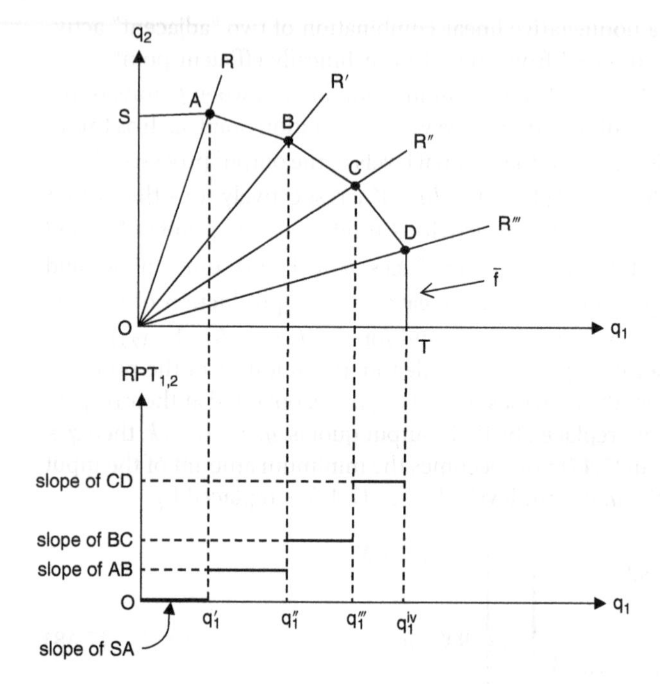

Figure 7.23 Output transformation curve.

So for this joint output linear production model, the rate of transformation of one product for another is discontinuous and increasing (or at least nondecreasing) along any "well-behaved" product transformation curve, as Figure 7.23 indicates. If output q_2 is held fixed at the $\bar{q}_2$ level, then the total (real resource) cost function of output q_1 given $q_2 = \bar{q}_2$ along with the associated marginal and average (real resource) cost functions for q_1, determined, respectively, as $MC_1 = df/dq_1, AC_1 = f/q_1$, are illustrated in Figure 7.24. Note that MC_1 is discontinuous and increasing, with this latter property indicating that the model exhibits **increasing (real resource) opportunity cost**, i.e. the marginal input requirement resulting from increased production of one output increases given that the remaining output level is held constant.

Example 7.5 Given the output activity vectors $\boldsymbol{b}_1' = (2,9)$, $\boldsymbol{b}_2' = (5,7)$, and $\boldsymbol{b}_3' = (8,4)$, determine the total, marginal, and average (real resource) cost functions for q_1 when $q_2 = \bar{q}_2 = 25$. As q_1 increases from the zero level, the first output activity encountered is $\boldsymbol{b}_1$ and thus the system

$$
\begin{aligned}
&f = x_1 + x_2 + x_3 \\
&x_1 \begin{bmatrix} 2 \\ 9 \end{bmatrix} + x_2 \begin{bmatrix} 5 \\ 7 \end{bmatrix} + x_3 \begin{bmatrix} 8 \\ 4 \end{bmatrix} = \begin{bmatrix} q_1 \\ 25 \end{bmatrix} \\
&x_j \geq 0, j = 1,2,3,
\end{aligned}
\tag{7.49}
$$

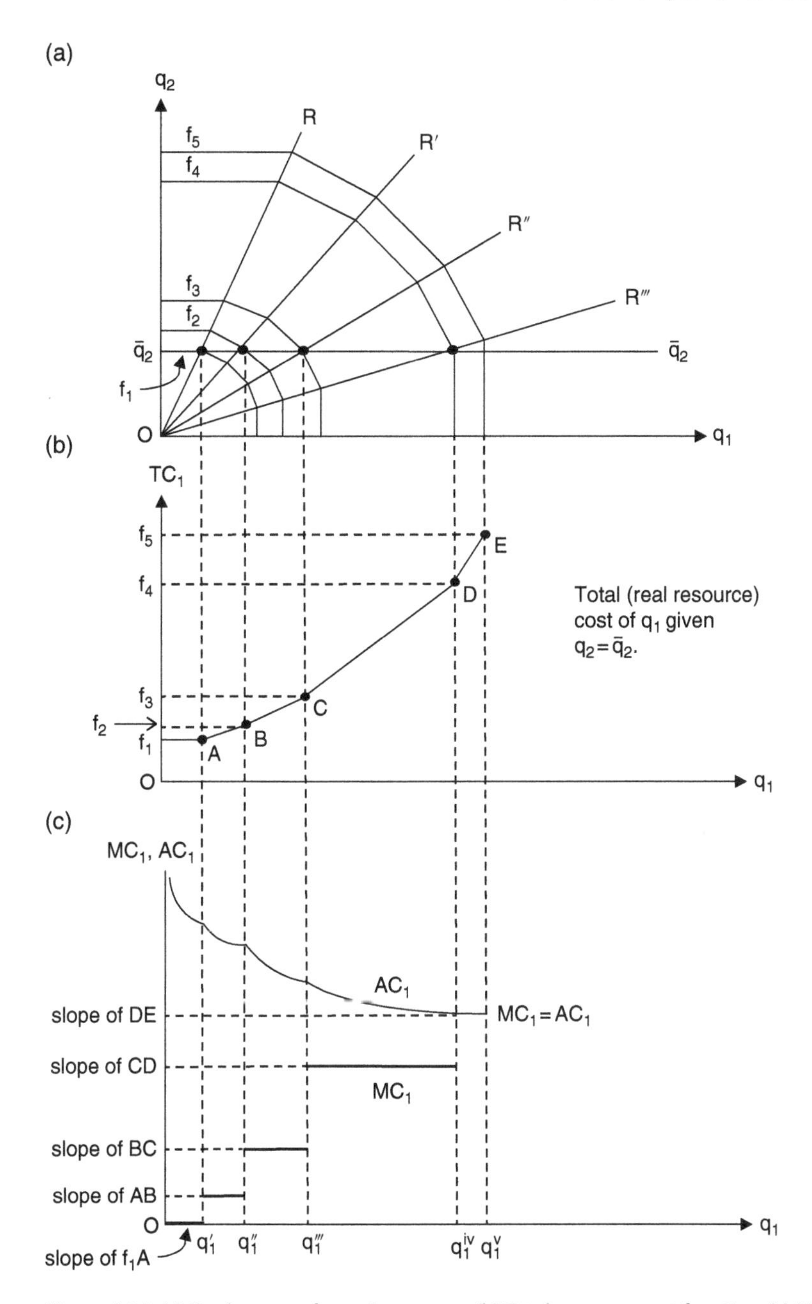

Figure 7.24 (a) Product transformation curves; (b) Total resource cost function; (c) Marginal and average cost functions.

becomes

$$\left.\begin{array}{l} f = x_1 \\ 2x_1 = q_1 \\ 9x_1 = 25 \end{array}\right\} \text{or } f = \frac{25}{9}, 0 \le q_i \le \frac{50}{9}. \tag{7.49.1}$$

Next, if activity $\boldsymbol{b}_2$ is brought on line and $\boldsymbol{b}_1$, $\boldsymbol{b}_2$ are run in combination, (7.49) appears as

$$\left.\begin{array}{l} f = x_1 + x_2 \\ 2x_1 + 5x_2 = q_1 \\ 9x_1 + 7x_2 = 25 \end{array}\right\} \text{or } f = \frac{75}{31} + \frac{2}{31}q_1, \frac{50}{9} \le q_i \le \frac{125}{7}. \tag{7.49.2}$$

If activity $\boldsymbol{b}_3$ replaces $\boldsymbol{b}_1$ so that now $\boldsymbol{b}_2$, $\boldsymbol{b}_3$ operate jointly, (7.49) yields

$$\left.\begin{array}{l} f = x_2 + x_3 \\ 5x_2 + 8x_3 = q_1 \\ 7x_2 + 4x_3 = 25 \end{array}\right\} \text{or } f = \frac{25}{12} + \frac{1}{12}q_1, \frac{125}{7} \le q_i \le 50. \tag{7.49.3}$$

And if $\boldsymbol{b}_3$ is run alone, (7.49) reduces to

$$\left.\begin{array}{l} f = x_3 \\ 8x_3 = q_1 \\ 4x_3 = 25 \end{array}\right\} \text{or } f = \frac{q_1}{8}, q_1 \ge 50. \tag{7.49.4}$$

Upon combining (7.49.1)–(7.49.4), the total (real resource) cost function for output q_1 given $\bar{q}_2 = 25$ is

$$f = \begin{cases} 25/9, & 0 \le q_1 \le 50/9, \\ \frac{75}{31} + \frac{2}{31}q_1, & 50/9 \le q_1 \le 125/7, \\ \frac{25}{12} + \frac{1}{12}q_1, & 125/7 \le q_1 \le 50, \\ q_1/8, & q_1 \ge 50 \end{cases} \tag{7.50}$$

(see Figure 7.25).

From (7.50), the marginal and average (real resource) cost functions for q_1 given $\bar{q}_2 = 25$ are, respectively,

$$MC_1 = \frac{df}{dq_1} = \begin{cases} 0, & 0 \le q_1 \le 50/9, \\ 2/31, & 50/9 \le q_1 \le 125/7, \\ 1/12, & 125/7 \le q_1 \le 50, \\ 1/8, & q_1 \ge 50; \end{cases} \tag{7.51}$$

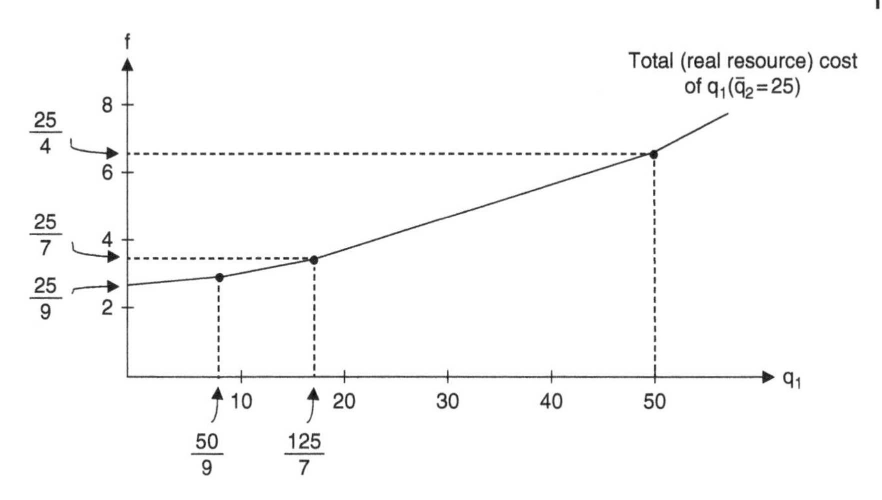

Figure 7.25 Total (real resource) cost function for q_1 given $\bar{q}_2 = 25$.

$$AC_1 = \frac{f}{q_1} = \begin{cases} 25/9q_1, & 0 \leq q_1 \leq 50/9, \\ \dfrac{75}{31q_1} + \dfrac{2}{31}, & 50/9 \leq q_1 \leq 125/7, \\ \dfrac{25}{12q_1} + \dfrac{1}{12}, & 125/7 \leq q_1 \leq 50, \\ 1/8, & q_1 \geq 50 \end{cases} \tag{7.52}$$

(see Figure 7.26).

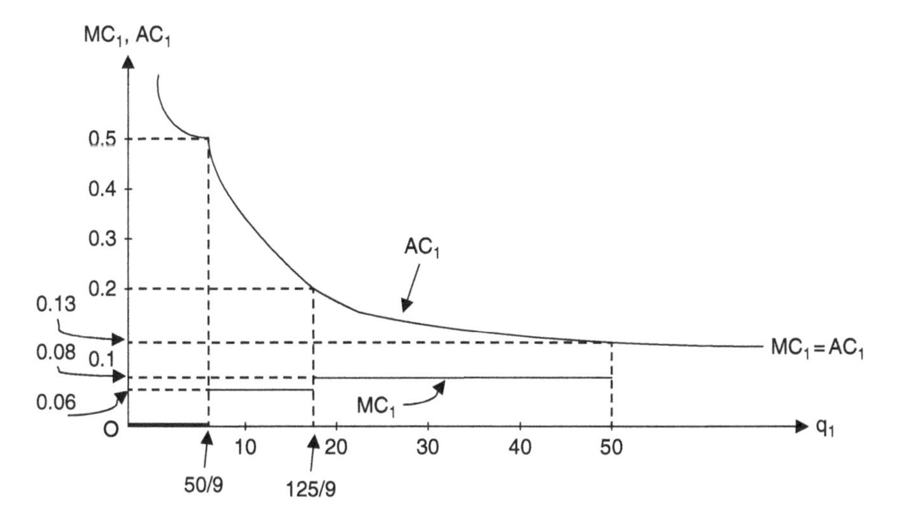

Figure 7.26 Marginal and average (real resource) cost functions for q_1 given $\bar{q}_2 = 25$.

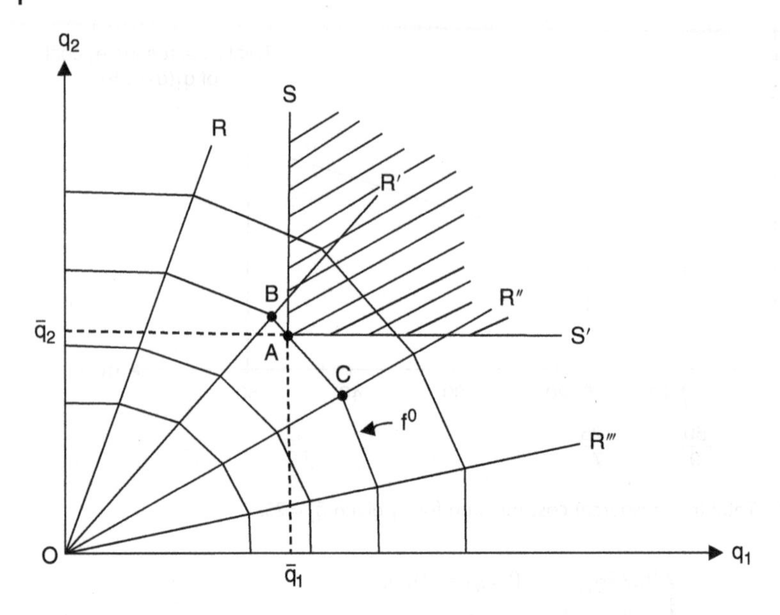

Figure 7.27 Given $\bar{q}_1$, $\bar{q}_2$, the optimal levels of the second and third activities are at Point A.

If in Figure 7.27 $\bar{q}_1$, $\bar{q}_2$ represent fixed output quotas for the products q_1, q_2, respectively, then the minimum amount of the single scarce input needed to meet these quotas is f^o, the value of f corresponding to point A where the (shaded) region of feasible solutions (the convex polyhedral cone SAS' of Figure 7.27) has a point in common with the lowest possible output transformation curve attainable. Here the optimal levels of the second and third activities are, respectively, $f^o x_2 = f^0 AC/BC$, $f^o x_3 = f^o(1 - x_2) = f^0 BA/BC$ while the amounts of the products produced exactly fulfil the quotas. ■

7.15 Joint Production and Cost Minimization

In the preceding section, the objective $f = \mathbf{1}'\mathbf{X}$ represented the total amount of the single scarce primary input used to meet the production quotas q_i, $i = 1, \ldots, k$. Let us now assume that the total cost of the primary input is to be minimized subject to a set of output constraints and nonnegativity conditions and that r additional **shadow inputs** are to be employed, i.e. the amounts of these supplemental inputs vary proportionately with the levels at which the various output activities are operated.[12] In this regard, let the r shadow inputs be denoted as $s_1,\ldots,s_r$, where

$$s_l = \sum_{j=1}^{p} d_{lj}x_j,\ l = 1,\ldots,r,$$

and d_{lj} is the quantity of the lth shadow input employed per unit of x_j, $j = 1, \ldots, p$. So in addition to the aforementioned set of structural constraint inequalities, we must now also incorporate into the model the system of equations

$$\begin{aligned}
&d_{11}x_1 + d_{12}x_2 + \cdots + d_{1p}x_p = s_1\\
&d_{21}x_1 + d_{22}x_2 + \cdots + d_{2p}x_p = s_2\\
&\ldots\ldots\ldots\ldots\ldots\ldots\ldots\ldots\ldots\ldots\\
&d_{r1}x_1 + d_{r2}x_2 + \cdots + d_{rp}x_p = s_r
\end{aligned}$$

or

$$\boldsymbol{DX} = \boldsymbol{S}, \tag{7.53}$$

where $\boldsymbol{D} = [\boldsymbol{d}_1, \boldsymbol{d}_2, \ldots, \boldsymbol{d}_p]$ is a $(r \times p)$ matrix of shadow input coefficients whose jth column is the **shadow activity** $\boldsymbol{d}_j, \boldsymbol{d}_j' = (d_{1j}, d_{2j}, \ldots, d_{rj}), j = 1, \ldots, p$, and $\boldsymbol{S}$ is a $(r \times 1)$ vector of shadow input levels.

If c is the unit price of the single primary input and

$$\boldsymbol{C}_s = \begin{bmatrix} c_{s1} \\ c_{s2} \\ \vdots \\ c_{sr} \end{bmatrix}$$

is an $(r \times 1)$ vector whose lth component c_{sl} is the cost per unit of the lth shadow factor, $l = 1, \ldots, r$, then the constrained cost minimization model at hand is

$$\begin{aligned}
&\min f = c\sum_{j=1}^{p} x_j + \sum_{l=1}^{r} c_{sl}s_l \quad \text{s.t.}\\
&\sum_{j=1}^{p} b_{ij}x_j \geq q_i,\ i = 1,\ldots,k\\
&\sum_{j=1}^{p} d_{lj}x_j \geq s_l,\ l = 1,\ldots,r\\
&x_j \geq 0,\ j = 1,\ldots,p
\end{aligned} \tag{7.54}$$

12 The concept of a shadow input was developed by Frisch (1965), p. 22. Suffice it to say that as any of the output process levels x_j, $j = 1, \ldots, p$, increases, more of each supplemental input is concomitantly required, i.e. they follow (indirect proportion) behind the x_j's "like a shadow."

or

$$\begin{aligned} \min f &= c\mathbf{1}'\mathbf{X} + \mathbf{C}_s'\mathbf{S} \text{ s.t.} \\ \mathbf{BX} &\geq \mathbf{q} \\ \mathbf{DX} &= \mathbf{S} \\ \mathbf{X} &\geq \mathbf{O}. \end{aligned} \tag{7.54.1}$$

Upon substituting s_l, $l = 1, \ldots, r$, into the objective form of (7.54) we ultimately seek to

$$\begin{aligned} \min f &= c\sum_{j=1}^{p} x_j + \sum_{l=1}^{r} c_{sl}\left(\sum_{j=1}^{p} d_{lj}x_j\right) = \sum_{j=1}^{p}\left(c + \sum_{l=1}^{r} c_{sl}d_{lj}\right)x_j \text{ s.t.} \\ &\sum_{j=1}^{p} b_{ij}x_j \geq q_i,\ i = 1,\ldots,k \\ &x_j \geq 0,\ j = 1,\ldots,p \end{aligned} \tag{7.55}$$

or, from (7.54.1),

$$\begin{aligned} \min f &= (c\mathbf{1} + \mathbf{D}'\mathbf{C}_s)'\mathbf{X} \text{ s.t.} \\ \mathbf{BX} &\geq \mathbf{q} \\ \mathbf{X} &\geq \mathbf{O}. \end{aligned} \tag{7.55.1}$$

The net effect of this substitution is on the objective function coefficient associated with each output process level, i.e. it adds to the unit cost of the primary input the unit cost of using shadow inputs. Thus, the objective function coefficient on x_j is

$$c + \sum_{l=1}^{r} c_{sl}d_{lj} = c + \mathbf{C}_s'\mathbf{d}_j,\ j = 1,\ldots,p. \tag{7.56}$$

Clearly (7.56) represents the total cost of all resources used to operate the jth output process at the unit level.

If in (7.55) we define the primal surplus variable x_{p+i} as the amount of **overproduction** of the ith product – it represents the amount by which its output quota is exceeded, then

$$x_{p+i} = \sum_{j=1}^{p} b_{ij}x_j - q_i \geq 0,\ i = 1,\ldots,k.$$

Looking to the symmetric dual of (7.55.1) we have

$$\begin{aligned} \max h = \boldsymbol{q}'\boldsymbol{U} \ \text{ s.t.} \\ \boldsymbol{B}'\boldsymbol{U} \leq c\boldsymbol{1} + \boldsymbol{D}'\boldsymbol{C}_s \\ \boldsymbol{U} \geq \boldsymbol{O} \end{aligned} \tag{7.57}$$

or

$$\begin{aligned} \max h = \sum_{j=1}^{k} q_i u_i \ \text{ s.t.} \\ \sum_{i=1}^{k} b_{ij} u_i \leq c + \sum_{l=1}^{r} c_{sl} d_{lj}, \ j = 1, \ldots, p \\ u_i \geq 0, \ i = 1, \ldots, k, \end{aligned} \tag{7.57.1}$$

where the ith dual variable $u_i = \partial f / \partial q_i$, $i = 1, \ldots, k$, represents the rate of change of total cost with respect to the ith production quota q_i; it is the **marginal cost** of the ith product, i.e. if the production of q_i is increased by one unit, total cost increases by $\$u_i$. Alternatively, u_i represents the potential cost reduction to the firm if the ith output quota is decreased by one unit. Thus $\$u_i$ is the maximum amount the firm would be willing to pay to decrease the ith output quota by one unit. Here, too, the dual variables associated with the primal structural constraints serve as internal (accounting) or imputed (shadow) prices; since the output quotas set a lower limit to total cost, their fulfillment is valued on an opportunity cost basis.

Looking to the dual objective in (7.57) (or in 7.57.1) we see that h represents the total "imputed" cost of the firm's minimum output requirements. Hence, the firm seeks to make this quantity as large as possible, i.e. it desires to maximize its total potential cost reduction. Moreover, a glance at the jth dual structural constraint reveals that $\sum_{i=1}^{k} b_{ij} u_i$, the "imputed" cost of the outputs produced by operating the jth activity at the unit level (i.e. one unit of the single scarce primary input is utilized), cannot exceed $c + \sum_{l=1}^{r} c_{sl} d_{lj}$, the total cost of "all" inputs used to operate the jth output process at the unit level. Based on this discussion, it is evident that the dual problem requires the firm to determine the largest imputed cost of its set of output quotas that fully accounts for all of the resource costs incurred by the p output activities.

If in (7.57.1) we define the dual slack variable u_{k+j} as the **accounting loss** associated with the jth output activity – it represents the amount by which the total resource cost per unit of activity j exceeds the total imputed cost of the outputs produced by running activity j at the unit level, then

$$u_{k+j} = c + \sum_{l=1}^{r} c_{sl} d_{lj} - \sum_{i=1}^{k} b_{ij} u_i \geq 0, \ j = 1, \ldots, p.$$

If at an optimal solution to the dual problem a dual structural constraint holds as an equality, the imputed cost of the outputs produced by operating activity j at the unit level exactly equals the total cost of the inputs used to run the process at that level. In this instance output activity j is operated at a positive level (its accounting loss is zero) and thus $x_j > 0$. If, however, strict inequality prevails in a dual structural constraint at the optimum, activity j is unprofitable and will not be operated ($x_j = 0$) since the total input cost incurred per unit of the jth process exceeds its imputed output cost per unit (the accounting loss for activity j is positive). These observations may be conveniently summarized by the complementary slackness conditions that hold at an optimal solution to the primal-dual pair of problems. Specifically, if at any such solution $\boldsymbol{X}_o$, $(\boldsymbol{X}_s)_o$ depict $(p \times 1)$ and $(k \times 1)$ vectors of primal activity levels and surplus (overproduction) variables, respectively, while $\boldsymbol{U}_o$, $(\boldsymbol{U}_s)_o$ represent $(k \times 1)$ and $(p \times 1)$ vectors of dual imputed (marginal) costs and accounting loss (slack) variables, respectively, then

$$(\boldsymbol{U}_s)_o' \boldsymbol{X}_o + \boldsymbol{U}_o' (\boldsymbol{X}_s)_o = 0 \text{ or } \sum_{j=1}^{p} x_j^o u_{k+j}^o + \sum_{i=1}^{k} u_i^o x_{p+i}^o = 0$$

so that these conditions apply:

a) If $x_j^o > 0$, then $u_{k+j}^o = 0$ – if an activity is undertaken at a positive level in the optimal primal solution, its corresponding accounting loss figure is zero.
b) If $u_{k+j}^o > 0$, then $x_j^o = 0$ – whenever a surplus variable appears in the optimal dual solution, the activity to which it refers is not undertaken.
c) If $u_i^o > 0$, then $x_{p+i}^o = 0$ – if u_i^o appears in the optimal dual solution, the ith output quota q_i is met exactly so that if q_i is reduced by one unit, the cost savings to the firm is $\$u_i^o$.
d) If $x_{p+j}^o > 0$, then $u_i^o = 0$ – if overproduction of output i occurs in the optimal primal solution, its corresponding marginal cost is zero, i.e. if the ith output quota q_i has been exceeded, the potential reduction in cost from a one unit decrease in q_i must be zero.

7.16 Cost Indifference Curves

To convert a given output transformation curve (depicted by ABDC in Figure 7.28 and corresponding to $\bar{f}$ units of the single scarce primary input) into a **cost indifference** or **isocost curve**, we must answer the following question. Specifically, what points yield $\$\bar{f}$ in total cost? For any output activity level x_j, unit cost is $c + \sum_{l=1}^{r} c_{sl} d_{lj}, j = 1, \ldots, p$. So for the four output activities given in Figure 7.28, their unit cost levels are, respectively,

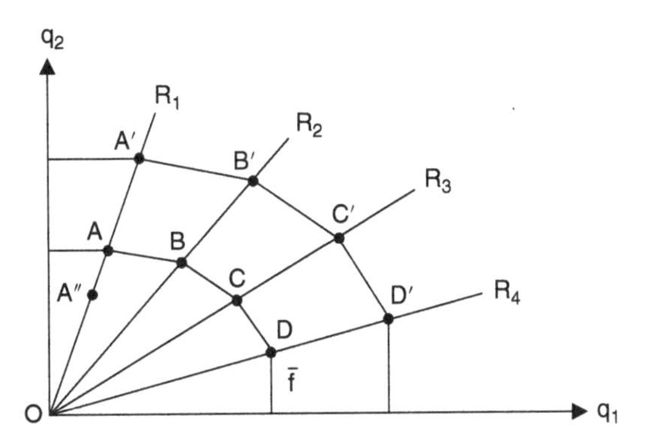

Figure 7.28 A′ B′ C′ D′ is an isocost curve.

$$c_1 = c + \sum_{l=1}^{r} c_{sl} d_{l1}, \qquad c_3 = c + \sum_{l=1}^{r} c_{sl} d_{l3},$$

$$c_2 = c + \sum_{l=1}^{r} c_{sl} d_{l2}, \qquad c_4 = c + \sum_{l=1}^{r} c_{sl} d_{l4}.$$

We note first that if $c_1 = c_2 = c_3 = c_4 = \$1$, then each of the points A, B, C, and D represents both $\bar{f}$ units of the primary input and $\$\bar{f}$ in total cost (e.g. at A, $\bar{f}\ (\$1) = \$\bar{f}$). Next, if, say, $c_1 < \$1$, then point A involves less than $\$\bar{f}$ in total cost. That is, since every unit of the primary input used by activity R_1 costs $\$c_1$, point A represents a total cost of $\$\bar{f}c_1 < \$\bar{f}$. So for the firm to spend $\$\bar{f}$ using activity $R_1, f' = \bar{f}/c_1\left(>\bar{f}\right)$ units of the primary input must be used at point A′ of Figure 7.28 (note that $f' \cdot c_1 = \bar{f}$) so that the utilization of the single primary input is increased by the multiple $\frac{1-c_1}{c_1}$, i.e., $f' - \bar{f} = \bar{f}\left(\frac{1-c_1}{c_1}\right)$. However, if $c_1 > \$1$, point A now involves more than $\$\bar{f}$ in total cost. Hence every unit of the primary input used by activity R_1 costs $\$c_1$ so that point A depicts a total cost of $\$\bar{f}c_1 > \$\bar{f}$. Thus, for the firm to spend $\$\bar{f}$ on process R_1, $f'' = \bar{f}/c_1\left(<\bar{f}\right)$ units of the primary input must be used (see Point A″ of Figure 7.28). In this instance, resource utilization decreases by the factor $\frac{c_1-1}{c_1}$ since $\bar{f} - f'' = \bar{f} - \frac{\bar{f}}{c_1} = \bar{f}\left(\frac{c_1-1}{c_1}\right)$. If the remaining output activities R_2, R_3, and R_4 are viewed in a similar fashion, then a hypothetical isocost curve depicting a total cost of $\$\bar{f}$ is A′ B′ C′ D′ (implicit in this discussion is the assumption that $\$1 > c_1 > c_2 > c_3 > c_4$).

As was the case with output transformation curves, successive isocost curves are radial projections of the one corresponding to a total cost level of $\$\bar{f}$ (or, more precisely, of the unit cost or \$1 isocost curve) and exhibit constant (cost) returns to scale, i.e. they reflect the linear homogeneity property of the fixed coefficients linear technology. In this regard, if the production of all outputs is increased by λ%, total cost also increases by 'λ%. Moreover, the more costly a particular activity is, the further out will be the shift along the activity ray of a

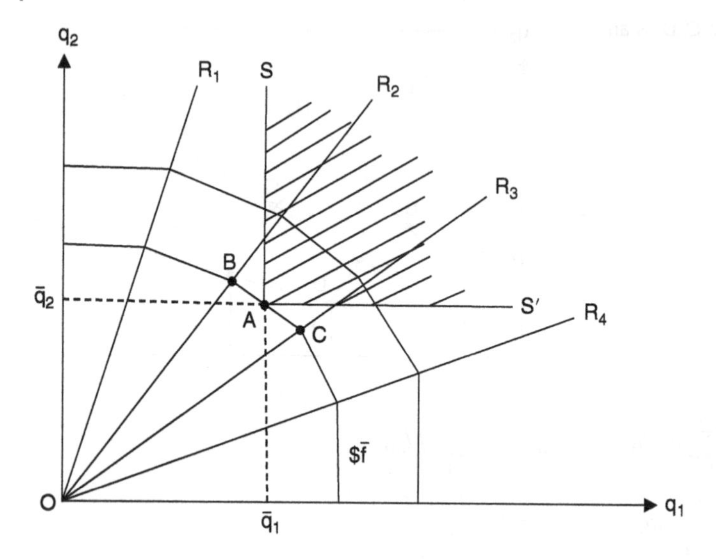

Figure 7.29 Cost minimizing output combination at point A.

point representing $\bar{f}$ units of the primary input to a point representing $\$\bar{f}$ of total cost. For instance, since in Figure 7.28 it was assumed that $c_1 > c_2 > c_3 > c_4$, it follows that distance AA′ < distance BB′ < distance CC′ < distance DD′ or that $\frac{1-c_1}{c_1} < \frac{1-c_2}{c_2} < \frac{1-c_3}{c_3} < \frac{1-c_4}{c_4}$.

If from a set of output transformation curves we derive a set of isocost curves, then the graphical solution to a primal problem such as (7.55) is indicated, for $q \, \varepsilon \, \mathcal{E}^2$, in Figure 7.29. Here the minimum quantities of q_1, q_2 that are to be produced are $\bar{q}_1, \bar{q}_2$, respectively, so that the (shaded) region of admissible solution is the convex polyhedral cone SAS'. Clearly the cost minimizing ($\$\bar{f}$) output combination occurs at point A where the $\$\bar{f}$ isocost curve is the lowest one attainable that also preserves feasibility. At point A activity 2 contributes $(\$\bar{f})x_2 = (\$\bar{f})\frac{AC}{BC}$ to the $\$\bar{f}$ total cost level while activity 3 contributes $(\$\bar{f})x_3 = (\$\bar{f})(1-x_2) = (\$\bar{f})\frac{BA}{BC}$ to the same. In addition, the output quotas are exactly fulfilled in the determination of the $\$\bar{f}$ cost level.

7.17 Activity Levels Interpreted as Individual Resource Levels

If the firm does not employ a single scarce primary input but actually employs p different (nonshadow) primary inputs, one for each individual output activity, then each such activity has associated with it a separate constant price per unit $c_j, j = 1, \ldots, p$. Letting the unit resource cost of the jth output activity correspond to the jth component of the $(p \times 1)$ cost vector $\boldsymbol{C}$, the total cost generated by all p activities is $TC = \sum_{j=1}^{p} c_j x_j = \boldsymbol{C}'\boldsymbol{X}$. In view of this modification, (7.55), (7.55.1) may be rewritten, respectively, as

$$\begin{aligned} \min f &= \sum_{j=1}^{p}\left(c_j + \sum_{l=1}^{r} c_{sl} d_{lj}\right) x_j \quad \text{s.t.} \\ &\sum_{j=1}^{p} b_{ij} x_j \geq q_i,\ i = 1, \ldots, k \\ &x_j \geq 0,\ j = 1, \ldots, p \end{aligned} \tag{7.58}$$

or

$$\begin{aligned} \min f &= (\boldsymbol{C} + \boldsymbol{D}'\boldsymbol{C}_s)'\boldsymbol{X} \quad \text{s.t.} \\ \boldsymbol{B}\boldsymbol{X} &\geq \boldsymbol{q} \\ \boldsymbol{X} &\geq \boldsymbol{O}. \end{aligned} \tag{7.58.1}$$

The symmetric dual thus becomes

$$\begin{aligned} \max h &= \boldsymbol{q}'\boldsymbol{U} \quad \text{s.t.} \\ \boldsymbol{B}'\boldsymbol{U} &\leq \boldsymbol{C} + \boldsymbol{D}'\boldsymbol{C}_s \\ \boldsymbol{U} &\geq \boldsymbol{O} \end{aligned} \tag{7.59}$$

or

$$\begin{aligned} \max h &= \sum_{i=1}^{k} q_i u_i \quad \text{s.t.} \\ &\sum_{i=1}^{k} b_{ij} u_i \leq c_j + \sum_{l=1}^{r} c_{sl} d_{lj},\ j = 1, \ldots, p \\ &u_i \geq 0,\ i = 1, \ldots, k. \end{aligned} \tag{7.59.1}$$

Example 7.6 Given that a cost minimizing firm employs a single scarce primary input to support the output activities

$$\boldsymbol{b}_1 = \begin{bmatrix} 10 \\ 3 \\ 1 \end{bmatrix}, \boldsymbol{b}_2 = \begin{bmatrix} 8 \\ 1 \\ 5 \end{bmatrix}, \text{and } \boldsymbol{b}_3 = \begin{bmatrix} 4 \\ 8 \\ 3 \end{bmatrix},$$

let us assume that the (minimum) output quotas imposed on the firm's operations are $q_1 = 40$, $q_2 = 30$, and $q_3 = 35$. In addition, each output activity has associated with it the shadow activities

$$d_1 = \begin{bmatrix} 2 \\ 2 \end{bmatrix}, d_2 = \begin{bmatrix} 4 \\ 1 \end{bmatrix}, \text{and } d_3 = \begin{bmatrix} 2 \\ 5 \end{bmatrix}.$$

If the price per unit of the primary factor is \$20 and the unit prices of the two shadow factors are $c_{s1} = \$6$, $c_{s2} = \$4$, respectively, then, from (7.29), the objective function coefficients are

$$c + C_s' d_1 = 20 + (6,4)\begin{bmatrix} 2 \\ 2 \end{bmatrix} = 40,$$

$$c + C_s' d_2 = 20 + (6,4)\begin{bmatrix} 4 \\ 1 \end{bmatrix} = 48,$$

$$c + C_s' d_3 = 20 + (6,4)\begin{bmatrix} 2 \\ 5 \end{bmatrix} = 52.$$

Hence, from (7.55), the firm desires to

$$\min f = 40x_1 + 48x_2 + 52x_3 \text{ or}$$

$$\max h = -f = -40x_1 - 48x_2 - 52x_3 \text{ s.t.}$$

$$10x_1 + 8x_2 + 4x_3 \geq 40$$

$$3x_1 + x_2 + 8x_3 \geq 30$$

$$x_1 + 5x_2 + 3x_3 \geq 35$$

$$x_1, x_2, x_3 \geq 0.$$

If x_4, x_5, and x_6 represent nonnegative surplus variables, the optimal simplex matrix corresponding to this problem is

$$\begin{bmatrix} -\frac{1}{37} & 1 & 0 & 0 & \frac{3}{37} & -\frac{8}{37} & 0 & \frac{190}{37} \\ \frac{14}{37} & 0 & 1 & 0 & -\frac{5}{37} & \frac{1}{37} & 0 & \frac{115}{37} \\ -\frac{322}{37} & 0 & 0 & 1 & \frac{4}{37} & -\frac{60}{37} & 0 & \frac{500}{37} \\ \frac{800}{37} & 0 & 0 & 0 & \frac{116}{37} & \frac{332}{37} & 1 & -\frac{15100}{37} \end{bmatrix}$$

with $\frac{190}{37}$ units of the primary input used by activity 2 and $\frac{115}{37}$ units of the same used by activity 3, thus requiring a total of $\frac{305}{37}$ units of the primary factor at a total cost of $\frac{\$15100}{37} = \$408.10.$[13]

From this solution we obtain four separate data sets that are of strategic importance to the firm:

i) *The output activity mix* $(x_1 = 0, x_2 = 190/37, \text{and } x_3 = 115/37\ )$;[14]
ii) *Overproduction levels.* The primal surplus variables are: $x_4 = 500/37$ and $x_5 = x_6 = 0$, thus indicating that the amount of product one produced exceeds its quota by 500/37 units while the output of products two and three exactly fulfill their quotas.
iii) *The product marginal costs.* $u_1 = 0$ while $u_2 = \$116/37$, $u_3 = \$332/37$, i.e. since the minimum output requirement for product one is exceeded, there is no potential cost savings if the quota for this product is reduced by one unit; and since the second and third output quotas are exactly met, the potential cost savings accruing to the firm by reducing these restrictions by one unit each are \$116/37, \$332/37, respectively.
iv) *Accounting loss figures.* The dual slack variables are $u_4 = \$800/37$, $u_5 = u_6 = 0$, thus indicating that activities 2 and 3 just break even and, consequently, are operated at an optimal level while activity 1 is unprofitable and thus stands idle – its total resource cost per unit exceeds the total imputed cost of the outputs produced by operating this activity at the unit level by \$800/37). All of the preceding information sets can be conveniently summarized by utilizing the complementary slackness conditions:

---

13 Note that only $\$20\left(\frac{305}{37}\right) = \$164.86$ of the total cost figure is attributed to the employment of the primary factor; with the remaining \$243.24 due to the use of shadow factors.

14 The amounts of the three products produced by activity two are

$$x_2\boldsymbol{b}_2 = \frac{190}{37}\begin{bmatrix}8\\1\\5\end{bmatrix} = \begin{bmatrix}1520/37\\190/37\\950/37\end{bmatrix}.$$

(Activity 2 produces 1520/37 units of the first output, 190/37 units of the second output, and 950/37 units of the third output); the amounts produced by activity 3 are

$$x_3\boldsymbol{b}_3 = (1-x_2)\boldsymbol{b}_3 = \frac{115}{37}\begin{bmatrix}4\\8\\3\end{bmatrix} = \begin{bmatrix}460/37\\920/37\\345/37\end{bmatrix}.$$

(Activity 3 produces 460/37 units of the first output one, 920/37 units of the second output, and 345/37 units of the third output).

Activity mix vs. Accounting loss values	**Overproduction vs. Marginal costs**
$x_1^o u_4^o = 0 \cdot \frac{800}{37} = 0$	$u_1^o x_4^o = 0 . \frac{500}{37} = 0$
$x_2^o u_5^o = \frac{190}{37} \cdot 0 = 0$	$u_2^o x_5^o = \frac{116}{37} \cdot 0 = 0$
$x_3^o u_6^o = \frac{115}{37} \cdot 0 = 0$	$u_3^o x_6^o = \frac{332}{37} \cdot 0 = 0$

■

Let us next determine the isocost curve corresponding to $f^o = 305/37$ units of the single primary input by first finding the $f^o = 305/37$ transformation curve in output space (i.e. the $f^o = 305/37$ transformation curve is to be converted into the $\$f^o = \$305/37$ isocost curve). To do so, let us first answer the following question. Specifically, how much of each of the three products $\bar{q}_1, \bar{q}_2$ and $\bar{q}_3$ can be produced if each output activity is operated at the 305/37 level? To solve this problem, we simply need to employ the set of technological relationships $b_{ij}x_j = q_i$ with $x_j = 305/37$, where $i, j = 1, 2, 3$. Using the output activities specified above, we obtain, respectively,

$$\begin{bmatrix} 10 \\ 3 \\ 1 \end{bmatrix} \frac{305}{37} = \begin{bmatrix} q_1 \\ q_2 \\ q_3 \end{bmatrix} = \begin{bmatrix} 82.43 \\ 24.73 \\ 8.24 \end{bmatrix},$$

$$\begin{bmatrix} 8 \\ 1 \\ 5 \end{bmatrix} \frac{305}{37} = \begin{bmatrix} q_1 \\ q_2 \\ q_3 \end{bmatrix} = \begin{bmatrix} 65.95 \\ 8.24 \\ 41.21 \end{bmatrix},$$

$$\begin{bmatrix} 4 \\ 8 \\ 3 \end{bmatrix} \frac{305}{37} = \begin{bmatrix} q_1 \\ q_2 \\ q_3 \end{bmatrix} = \begin{bmatrix} 32.97 \\ 65.95 \\ 24.73 \end{bmatrix}.$$

These three sets of output combinations correspond to the points A, B, and C, respectively, of Figure 7.30 and form a convex polyhedral cone emanating from the origin. The $f^o = 305/37$ *transformation surface* (since each output activity has three components) is then constructed by joining these points with straight line segments, i.e. the transformation surface is the triangle ABC.

If $c_j = c + \mathbf{C}_s' \mathbf{d}_j, j = 1,2,3,$ and if $c_1 = c_2 = c_3 = \$1$, each point on the transformation surface ABC above would represent both $f^o = 305/37$ units of the single primary input as well as the $\$f^o = \frac{\$305}{37}$ total cost level. With $c_1 = \$40 > 1$, point

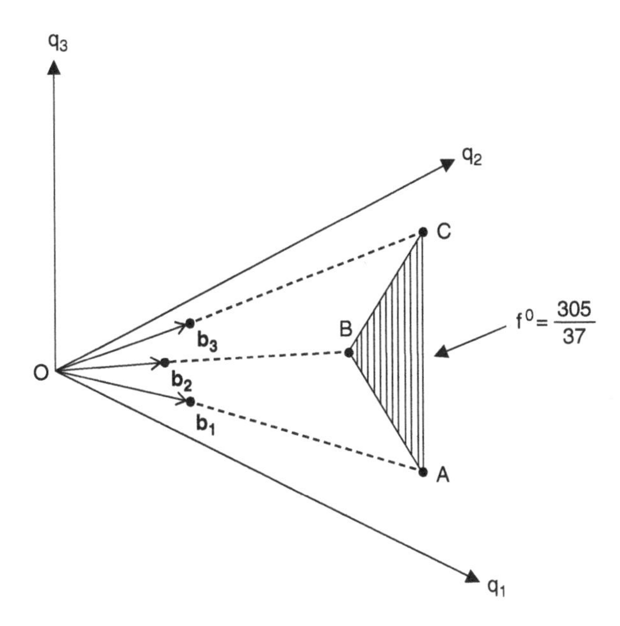

Figure 7.30 Output transformation surface.

A involves more than the total cost of $\frac{\$305}{37}$. Hence, the amount of the primary input used by activity 1 must decrease by the factor $\frac{c_1 - 1}{c_1} = 39/40$ or a total input level of $f^o/c_1 = \frac{305}{1480} = 0.206$ must be used by activity 1. Similarly, for $c_2 = \$48 > 1$, point B also involves more than $\frac{\$305}{37}$ in total cost so that the amount of the primary input utilized by activity 2 must be decreased by the factor $\frac{c_2 - 1}{c_2} = \frac{47}{48}$ or a total input level of $f^o/c_2 = \frac{305}{1776} = 0.172$ must be employed by activity 2. Finally, with $c_3 = \$52 > 1$, point C corresponds to more than $\frac{\$305}{37}$ in total cost so that the amount of the input used by activity 3 must be decreased by the factor $\frac{c_3 - 1}{c_3} = \frac{51}{52}$ or a total input level of $f^o/c_3 = \frac{305}{1924} = 0.159$ is warranted for activity 3. We may plot the $\$f^o = \frac{\$305}{37}$ "isocost surface" in output space by performing a set of calculations similar to those undertaken to derive the $f^o = \frac{\$305}{37}$ transformation surface, i.e. again using the output activities $\boldsymbol{b}_j$, $j = 1, 2, 3$, we have, respectively,

$$\begin{bmatrix}10\\3\\1\end{bmatrix}0.206 = \begin{bmatrix}q_1\\q_2\\q_3\end{bmatrix} = \begin{bmatrix}2.060\\0.618\\0.206\end{bmatrix},$$

$$\begin{bmatrix}8\\1\\5\end{bmatrix}0.172 = \begin{bmatrix}q_1\\q_2\\q_3\end{bmatrix} = \begin{bmatrix}1.376\\0.172\\0.860\end{bmatrix},$$

$$\begin{bmatrix}4\\8\\3\end{bmatrix}0.159 = \begin{bmatrix}q_1\\q_2\\q_3\end{bmatrix} = \begin{bmatrix}0.636\\1.272\\0.477\end{bmatrix}.$$

These latter sets of output combinations correspond to points A′, B′, and C′, respectively, in Figure 7.31. The $\$f^o = \frac{\$305}{37}$ isocost surface is then drawn by joining these new points by using straight line segments.

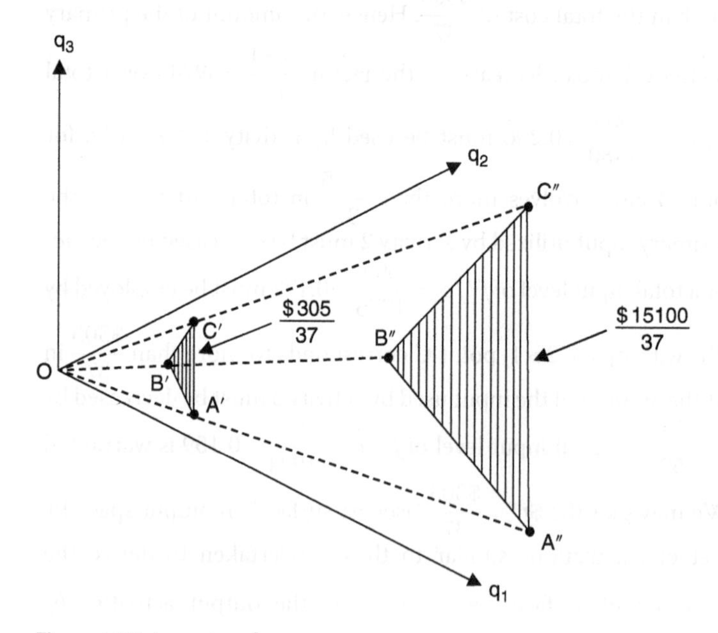

Figure 7.31 Isocost surfaces.

We may also find the optimal isocost surface corresponding to a $\frac{\$15100}{37}$ total cost level by forming the radial projection of the $\frac{\$305}{37}$ isocost surface, i.e. since $\frac{\$15100}{37} = 49.52\left(\frac{\$305}{37}\right)$, we may increase each component of the points A′, B′, and C′ by the scalar multiple 49.52 to generate a new set of points A″, B″, and C″ found on the $\frac{\$15100}{37}$ isocost surface (Figure 7.31). The coordinates of these points are, respectively,

$$49.52\begin{bmatrix}2.060\\0.618\\0.206\end{bmatrix} = \begin{bmatrix}102.01\\30.60\\10.20\end{bmatrix},$$

$$49.52\begin{bmatrix}1.376\\0.172\\0.860\end{bmatrix} = \begin{bmatrix}68.14\\8.52\\42.59\end{bmatrix},$$

$$49.52\begin{bmatrix}0.636\\1.272\\0.477\end{bmatrix} = \begin{bmatrix}31.49\\62.99\\23.62\end{bmatrix}.$$ [15]

15 If we were initially interested in only the \$15100/37 isocost surface, then we could have derived it in a more straightforward fashion by utilizing the structure of the initial problem itself, e.g. if from the objective function we set $f = 151100/37 = 40x_1$ then $x_1 = 10.203$ and thus, from $\boldsymbol{b}_1$, we obtain

$$10.203\begin{bmatrix}10\\3\\1\end{bmatrix} = \begin{bmatrix}q_1\\q_2\\q_3\end{bmatrix} = \begin{bmatrix}102.03\\30.61\\10.20\end{bmatrix}$$

as the coordinates of point A″ in Figure 7.31 (any differences between this set of values and the preceding ones are due to rounding). Similarly, to get the coordinates of point B″, setting $f = 151100/37 = 48x_2$ yields $x_2 = 8.502$ so that

$$8.502\begin{bmatrix}8\\1\\5\end{bmatrix} = \begin{bmatrix}q_1\\q_2\\q_3\end{bmatrix} = \begin{bmatrix}68.02\\8.50\\42.51\end{bmatrix}.$$

The coordinates of point C″ are found in like fashion.

8

Sensitivity Analysis

8.1 Introduction

Throughout this chapter we shall invoke the assumption that we have already obtained an optimal basic feasible solution to a linear programming problem in standard form, i.e. to a problem of the form

$$\begin{aligned} &max\, f(\bar{X}) = \bar{C}'\bar{X} \qquad \text{s.t.} \\ &\bar{A}\bar{X} = \boldsymbol{b}, \bar{X} \geq \boldsymbol{O}. \end{aligned}$$

We then seek to employ the information contained within the optimal simplex matrix to gain some insight into the solution of a variety of linear programming problems that are essentially slight modifications of the original one. Specifically, we shall undertake a **sensitivity analysis** of the initial linear programming problem, i.e. this sort of **post-optimality analysis** involves the introduction of discrete changes in any of the components of the matrices $\bar{\boldsymbol{C}}, \boldsymbol{b}, \text{or}\, \bar{\boldsymbol{A}}$, in which case the values of c_j, b_i, or a_{ij}, $i = 1, \ldots, m; j = 1, \ldots, n - m$, respectively, are altered (increased or decreased) in order to determine the extent to which the original problem may be modified without violating the feasibility or optimality of the original solution. (It must be remembered, however, that any change in, say, the objective function coefficient c_k is a ceteris paribus change, e.g. all other c_j's are held constant. This type of change also holds for the components of $\boldsymbol{b}, \bar{\boldsymbol{A}}$.)

8.2 Sensitivity Analysis

Turning again to our linear programming problem in standard form, let us

$$\begin{aligned} &max\, f(\bar{X}) = \boldsymbol{C}_B'\boldsymbol{X}_B + \boldsymbol{C}_R'\boldsymbol{X}_R \qquad \text{s.t.} \\ &\quad \boldsymbol{B}\boldsymbol{X}_B + \boldsymbol{R}\boldsymbol{X}_R = \boldsymbol{b}, \text{ with } \boldsymbol{X}_B, \boldsymbol{X}_R \geq \boldsymbol{O}. \end{aligned}$$

Linear Programming and Resource Allocation Modeling, First Edition. Michael J. Panik.

Then, as previously determined, with

$$X_B = B^{-1}b - B^{-1}RX_R,$$
$$f(\bar{X}) = C'_B B^{-1}b + \left(C'_R - C'_B B^{-1}R\right)X_R,$$

It follows that if $X_R = O$, the resulting solution is basic (primal) feasible if $X_B = B^{-1}b \geq O$ and optimal (dual feasible) if $C'_R - C'_B B^{-1}R \leq O'$.

8.2.1 Changing an Objective Function Coefficient

If we now change $C' = \left(C'_B, C'_R\right)$ to $C' + \Delta C' = \left(C'_B + \Delta C'_B, C'_R + \Delta C'_R\right)$, then

$$\begin{aligned} f(\bar{X}) &= (C_B + \Delta C_B)' X_B + (C_R + \Delta C_R)' X_R \\ &= (C_B + \Delta C_B)'\left(B^{-1}b - B^{-1}RX_R\right) + (C_R + \Delta C_R)' X_R \\ &= C'_B B^{-1}b + \Delta C'_B B^{-1}b + \left(C'_R - C'_B B^{-1}R + \Delta C'_R - \Delta C'_B B^{-1}R\right)X_R. \end{aligned}$$

Since $X_B = B^{-1}b$ is independent of $\bar{C}$, this new basic solution remains primal feasible while, if $\left(C'_R - C'_B B^{-1}b\right) + \left(\Delta C'_R - \Delta C'_B B^{-1}R\right) \leq O'$, dual feasibility or primal optimality holds. Since our discussion from this point on will be framed in terms of examining selected components within the optimal simplex matrix, it is evident from the last row of the same that primal optimality or dual feasibility is preserved if $C'_B B^{-1}R - C'_R + \left(\Delta C'_B B^{-1}R - \Delta C'_R\right) \geq O'$. Let us now consider the various subcases that may obtain.

If $\Delta C_B = O$ (so that an objective function coefficient associated with a nonbasic variable changes), it follows that optimality is preserved if $C'_B B^{-1}R - C'_R - \Delta C'_R \geq O'$, i.e. if

$$-\bar{c}_j - \Delta c_{Rj} \geq 0, j = 1, \ldots, n-m. \tag{8.1}$$

How may changes in c_{Rj} affect (8.1)? First, if c_{Rj} decreases in value so that $\Delta c_{Rj} < 0$, then (8.1) is still satisfied. In fact, c_{Rj} may decrease without bound and still not violate (8.1). Hence the lower limit to any decrease in c_{Rj} is $-\infty$. Next, if c_{Rj} increases so that $\Delta c_{Rj} > 0$, then (8.1) will hold so long as c_{Rj} increases by no more than $-\bar{c}_j$. Hence the only circumstance in which x_{Rj} becomes basic is if its objective function coefficient c_{Rj} increases by more than $-\bar{c}_j$. If $\Delta c_{Rj} = \hat{c}_{Rj} - c_{Rj}$, where $\hat{c}_{Rj}$ is the new level of c_{Rj}, then the upper limit on $\hat{c}_{Rj}$ to which c_{Rj} can increase without violating primal optimality is $c_{Rj} - \bar{c}_j$.

If $\Delta C_R = O$ (so that an objective function coefficient associated with a basic variable changes), it follows that the optimality criterion is not violated if $C'_B B^{-1}R - C'_R + \Delta C'_B B^{-1}R \geq O'$, i.e. if

$$-\bar{c}_j + \sum_{i=1}^{m} y_{ij}\Delta c_{Bi} \geq 0, j = 1, \ldots, n-m. \tag{8.2}$$

Let us assume that c_{Bk} changes. Hence (8.2) becomes

$$-\bar{c}_j + y_{kj}\Delta c_{Bk} \geq 0, j = 1,\ldots,n-m. \tag{8.2.1}$$

We now consider the circumstances under which changes in c_{Bk} preserve the sense of this inequality. First, if $y_{kj} = 0$, then (8.2.1) holds regardless of the direction or magnitude of the change in c_{Bk}. Second, if $y_{kj} > 0$, then (8.2.1) is satisfied if $\Delta c_{Bk} \geq 0$, i.e. Δc_{Bk} may increase without bound. Hence, (8.2.1) is affected only when c_{Bk} decreases or $\Delta c_{Bk} < 0$. Since (8.2.1) implies that $\bar{c}_j / y_{kj} \leq \Delta c_{Bk}, j = 1,\ldots,n-m,$ it is evident that the lower limit to which Δc_{Bk} can decrease without violating optimality is $\bar{c}_j / y_{kj}, y_{kj} > 0$. Hence, x_{Rj} remains nonbasic as long as its objective function coefficient does not decrease by more than $\bar{c}_j / y_{kj}$. Now, it may be the case that $y_{kj} > 0$ for two or more nonbasic variables. To ensure that (8.2.1) is satisfied for all such j, the lower limit to which Δc_{Bk} can decrease without violating the optimality of the current solution is

$$\underline{lim}\, \Delta c_{Bk} = \max_j \left\{ \bar{c}_j / y_{kj}, y_{kj} > 0 \right\}. \tag{8.3}$$

Finally, if $y_{kj} < 0$, (8.2.1) holds if $\Delta c_{Bk} \leq 0$, i.e. Δc_{Bk} may decrease without bound without affecting primal optimality. Thus, the optimality criterion is affected by changes in c_{Bk} only when $\Delta c_{Bk} > 0$. In this instance (8.2.1) may be written as $\bar{c}_j / y_{kj} \geq \Delta c_{Bk}, j = 1,\ldots,n-m,$ so that the upper limit to any change in c_{Bk} is $\bar{c}_j / y_{kj}, y_{kj} < 0$. Hence x_{Rj} maintains its nonbasic status provided that the increase in c_{Bk} does not exceed $\bar{c}_j / y_{kj}$. If $y_{kj} < 0$ for two or more nonbasic variables, then (8.2.1) holds for all relevant j if the upper limit to which Δc_{Bk} can increase is specified as

$$\overline{lim}\, \Delta c_{Bk} = \min_j \left\{ \bar{c}_j / y_{kj}, y_{kj} < 0 \right\}. \tag{8.4}$$

Actually, the optimality criterion (8.2.1) will be maintained if both the lower and upper limits (8.3), (8.4), respectively, are subsumed under the inequality

$$\underline{lim}\, \Delta c_{Bk} = \max_j \left\{ \bar{c}_j / y_{kj}, y_{kj} > 0 \right\} \leq \Delta c_{Bk} \leq \overline{lim}\, \Delta c_{Bk} = \min_j \left\{ \bar{c}_j / y_{kj}, y_{kj} < 0 \right\}. \tag{8.5}$$

Hence (8.5) provides the bounds within which Δc_{Bk} may decrease or increase without affecting the optimality of the current solution. As an alternative to the approach pursued by defining (8.3), (8.4), we may express these limits in terms of the new level of c_{Bk}, call it $\hat{c}_{Bk}$, by noting that, for $\Delta c_{Bk} = \hat{c}_{Bk} - c_{Bk}$, (8.3) becomes

$$\underline{lim}\, \hat{c}_{Bk} = \max_j \left\{ c_{Bk} + \bar{c}_j / y_{kj}, y_{kj} > 0 \right\}; \tag{8.3.1}$$

while (8.4) may be rewritten as

$$\overline{lim}\,\hat{c}_{Bk} = \min_j \{c_{Bk} + \bar{c}_j / y_{kj}, y_{kj} < 0\}. \tag{8.4.1}$$

Hence (8.3.1) (8.4.1) specifies the lower (upper) limit to which c_{Bk} can decrease (increase) without violating the optimality criterion (8.2.1). In view of this modification, (8.5) appears as

$$\underline{lim}\,\hat{c}_{Bk} = \max_j \{c_{Bk} + \bar{c}_j / y_{kj}, y_{kj} > 0\} \le \hat{c}_{Bk} \le \overline{lim}\,\hat{c}_{Bk} = \min_j \{c_{Bk} + \bar{c}_j / y_{kj}, y_{kj} < 0\}. \tag{8.5.1}$$

One final point is in order. If c_{Bk} changes in a fashion such that (8.5) and (8.5.1) are satisfied, then the current basic feasible solution remains optimal but the value of the objective function changes to $f = \boldsymbol{C}_B' \boldsymbol{B}^{-1} \boldsymbol{b} + \Delta c_{Bk} x_{Bk}$.

Example 8.1 We found in Chapter 4 that the optimal simplex matrix associated with the problem:

$$\begin{aligned} \max f &= 2x_1 + 3x_2 + 4x_3 \quad \text{s.t.} \\ 2x_1 + x_2 + 4x_3 &\le 100 \\ x_1 + 3x_2 + x_3 &\le 80 \\ x_1, x_2, x_3 &\ge 0 \end{aligned}$$

appears as

$$\begin{bmatrix} \frac{5}{11} & 0 & 1 & \frac{3}{11} & -\frac{1}{11} & 0 & 20 \\ \frac{2}{11} & 1 & 0 & -\frac{1}{11} & \frac{4}{11} & 0 & 20 \\ \frac{4}{11} & 0 & 0 & \frac{9}{11} & \frac{8}{11} & 1 & 140 \end{bmatrix}.$$

A glance at the objective function row of this matrix reveals that $-\bar{c}_1 = \frac{4}{11}$, $-\bar{c}_2 = \frac{9}{11}$, and $-\bar{c}_3 = \frac{8}{11}$. Hence the above discussion indicates that $x_{R1} = x_1$, $x_{R2} = x_4$, and $x_{R3} = x_5$ should each maintain their nonbasic status provided that their objective function coefficients do not increase by more than $\Delta c_{R1} = \frac{4}{11}$, $\Delta c_{R2} = \frac{9}{11}$, and $\Delta c_{R3} = \frac{8}{11}$ units, respectively. Additionally, the upper limits to which $c_{R1} = c_1$, $c_{R2} = c_4$, and $c_{R3} = c_5$ can individually increase without violating primal optimality are provided in Table 8.1.

Table 8.1 Limits to which c_{R1}, c_{R2}, and c_{R3} can increase or decrease without violating primal optimality.

Nonbasic variables	Objective function coefficients	Lower limit	Upper limit
$x_{R1} = x_1$	$c_{R1} = c_1 = 2$	$-\infty$	$c_{R1} - \bar{c}_1 = \frac{26}{11}$
$x_{R2} = x_4$	$c_{R2} = c_4 = 0$	$-\infty$	$c_{R2} - \bar{c}_2 = \frac{9}{11}$
$x_{R3} = x_5$	$c_{R3} = c_5 = 0$	$-\infty$	$c_{R3} - \bar{c}_3 = \frac{8}{11}$

Next, from (8.5), we see that the limits within which $\Delta c_{B1} = \Delta c_3, \Delta c_{B2} = \Delta c_2$ can vary without affecting the optimality of the current solution are

$$\underline{lim}\,\Delta c_{B1} = max\left\{\frac{\bar{c}_1}{y_{11}} = \frac{-4/11}{5/11} = -\frac{4}{5}, \frac{\bar{c}_2}{y_{12}} = \frac{-9/11}{3/11} = -3\right\} = -\frac{4}{5} \leq \Delta c_{B1}$$
$$\leq \overline{lim}\,\Delta c_{B1} = min\left\{\frac{\bar{c}_3}{y_{13}} = \frac{-8/11}{4/11} = 8\right\} = 8;$$

$$\underline{lim}\,\Delta c_{B2} = max\left\{\frac{\bar{c}_1}{y_{21}} = \frac{-4/11}{2/11} = -2, \frac{\bar{c}_3}{y_{23}} = \frac{-8/11}{4/11} = -2\right\} = -2 \leq \Delta c_{B2}$$
$$\leq \overline{lim}\,\Delta c_{B2} = min\left\{\frac{\bar{c}_2}{y_{22}} = \frac{-8/11}{-1/11} = 9\right\} = 9.$$

Moreover, we may utilize (8.5.1) to specify the limits within which $c_{B1} = c_3$, $c_{B2} = c_2$ can individually decrease or increase while preserving optimality (Table 8.2). ■

Table 8.2 Limits to which c_{B1}, c_{B2} can increase or decrease while preserving primal optimality.

Basic variables	Objective function coefficients	Lower limit	Upper limit
$x_{B1} = x_3$	$c_{B1} = c_3 = 4$	$c_{B1} + \frac{\bar{c}_1}{y_{11}} = \frac{16}{5}$	$c_{B1} + \frac{\bar{c}_3}{y_{13}} = 12$
$x_{B2} = x_2$	$c_{B2} = c_2 = 3$	$c_{B2} + \frac{\bar{c}_2}{y_{21}} = c_{B2} + \frac{\bar{c}_3}{y_{23}} = 1$	$c_{B2} + \frac{\bar{c}_2}{y_{22}} = 12$

8.2.2 Changing a Component of the Requirements Vector

If we next change $\boldsymbol{b}$ to $\boldsymbol{b}+\Delta\boldsymbol{b}$, the new vector of basic variables becomes

$$\begin{aligned}\hat{X}_B &= B^{-1}(b+\Delta b)\\ &= B^{-1}b + B^{-1}\Delta b.\end{aligned} \tag{8.6}$$

If this new basic solution is primal feasible $(\hat{X}_B \geq O)$, then, since $\hat{X}_B$ is independent of $\bar{C}$, primal optimality or dual feasibility is also preserved. For $\Delta b' = (\Delta b_1, \ldots, \Delta b_m)$, (8.6) becomes

$$\hat{X}_B = B^{-1}b + \begin{bmatrix}\beta_1\Delta b\\ \vdots\\ \beta_m\Delta b\end{bmatrix} = X_B + \begin{bmatrix}\sum_{j=1}^{m}\beta_{ij}\Delta b_j\\ \vdots\\ \sum_{j=1}^{m}\beta_{mj}\Delta b_j\end{bmatrix} \text{ or} \tag{8.7}$$

$$\hat{x}_{Bi} = x_{Bi} + \sum_{j=1}^{m}\beta_{ij}\Delta b_j, i = 1,\ldots,m, \tag{8.7.1}$$

where $\beta_i = (\beta_{i1}, \ldots, \beta_{im})$, $i = 1, \ldots, m$, denotes the ith row of B^{-1}. Let us assume that b_k changes. In order for this change to ensure the feasibility of the resulting basic solution, we require, by virtue of (8.7.1), that

$$x_{Bi} + \beta_{ik}\Delta b_k \geq 0, i = 1,\ldots,m. \tag{8.7.2}$$

Under what conditions do changes in b_k preserve the sense of this inequality? First, if $\beta_{ik} = 0$, then (8.7.2) holds for any Δb_k. Second, for $\beta_{ik} > 0$, (8.7.2) is satisfied if $\Delta b_k \geq 0$, i.e. Δb_k may increase without bound so that its upper limit is $+\infty$. Thus, the feasibility of the new solution is affected only when $\Delta b_k < 0$. Since (8.7.2) implies that $-x_{Bi}/\beta_{ik} \leq \Delta b_k$, $i = 1, \ldots, m$, it follows that the lower limit to which Δb_k can decrease without violating primal feasibility is $-x_{Bi}/\beta_{ik}$, $\beta_{ik} > 0$. Hence x_{Bi} remains in the set of basic variables as long as b_k does not decrease by more than $-x_{Bi}/\beta_{ik}$. If $\beta_{ik} > 0$ for two or more basic variables x_{Bi}, then, to ensure that (8.7.2) is satisfied for all such i, the lower limit to which Δb_k can decrease without violating the feasibility of the new primal solution is

$$\underline{lim}\,\Delta b_k = \max_i\{-x_{Bi}/\beta_{ik}, \beta_{ik} > 0\}. \tag{8.8}$$

Finally, if $\beta_{ik} < 0$, (8.7.2) holds if $\Delta b_k < 0$, i.e. Δb_k can decrease without bound so that its lower limit is $-\infty$. Thus, the feasibility of the new basic solution is affected by changes in b_k only when $\Delta b_k > 0$. In this instance (8.7.2) may be rewritten as $-x_{Bi}/\beta_{ik} \geq \Delta b_k$, $i = 1, \ldots, m$, so that the upper limit to Δb_k is $-x_{Bi}/\beta_{ik}$, $\beta_{ik} < 0$. Hence x_{Bi} maintains its basic status provided that b_k does not increase by more than $-x_{Bi}/\beta_{ik}$. And if $\beta_{ik} < 0$ for more than one basic variable

x_{Bi}, then (8.7.2) holds for all relevant i if the upper limit to which Δb_k can increase is specified as

$$\overline{lim}\,\Delta b_k = \min_i\{-x_{Bi}/\beta_{ik}, \beta_{ik} < 0\}. \tag{8.9}$$

If we choose to combine the lower and upper limits (8.8), (8.9), respectively, into a single expression, then (8.7.2) will be maintained if

$$\underline{lim}\,\Delta b_k = \max_i\{-x_{Bi}/\beta_{ik}, \beta_{ik} > 0\} \leq \overline{lim}\,\Delta b_k = \min_i\{-x_{Bi}/\beta_{ik}, \beta_{ik} < 0\}. \tag{8.10}$$

Hence (8.10) provides the bounds within which Δb_k may decrease or increase without affecting the feasibility of $\hat{X}_B$. As an alternative to (8.10), we may express the limits contained therein in terms of the new level of b_k, call it $\hat{b}_k$, by noting that, since $\Delta b_k = \hat{b}_k - b_k$,

$$\begin{aligned} \underline{lim}\,\Delta\hat{b}_k &= \max_i\{b_k - x_{Bi}/\beta_{ik}, \beta_{ik} > 0\} \leq \hat{b}_k \leq \overline{lim}\,b_k \\ &= \min_i\{b_k - x_{Bi}/\beta_{ik}, \beta_{ik} < 0\}. \end{aligned} \tag{8.10.1}$$

What about the effect of a change in a component of the requirements vector on the optimal value of the objective function? By virtue of the preceding discussion, if (8.10) or (8.10.1) is satisfied, those variables that are currently basic remain so. However, a change in any of the requirements b_i, $i = 1. \ldots, m$, alters the optimal values of these variables, which in turn affect the optimal value of f. If b_k alone changes, the magnitude of the change in f precipitated by a change in b_k is simply $\Delta f = u_k \Delta b_k$ since, as noted in Chapter 6, $u_k = \Delta f / \Delta b_k$ represents a measure of the amount by which the optimal value of the primal objective function changes per unit change in the kth requirement b_k.

Example 8.2 Let us again look to the optimal simplex matrix provided at the outset of the previous example problem. In the light of (6.4), if we examine the columns of this final primal simplex matrix that were, in the original matrix, the columns of the identity matrix, we find that the ith column of $\boldsymbol{B}^{-1}$ appears under the column that was initially the ith column of $\boldsymbol{I}_2$. Hence the inverse of the optimal basis matrix is

$$\boldsymbol{B}^{-1} = \begin{bmatrix} {}^{3}/_{11} & -{}^{1}/_{11} \\ -{}^{1}/_{11} & {}^{4}/_{11} \end{bmatrix}.$$

Table 8.3 Limits within which b_1, b_2 can decrease or increase without violating primal optimality.

Basic variables	Original requirements	Lower limit	Upper limit
$x_{B1} = x_3$	$b_1 = 100$	$b_1 - x_{B1}/\beta_{11} = 26.67$	$b_1 - x_{B2}/\beta_{21} = 320$
$x_{B2} = x_2$	$b_2 = 80$	$b_2 - x_{B2}/\beta_{22} = 25$	$b_2 - x_{B1}/\beta_{12} = 300$

By utilizing the components of this inverse, we may determine the limits within which Δb_1, Δb_2 can vary without affecting the feasibility of the new vector of basic variables according to

$$\underline{lim}\,\Delta b_1 = max\left\{\frac{-x_{B1}}{\beta_{11}} = \frac{-20}{3/11} = -73.33\right\} = -73.33 \leq \Delta b_1 \leq \overline{lim}\,\Delta b_1 =$$

$$min\left\{\frac{-x_{B2}}{\beta_{21}} = \frac{-20}{-1/11} = 220\right\} = 220;$$

$$\underline{lim}\,\Delta b_2 = max\left\{\frac{-x_{B2}}{\beta_{22}} = \frac{-20}{4/11} = -55\right\} = -55 \leq \Delta b_2 \leq \overline{lim}\,\Delta b_2 =$$

$$min\left\{\frac{-x_{B1}}{\beta_{12}} = \frac{-20}{-1/11} = 220\right\} = 220.$$

Additionally, we may utilize (8.10.1) to specify the limits within which b_1, b_2 can individually decrease or increase without violating the feasibility of $\hat{X}_B$ (Table 8.3). ■

8.2.3 Changing a Component of the Coefficient Matrix

We now look to the effect on the optimal solution of a change in a component of the coefficient matrix $\bar{A}$ associated with the structural constraint system $\bar{A}\bar{X} = b$. Since $\bar{A}$ was previously partitioned as $\bar{A} = [B \vdots R]$, it may be the case that some component r_{ij}, $i = 1, \ldots, m$; $j = 1, \ldots, n - m$, of a nonbasic vector r_j changes; or a component b_{ki}, $i, k = 1, \ldots, m$, of a basic vector b_i changes.

Let us begin by assuming that the lth component of the nonbasic vector $r'_k = (r_{1k}, \ldots, r_{lk}, \ldots, r_{mk})$ changes to $\hat{r}_{lk} = r_{lk} + \Delta r_{lk}$ so that r_k is replaced by $\hat{r}_k = r_k + \Delta r_{lk} e_l$ in $R = [r_1, \ldots, r_k, \ldots, r_{n-m}]$. Hence the matrix of nonbasic vectors may now be expressed as $\hat{R} = [r_1, \ldots, \hat{r}_k, \ldots, r_{n-m}]$. With $\hat{X}_B = B^{-1}b - B^{-1}RX_{\hat{R}}$ independent of Δr_{lk} at a basic feasible solution (since $X_{\hat{R}} = O$), it follows that this nonbasic solution is primal feasible while, if $-C'_{\hat{R}} + C'_B B^{-1}\hat{R} = -C'_{\hat{R}} + U'\hat{R} \geq O'$, dual feasibility or primal optimality is also preserved. Since

$$\begin{aligned} -C'_{\hat{R}} + U'\hat{R} &= (-c_{R1} + U'r_1, \ldots, -c_{Rk} + U'r_k + u_l \Delta r_{lk}, \ldots, -c_{R,n-m} + U'r_{n-m}) \\ &= (-\bar{c}_1, \ldots, -\bar{c}_k + u_l \Delta r_{lk}, \ldots, -\bar{c}_{n-m}) \end{aligned} \quad (8.11)$$

and $-\bar{c}_j \geq 0, j = 1, \ldots, k, \ldots, n-m$, it is evident that primal optimality is maintained if

$$-\bar{c}_k + u_l \Delta r_{lk} \geq 0. \quad (8.12)$$

Under what circumstances will changes in r_{lk} preserve the sense of this inequality? First, if $u_l = 0$, then (8.12) holds regardless of the direction or magnitude of the change in r_{lk}. Second, for $u_l > 0$, (8.12) is satisfied if $\Delta r_{lk} \geq 0$, i.e. Δr_{lk} may increase without bound so that its upper limit becomes $+\infty$. Hence (8.12) is affected only when r_{lk} decreases or $\Delta r_{lk} < 0$. Since (8.12) implies that $\bar{c}_k / u_l \leq \Delta r_{lk}$, it is evident that the lower limit to which Δr_{lk} can decrease without violating optimality is $\bar{c}_k / u_l, u_l > 0$. Thus, $\hat{r}_k$ enters the basis (or $x_{\hat{R}k}$ enters the set of basic variables) only if the decrease in r_{lk} is greater than $\bar{c}_k / u_l, u_l > 0$. If we express this lower limit in terms of the new level of $r_{lk}, \hat{r}_{lk}$, then, for $\Delta r_{lk} = \hat{r}_{lk} - r_{lk}$,

$$\underline{lim}\, \hat{r}_{lk} = r_{lk} + \bar{c}_k / u_l, u_l > 0. \quad (8.13)$$

Next, let us assume that the lth component of the basis vector $b'_k = (b_{1k}, \ldots, b_{lk}, \ldots, b_{mk})$ changes to $\hat{b}_{lk} = b_{lk} + \Delta b_{lk}$ so that b_k is replaced in the basis matrix $B = [b_1, \ldots, b_k, \ldots, b_m]$ by

$$\hat{b}_k = \begin{bmatrix} b_{1k} \\ \vdots \\ \hat{b}_{lk} \\ \vdots \\ b_{mk} \end{bmatrix} = \begin{bmatrix} b_{1k} \\ \vdots \\ b_{lk} + \Delta b_{lk} \\ \vdots \\ b_{mk} \end{bmatrix} = b_k + \Delta b_{lk} e_l.$$

Hence our initial problem becomes one of obtaining the inverse of the new basis matrix $\hat{B} = \left[b_1, \ldots, \hat{b}_k, \ldots, b_m\right]$. Then once $\hat{B}^{-1}$ is known, it can be shown that

$$X_{\hat{B}} = \hat{B}^{-1} b = \begin{bmatrix} x_{B1} - \dfrac{\beta_{1l} \Delta b_{lk}}{1 + \beta_{kl} \Delta b_{lk}} x_{Bk} \\ \vdots \\ \dfrac{x_{Bk}}{1 + \beta_{kl} \Delta b_{lk}} \\ \vdots \\ x_{Bm} - \dfrac{\beta_{ml} \Delta b_{lk}}{1 + \beta_{kl} \Delta b_{lk}} x_{Bk} \end{bmatrix}.$$

For $X_{\hat{B}}$ primal feasible, we require that

$$1+\beta_{kl}\Delta b_{lk}>0,$$
$$x_{Bi}-\frac{\beta_{il}\Delta b_{lk}}{1+\beta_{kl}\Delta b_{lk}}x_{Bk}\geq 0, i=1,\ldots,m.$$

If $1+\beta_{kl}\Delta b_{lk}>0$, then upon solving the second inequality for Δb_{lk},

$$\begin{aligned}&\text{(a)}\ \frac{x_{Bi}}{\beta_{il}x_{Bk}-\beta_{kl}x_{Bi}}\leq \Delta b_{lk} \text{ for } \beta_{il}x_{Bk}-\beta_{kl}x_{Bi}<0;\\ &\text{(b)}\ \frac{x_{Bi}}{\beta_{il}x_{Bk}-\beta_{kl}x_{Bi}}\geq \Delta b_{lk} \text{ for } \beta_{il}x_{Bk}-\beta_{kl}x_{Bi}>0, i=1,\ldots,m.\end{aligned} \tag{8.14}$$

Since these inequalities must hold for all $i\neq k$ when $1+\beta_{kl}\Delta b_{lk}>0$, the lower and upper limits to which Δb_{lk} can decrease or increase without violating primal feasibility may be expressed, from (8.14 a, b), respectively, as

$$\begin{aligned}\underline{lim}\,\Delta b_{lk}&=\max_{i\neq k}\left\{\frac{x_{Bi}}{\beta_{il}x_{Bk}-\beta_{kl}x_{Bi}}\middle|\beta_{il}x_{Bk}-\beta_{kl}x_{Bi}<0\right\}\leq \Delta b_{lk}\\ &\leq \overline{lim}\,\Delta b_{lk}=\min_{i\neq k}\left\{\frac{x_{Bi}}{\beta_{il}x_{Bk}-\beta_{kl}x_{Bi}}\middle|\beta_{il}x_{Bk}-\beta_{kl}x_{Bi}>0\right\}.\end{aligned} \tag{8.15}$$

If we re-express these limits in terms of the new level of b_{lk}, $\hat{b}_{lk}$, then, for $\Delta b_{lk}=\hat{b}_{lk}-b_{lk}$, (8.15) becomes

$$\begin{aligned}\underline{lim}\,\hat{b}_{lk}&=\max_{i\neq k}\left\{b_{lk}+\frac{x_{Bi}}{\beta_{il}x_{Bk}-\beta_{kl}x_{Bi}}\middle|\beta_{il}x_{Bk}-\beta_{kl}x_{Bi}<0\right\}\leq \hat{b}_{lk}\\ &\leq \overline{lim}\,\hat{b}_{lk}=\min_{i\neq k}\left\{b_{lk}+\frac{x_{Bi}}{\beta_{il}x_{Bk}-\beta_{kl}x_{Bi}}\middle|\beta_{il}x_{Bk}-\beta_{kl}x_{Bi}>0\right\}.\end{aligned} \tag{8.15.1}$$

What is the effect of a change in b_{lk} upon the primal optimality (dual feasibility) requirement $C'_{\hat{B}}\hat{B}^{-1}R-C'_R\geq O'$? Here, too, it can be shown that the lower and upper limits to which Δb_{lk} can decrease or increase without violating primal optimality may be written as

$$\begin{aligned}\underline{lim}\,\Delta b_{lk}&=\max_{j}\left\{\frac{-\bar{c}_j}{y_{kj}u_l+\bar{c}_j\beta_{kl}}\middle|y_{kj}u_l+\bar{c}_j\beta_{kl}<0\right\}\leq \Delta b_{lk}\\ &\leq \overline{lim}\,\Delta b_{lk}=\min_{j}\left\{\frac{-\bar{c}_j}{y_{kj}u_l+\bar{c}_j\beta_{kl}}\middle|y_{kj}u_l+\bar{c}_j\beta_{kl}>0\right\}.\end{aligned} \tag{8.16}$$

Alternatively, we may write these limits in terms of the new level of b_{lk}, $\hat{b}_{lk}$, since, if $\Delta b_{lk} = \hat{b}_{lk} - b_{lk}$, (8.16) becomes

$$\begin{aligned} \underline{lim}\, \hat{b}_{lk} &= \max_j \left\{ b_{lk} + \frac{-\bar{c}_j}{y_{kj}u_l + \bar{c}_j\beta_{kl}} \middle| y_{kj}u_l + \bar{c}_j\beta_{kl} < 0 \right\} \leq \hat{b}_{lk} \\ &\leq \overline{lim}\, \hat{b}_{lk} = \min_j \left\{ b_{lk} + \frac{-\bar{c}_j}{y_{kj}u_l + \bar{c}_j\beta_{kl}} \middle| y_{kj}u_l + \bar{c}_j\beta_{kl} > 0 \right\}. \end{aligned} \tag{8.16.1}$$

We next determine the effect of a change in the element b_{lk} upon the optimal value of the objective function. Again dispensing with the derivation we have the result that the new value of f is

$$f\left(X_{\hat{B}}\right) = f(X_B) - \frac{\Delta b_{lk} x_{Bk} u_l}{1 + \beta_{kl}\Delta b_{lk}} \tag{8.17}$$

Example 8.3 It was indicated at the beginning of Example 8.1 that the optimal simplex matrix corresponding to the problem:

$$\begin{aligned} \max f = 2x_1 + 3x_2 + 4x_3 \quad \text{s.t.} \\ 2x_1 + x_2 \quad + 4x_3 \leq 100 \\ x_1 + 3x_2 + \quad x_3 \leq 80 \\ x_1, x_2, x_3 \geq 0 \end{aligned}$$

is

$$\begin{bmatrix} \frac{5}{11} & 0 & 1 & \frac{3}{11} & -\frac{1}{11} & 0 & 20 \\ \frac{2}{11} & 1 & 0 & -\frac{1}{11} & \frac{4}{11} & 0 & 20 \\ \frac{4}{11} & 0 & 0 & \frac{9}{11} & \frac{8}{11} & 1 & 140 \end{bmatrix}.$$

Moreover, since x_2, x_3 are basic variables,

$$\boldsymbol{B} = \begin{bmatrix} b_{11} & b_{12} \\ b_{21} & b_{22} \end{bmatrix} = \begin{bmatrix} 4 & 1 \\ 1 & 3 \end{bmatrix}, \boldsymbol{B}^{-1} = \begin{bmatrix} \beta_{11} & \beta_{12} \\ \beta_{21} & \beta_{22} \end{bmatrix} = \begin{bmatrix} \frac{3}{11} & -\frac{1}{11} \\ -\frac{1}{11} & \frac{4}{11} \end{bmatrix},$$

$$\boldsymbol{R} = \begin{bmatrix} r_{11} & r_{12} & r_{13} \\ r_{21} & r_{22} & r_{23} \end{bmatrix} = \begin{bmatrix} 2 & 1 & 0 \\ 1 & 0 & 1 \end{bmatrix}.$$

From the objective function row of the optimal simplex matrix we find, for instance, that $\bar{c}_1 = -\frac{4}{11}$. Hence $x_{R1} = x_1$ will maintain its nonbasic status provided that r_{11}, r_{21} do not decrease by more than $\Delta r_{11} = \bar{c}_1/u_1 = -\frac{4}{9}, \Delta r_{21} = \bar{c}_1/u_2 = -\frac{1}{2}$, respectively. In addition, (8.13) may be utilized to depict the lower limits to which r_{11}, r_{21}, can individually decrease without violating primal optimality (Table 8.4).

Next, from (8.15), we see that the limits within which the components of the basis vectors $\boldsymbol{b}_1' = (b_{11}, b_{21}), \boldsymbol{b}_2' = (b_{12}, b_{22})$ can vary without violating primal feasibility are:

$$\underline{lim}\,\Delta b_{11} = \max_{i \neq 1}\left\{\frac{x_{B2}}{\beta_{21}x_{B1} - \beta_{11}x_{B2}} = \frac{20}{\left(-\frac{1}{11}\right)(20) - \left(\frac{3}{11}\right)(20)}\right\}$$

$$= -2.75 \leq \Delta b_{11} \leq \overline{lim}\,\Delta b_{11} = +\infty;$$

$$\underline{lim}\,\Delta b_{21} = -\infty \leq \Delta b_{21} \leq \overline{lim}\,\Delta b_{21}$$

$$= \min_{i \neq 1}\left\{\frac{x_{B2}}{\beta_{22}x_{B1} - \beta_{12}x_{B2}} = \frac{20}{\left(\frac{4}{11}\right)(20) - \left(-\frac{1}{11}\right)(20)}\right\} = 2.20;$$

$$\underline{lim}\,\Delta b_{12} = -\infty \leq \Delta b_{12} \leq \overline{lim}\,\Delta b_{12}$$

$$= \min_{i \neq 2}\left\{\frac{x_{B1}}{\beta_{11}x_{B2} - \beta_{21}x_{B1}} = \frac{20}{\left(\frac{3}{11}\right)(20) - \left(-\frac{1}{11}\right)(20)}\right\} = 2.75;$$

$$\underline{lim}\,\Delta b_{22} = \max_{i \neq 2}\left\{\frac{x_{B1}}{\beta_{12}x_{B2} - \beta_{22}x_{B1}} = \frac{20}{\left(-\frac{1}{11}\right)(20) - \left(\frac{4}{11}\right)(20)}\right\}$$

$$= -2.20 \leq \Delta b_{22} \leq \overline{lim}\,\Delta b_{22} = +\infty.$$

Table 8.4 Limits to which r_{11}, r_{21} can decrease or increase without violating primal optimality.

Component of r_1	Lower limit	Upper limit
$r_{11} = 2$	$r_{11} + \bar{c}_1/u_1 = 14/9$	$+\infty$
$r_{21} = 1$	$r_{21} + \bar{c}_1/u_2 = \frac{1}{2}$	$+\infty$

Moreover, we may employ (8.15.1) to specify the limits to which b_{11}, b_{21}, b_{12} and b_{22} can individually decrease or increase without affecting the feasibility of the current solution (Table 8.5).

We now turn to the specification of the limits within which the components of $\boldsymbol{b}_1, \boldsymbol{b}_2$ can vary without affecting primal optimality. From (8.16):

$$\underline{lim}\,\Delta b_{11} = max\left\{\frac{-\bar{c}_3}{y_{13}u_1 + \bar{c}_3\beta_{11}} = \frac{8/11}{\left(-\frac{1}{11}\right)\left(\frac{9}{11}\right)-\left(\frac{8}{11}\right)\left(\frac{3}{11}\right)}\right\}$$

$$= -2.66 \le \Delta b_{11} \le \overline{lim}\,\Delta b_{11}$$

$$= min\left\{\frac{-\bar{c}_1}{y_{11}u_1 + \bar{c}_1\beta_{11}} = \frac{4/11}{\left(\frac{5}{11}\right)\left(\frac{9}{11}\right)-\left(\frac{4}{11}\right)\left(\frac{3}{11}\right)}\right\} = 1.33;$$

$$\underline{lim}\,\Delta b_{21} = -\infty \le \Delta b_{21} \le \overline{lim}\,\Delta b_{21}$$

$$= min\left\{\frac{-\bar{c}_1}{y_{11}u_2 + \bar{c}_1\beta_{12}} = \frac{4/11}{\left(\frac{5}{11}\right)\left(\frac{8}{11}\right)-\left(\frac{4}{11}\right)\left(-\frac{1}{11}\right)} = 1,\right.$$

$$\left.\frac{-\bar{c}_2}{y_{12}u_2 + \bar{c}_2\beta_{12}} = \frac{9/11}{\left(\frac{3}{11}\right)\left(\frac{8}{11}\right)-\left(\frac{9}{11}\right)\left(-\frac{1}{11}\right)} = 3\right\} = 1;$$

Table 8.5 Limits to which b_{11}, b_{21}, b_{12} and b_{22} can decrease or increase without violating primal feasibility.

Component of b_i, $i = 1, 2$	Lower limit	Upper limit
$b_{11} = 4$	$b_{11} + \frac{x_{B2}}{\beta_{21}x_{B1} - \beta_{11}x_{B2}} = 1.25$	$+\infty$
$b_{21} = 1$	$-\infty$	$b_{21} + \frac{x_{B2}}{\beta_{22}x_{B1} - \beta_{12}x_{B2}} = 3.20$
$b_{12} = 1$	$-\infty$	$b_{12} + \frac{x_{B2}}{\beta_{11}x_{B2} - \beta_{21}x_{B1}} = 3.75$
$b_{22} = 3$	$b_{22} + \frac{x_{B1}}{\beta_{12}x_{B2} - \beta_{22}x_{B1}} = 0.80$	$+\infty$

$$\underline{lim}\,\Delta b_{12} = -\infty \le \Delta b_{12} \le \overline{lim}\,\Delta b_{12}$$

$$= min\left\{\frac{-\bar{c}_1}{y_{21}u_1 + \bar{c}_1\beta_{21}} = \frac{4/11}{\left(\frac{2}{11}\right)\left(\frac{9}{11}\right) - \left(\frac{4}{11}\right)\left(-\frac{1}{11}\right)} = 2,\right.$$

$$\left.\frac{-\bar{c}_3}{y_{23}u_1 + \bar{c}_3\beta_{21}} = \frac{8/11}{\left(\frac{4}{11}\right)\left(\frac{9}{11}\right) - \left(\frac{8}{11}\right)\left(-\frac{1}{11}\right)} = 2\right\} = 2;$$

$$\underline{lim}\,\Delta b_{22} = max\left\{\frac{-\bar{c}_2}{y_{22}u_2 + \bar{c}_2\beta_{22}} = \frac{9/11}{\left(-\frac{1}{11}\right)\left(\frac{8}{11}\right) - \left(\frac{9}{11}\right)\left(\frac{4}{11}\right)}\right\}$$

$$= -2.25 \le \Delta b_{22} \le \overline{lim}\,\Delta b_{22} = +\infty.$$

Additionally, from (8.16.1), we may determine the lower and upper limits to which b_{11}, b_{21}, b_{12} and b_{22} can individually decrease or increase while still preserving primal optimality (Table 8.6).

Since changes in the components of $\boldsymbol{b}_1$, $\boldsymbol{b}_2$ may affect both primal feasibility and optimality, we may combine Tables 8.5 and 8.6 to obtain the lower and upper limits to which b_{11}, b_{21}, b_{12}, and b_{22} can individually decrease or increase without affecting either primal feasibility or optimality (Table 8.7). ■

Table 8.6 Limits to which b_{11}, b_{21}, b_{12}, and b_{22} can decrease or increase without violating primal optimality.

Component of b_i, $i = 1, 2$	Lower limit	Upper limit
$b_{11} = 4$	$b_{11} + \frac{-\bar{c}_3}{y_{13}u_1 + \bar{c}_3\beta_{11}} = 1.34$	$b_{11} + \frac{-\bar{c}_1}{y_{11}u_1 + \bar{c}_1\beta_{11}} = 5.33$
$b_{21} = 1$	$-\infty$	$b_{21} + \frac{-\bar{c}_1}{y_{11}u_2 + \bar{c}_1\beta_{12}} = 2$
$b_{12} = 1$	$-\infty$	$b_{12} + \frac{-\bar{c}_j}{y_{2j}u_1 + \bar{c}_j\beta_{21}} = 2, j = 1,3$
$b_{22} = 3$	$b_{22} + \frac{-\bar{c}_2}{y_{22}u_2 + \bar{c}_2\beta_{22}} = 0.75$	$+\infty$

Table 8.7 Limits within which b_{11}, b_{21}, b_{12}, and b_{22} can decrease or increase, without violating either primal feasibility or optimality.

Component of b_i, $i = 1, 2$	Lower limit	Upper limit
$b_{11} = 4$	1.34	5.33
$b_{21} = 1$	$-\infty$	2
$b_{12} = 1$	$-\infty$	2
$b_{22} = 3$	0.80	$+\infty$

8.3 Summary of Sensitivity Effects

I) **Changing an Objective Function Coefficient.**
  A) Primal feasibility. Always preserved.
  B) Primal optimality.
    1) Changing the coefficient c_{Rj} of a nonbasic variable x_{Rj}:
      a) preserved if $\Delta c_{Rj} < 0$.
      b) preserved if $\Delta c_{Rj} > 0$ and $\Delta c_{Rj} \leq -\bar{c}_j$.
    2) Changing the coefficient c_{Bk} of a basic variable x_{Bk}:
      a) preserved if $y_{kj} = 0$.
      b) $y_{kj} > 0 \begin{cases} \text{preserved if } \Delta c_{Bk} \geq 0 \\ \text{preserved if } \Delta c_{Bk} < 0 \text{ and } \Delta c_{Bk} \geq \bar{c}_j / y_{kj}. \end{cases}$
      c) preserved if more than one $y_{kj} > 0, \Delta c_{Bk} < 0$, and $\Delta c_{Bk} \geq max_j \{\bar{c}_j / y_{kj}\}$.
      d) $y_{kj} < 0 \begin{cases} \text{preserved if } \Delta c_{Bk} \leq 0. \\ \text{preserved if } \Delta c_{Bk} > 0 \text{ and } \Delta c_{Bk} \leq \bar{c}_j / y_{kj} \end{cases}$
      e) preserved if more than one $y_{kj} < 0, \Delta c_{Bk} > 0$, and $\Delta c_{Bk} \leq min_j \{\bar{c}_j / y_{kj}\}$.

II) **Changing a Component of the Requirement Vector.**
  A) Primal optimality. Always preserved.
  B) Primal feasibility, i.e. conditions under which all $x_{Bi} \geq 0$. Changing the requirement b_k:
    a) preserved if $\beta_{ik} = 0$.
    b) $\beta_{ik} > 0 \begin{cases} \text{preserved if } \Delta b_k \geq 0; \\ \text{preserved if } \Delta b_k < 0 \text{ and } \Delta b_k \geq -x_{Bi} / \beta_{ik}. \end{cases}$
    c) preserved if more than one $\beta_{ik} > 0, \Delta b_k < 0$, and $\Delta b_k \geq max_j \{-x_{Bi} / \beta_{ik}\}$.

d) $\beta_{ik} < 0 \begin{cases} \text{preserved if } \Delta b_k < 0; \\ \text{preserved if } \Delta b_k > 0 \text{ and } \Delta b_k \le -x_{Bi}/\beta_{ik}. \end{cases}$

e) preserved if more than one
$\beta_{ik} < 0, \Delta b_k > 0, \text{and } \Delta b_k \le min_j\{-x_{Bi}/\beta_{ik}\}.$

III) **Changing a Component of an Activity Vector.**

A) Changing a Component r_{lk} of a Nonbasic Activity Vector $\boldsymbol{r}_k$.

1) Primal feasibility. Always preserved.

2) Primal optimality:

a) Preserved if $u_l = 0$.

b) $u_l > 0 \begin{cases} \text{preserved if } \Delta r_{lk} \ge 0; \\ \text{preserved if } \Delta r_{lk} < 0 \text{ and } \Delta r_{lk} \ge \bar{c}_k / u_l. \end{cases}$

B) Changing a Component b_{lk} of a Basic Activity Vector $\boldsymbol{b}_k$.

1) Primal feasibility, i.e. conditions under which all $x_{Bi} \ge 0$.

$$1 + \beta_{kl}\Delta b_{lk} > 0 \begin{cases} \text{preserved if } \beta_{il}x_{Bk} - \beta_{kl}x_{Bi} < 0 \text{ and} \\ \Delta b_{lk} \ge \max_{i \ne k}\left\{\dfrac{x_{Bi}}{\beta_{il}x_{Bk} - \beta_{kl}x_{Bi}}\right\}; \\ \text{preserved if } \beta_{il}x_{Bk} - \beta_{kl}x_{Bi} > 0 \text{ and} \\ \Delta b_{lk} \le \min_{i \ne k}\left\{\dfrac{x_{Bi}}{\beta_{il}x_{Bk} - \beta_{kl}x_{Bi}}\right\}. \end{cases}$$

2) Primal optimality, i.e. conditions under which all $-\bar{c}_j \ge 0$.

$$1 + \beta_{kl}\Delta b_{lk} > 0 \begin{cases} \text{preserved if } y_{kj}u_l + \bar{c}_j\beta_{kl} < 0 \text{ and} \\ \Delta b_{lk} \ge \max_{j}\left\{\dfrac{-\bar{c}_j}{y_{kj}u_l + \bar{c}_j\beta_{kl}}\right\}; \\ \text{preserved if } y_{kj}u_l + \bar{c}_j\beta_{kl} > 0 \text{ and} \\ \Delta b_{lk} \le \min_{j}\left\{\dfrac{-\bar{c}_j}{y_{kj}u_l + \bar{c}_j\beta_{kl}}\right\}. \end{cases}$$

Example 8.4 Given that a profit-maximizing firm produces two products x_1, x_2 using the activities

$$\boldsymbol{a}_1 = \begin{bmatrix} 10 \\ 2 \\ 3 \\ 0 \end{bmatrix}, \boldsymbol{a}_2 = \begin{bmatrix} 8 \\ 3 \\ 0 \\ 5 \end{bmatrix},$$

let us assume that the first two inputs are fixed at the levels $b_1 = 40$, $b_2 = 30$, respectively, while the third and fourth factors are freely variable. (Note the special structure of these activities, i.e. input three is used only in activity one while input four is used exclusively by activity two.) If the unit price of product one (two) is $p_1 = \$25$ ($p_2 = \$30$) and the price per unit of factor three (four) is $q_3 = \$3$ ($q_4 = \$4$), then

$$\begin{aligned} TR &= 25x_1 + 30x_2, \\ TVC_1 &= 3\cdot 3 + 4\cdot 0 = 9, \\ TVC_2 &= 3\cdot 0 + 4\cdot 5 = 20, \\ TVC &= 9x_1 + 20x_2 \end{aligned}$$

and thus total profit appears as $f = (25 - 9)x_1 + (30 - 20)x_2 = 16x_1 + 10x_2$. Hence the profit maximization problem to be solved is

$$\begin{aligned} max\, f = 16x_1 + 10x_2 \quad & \text{s.t.} \\ 10x_1 + \ 8x_2 &\le 40 \\ 2x_1 + \ 3x_2 &\le 30 \\ x_1, x_2 &\ge 0. \end{aligned}$$

It is easily shown that the optimal simplex matrix associated with this problem is

$$\begin{bmatrix} 1 & \frac{4}{5} & \frac{1}{10} & 0 & 0 & 4 \\ 0 & \frac{7}{5} & -\frac{1}{5} & 1 & 0 & 22 \\ 0 & \frac{14}{5} & \frac{8}{5} & 0 & 1 & 64 \end{bmatrix},$$

where columns 3 and 4 correspond to nonnegative slack variables. Here the firm produces 4 units of x_1 at a profit of \$64. Given this optimal basic feasible solution, let us subject it to a complete sensitivity analysis so as to determine how changes in product and factor prices, resource availability, and the firm's technology affect the feasibility and optimality of the original solution.

A) Changing Product and Factor Prices.

From above, $c_{B1} = p_1 - TVC_1 = 25 - 9 = 16$. Since both $y_{11} = \frac{4}{5}, y_{12} = \frac{1}{10}$ are positive, primal optimality is preserved if p_1 increases (TVC_1 constant) or TVC_1 decreases (p_1 constant) or p_1 increases and, at the same time, TVC_1 decreases. In addition, if p_1 decreases (TVC_1 constant), primal optimality is preserved if the decrease does not exceed

$$\max_j \left\{ \frac{\bar{c}_1}{y_{11}} = \frac{-14/5}{4/5} = -\frac{14}{4}, \frac{\bar{c}_2}{y_{12}} = \frac{-8/5}{1/10} = -16 \right\} = \frac{14}{4}.$$

And if TVC_1 increases (p_1 constant), primal optimality is maintained if the increase does not exceed

$$\min_j\left\{-\frac{\bar{c}_1}{y_{11}}=\frac{14/5}{4/5}=\frac{14}{4},-\frac{\bar{c}_2}{y_{12}}=\frac{8/5}{1/10}=16\right\}=\frac{14}{4}.$$[1]

(Note that the increase in the TVC_1 that preserved primal optimality is the negative of the decrease in p_1, which does the same.)

If p_1 decreases and, simultaneously, TVC_1 increases, then, from (8.20), to preserve primal optimality, p_1 cannot decrease by more than

$$\begin{aligned}&\max_j\left\{\frac{\bar{c}_j}{y_{1j}}+\Delta TVC_1<0, y_{1j}>0\right\}=\\&\max_j\left\{-\frac{14}{4}+\Delta TVC_1<0,-16+\Delta TVC_1<0\right\}.\end{aligned}\tag{8.18}$$

For instance, if TVC_1 increases by \$2, then $\Delta TVC_1=2$ and thus (8.18) indicates that, to maintain primal optimality, p_1 cannot decrease by more than $\max\left\{-\frac{3}{2},-14\right\}=-\frac{3}{2}$. Similarly, TVC_1 cannot increase by more than

$$\min_j\left\{\Delta p_1-\frac{\bar{c}_j}{y_{ij}}>0, y_{ij}>0\right\}=\min_j\left\{\Delta p_1+\frac{14}{4}>0,\Delta p_1+16>0\right\}\tag{8.19}$$

in order for primal optimality to be preserved. If p_1 decreases by \$1, $\Delta p_1=-1$ and thus (8.19) yields $\min\left\{\frac{5}{2},15\right\}=\frac{5}{2}$.

Again looking to the primal objective, we see that $c_{R1}=p_2-TVC_2=30-20=10$. Here primal optimality is preserved if p_2 decreases (TVC_2 constant) or TVC_2 increases (p_2 constant) or if p_2 decreases and, at the same time, TVC_2 increases. In general, since $\Delta c_{R1}=\Delta p_2-\Delta TVC_2$, primal optimality is maintained if both p_2, TVC_2 change in a fashion such that $\Delta p_2<\Delta TVC_2$, e.g. even though p_2

---

1 To see this latter point let us use $c_{Bk}=p_k-TVC_k$ to obtain $\Delta c_{Bk}=\Delta p_k-\Delta TVC_k$. Hence (8.2) becomes

$$-\bar{c}_j+y_{kj}\Delta c_{Bk}=-\bar{c}_j+y_{kj}(\Delta p_k-\Delta TVC_k)\geq 0, j=1,\ldots,n-m.\tag{8.20}$$

If $\Delta p_k=0$,this expression may be rewritten as $-\bar{c}_j/y_{kj}\geq\Delta TVC_k$. For $\Delta TVC_k>0$, the upper limit to which ΔTVC_k can increase without violating primal optimality is $-\bar{c}_j/y_{kj}$. With $y_{kj}>0$ for two or more nonbasic variables, the upper limit on any increase in ΔTVC_k is thus

$$\min_j\left\{\frac{-\bar{c}_j}{y_{kj}}, y_{kj}>0\right\}.$$

may actually increase, primal optimality holds so long as the increase in TVC_2 exceeds that of p_2.

If p_2 increases (TVC_2 constant) primal optimality is preserved if $\Delta p_2 \le -\bar{c}_2 = \frac{8}{5}$; and if TVC_2 decreases (p_2 constant), primal optimality holds if $\Delta\, TVC_2 \ge -\frac{8}{5}$. If p_2, TVC_2 change simultaneously, primal optimality is preserved if $\Delta p_2 > \Delta TVC_2$ and $\Delta p_2 - \Delta TVC_2 \le -\bar{c}_2 = \frac{8}{5}$, e.g. if TVC_2 decreases by \$1, $\Delta TVC_2 = -1$ and thus primal optimality holds if $\Delta p_2 \le \frac{3}{5}$ (p_2 cannot increase by more than \$3/5); if p_2 increases by \$2, $\Delta p_2 = 2$ and thus, primal optimality is preserved if $\Delta TVC_2 \ge \frac{2}{5}$.

(TVC_2 cannot increase by more than \$2/5).

B) Changing Resource Requirements.

From the preceding simplex matrix, we have

$$\boldsymbol{B}^{-1} = \begin{bmatrix} \beta_{11} & \beta_{12} \\ \beta_{21} & \beta_{22} \end{bmatrix} = \begin{bmatrix} 1/10 & 0 \\ -1/5 & 1 \end{bmatrix}.$$

How does changing $b_1 = 40$ affect primal feasibility? With $\beta_{11} = 1/10 > 0$, $x_{B1} = x_1$ remains feasible for any increase in b_1. If b_1 is decreased, x_1 remains feasible so long as $\Delta b_1 \ge -x_{B1}/\beta_{11} = -4/(1/10) = -40$. And with $\beta_{21} = -1/5 < 0$, the feasibility of $x_{B2} = x_4$ will be maintained for any decrease in b_1 while if b_1 increases, the feasibility of x_4 will not be violated if $\Delta b_1 \le -22/(-1/5) = 110$. In sum, primal feasibility holds if $-40 \le \Delta b_1 \le 110$.

How is primal feasibility affected by perturbing $b_2 = 30$? With $\beta_{12} = 0$, the feasibility of $x_{B1} = x_1$ is always preserved. And with $\beta_{22} = 1 > 0$, the feasibility of $x_{B2} = x_4$ holds for any increase in b_2, while if b_2 decreases, x_4 remains feasible if $\Delta b_2 \ge -x_{B2}/\beta_{22} = -22$. Thus, to preserve primal feasibility, $-22 \le \Delta b_2 < +\infty$.

C) Changes in Technology

Since x_1, x_4 are basic variables in the optimal simplex matrix above,

$$\boldsymbol{B} = \begin{bmatrix} b_{11} & b_{12} \\ b_{21} & b_{22} \end{bmatrix} = \begin{bmatrix} 10 & 0 \\ 2 & 1 \end{bmatrix}, \boldsymbol{B}^{-1} = \begin{bmatrix} \beta_{11} & \beta_{12} \\ \beta_{21} & \beta_{22} \end{bmatrix} = \begin{bmatrix} 1/10 & 0 \\ -1/5 & 1 \end{bmatrix}, \text{ and}$$

$$\boldsymbol{R} = \begin{bmatrix} r_{11} & r_{12} \\ r_{21} & r_{22} \end{bmatrix} = \begin{bmatrix} 8 & 1 \\ 3 & 0 \end{bmatrix}.$$

Again looking to the optimal simplex matrix, we see that $\bar{c}_1 = -14/5, \bar{c}_2 = -8/5$. Since $u_1 = 8/5 > 0$, primal optimality is preserved if r_{11} increases or, if decreased, $\Delta r_{11} \geq \bar{c}_1/u_1 = (-14/5)/(8/5) = -\frac{7}{4}$. With $u_2 = 0$, primal optimality is maintained for any change in r_{21}(as well as in r_{22}). Next, it is easily seen that primal optimality holds if r_{12} increases in value; if r_{12} decreases, primal optimality holds if $\Delta r_{12} \geq \bar{c}_2/u_1 = (-8/5)/(8/5) = -1$.

How may we alter b_{11} so as to maintain primal feasibility? Since $\beta_{21}x_{B1} - \beta_{11}x_{B2} = \left(-\frac{1}{5}\right)4 - \left(\frac{1}{10}\right)22 = -3 < 0$, the decrease in b_{11} cannot exceed

$$\max_{i \neq 1} \left\{ \frac{x_{B2}}{\beta_{21}x_{B1} - \beta_{11}x_{B2}} \right\} = -\frac{22}{3};$$

while b_{11} can increase without bound without violating primal feasibility. What about the effect of changes in b_{21} on primal feasibility? With $\beta_{22}x_{B1} - \beta_{12}x_{B2} = 1 \cdot 4 - 0 \cdot 22 = 4 > 0$, the increase in b_{21} cannot exceed

$$\min_{i \neq 1} \left\{ \frac{x_{B2}}{\beta_{22}x_{B1} - \beta_{12}x_{B2}} \right\} = \frac{11}{2};$$

b_{21} can decrease without bound and not affect primal feasibility. Next, since $\beta_{11}x_{B2} - \beta_{21}x_{B1} = \left(\frac{1}{10}\right)22 - \left(-\frac{1}{5}\right)4 = 3 > 0$, the increase in b_{21} cannot exceed

$$\min_{i \neq 2} \left\{ \frac{x_{B1}}{\beta_{11}x_{B2} - \beta_{21}x_{B1}} \right\} = \frac{4}{3}$$

if primal feasibility is to be maintained; b_{21} may decrease to $-\infty$ without affecting the feasibility of the optimal primal solution. Finally, with $\beta_{12}x_{B2} - \beta_{22}x_{B1} = 0 \cdot 22 - 1 \cdot 4 = -4 < 0$, primal feasibility holds if the decrease in b_{22} does not exceed

$$\max_{i \neq 2} \left\{ \frac{x_{B1}}{\beta_{12}x_{B2} - \beta_{22}x_{B1}} \right\} = -1;$$

and b_{11} can increase without bound and not affect primal feasibility.

Let us now determine the limits within which the components of the basic activities can vary without affecting primal optimality. For b_{11}, since $y_{11}u_1 + \bar{c}_1\beta_{11} = (4/5)(8/5) - (14/5)(1/10) = 1, y_{12}u_1 + \bar{c}_2\beta_{11} = (1/10)(8/5) - (8/5)(1/10) = 0$, it is evident that the upper limit on Δb_{11} that preserved primal optimality is

$$\min_{j} \left\{ \frac{-\bar{c}_1}{y_{11}u_1 + \bar{c}_1\beta_{11}} \right\} = 14/5;$$

b_{11} can decrease without bound without affecting the optimality of the current solution. Next, looking to b_{21}, with $y_{11}u_2 + \bar{c}_1\beta_{12} = (4/5)\cdot 0 - (14/5)\cdot 0 = y_{12}u_2 + \bar{c}_2\beta_{12} = (1/10)\cdot 0 - (8/5)\cdot 0 = 0$, we conclude that b_{21} may either increase or decrease without bound and not breach primal optimality. If we now consider b_{12}, since

$$y_{21}u_1 + \bar{c}_1\beta_{21} = (7/5)(8/5) - (14/5)(-1/5) = 14/5, y_{22}u_1 + \bar{c}_2\beta_{21}$$
$$= (-1/5)(8/5) - (8/5)(-1/5) = 0,$$

it follows that the upper bound on the increase in Δb_{12} is

$$\min_j \left\{ \frac{-\bar{c}_1}{y_{21}u_1 + \bar{c}_1\beta_{21}} \right\} = 1;$$

while primal optimality also holds if b_{12} decreases without limit. Finally, for b_{22}, with

$$y_{21}u_2 + \bar{c}_1\beta_{22} = (7/5)\cdot 0 - (14/5)\cdot 1 = -\frac{14}{5}, y_{22}u_2 + \bar{c}_2\beta_{22}$$
$$= (-1/5)\cdot 0 - (8/5)\cdot 1 = -\frac{8}{5},$$

primal optimality holds if the decrease in Δb_{22} does not exceed

$$\max_j \left\{ \frac{-\bar{c}_1}{y_{21}u_2 + \bar{c}_1\beta_{22}} = -1, \frac{-\bar{c}_2}{y_{22}u_2 + \bar{c}_2\beta_{22}} = -1 \right\} = -1;$$

primal optimality holds if b_{22} increases without bound. ■

9

Analyzing Structural Changes

9.1 Introduction

Oftentimes, one is faced with the task of adding or deleting certain variables or structural constraints in a problem for which an optimal basic feasible solution has been found. As in the preceding chapter, rather than resolve completely the adjusted problem, we shall use the information contained within the optimal simplex matrix of the original linear program to determine the effects of the said changes upon the feasibility or optimality of the solution at hand.

9.2 Addition of a New Variable

We first look to the effect on the optimal solution of a given linear programming problem of the introduction of a new variable x_{n+1}, with c_{n+1} and $\boldsymbol{a}_{n+1}$ representing its associated objective function coefficient and vector of structural constraint coefficients, respectively. In particular, we are now confronted with solving a problem of the form

$$\max\ f(\bar{X}, x_{n+1}) = \bar{C}'\bar{X} + c_{n+1}x_{n+1} \quad \text{s.t.}$$

$$\left[\bar{A} \vdots \boldsymbol{a}_{n+1}\right]\begin{bmatrix} \bar{X} \\ \cdots\cdots \\ x_{n+1} \end{bmatrix} = \boldsymbol{b}, \bar{X} \geq \boldsymbol{O}, x_{n+1} \geq 0.$$

If $\boldsymbol{X_B} = \boldsymbol{B}^{-1}\boldsymbol{b}$ represents an optimal basic feasible solution to the problem

$$\begin{aligned} &\max\ f(\bar{X}) = \bar{C}'\bar{X} \quad \text{s.t.} \\ &\bar{A}\bar{X} = \boldsymbol{b}, \bar{X} \geq \boldsymbol{O}, \end{aligned}$$

then we may also view it as a basic feasible solution to the former problem with x_{n+1} deemed nonbasic or zero. Moreover, this solution will be optimal if

Linear Programming and Resource Allocation Modeling, First Edition. Michael J. Panik.

$\bar{c}_{n+1-m} = c_{R,n+1-m} - C_B' B^{-1} r_{n+1-m} = c_{n+1} - u' a_{n+1} \leq 0$. Obviously then, if $\bar{c}_{n+1-m} > 0$, the current solution is no longer optimal so that a_{n+1} must be brought into the basis, in which case we must insert into the optimal simplex matrix an additional column of the form

$$\begin{bmatrix} Y_{n+1-m} \\ \cdots\cdots\cdots \\ -\bar{c}_{n+1-m} \end{bmatrix} = \begin{bmatrix} B^{-1} a_{n+1} \\ \cdots\cdots\cdots \\ -\bar{c}_{n+1-m} \end{bmatrix},$$

where $-\bar{c}_{n+1-m} < 0$.

Example 9.1 Given that the optimal simplex matrix associated with the problem presented at the outset of Example 8.1 appears as

$$\begin{bmatrix} \frac{5}{11} & 0 & 1 & \frac{3}{11} & -\frac{1}{11} & 0 & 20 \\ \frac{2}{11} & 1 & 0 & -\frac{1}{11} & \frac{4}{11} & 0 & 20 \\ \frac{4}{11} & 0 & 0 & \frac{9}{11} & \frac{8}{11} & 1 & 140 \end{bmatrix}$$

with

$$B^{-1} = \begin{bmatrix} \frac{3}{11} & -\frac{1}{11} \\ -\frac{1}{11} & \frac{4}{11} \end{bmatrix},$$

let us introduce a new variable x_6 with $c_6 = 3$ and $a_6' = (^1/_2, 2)$. Since $\bar{c}_4 = c_6 - u' a_6 = 25/22 > 0$, the current solution is not an optimal basic feasible solution to this modified problem. Let us therefore compute

$$Y_4 = B^{-1} a_6 = \begin{bmatrix} \frac{3}{11} & -\frac{1}{11} \\ -\frac{1}{11} & \frac{4}{11} \end{bmatrix} \begin{bmatrix} \frac{1}{2} \\ 2 \end{bmatrix} = \begin{bmatrix} -\frac{1}{22} \\ \frac{15}{22} \end{bmatrix}.$$

Inserting

$$\begin{bmatrix} Y_4 \\ \cdots\cdots \\ -\bar{c}_4 \end{bmatrix} = \begin{bmatrix} -\frac{1}{22} \\ \frac{15}{22} \\ -\frac{25}{22} \end{bmatrix}$$

as an additional (sixth) column into the previous simplex matrix yields

$$\begin{bmatrix} \frac{5}{11} & 0 & 1 & \frac{3}{11} & -\frac{1}{11} & -\frac{1}{22} & 0 & 20 \\ \frac{2}{11} & 1 & 0 & -\frac{1}{11} & \frac{4}{11} & \frac{15}{22} & 0 & 20 \\ \frac{4}{11} & 0 & 0 & \frac{9}{11} & \frac{8}{11} & -\frac{25}{22} & 1 & 140 \end{bmatrix}.$$

An additional iteration on this adjusted simplex matrix yields a new optimal basic feasible solution. ∎

9.3 Addition of a New Structural Constraint

Let us now assume that we are faced with the problem of introducing an additional structural constraint into a linear programming problem for which an optimal basic feasible solution has already been obtained. We may write this new $(m + 1)$st structural constraint as

$$\sum_{j=1}^{n} a_{m+1,j} x_j \leq, \geq, = b_{m+1} \text{ or } \boldsymbol{\alpha}\bar{\boldsymbol{X}} \leq, \geq, = b_{m+1},$$

where $\boldsymbol{\alpha} = (a_{m+1,1}, \ldots, a_{m+1,n})$ is a $(1 \times n)$ vector of coefficients on the variables x_j, $j = 1, \ldots, n$.

First, for $\boldsymbol{\alpha}\bar{\boldsymbol{X}} \leq b_{m+1}$, if x_{n+1} represents a nonnegative slack variable, then we seek a solution to a problem of the form

$$\begin{aligned} & max\, f(\bar{\boldsymbol{X}}, x_{n+1}) = \bar{\boldsymbol{C}}'\bar{\boldsymbol{X}} + 0x_{n+1} \quad \text{s.t.} \\ & \bar{\boldsymbol{A}}\bar{\boldsymbol{X}} + x_{n+1}\boldsymbol{O} = \boldsymbol{b} \\ & \boldsymbol{\alpha}\bar{\boldsymbol{X}} + x_{n+1} = b_{m+1} \\ & \bar{\boldsymbol{X}} > \boldsymbol{O}, x_{n+1} \geq 0. \end{aligned}$$

The simplex matrix associated with this problem may be written as

$$\begin{array}{c} \quad \boldsymbol{X}_B \quad \boldsymbol{X}_R \quad x_{n+1} \\ \begin{bmatrix} \boldsymbol{B} & \boldsymbol{R} & \boldsymbol{O} & \boldsymbol{O} & \boldsymbol{b} \\ \boldsymbol{\alpha}_B & \boldsymbol{\alpha}_R & 1 & 0 & b_{m+1} \\ -\boldsymbol{C}_B' & -\boldsymbol{C}_R' & 0 & 1 & 0 \end{bmatrix} \end{array},$$

where $\boldsymbol{\alpha} = (\boldsymbol{\alpha}_B, \boldsymbol{\alpha}_r)$. Here $\boldsymbol{\alpha}_B$ is a $(1 \times m)$ vector containing those components $a_{m+1,j}$, which correspond to the columns of $\boldsymbol{B}$ and $\boldsymbol{\alpha}_r$ is a $(1 \times n-m)$ vector whose components $a_{m+1,j}$ correspond to the columns of $\boldsymbol{R}$. If this simplex matrix is premultiplied by the inverse of the $m+1$ s.t. order basis matrix

$$\bar{\boldsymbol{B}} = \begin{bmatrix} \boldsymbol{B} & \boldsymbol{O} & \boldsymbol{O} \\ \boldsymbol{\alpha_B} & 1 & 0 \\ -\boldsymbol{C}'_B & 0 & 1 \end{bmatrix}, \bar{\boldsymbol{B}}^{-1} = \begin{bmatrix} \boldsymbol{B}^{-1} & \boldsymbol{O} & \boldsymbol{O} \\ -\boldsymbol{\alpha_B}\boldsymbol{B}^{-1} & 1 & 0 \\ \boldsymbol{C}'_B\boldsymbol{B}^{-1} & 0 & 1 \end{bmatrix},$$

the resulting product

$$\begin{bmatrix} \boldsymbol{I}_m & \boldsymbol{B}^{-1}\boldsymbol{R} & \boldsymbol{O} & \boldsymbol{O} & \boldsymbol{B}^{-1}\boldsymbol{b} \\ \boldsymbol{O}' & \boldsymbol{\alpha_R} - \boldsymbol{\alpha_B}\boldsymbol{B}^{-1}\boldsymbol{R} & 1 & 0 & b_{m+1} - \boldsymbol{\alpha_B}\boldsymbol{B}^{-1}\boldsymbol{b} \\ \boldsymbol{O}' & \boldsymbol{C}'_B\boldsymbol{B}^{-1}\boldsymbol{R} - \boldsymbol{C}'_R & 0 & 1 & \boldsymbol{C}'_B\boldsymbol{B}^{-1}\boldsymbol{b} \end{bmatrix}$$

has, with $\boldsymbol{X_R} = \boldsymbol{O}$, the partitioned rows

$$\begin{aligned} &\boldsymbol{X_B} = \boldsymbol{B}^{-1}\boldsymbol{b}, \\ &x_{n+1} = b_{m+1} - \boldsymbol{\alpha_B}\boldsymbol{B}^{-1}\boldsymbol{b} = b_{m+1} - \boldsymbol{\alpha_B}\boldsymbol{X_B}, \text{and} \\ &f = \boldsymbol{C}'_B\boldsymbol{B}^{-1}\boldsymbol{b} = \boldsymbol{C}'_B\boldsymbol{X_B}, \end{aligned}$$

respectively. That is

$$\boldsymbol{X}_{\bar{B}} = \begin{bmatrix} \boldsymbol{X_B} \\ \cdots \\ x_{n+1} \end{bmatrix} = \begin{bmatrix} \boldsymbol{X_B} \\ \cdots\cdots\cdots\cdots \\ b_{m+1} - \boldsymbol{\alpha_B}\boldsymbol{X_B} \end{bmatrix} \text{ or } \begin{cases} x_{\bar{B}i} = x_{Bi}, i = 1, \ldots, m; \\ x_{\bar{B},m+1} = x_{n+1} = b_{m+1} - \boldsymbol{\alpha_B}\boldsymbol{X_B}, \end{cases}$$

$$f(\boldsymbol{X}_{\bar{B}}) = f(\boldsymbol{X_B}).$$

Now, if $x_{\bar{B},m+1} > 0$, this basic solution to the enlarged problem is feasible, i.e. the optimal basic feasible solution to the original problem also satisfies the new structural constraint or $\boldsymbol{\alpha}\boldsymbol{X}_{\bar{B}} \leq b_{m+1}$. Moreover, if the original basic solution was optimal, then an examination of the last row of the preceding simplex matrix reveals that the solution to the extended problem is also optimal (since the optimality criterion for each is the same) with equality holding between the optimal values of the objective functions for the original and enlarged problems. If $x_{\bar{B},m+1} < 0$ (in which case $\boldsymbol{\alpha}\boldsymbol{X}_{\bar{B}} > b_{m+1}$) so that primal feasibility is violated, then, since the optimality criterion is satisfied, the dual simplex method may be employed to obtain a basic feasible solution to the modified problem. In this regard, it is obvious that the optimal objective value for the original problem provides us with an upper bound on the same for the enlarged problem.

Example 9.2 Again returning to the problem presented in Example 8.1 with optimal simplex matrix

$$\begin{bmatrix} \frac{5}{11} & 0 & 1 & \frac{3}{11} & -\frac{1}{11} & 0 & 20 \\ \frac{2}{11} & 1 & 0 & -\frac{1}{11} & \frac{4}{11} & 0 & 20 \\ \frac{4}{11} & 0 & 0 & \frac{9}{11} & \frac{8}{11} & 1 & 140 \end{bmatrix},$$

let us incorporate the additional structural constraint $3x_1 + 5x_2 + 2x_3 \leq 60$ into the original problem. Since the current optimal basic feasible solution violates this new constraint (i.e. $3(0) + 5(20) + 2(20) > 60$), let us introduce the nonnegative slack variable x_6 and proceed to obtain an optimal basic feasible solution to this modified problem. Since

$$\boldsymbol{\alpha} = (3,5,2,0,0), \boldsymbol{\alpha}_B = (2,5), \boldsymbol{\alpha}_R = (3,0,0),$$

$$\boldsymbol{\alpha}_R - \boldsymbol{\alpha}_B \boldsymbol{B}^{-1}\boldsymbol{R} = (3,0,0) - (2,5)\begin{bmatrix} \frac{3}{11} & -\frac{1}{11} \\ -\frac{1}{11} & \frac{4}{11} \end{bmatrix}\begin{bmatrix} 2 & 1 & 0 \\ 1 & 0 & 1 \end{bmatrix} = \left(\frac{13}{11}, -\frac{1}{11}, -\frac{18}{11}\right), \text{ and}$$

$$b_{m+1} - \boldsymbol{\alpha}_B \boldsymbol{X}_B = 60 - (2,5)\begin{bmatrix} 20 \\ 20 \end{bmatrix} = -80,$$

it follows that the enlarged simplex matrix appears as

$$\begin{bmatrix} \frac{5}{11} & 0 & 1 & \frac{3}{11} & -\frac{1}{11} & 0 & 0 & 20 \\ \frac{2}{11} & 1 & 0 & -\frac{1}{11} & \frac{4}{11} & 0 & 0 & 20 \\ \frac{13}{11} & 0 & 0 & -\frac{1}{11} & -\frac{18}{11} & 1 & 0 & -80 \\ \frac{4}{11} & 0 & 0 & \frac{9}{11} & \frac{8}{11} & 0 & 1 & 140 \end{bmatrix}.$$

An application of the dual simplex method to this adjusted matrix will yield a new optimal basic feasible solution. ■

Next, if $\boldsymbol{\alpha}\bar{X} \geq b_{m+1}$, then x_{n+1} represents a nonnegative surplus variable. In this circumstance, let us add $\boldsymbol{\alpha}\bar{X} - x_{n+1} = b_{m+1}$ or $\bar{\boldsymbol{\alpha}}\bar{X} + x_{n+1} = \bar{b}_{m+1}$ to the original problem, where $\bar{\boldsymbol{\alpha}} = -\boldsymbol{\alpha}$ and $\bar{b}_{m+1} = -b_{m+1}.$ The calculations then proceed exactly as above with $\bar{\boldsymbol{\alpha}}$ and $\bar{b}_{m+1}$ replacing $\boldsymbol{\alpha}$ and b_{m+1}, respectively.

Example 9.3 If into the preceding example we now introduce the new structural constraint $x_1 - 2x_2 + x_3 \geq 30$, we see that the optimal basic feasible solution to the original problem also violates this constraint since −2 (20) + 20 < 30. For x_6 a nonnegative slack variable,

$$\bar{\alpha} = (-1,2,-1,0,0), \bar{\alpha}_B = (-1,2), \bar{\alpha}_R = (-1,0,0),$$

$$\bar{\alpha}_R - \bar{\alpha}_B B^{-1} R = (-1,0,0) - (-1,2) \begin{bmatrix} \frac{3}{11} & -\frac{1}{11} \\ -\frac{1}{11} & \frac{4}{11} \end{bmatrix} \begin{bmatrix} 2 & 1 & 0 \\ 1 & 0 & 1 \end{bmatrix} = \left(-\frac{10}{11}, \frac{5}{11}, -\frac{9}{11}\right),$$

$$\bar{b}_{m+1} - \bar{\alpha}_B X_B = -30 - (-1,2) \begin{bmatrix} 20 \\ 20 \end{bmatrix} = -50,$$

and thus we have the enlarged simplex matrix

$$\begin{bmatrix} \frac{5}{11} & 0 & 1 & \frac{3}{11} & -\frac{1}{11} & 0 & 0 & 20 \\ \frac{2}{11} & 1 & 0 & -\frac{1}{11} & \frac{4}{11} & 0 & 0 & 20 \\ -\frac{10}{11} & 0 & 0 & \frac{5}{11} & -\frac{9}{11} & 1 & 0 & -50 \\ \frac{4}{11} & 0 & 0 & \frac{9}{11} & \frac{8}{11} & 0 & 1 & 140 \end{bmatrix}.$$

The dual simplex method can then be applied to this adjusted matrix to generate a new optimal basic feasible solution. ■

Finally, if $\alpha\bar{X} = b_{m+1}$, then x_{n+1} will represent a nonnegative artificial variable. If in the solution to the enlarged problem we find that $x_{\bar{B},m+1} = x_{n+1} = b_{m+1} - \alpha_B X_B = 0$, then it is evident that the optimal basic feasible solution for the original problem exactly satisfies the new constraint, in which case the original solution is also the optimal solution for the enlarged problem. The implication of this degenerate solution to the enlarged problem is that the added structural constraint is redundant. If $x_{\bar{B},m+1} < 0$, then the artificial variable x_{n+1} may be treated as if it were a slack variable, in which case the dual simplex method may be applied as in Example 9.2. And if $x_{\bar{B},m+1} > 0$, then the artificial variable x_{n+1} may be handled as if it represented a surplus variable. Again the dual simplex method may be applied per Example 9.3.

9.4 Deletion of a Variable

The deletion of a positive variable is tantamount to driving it to zero so that its contribution to optimality is withdrawn. This being the case, it is obvious that the deletion of a nonbasic variable is a trivial process. To delete a positive basic variable, we may simply pivot its associated basis vector out of the basis, where the pivotal column is determined by the dual simplex entry criterion. This procedure will result in a basic solution that is primal optimal but not necessarily primal feasible, at which point the column corresponding to the outgoing basis vector may be deleted from the simplex matrix. If warranted, reoptimization of the primal objective function may then be carried out using the dual simplex routine.

9.5 Deletion of a Structural Constraint

If a structural constraint is to be deleted from the optimal simplex matrix, consideration must first be given to whether or not that constraint is binding at the optimal basic feasible solution. In this regard, if the original structural constraint was of the "≤" or "≥" variety and its associated slack or surplus variable appears in the optimal set of basic variables at a positive level, thus indicating that the constraint is not binding at the optimal extreme point, then clearly the current solution will also be an optimal basic feasible solution to the modified problem. For instance, if in Figure 9.1 point A represents an optimal basic feasible

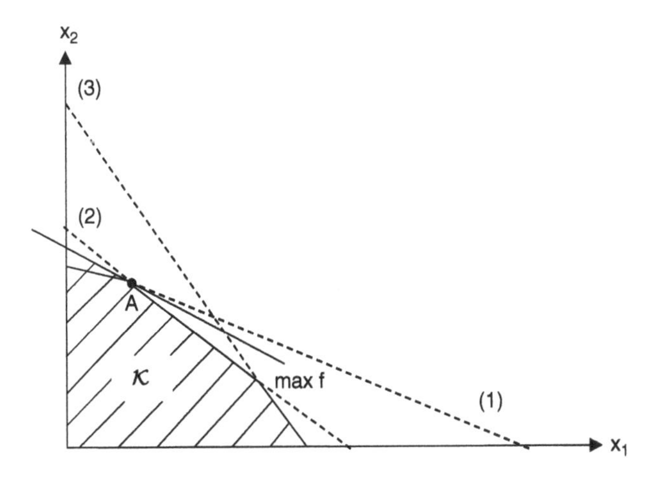

Figure 9.1 Only structural constraints (1), (2) binding at the optimal solution Point A.

solution to some linear programming problem and the "≤" structural constraint (3) is deleted from the constraint system, then point A also represents an optimal basic feasible solution to the adjusted problem since only structural constraints (1), (2) are binding there for each problem.

If, on the one hand, the structural constraint to be deleted was of the "≤" type and its corresponding slack variable is zero at the optimal basic feasible solution, thus indicating that the constraint is binding there, then the constraint may be deleted from the optimal simplex matrix by subtracting a nonnegative surplus variable from its left-hand side. To justify this operation let us assume that the rth structural constraint is binding at an optimal basic feasible solution so that $a_{r1}x_1 + \cdots + a_{rn}x_n = b_r$. Given that this constraint is to be deleted from the optimal simplex matrix, the introduction of the surplus variable x_{n+1} allows for the possibility that $a_{r1}x_1 + \cdots + a_{rn}x_n > b_r$ once this constraint is deleted, i.e. $a_{r1}x_1 + \cdots + a_{rn}x_n - b_r = x_{n+1} \geq 0$. A glance back at Section 9.2 involving the addition of a new variable to a given linear programming problem reveals that in this instance we seek to generate a solution to a problem of the form

$$max f(\bar{X}, x_{n+1}) = \bar{C}'\bar{X} + 0x_{n+1} \text{ s.t.}$$

$$\left[\bar{A} \vdots -\mathbf{e}_r\right]\begin{bmatrix} \bar{X} \\ \cdots \\ x_{n+1} \end{bmatrix} = \mathbf{b}, \bar{X} \geq \mathbf{O}, x_{n+1} \geq 0.$$

Since $\bar{c}_{n+1-m} = c_{R,n+1-m} - C_B'B^{-1}r_{n+1-m} = \mathbf{u}'\mathbf{e}_r = u_r > 0$, the current solution is no longer optimal so that we must insert into the optimal simplex matrix an additional column of the form

$$\begin{bmatrix} Y_{n+1-m} \\ \cdots\cdots\cdots \\ -\bar{c}_{n+1-m} \end{bmatrix} = \begin{bmatrix} -B^{-1}\mathbf{e}_r \\ \cdots\cdots\cdots \\ -u_r \end{bmatrix}.$$

The primal simplex routine then yields an optimal basic feasible solution to the modified problem. On the other hand, if the structural constraint to be deleted was of the "≥" variety and its associated surplus variable is zero at the optimal basic feasible solution so that the constraint is binding there, then the constraint may be deleted from the optimal simplex matrix by adding to its left-hand side a nonnegative slack variable. Again looking to the rth (binding) structural constraint, the introduction of the slack variable x_{n+1} provides for the likelihood that $a_{r1}x_1 + \cdots + a_{rn}x_n < b_r$ when the constraint is deleted, i.e. $x_{n+1} = b_r - a_{r1}x_1 - \cdots - a_{rn}x_n \geq 0$. Proceeding as above, we wish to determine a solution to

$$max\, f(\bar{X}, x_{n+1}) = \bar{C}'\bar{X} + 0x_{n+1} \quad \text{s.t.}$$

$$\left[\bar{A} \vdots e_r\right]\begin{bmatrix} \bar{X} \\ \cdots \\ x_{n+1} \end{bmatrix} = b, \bar{X} \geq O, x_{n+1} \geq 0.$$

And if the (rth) structural constraint to be deleted was initially a strict equality, then we may simultaneously add and subtract nonnegative slack and surplus variables x_{n+1}, x_{n+2}, respectively, on its left-hand side, i.e. we must now solve

$$max\, f(\bar{X}, x_{n+1}, x_{n+2}) = \bar{C}'\bar{X} + 0x_{n+1} + 0x_{n+2} \quad \text{s.t.}$$

$$\left[\bar{A} \vdots e_r \vdots - e_r\right]\begin{bmatrix} \bar{X} \\ x_{n+1} \\ x_{n+2} \end{bmatrix} = b, \bar{X} \geq O; x_{n+1}, x_{n+2} \geq 0.$$

Example 9.4 The optimal simplex matrix associated with the problem

$$max\, f = \frac{1}{4}x_1 + x_2 \quad \text{s.t.}$$

$$\begin{aligned} x_1 \quad\quad & \leq 30 \\ x_1 + x_2 & \leq 50 \\ x_2 & \leq 40 \\ x_1, x_2 & \geq 0 \end{aligned}$$

appears as

$$\begin{bmatrix} 0 & 0 & 1 & -1 & 1 & 0 & 20 \\ 1 & 0 & 0 & 1 & -1 & 0 & 10 \\ 0 & 1 & 0 & 0 & 1 & 0 & 40 \\ 0 & 0 & 0 & \frac{1}{4} & \frac{3}{4} & 1 & 85/2 \end{bmatrix}$$

with $x_1 = 10$, $x_2 = 40$, $x_3 = 20$, and $f = 85/2$. To delete the third (binding) structural constraint $x_2 \leq 40$, let us compute, from above, the vector

$$\begin{bmatrix} Y_3 \\ \cdots \\ -\bar{c}_3 \end{bmatrix} = \begin{bmatrix} -B^{-1}e_3 \\ \cdots\cdots\cdots \\ -u_3 \end{bmatrix} = \begin{bmatrix} -1 \\ 1 \\ -1 \\ -\frac{3}{4} \end{bmatrix}$$

and insert it as an additional (sixth) column into the above matrix. Then we obtain

$$\begin{bmatrix} 0 & 0 & 1 & -1 & 1 & -1 & 0 & 20 \\ 1 & 0 & 0 & 1 & -1 & 1 & 0 & 10 \\ 0 & 1 & 0 & 0 & 1 & -1 & 0 & 40 \\ 0 & 0 & 0 & 1 & 1 & -\frac{3}{4} & 1 & 85/2 \end{bmatrix} \rightarrow \begin{bmatrix} 1 & 0 & 1 & 0 & 0 & 0 & 0 & 30 \\ 1 & 0 & 0 & 1 & -1 & 1 & 0 & 10 \\ 1 & 1 & 0 & 1 & 0 & 0 & 0 & 50 \\ \frac{3}{4} & 0 & 0 & \frac{7}{4} & \frac{1}{4} & 0 & 1 & 50 \end{bmatrix}$$

with $x_2 = 50, x_3 = 30, x_6 = 10$, and $f = 50$. Hence, we have allowed x_2 to increase in value beyond 40 (remember that initially $x_2 \leq 40$) by introducing the surplus variable $x_6 = 10$ into the set of basic variables in order to absorb the excess (surplus) of x_2 above 40 units. ■

10

Parametric Programming

10.1 Introduction

In Chapter 8 we considered an assortment of post-optimality problems involving discrete changes in only selected components of the matrices $\bar{C}, b, \text{or}\, \bar{A}$. Therein emphasis was placed on the extent to which a given problem may be modified without breaching its feasibility or optimality. We now wish to extend this sensitivity analysis a bit further to what is called **parametric analysis**. That is, instead of just determining the amount by which a few individual components of the aforementioned matrices may be altered in some particular way before the feasibility or optimality of the current solution is violated, let us generate a sequence of basic solutions that, in turn, become optimal, one after the other, as all of the components of $\bar{C}$, b, or a column of $\bar{A}$ vary continuously in some prescribed direction. In this regard, the following parametric analysis will involve a marriage between sensitivity analysis and simplex pivoting.

10.2 Parametric Analysis

If we seek to

$$\begin{aligned} \max f(\bar{X}) &= \bar{C}'\bar{X} \quad \text{s.t.} \\ \bar{A}\bar{X} &= b, \bar{X} \geq O, \end{aligned}$$

then an optimal basic feasible solution emerges if $X_B = B^{-1}b \geq O$ and $C'_R - C'_B B^{-1}R \leq O'$ (or, in terms of the optimal simplex matrix $C'_B B^{-1}R - C'_R \geq O'$) with $f(\bar{X}) = C'_B X_B = C'_B B^{-1}b$. Given this result as our starting point, let us examine the process of parameterizing the objective function.

Linear Programming and Resource Allocation Modeling, First Edition. Michael J. Panik.

10.2.1 Parametrizing the Objective Function

Given the above optimal basic feasible solution, let us replace $\bar{C}$ by $C^* = \bar{C} + \theta S$, where θ is a nonnegative scalar parameter and S is a specified, albeit arbitrary, $(n \times 1)$ vector that determines a given direction of change in the coefficients c_j, $j = 1, \ldots, n$. In this regard, the c^*_j, $j = 1, \ldots, n$, are specified as linear functions of the parameter θ. Currently, $-C'_R + C'_B B^{-1} R \geq O'$ or $-\bar{c}_j = -c_{Rj} + C'_B Y_j \geq 0$, $j = 1, \ldots, n-m$. If C^* is partitioned as

$$\begin{bmatrix} C^*_B \\ C^*_R \end{bmatrix} = \begin{bmatrix} C_B \\ C_R \end{bmatrix} + \theta \begin{bmatrix} S_B \\ S_R \end{bmatrix},$$

where S_B (S_R) contains the components of S corresponding to the components of $\bar{C}$ within $C_B(C_R)$, then, when C^* replaces $\bar{C}$. The *revised optimality condition* becomes

$$\begin{aligned} -\left(C^*_R\right)' + \left(C^*_B\right)' B^{-1} R &= -\left(C'_R + \theta S'_R\right) + \left(C'_B + \theta S'_B\right) B^{-1} R \\ &= \left(-C'_R + C'_B B^{-1} R\right) + \theta\left(-S'_R + S'_B B^{-1} R\right) \geq O' \end{aligned} \tag{10.1}$$

or, in terms of the individual components of (10.1),

$$-\bar{c}^*_j = -\bar{c}_j + \theta\left(-S_{Rj} + S'_B Y_j\right) \geq 0, j = 1, \ldots, n-m. \tag{10.1.1}$$

(Note that the parametrization of f affects only primal optimality and not primal feasibility since X_B is independent of $\bar{C}$.) Let us now determine the largest value of θ (known as its **critical value,** θ_c) for which (10.1.1) holds. Upon examining this expression, it is evident that the critical value of θ is that for which any increase in θ beyond θ_c makes at least one of the $-\bar{c}^*_j$ values negative, thus violating optimality.

How large of an increase in θ preserves optimality? First, if $-S'_R + S'_B B^{-1} R \geq O'$ or $-S_{Rj} + S'_B Y_j \geq 0$, the θ can be increased without bound while still maintaining the revised optimality criterion since, in this instance, (10.1.1) reveals that $-\bar{c}^*_j \geq -\bar{c}_j \geq 0, j = 1, \ldots, n-m$. Next, if $-S_{Rj} + S'_B Y_j < 0$ for some particular value of j, then $-\bar{c}^*_j \geq 0$ for

$$\theta \leq \frac{-c_j}{-S_{Rj} + S'_B Y_j = \theta_c}.$$

Hence, the revised optimality criterion is violated when $\theta > \theta_c$ for some nonbasic variable x_{Rj}. Moreover, if $-S_{Rj} + S'_B Y_j < 0$ for two or more nonbasic variables, then

$$\theta_c = \min_j \left\{ \frac{\bar{c}_j}{-S_{Rj} + S'_B Y_j} \middle| -S_{Rj} + S'_B Y_j < 0 \right\}. \tag{10.2}$$

So if θ increases and the minimum in (10.2) is attained for $j = k$, it follows that $\theta_c = \bar{c}_k / \left(-S_{Rk} + S'_B Y_k\right)$ or $-\bar{c}_k^* = 0$. With $-\bar{c}_k^* = 0$, the case of multiple optimal basic feasible solutions obtains, i.e. the current basic feasible solution (call it $X_B^{(1)}$) remains optimal and the alternative optimal basic feasible solution (denoted $X_B^{(2)}$) emerges when r_k is pivoted into the basis (or x_{Rk} enters the set of basic variables). And as θ increases slightly beyond θ_c, $-\bar{c}_k^* < 0$ so that $X_B^{(2)}$ becomes the unique optimal basic feasible solution.

What about the choice of the direction vector S? As indicated above, S represents the direction in which the objective function coefficients are varied. So while the components of S are quite arbitrary (e.g. some of the c_j values may be increased while others are decreased), it is often the case that only a single objective function coefficient c_j is to be changed, i.e. $S = e_j$ (or $-e_j$). Addressing ourselves to this latter case, if $C^* = \bar{C} + \theta e_k$ and x_k is the nonbasic variable x_{Rk}, then (10.1.1) reduces to $-\bar{c}_j^* = -\bar{c}_j \geq 0$ for $j \neq k$; while for $j = k$, $-\bar{c}_k^* = -\bar{c}_k + \theta(-S_{Rk}) = -\bar{c}_k - \theta \geq 0$. If θ increases so that $\theta_c = -\bar{c}_k$, or $-c_k^* = 0$, then the preceding discussion dictates that the column within the simplex matrix associated with x_k is to enter the basis, with no corresponding adjustment in the current value of f warranted. Next, if x_k is the basic variable x_{Bk}, (10.1.1) becomes $-\bar{c}_j^* = -\bar{c}_j + \theta S_{Bk} y_{kj} = -\bar{c}_j + \theta y_{kj} \geq 0, j = 1, \ldots, n-m$. In this instance (10.2) may be rewritten as

$$\theta_c = \min_j \left\{ \frac{\bar{c}_j}{y_{kj}} \middle| y_{kj<0} \right\}. \tag{10.2.1}$$

If the minimum in (10.2.1) is attained for $j = r$, then $\theta_c = \bar{c}_r / y_{kr}$ or $-\bar{c}_r^* = 0$. Then r_r may be pivoted into the basis so that, again, the case of multiple optimal basic feasible solutions holds. Furthermore, when x_{Rr} is pivoted into the set of basic variables, the current value of f must be adjusted by an amount $\Delta f = \Delta c_{Bk} x_k = \theta_c x_{Bk}$ as determined in Chapter 8.

Example 10.1 Given that we desire to

$$\begin{aligned}
&\max f = x_1 + 4x_2 \quad \text{s.t.} \\
&\frac{1}{2}x_1 + x_2 \leq 50 \\
&x_2 \leq 40 \\
&x_1 - x_2 \leq 20 \\
&x_1 + \frac{1}{2}x_2 \leq 50 \\
&x_1, x_2 \geq 0,
\end{aligned}$$

the optimal simplex matrix appears as

$$\begin{bmatrix} 1 & 0 & 2 & -2 & 0 & 0 & 0 & 20 \\ 0 & 1 & 0 & 1 & 0 & 0 & 0 & 40 \\ 0 & 0 & -2 & 3 & 1 & 0 & 0 & 40 \\ 0 & 0 & -2 & 3/2 & 0 & 1 & 0 & 10 \\ 0 & 0 & 2 & 2 & 0 & 0 & 1 & 180 \end{bmatrix}$$

with $x_1 = 20, x_2 = 40, x_5 = 40, x_6 = 10$, and $f = 180$. As indicated in Figure 10.1, the optimal extreme point is A. Let us parametrize $\bar{C}$ by choosing $S = -e_2$ so that

$$C^* = \bar{C} + \theta S = \begin{bmatrix} 1 \\ 4 \\ 0 \\ 0 \\ 0 \\ 0 \end{bmatrix} + \theta \begin{bmatrix} 0 \\ -1 \\ 0 \\ 0 \\ 0 \\ 0 \end{bmatrix} = \begin{bmatrix} 1 \\ 4-\theta \\ 0 \\ 0 \\ 0 \\ 0 \end{bmatrix}.$$

Since $C'_B = (c_1, c_2, c_5, c_6) = (1,4,0,0), C'_R = (c_3, c_4) = (0,0)$, it follows that $S'_B = (S_{B1}, S_{B2}, S_{B3}, S_{B4}) = (0,-1,0,0)$, and $S'_R = (S_{R1}, S_{R2}) = (0,0)$. Then for $j = 1, 2$,

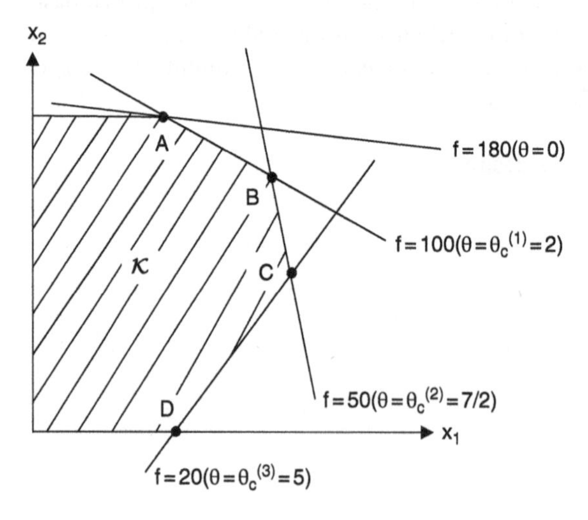

Figure 10.1 Generating the sequence A, B, C, and D of optimal extreme points parametrically.

$$S_B'Y_1 - S_{R1} = (0,-1,0,0)\begin{bmatrix} 2 \\ 0 \\ -2 \\ -2 \end{bmatrix} - 0 = 0,$$

$$S_B'Y_2 - S_{R2} = (0,-1,0,0)\begin{bmatrix} -2 \\ 1 \\ 3 \\ 3/2 \end{bmatrix} - 0 = -1,$$

so that, from (10.2), $\theta_c = \theta_c^{(1)} = \bar{c}_2/\left(S_B'Y_2 - S_{R2}\right) = -2/-1 = 2$. (Note that since the minimum in (10.2) occurred for $j = 2$, $\boldsymbol{r}_2$ will eventually be pivoted into the basis.) In this regard, for $0 \leq \theta < 2$, the current basic feasible solution remains optimal with $max\, f = (\boldsymbol{C_B} + \theta \boldsymbol{S_B})' \boldsymbol{X_B}$. As indicated in Figure 10.1, when $\theta = \theta_c^{(1)}$, the adjusted problem

$$max f(\bar{X}) = \left(\bar{C} + \theta_c^{(1)} S\right)' \bar{X} = (1,2,0,0,0,0)\bar{X} \text{ s.t.}$$

the original constraints has two optimal extreme point solutions: (i) point A; and (ii) point B. To find the optimal simplex matrix for the modified problem that corresponds to point A, let us make the following adjustments in the above simplex matrix. First, let us determine from (10.1.1),

$$-\bar{c}_1^* = -\bar{c}_1 + \theta_c\left(S_B'Y_1 - S_{R1}\right) = 2 + 2(0) = 2,$$
$$-\bar{c}_2^* = -\bar{c}_2 + \theta_c\left(S_B'Y_2 - S_{R2}\right) = 2 + 2(-1) = 0.$$

Next, since we have decreased the objective function coefficient associated with a basic variable (with $\theta = \theta_c^{(1)} = 2, c_2^* = 4 - \theta$ decreases to 2), the appropriate adjustment in f is $\Delta f = \Delta c_{B2} x_{B2} = \Delta c_2 x_2 = -\theta_c x_2 = (-2)40 = -80$. Hence, the adjusted value of f is 100. Once these values replace those already appearing in the last row of the preceding optimal simplex matrix, we obtain

$$\begin{bmatrix} 1 & 0 & 2 & -2 & 0 & 0 & 0 & 20 \\ 0 & 1 & 0 & 1 & 0 & 0 & 0 & 40 \\ 0 & 0 & -2 & 3 & 1 & 0 & 0 & 40 \\ 0 & 0 & -2 & \frac{3}{2} & 0 & 1 & 0 & 10 \\ 0 & 0 & 2 & 0 & 0 & 0 & 1 & 100 \end{bmatrix}.$$

So for $\theta = \theta_c^{(1)} = 2$, point A yields the optimal basic feasible solution consisting of $x_1 = 20, x_2 = 40, x_5 = 40, x_6 = 10$, and now $f = 100$. If we utilize the present matrix to pivot to the implied alternative optimal basic feasible solution at point B, the corresponding optimal simplex matrix is

$$\begin{bmatrix} 1 & 0 & -\frac{2}{3} & 0 & 0 & \frac{4}{3} & 0 & \frac{100}{3} \\ 0 & 1 & \frac{4}{3} & 0 & 0 & -\frac{2}{3} & 0 & \frac{100}{3} \\ 0 & 0 & 2 & 0 & 1 & -2 & 0 & 20 \\ 0 & 0 & -\frac{4}{3} & 1 & 0 & \frac{2}{3} & 0 & \frac{20}{3} \\ 0 & 0 & 2 & 0 & 0 & 0 & 1 & 100 \end{bmatrix}$$

and thus the said solution appears as $x_1 = \frac{100}{3}, x_2 = \frac{100}{3}, x_4 = \frac{20}{3}, x_5 = 20$, and $f = 100$.

If θ is increased slightly beyond $\theta_c^{(1)} = 2$, we obtain a unique optimal basic feasible solution at point B. In this regard, how large of an increase in θ beyond $\theta_c^{(1)}$ is tolerable before this extreme point ceases to be optimal? Since $\boldsymbol{C}_{\boldsymbol{B}}' = (c_1, c_2, c_5, c_4) = (1,4,0,0), \boldsymbol{C}_{\boldsymbol{R}}' = (c_3, c_6) = (0,0)$, it follows that $\boldsymbol{S}_{\boldsymbol{B}}' = (0,-1,0,0)$, $\boldsymbol{S}_{\boldsymbol{R}}' = (0,0)$. Again, for $j = 1, 2$, we have, from the latest optimal simplex matrix,

$$\boldsymbol{S}_{\boldsymbol{B}}'\boldsymbol{Y}_1 - S_{R1} = (0,-1,0,0)\begin{bmatrix} -2/3 \\ 4/3 \\ 2 \\ -4/3 \end{bmatrix} - 0 = -4/3,$$

$$\boldsymbol{S}_{\boldsymbol{B}}'\boldsymbol{Y}_2 - S_{R2} = (0,-1,0,0)\begin{bmatrix} 4/3 \\ -2/3 \\ -2 \\ 2/3 \end{bmatrix} - 0 = -2/3.$$

And from (10.2), $\theta_c = \bar{c}_1 / (\boldsymbol{S}_{\boldsymbol{B}}'\boldsymbol{Y}_1 - S_{R1}) = -2/(-4/3) = 3/2$. Then $\theta_c^{(2)} = \theta_c^{(1)} + \theta_c = 7/2$.(Since the minimum in (10.2) occurs for $j = 1$, $\boldsymbol{r}_1$ will be pivoted into the basis when an additional iteration of the simplex routine is undertaken.) Hence $\theta_c^{(1)} = 2 < \theta < \frac{7}{2} = \theta_c^{(2)}$, the current basic feasible solution remains optimal with $max\, f = \left(\boldsymbol{C}_{\boldsymbol{B}} + \left(\theta_c^{(1)} + \theta\right)\boldsymbol{S}\right)'\boldsymbol{X}_{\boldsymbol{B}}$. When $\theta = \theta_c^{(2)}$, the modified problem

$$max\, f(\bar{X}) = \left(\bar{C} + \theta_c^{(2)} S\right)' \bar{X} = (1,1/2,0,0,0,0)\bar{X} \text{ s.t.}$$

the original constraints again exhibits multiple optimal basic feasible solutions: (i) point B; and (ii) point C. To find the optimal simplex matrix for the modified problem that is associated with extreme point B, let us find, using (10.1.1),

$$-\bar{c}_1^* = -\bar{c}_1 + \theta_c\left(S_B' Y_1 - S_{R1}\right) = 2 + \frac{3}{2}\left(-\frac{4}{3}\right) = 0,$$

$$-\bar{c}_2^* = -\bar{c}_2 + \theta_c\left(S_B' Y_2 - S_{R2}\right) = 0 + \frac{3}{2}\left(\frac{2}{3}\right) = 1.$$

Additionally, $\Delta f = \Delta c_{B2} x_2 = -\theta_c x_2 = \left(-\frac{3}{2}\right)\left(\frac{100}{3}\right) = -50$ since, for $\theta = \theta_c^{(2)} = \frac{7}{2}$, $c_2^* = 4 - \theta$ decreases to $^1/_2$, to wit, the modified value of f is 50. An insertion of these values into the last row of the latest simplex matrix thus yields

$$\begin{bmatrix} 1 & 0 & -\frac{2}{3} & 0 & 0 & \frac{4}{3} & 0 & \frac{100}{3} \\ 0 & 1 & \frac{4}{3} & 0 & 0 & -\frac{2}{3} & 0 & \frac{100}{3} \\ 0 & 0 & 2 & 0 & 1 & -2 & 0 & 20 \\ 0 & 0 & -\frac{4}{3} & 1 & 0 & \frac{2}{3} & 0 & \frac{20}{3} \\ 0 & 0 & 0 & 0 & 0 & 1 & 1 & 50 \end{bmatrix}.$$

So for $\theta = \theta_c^{(2)} = \frac{7}{2}$, point B represents the optimal basic feasible solution $x_1 = \frac{100}{3}, x_2 = \frac{100}{3}, x_4 = \frac{20}{3}, x_5 = 20$, and $f = 50$. If we now employ the current simplex matrix to generate the alternative optimal basic feasible solution at point C, we have

$$\begin{bmatrix} 1 & 0 & 0 & 0 & \frac{1}{3} & \frac{2}{3} & 0 & 40 \\ 0 & 1 & 0 & 0 & -\frac{2}{3} & \frac{2}{3} & 0 & 20 \\ 0 & 0 & 1 & 0 & \frac{1}{2} & -1 & 0 & 10 \\ 0 & 0 & 0 & 1 & \frac{2}{3} & -\frac{2}{3} & 0 & 20 \\ 0 & 0 & 0 & 0 & 0 & 1 & 1 & 50 \end{bmatrix}$$

wherein $x_1 = 40$, $x_2 = 20$, $x_3 = 10$, $x_4 = 20$, and $f = 50$.

If θ is made to increase a bit above $\theta_c^{(2)} = \frac{7}{2}$, point C becomes the unique optimal basic feasible solution. Relative to such an increase, what is the largest allowable increment in θ beyond $\theta_c^{(2)}$ that does not violate the optimality of extreme point C? With $\boldsymbol{C}_B' = (c_1, c_2, c_3, c_4) = (1,4,0,0), \boldsymbol{C}_R' = (c_5, c_6) = (0,0)$, we have $\boldsymbol{S}_B' = (0,-1,0,0), \boldsymbol{S}_R' = (0,0)$. Then for $j = 1, 2$,

$$\boldsymbol{S}_B'Y_1 - S_{R1} = (0,-1,0,0)\begin{bmatrix} 1/3 \\ -2/3 \\ 1/2 \\ 2/3 \end{bmatrix} - 0 = \frac{2}{3},$$

$$\boldsymbol{S}_B'Y_2 - S_{R2} = (0,-1,0,0)\begin{bmatrix} 2/3 \\ 2/3 \\ -1 \\ -2/3 \end{bmatrix} - 0 = -\frac{2}{3}.$$

And from (10.2), $\theta_c = \bar{c}_2/(\boldsymbol{S}_B'Y_2 - S_{R2}) = -1/(-2/3) = \dfrac{3}{2}$ so that $\theta_c^{(3)} = \theta_c^{(2)} + \theta_c = 5$. (Note also that since the minimum in (10.2) holds for $j = 2$, we see that when the next simplex iteration is executed, $\boldsymbol{r}_2$ will be pivoted into the basis.) Hence, for $\theta_c^{(2)} = \dfrac{7}{2} < \theta < 5 = \theta_c^{(3)}$, the current basic feasible solution is optimal with $max\, f = (\boldsymbol{C}_B + (\theta_c^{(2)} + \theta_c)\boldsymbol{S})'\boldsymbol{X}_B$. For $\theta = \theta_c^{(3)}$, the adjusted problem

$$max\, f(\bar{X}) = \left(\bar{C} + \theta_c^{(3)}\boldsymbol{S}\right)'\bar{X} = (1,-1,0,0,0,0)\bar{X} \text{ s.t.}$$

The original constraints as usual possess two optimal extreme point solutions: (i) point C; and (ii) point D. Proceeding as above, the optimal simplex matrix associated with extreme point C may be determined by first utilizing (10.1.1) to obtain

$$-\bar{c}_1^* = -\bar{c}_1 + \theta_c(\boldsymbol{S}_B'Y_1 - S_{R1}) = 0 + \frac{3}{2}\left(\frac{2}{3}\right) = 1,$$

$$-\bar{c}_2^* = -\bar{c}_2 + \theta_c\left(\boldsymbol{S}_B'Y_2 - S_{R2}\right) = 1 + \frac{3}{2}\left(-\frac{2}{3}\right) = 0.$$

Also, $\Delta f = \left(-\dfrac{3}{2}\right)(20) = -30$ since, with $\theta = \theta_c^{(3)} = 5, c_2^* = 4 - \theta$ becomes -1 so that the modified f value is 20. Then the desired matrix is

$$\begin{bmatrix} 1 & 0 & 0 & 0 & \frac{1}{3} & \frac{2}{3} & 0 & 40 \\ 0 & 1 & 0 & 0 & -\frac{2}{3} & \frac{2}{3} & 0 & 20 \\ 0 & 0 & 1 & 0 & \frac{1}{2} & -1 & 0 & 10 \\ 0 & 0 & 0 & 1 & \frac{2}{3} & -\frac{2}{3} & 0 & 20 \\ 0 & 0 & 0 & 0 & 1 & 0 & 1 & 20 \end{bmatrix}.$$

Hence, for $\theta = \theta_c^{(3)} = 5$, point C yields an optimal basic feasible solution consisting of $x_1 = 40$, $x_2 = 20$, $x_3 = 10$, $x_4 = 20$, and $f = 20$. And if we now pivot to the alternative optimal basic feasible solution found at extreme point D, the resulting simplex matrix

$$\begin{bmatrix} 1 & -1 & 0 & 0 & \frac{7}{6} & 0 & 0 & 20 \\ 0 & \frac{3}{2} & 0 & 0 & -1 & 1 & 0 & 30 \\ 0 & \frac{3}{2} & 1 & 0 & -\frac{1}{2} & 0 & 0 & 40 \\ 0 & 1 & 0 & 1 & 0 & 0 & 0 & 40 \\ 0 & 0 & 0 & 0 & 1 & 0 & 1 & 20 \end{bmatrix}$$

yields $x_1 = 20$, $x_3 = 40$, $x_4 = 40$, $x_6 = 30$, and $f = 20$.

If θ increases slightly above $\theta_c^{(3)} = 5$, point D represents a unique optimal basic feasible solution. Moreover, since now $C_B' = (c_1, c_6, c_3, c_4), C_R' = (c_2, c_5)$, $S_B' = (0,0,0,0), S_R' = (-1,0)$, and $S_B'Y_1 - S_{R1} = 1, S_B'Y_2 - S_{R2} = 0$, it follows that the current basic feasible solution remains optimal as θ increases without bound.

To summarize: the optimal basic feasible solution to the parametrized problem

$$\begin{aligned} max\, f(x_1, x_2, \theta) &= x_1 + (4-\theta)x_2 \quad \text{s.t.} \\ \frac{1}{2}x_1 + x_2 &\le 50 \\ x_2 &\le 40 \\ x_1 - x_2 &\le 20 \\ x_1 + \frac{1}{2}x_2 &\le 50 \\ x_1, x_2 &\ge 0 \end{aligned}$$

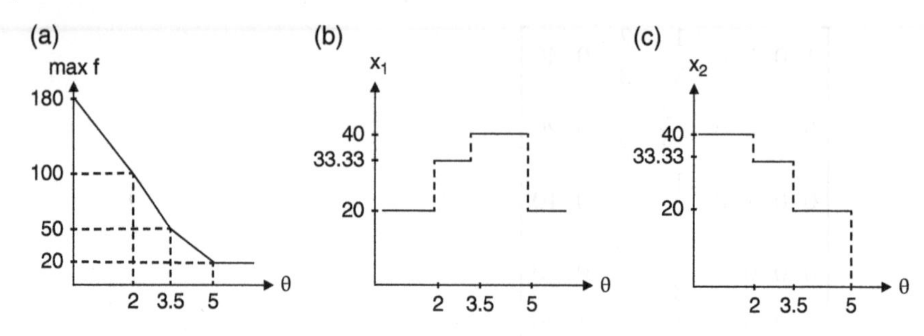

Figure 10.2 Parametric path of x_1, x_2, and *max f*.

is

$$\left(X_B^{(1)}\right)' = (x_1,x_2,x_5,x_6) = (20,40,40,10) \quad \text{for } 0 \le \theta \le 2;$$

$$\left(X_B^{(2)}\right)' = (x_1,x_2,x_5,x_4) = (100/3,100/3,20,20/3) \quad \text{for } 2 \le \theta \le \frac{7}{2};$$

$$\left(X_B^{(3)}\right)' = (x_1,x_2,x_3,x_4) = (40,20,10,20) \quad \text{for } \frac{7}{2} \le \theta \le 5;$$

$$\left(X_B^{(4)}\right)' = (x_1,x_6,x_3,x_4) = (20,30,40,40) \quad \text{for } 5 \le \theta \le +\infty.$$

In addition, Figure 10.2 traces out the path taken by x_1, x_2, and *max f* as θ increases through the above sequence of critical values. ■

10.2.2 Parametrizing the Requirements Vector

Given that our starting point is again an optimal basic feasible solution to a linear programming problem, let us replace $\boldsymbol{b}$ by $\boldsymbol{b}^* = \boldsymbol{b} + \phi \boldsymbol{t}$, where ϕ is a nonnegative scalar parameter and $\boldsymbol{t}$ is a given (arbitrary) $(m \times 1)$ vector that specifies the direction of change in the requirements b_i, $i = 1, \ldots, m$. Hence, the $b_i^*, i = 1,\ldots,m$, vary linearly with the parameter ϕ. Currently, primal feasibility holds for $\boldsymbol{X_B} = \boldsymbol{B}^{-1}\boldsymbol{b} \ge \boldsymbol{O}$ or $x_{\boldsymbol{B}i} \ge 0$, $i = 1, \ldots, m$. With $\boldsymbol{b}^*$ replacing $\boldsymbol{b}$, the *revised primal feasibility requirement* becomes

$$\begin{aligned} \boldsymbol{X}_{\boldsymbol{B}}^* = \boldsymbol{B}^{-1}\boldsymbol{b}^* &= \boldsymbol{B}^{-1}(\boldsymbol{b} + \phi \boldsymbol{t}) \\ &= \boldsymbol{X}_{\boldsymbol{B}} + \phi \boldsymbol{B}^{-1}\boldsymbol{t} \\ &= \boldsymbol{X}_{\boldsymbol{B}} + \phi \begin{bmatrix} \boldsymbol{\beta}_1 \boldsymbol{t} \\ \vdots \\ \boldsymbol{\beta}_m \boldsymbol{t} \end{bmatrix} \ge \boldsymbol{O}, \end{aligned} \tag{10.3}$$

where $\boldsymbol{\beta}_i$ devotes the ith row of $\boldsymbol{B}^{-1}$. In terms of the individual components of (10.3),

$$x_{Bi}^{*} = x_{Bi} + \phi\boldsymbol{\beta}_i\boldsymbol{t} \geq 0, i = 1,\ldots,m. \tag{10.3.1}$$

(Since the optimality criterion is independent of $\boldsymbol{b}$, the parametrization of the latter affects only primal feasibility and not primal optimality.) What is the largest value of ϕ for which (10.3.1) holds? Clearly the implied critical value of ϕ, ϕ_c, is that for which any increase in ϕ beyond ϕ_c violates feasibility, i.e. makes at least one of the x_{Bi}^{*} values negative.

To determine ϕ_c, let us first assume that $\boldsymbol{\beta}_i\boldsymbol{t} \geq 0$, $i = 1, \ldots, m$. Then ϕ can be increased without bound while still maintaining the revised feasibility criterion since, in this instance, $x_{Bi}^{*} \geq x_{Bi} \geq 0, i = 1,\ldots,m$. Next, if $\boldsymbol{\beta}_i\boldsymbol{t} < 0$ for some i, $x_{Bi}^{*} \geq 0$ for

$$\phi \leq \frac{-x_{Bi}}{\boldsymbol{\beta}_i\boldsymbol{t}} = \phi_c.$$

Hence, the revised feasibility criterion is violated when $\phi > \phi_c$ for some basic variable x_{Bi}. Moreover, if $\boldsymbol{\beta}_i\boldsymbol{t} < 0$ for two or more basic variables, then

$$\phi_c = \min_i \left\{ \frac{-x_{Bi}}{\boldsymbol{\beta}_i\boldsymbol{t}} \middle| \boldsymbol{\beta}_i\boldsymbol{t} < 0 \right\}. \tag{10.4}$$

Now, if ϕ increases and the minimum in (10.4) is attained for $i = k$, it follows that $\phi_c = -x_{Bk}/\boldsymbol{\beta}_k\boldsymbol{t}$ or $x_{Bk}^{*} = 0$. And as ϕ increases beyond $\phi_c, x_{Bk}^{*} < 0$ so that $\boldsymbol{b}_k$ must be removed from the basis to preserve primal feasibility, with its replacement vector determined by the dual simplex entry criterion.

Although the specification of the direction vector $\boldsymbol{t}$ is arbitrary, it is frequently the case that only a single requirement b_i is to be changed, so that $\boldsymbol{t} = \boldsymbol{e}_i$(or $-\boldsymbol{e}_i$). With respect to this particular selection of $\boldsymbol{t}$, if $\boldsymbol{b}^{*} = \boldsymbol{b} + \phi\boldsymbol{e}_k$, then (10.3.1) simplifies to $x_{Bi}^{*} = x_{Bi} + \phi\beta_{ik} \geq 0, i = 1,\ldots,m$. For this choice of $\boldsymbol{t}$ then, (10.4) may be rewritten as

$$\phi_c = \min_i \left\{ \frac{-x_{Bi}}{\beta_{ik}} \middle| \beta_{ik} < 0 \right\}. \tag{10.4.1}$$

If the minimum in (10.4.1) is attained for $i = l, \phi_c = -x_{Bl}/\beta_{lk}$ or $x_{Bl}^{*} = 0$. If ϕ then increases above this critical value, x_{Bl}^{*} turns negative so that b_l is pivoted out of the basis to maintain primal feasibility. Moreover, as b_k changes, f must be modified by an amount $\Delta f = u_k\Delta b_k = u_k\phi_c$.

Example 10.2 The optimal simplex matrix associated with the problem

$$\begin{aligned} max\, f &= x_1 + 4x_2 \quad \text{s.t.} \\ x_1 - x_2 &\leq 10 \\ \frac{1}{2}x_1 + x_2 &\leq 100 \\ x_2 &\leq 40 \\ x_1, x_2 &\geq 0, \end{aligned}$$

is

$$\begin{bmatrix} 1 & 0 & 1 & 0 & 1 & 0 & 50 \\ 0 & 0 & -\frac{1}{2} & 1 & -\frac{3}{2} & 0 & 35 \\ 0 & 1 & 0 & 0 & 1 & 0 & 40 \\ 0 & 0 & 1 & 0 & 5 & 1 & 210 \end{bmatrix}$$

with $x_{B1} = x_1 = 50$, $x_{B2} = x_4 = 35$, $x_{B3} = x_2 = 40$, and $f = 210$ (Figure 10.3). Here the feasible region corresponding to the stated problem is the area OABC while the optimal extreme point is B. Let us parametrize $\boldsymbol{b}$ by choosing $\boldsymbol{t} = \boldsymbol{e}_3$ so that

$$\boldsymbol{b}^* = \boldsymbol{b} + \phi\boldsymbol{t} = \begin{bmatrix} 10 \\ 100 \\ 40 \end{bmatrix} + \phi \begin{bmatrix} 0 \\ 0 \\ 1 \end{bmatrix} = \begin{bmatrix} 10 \\ 100 \\ 40 + \phi \end{bmatrix}.$$

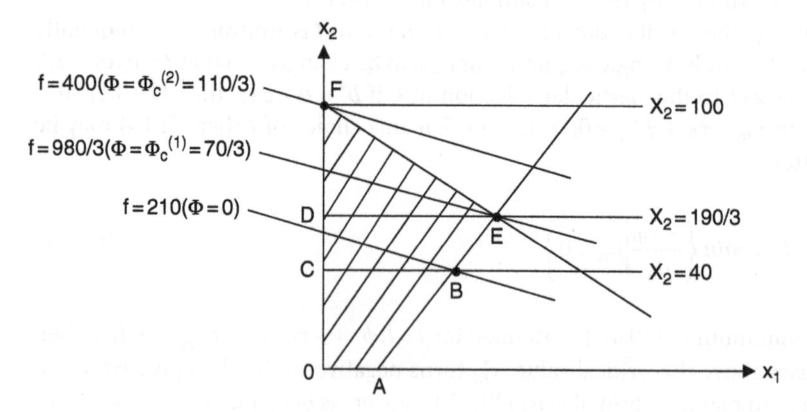

Figure 10.3 Generating the sequence B, E, and F of optimal extreme points parametrically.

From the optimal simplex matrix we may determine

$$X_B^* = X_B + \phi B^{-1} t = \begin{bmatrix} 50 \\ 35 \\ 40 \end{bmatrix} + \phi \begin{bmatrix} 1 & 0 & 1 \\ -\frac{1}{2} & 1 & -\frac{3}{2} \\ 0 & 0 & 1 \end{bmatrix} \begin{bmatrix} 0 \\ 0 \\ 1 \end{bmatrix} = \begin{bmatrix} 50+\phi \\ 35-\frac{3}{2}\phi \\ 40+\phi \end{bmatrix}.$$

Since $\beta_2 t = -\frac{3}{2} < 0$, it follows that $\phi_c = \phi_c^{(1)} = -x_{B2}/\beta_2 t = -35/(-3/2) = 70/3$ by virtue of (10.4). Then

$$\left(X_B^*\right)_{\phi_c^{(1)}} = \begin{bmatrix} 220/3 \\ 0 \\ 190/3 \end{bmatrix}, \text{and } \Delta f = u_3 \Delta b_3 = u_3 \phi_c = 350/3.$$

Once the adjusted values of the basic variables and the change in f are substituted into the last column of the above optimal simplex matrix, we obtain

$$\begin{bmatrix} 1 & 0 & 1 & 0 & 1 & 0 & \frac{220}{3} \\ 0 & 0 & -\frac{1}{2} & 1 & -\frac{3}{2} & 0 & 0 \\ 0 & 1 & 0 & 0 & 1 & 0 & \frac{190}{3} \\ 0 & 0 & 1 & 0 & 5 & 1 & \frac{980}{3} \end{bmatrix}.$$

Since the minimum in (10.4) was attained for $i = 2$, row two of this revised optimal simplex matrix is the pivotal row. Moreover, the dual simplex entry criterion selects the third column as the pivotal column so that the (circled) pivotal element is $-\frac{1}{2}$. Upon performing a pivot operation, we thus obtain

$$\begin{bmatrix} 1 & 0 & 0 & 2 & -2 & 0 & \frac{220}{3} \\ 0 & 0 & 1 & -2 & 3 & 0 & 0 \\ 0 & 1 & 0 & 0 & 1 & 0 & \frac{190}{3} \\ 0 & 0 & 0 & 2 & 2 & 1 & \frac{980}{3} \end{bmatrix}.$$

Hence our second optimal basic feasible solution occurs at extreme point E (since the extended feasible region is the area OAED) and has $x_{B1} = x_1 =$

$\frac{220}{3}$, $x_{B2} = x_3 = 0$, $x_{B3} = x_2 = \frac{190}{3}$, and $f = \frac{980}{3}$. (For this solution, the new requirements vector is

$$\boldsymbol{b}^*_{\phi_c^{(1)}} = \boldsymbol{b} + \phi_c^{(1)} \boldsymbol{t} = \begin{bmatrix} 10 \\ 100 \\ 190/3 \end{bmatrix},$$

i.e. this last simplex matrix represents an optimal basic feasible solution to

$$\begin{aligned} max\, f &= x_1 + 4x_2 \quad \text{s.t.} \\ x_1 - x_2 &\leq 10 \\ \frac{1}{2}x_1 + x_2 &\leq 100 \\ x_2 &\leq \frac{190}{3} \\ x_1, x_2 &\geq 0.) \end{aligned}$$

If ϕ is increased further, we now have

$$\boldsymbol{b}^* = \boldsymbol{b}^*_{\phi_c^{(1)}} + \phi \boldsymbol{t} = \begin{bmatrix} 10 \\ 100 \\ 190/3 \end{bmatrix} + \begin{bmatrix} 0 \\ 0 \\ 1 \end{bmatrix} = \begin{bmatrix} 10 \\ 100 \\ \frac{190}{3} + \phi \end{bmatrix}.$$

And from the latest optimal simplex matrix

$$\boldsymbol{X}_{\boldsymbol{B}}^* = \left(\boldsymbol{X}_{\boldsymbol{B}}^*\right)_{\phi_c^{(1)}} + \phi \boldsymbol{B}^{-1} \boldsymbol{t} = \begin{bmatrix} \frac{220}{3} \\ 0 \\ \frac{190}{3} \end{bmatrix} + \phi \begin{bmatrix} 0 & 2 & -2 \\ 1 & -2 & 3 \\ 0 & 0 & 1 \end{bmatrix} \begin{bmatrix} 0 \\ 0 \\ 1 \end{bmatrix} = \begin{bmatrix} \frac{220}{3} - 2\phi \\ 3\phi \\ \frac{190}{3} + \phi \end{bmatrix}.$$

With $\boldsymbol{\beta}_1 \boldsymbol{t} = -2 < 0$, (10.4) yields $\phi_c = \phi_c^{(2)} = x_{B1} / \boldsymbol{\beta}_1 \boldsymbol{t} = (-220/3)/(-2) = 110/3$. Then

$$\left(\boldsymbol{X}_{\boldsymbol{B}}^*\right)_{\phi_c^{(2)}} = \begin{bmatrix} 0 \\ 110 \\ 100 \end{bmatrix}, \text{and } \Delta f = u_3 \Delta b_3 = 220/3.$$

A substitution of these modified values into the preceding optimal simplex matrix thus yields

$$\begin{bmatrix} 1 & 0 & 0 & 2 & -2 & 0 & 0 \\ 0 & 0 & 1 & -2 & 3 & 0 & 110 \\ 0 & 1 & 0 & 0 & 1 & 0 & 100 \\ 0 & 0 & 0 & 2 & 2 & 1 & 400 \end{bmatrix} \rightarrow \begin{bmatrix} -\frac{1}{2} & 0 & 0 & -1 & 1 & 0 & 0 \\ \frac{3}{2} & 0 & 1 & 1 & 0 & 0 & 110 \\ \frac{1}{2} & 1 & 0 & 1 & 0 & 0 & 100 \\ 1 & 0 & 0 & 4 & 0 & 1 & 400 \end{bmatrix}.$$

Then the third optimal basic feasible solution corresponding to the modified feasible region OAEF occurs at extreme point F and consists of $x_{B1} = x_5 = 0$, $x_{B2} = x_3 = 110$, $x_{B3} = x_2 = 100$, and $f = 400$. Here the requirements vector for the problem just solved is

$$\boldsymbol{b}^*_{\phi_c^{(2)}} = \boldsymbol{b}^*_{\phi_c^{(1)}} + \phi_c^{(2)}\boldsymbol{t} = \begin{bmatrix} 10 \\ 100 \\ 100 \end{bmatrix},$$

i.e. we have generated an optimal basic feasible solution to

$$\begin{aligned} & max\, f = x_1 + 4x_2 \quad \text{s.t.} \\ & x_1 - x_2 \le 10 \\ & \frac{1}{2}x_1 + x_2 \le 100 \\ & x_2 \le 100 \\ & x_1, x_2 \ge 0. \end{aligned}$$

If ϕ is increased beyond $\phi_c^{(2)}$, it is easily demonstrated that

$$\boldsymbol{X}^*_{\boldsymbol{B}} = \left(\boldsymbol{X}^*_{\boldsymbol{B}}\right)_{\phi_c^{(2)}} + \phi\boldsymbol{B}^{-1}\boldsymbol{t} = \begin{bmatrix} \phi \\ 110 \\ 100 \end{bmatrix},$$

$$\boldsymbol{b}^* = \boldsymbol{b}^*_{\phi_c^{(2)}} + \phi\boldsymbol{t} = \begin{bmatrix} 10 \\ 100 \\ 100 + \phi \end{bmatrix}$$

so that, with $\boldsymbol{\beta}_i\boldsymbol{t} \ge 0$ for all i, ϕ may increase without bound without violating the feasibility of the current optimal solution.

In summary: the optimal basic feasible solution to the parametrized problem

$$\begin{aligned} max\, f &= x_1 + 4x_2 \quad \text{s.t.} \\ x_1 - x_2 &\le 10 \\ \frac{1}{2}x_1 + x_2 &\le 100 \\ x_2 &\le 40 + \phi \\ x_1, x_2 &\ge 0 \end{aligned}$$

is

$$\begin{aligned} \left(X_B^{(1)}\right)' &= (x_1, x_4, x_2) = (50, 35, 40) && \text{for } 0 \le \phi < 70/3; \\ \left(X_B^{(2)}\right)' &= (x_1, x_3, x_2) = (220/3, 0, 190/3) && \text{for } 70/3 \le \phi < 110/3; \\ \left(X_B^{(3)}\right)' &= (x_5, x_3, x_2) = (0, 110, 100) && \text{for } 110/3 \le \phi < +\infty. \end{aligned}$$

As far as the behavior of $max\, f$, x_1, x_2 and b_3 are concerned, Figure 10.4 charts the behavior of these quantities as ϕ increases through its sequence of critical values. As evidenced therein, $max\, f$ and b_3 are piecewise continuous linear functions of ϕ while x_1, x_2 are discontinuous functions of ϕ. For instance, if $70/3 \le \phi \le 110/3, max\, f = \boldsymbol{C}_B'\boldsymbol{B}^{-1}\boldsymbol{b}^* = \boldsymbol{U}'\boldsymbol{b}^* = \boldsymbol{U}'\boldsymbol{b} + \phi u_3$ is a linear function of ϕ with slope $d\, max\, f / d\phi = u_3 = 2$. ■

Example 10.3 If in the previous example we let $\boldsymbol{t} = -\boldsymbol{e}_2$, then

$$\boldsymbol{b}^* = \boldsymbol{b} + \phi \boldsymbol{t} = \begin{bmatrix} 10 \\ 100 \\ 40 \end{bmatrix} + \phi \begin{bmatrix} 0 \\ -1 \\ 0 \end{bmatrix}.$$

From the optimal simplex matrix for the original problem

$$\boldsymbol{X}_B^* = \boldsymbol{X}_B + \phi \boldsymbol{B}^{-1}\boldsymbol{t} = \begin{bmatrix} 50 \\ 35 \\ 40 \end{bmatrix} + \phi \begin{bmatrix} 1 & 0 & 1 \\ -\frac{1}{2} & 1 & -\frac{3}{2} \\ 0 & 0 & 1 \end{bmatrix} \begin{bmatrix} 0 \\ -1 \\ 0 \end{bmatrix} = \begin{bmatrix} 50 \\ 35 - \phi \\ 40 \end{bmatrix}.$$

With $\boldsymbol{\beta}_2\boldsymbol{t} = -1 < 0$,(10.4) yields $\phi_c = \phi_c^{(1)} = -x_{B2}/\boldsymbol{\beta}_2\boldsymbol{t} = -35/(-1) = 35$. Then

$$\left(\boldsymbol{X}_B^*\right)_{\phi_c^{(1)}} = \begin{bmatrix} 50 \\ 0 \\ 40 \end{bmatrix}, \left(\boldsymbol{b}^*\right)_{\phi_c^{(1)}} = \boldsymbol{b} + \phi_c^{(1)}\boldsymbol{t} = \begin{bmatrix} 10 \\ 65 \\ 40 \end{bmatrix}, \text{and } \Delta f = u_2 \Delta b_2 = 0.$$

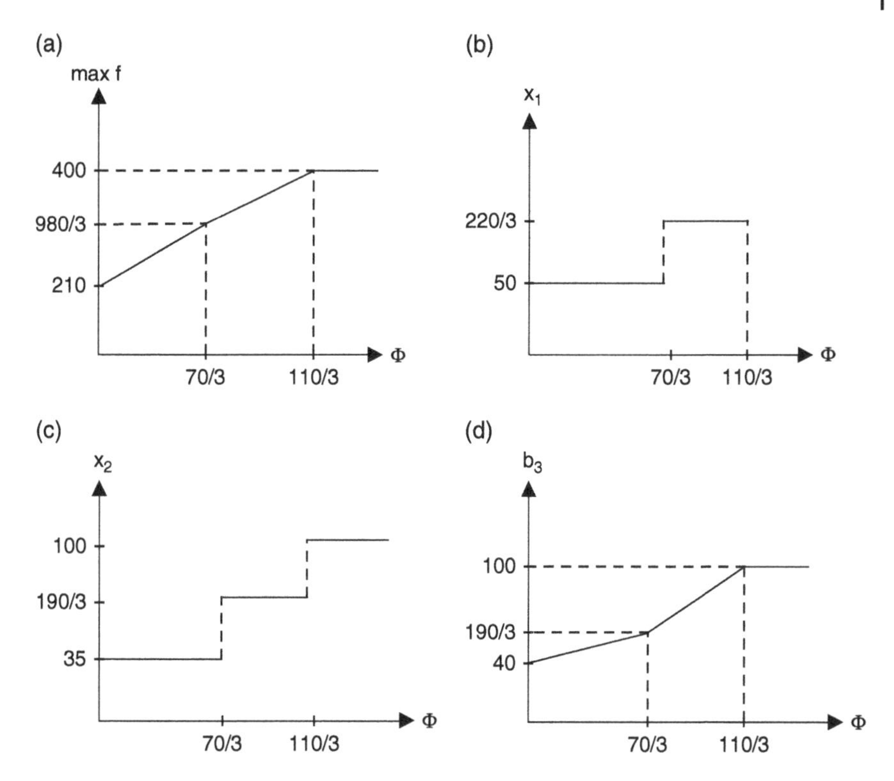

Figure 10.4 Parametric path of x_1, x_2, b_3, and *max f*.

Upon making the appropriate adjustments within the optimal simplex matrix we obtain

$$\begin{bmatrix} 1 & 0 & 1 & 0 & 1 & 0 & 50 \\ 0 & 0 & -\frac{1}{2} & 1 & -\frac{3}{2} & 0 & 0 \\ 0 & 1 & 0 & 0 & 1 & 0 & 40 \\ 0 & 0 & 1 & 0 & 5 & 1 & 210 \end{bmatrix} \rightarrow \begin{bmatrix} 1 & 0 & 0 & 2 & -2 & 0 & 50 \\ 0 & 0 & 1 & -2 & 3 & 0 & 0 \\ 0 & 1 & 0 & 0 & 1 & 0 & 40 \\ 0 & 0 & 0 & 2 & 2 & 1 & 210 \end{bmatrix},$$

where the second optimal basic feasible solution is $x_{B1} = x_1 = 50$, $x_{B2} = x_3 = 0$, $x_{B3} = x_2 = 40$, and $f = 210$. Increasing ϕ a bit further we have

$$\boldsymbol{b}^* = \boldsymbol{b}^*_{\phi_c^{(1)}} + \phi \boldsymbol{t} = \begin{bmatrix} 10 \\ 65 \\ 40 \end{bmatrix} + \phi \begin{bmatrix} 0 \\ -1 \\ 0 \end{bmatrix} = \begin{bmatrix} 10 \\ 65-\phi \\ 40 \end{bmatrix}.$$

Then, from the latest optimal simplex matrix

$$X_B^* = \left(X_B^*\right)_{\phi_c^{(1)}} + \phi B^{-1}t = \begin{bmatrix} 50 \\ 0 \\ 40 \end{bmatrix} + \phi \begin{bmatrix} 0 & 2 & -2 \\ 1 & -2 & 3 \\ 0 & 0 & 1 \end{bmatrix} \begin{bmatrix} 0 \\ -1 \\ 0 \end{bmatrix} = \begin{bmatrix} 50-2\phi \\ 2\phi \\ 40 \end{bmatrix}.$$

Since $\beta_1 t = -2 < 0$, we obtain from (10.4), $\phi_c = \phi_c^{(2)} = -x_{B1}/\beta_1 t = -50/(-2) = 25$. Hence

$$\left(X_B^*\right)_{\phi_c^{(2)}} = \begin{bmatrix} 0 \\ 50 \\ 40 \end{bmatrix}, b^*_{\phi_c^{(2)}} = b^*_{\phi_c^{(1)}} + \phi_c^{(2)} t = \begin{bmatrix} 10 \\ 40 \\ 40 \end{bmatrix}, \text{and } \Delta f = u_2 \Delta b_2 = -50.$$

Again, making the appropriate adjustments within the latest optimal simplex matrix, we have

$$\begin{bmatrix} 1 & 0 & 0 & 2 & -2 & 0 & 0 \\ 0 & 0 & 1 & -2 & 3 & 0 & 50 \\ 0 & 1 & 0 & 0 & 1 & 0 & 40 \\ 0 & 0 & 0 & 2 & 2 & 1 & 160 \end{bmatrix} \to \begin{bmatrix} -\frac{1}{2} & 0 & 0 & -1 & 1 & 0 & 0 \\ \frac{3}{2} & 0 & 1 & 1 & 0 & 0 & 50 \\ \frac{1}{2} & 1 & 0 & 1 & 0 & 0 & 40 \\ 1 & 0 & 0 & 4 & 0 & 1 & 160 \end{bmatrix},$$

with $x_{B1} = x_5 = 0$, $x_{B2} = x_3 = 50$, $x_{B3} = x_2 = 40$, and $f = 160$. If ϕ increases beyond $\phi_c^{(2)}$,

$$X_B^* = \left(X_B^*\right)_{\phi_c^{(2)}} + \phi B^{-1}t = \begin{bmatrix} 0 \\ 50 \\ 40 \end{bmatrix} + \phi \begin{bmatrix} 0 & -1 & 1 \\ 1 & 1 & 0 \\ 0 & 1 & 0 \end{bmatrix} \begin{bmatrix} 0 \\ -1 \\ 0 \end{bmatrix} = \begin{bmatrix} \phi \\ 50-\phi \\ 40-\phi \end{bmatrix}.$$

In this instance, $\beta_2 t = -1 < 0, \beta_3 t = -1 < 0$ so that

$$\phi_c = \phi_c^{(3)} = 40 = \min_i \left\{ \frac{-x_{B2}}{\beta_2 t} = \frac{-50}{-1} = 50, \frac{-x_{B3}}{\beta_3 t} = \frac{-40}{-1} = 40 \right\}$$

and thus

$$\left(X_B^*\right)_{\phi_c^{(3)}} = \begin{bmatrix} 40 \\ 10 \\ 0 \end{bmatrix} \text{ and } \Delta f = u_2 \Delta b_2 = 160.$$

Using this information to construct the modified optimal simplex matrix, we have

$$\begin{bmatrix} -\frac{1}{2} & 0 & 0 & -1 & 1 & 0 & 40 \\ \frac{3}{2} & 0 & 1 & 1 & 0 & 0 & 10 \\ \frac{1}{2} & 1 & 0 & 1 & 0 & 0 & 0 \\ 1 & 0 & 0 & 4 & 0 & 1 & 0 \end{bmatrix}.$$

And since both $y_{31}, y_{32} > 0$, the dual simplex entry criterion cannot find any vector to insert into the basis. In this instance, if ϕ increases beyond $\phi_c^{(3)}$, no feasible solution exists since $x_{B3} = x_2$ turns negative. ∎

One final point is in order. In the preceding two example problems, we parametrized $\boldsymbol{b}$ directly. However, if we consider the associated dual problems, then parametrizing $\boldsymbol{b}$ can be handled in the same fashion as parametrizing the objective function.

10.2.3 Parametrizing an Activity Vector

Given that we have obtained an optimal basic feasible solution to the problem

$$\begin{aligned} &max\, f(\bar{X}) = \bar{C}\bar{X} \quad \text{s.t.} \\ &\bar{A}\bar{X} = \boldsymbol{b}, \bar{X} \geq \boldsymbol{O}, \end{aligned}$$

let us undertake the parametrization of a column of the activity matrix $\bar{A}$. With $\bar{A}$ previously partitioned as $\bar{A} = [\boldsymbol{B} \vdots \boldsymbol{R}]$, it may be the case that: some nonbasic vector $\boldsymbol{r}_j$ is replaced by $\boldsymbol{r}_j^* = \boldsymbol{r}_j + \tau V, j = 1, \ldots, n-m$; or a basic vector $\boldsymbol{b}_i$ is replaced by $\boldsymbol{b}_i^* = \boldsymbol{b}_i + \tau V, i = 1, \ldots, m$, where τ represents a nonnegative scalar parameter and V is an arbitrary $(m \times 1)$ vector that determines the direction of change in the components of either $\boldsymbol{r}_j$ or $\boldsymbol{b}_i$. In each individual case, the said components are expressed as linear functions of τ.

Let us initially assume that $\boldsymbol{r}_k$ is replaced by $\boldsymbol{r}_k^* = \boldsymbol{r}_k + \tau V$ in $\boldsymbol{R} = [\boldsymbol{r}_1, \ldots, \boldsymbol{r}_k, \ldots, \boldsymbol{r}_{n-m}]$ so that the new matrix of nonbasic vectors appears as $\boldsymbol{R}^* = [\boldsymbol{r}_1, \ldots, \boldsymbol{r}_k^*, \ldots, \boldsymbol{r}_{n-m}]$. Since the elements within the basis matrix $\boldsymbol{B}$ are independent of the parametrization of $\boldsymbol{r}_k$, primal feasibility is unaffected (we still have $\boldsymbol{X_B} = \boldsymbol{B}^{-1}\boldsymbol{b} \geq \boldsymbol{O}$) but primal optimality may be compromised. In this regard, for $\tau = 0, \boldsymbol{C}_{\boldsymbol{B}}'\boldsymbol{B}^{-1}\boldsymbol{R} - \boldsymbol{C}_{\boldsymbol{R}}' \geq \boldsymbol{O}'$ or $-\bar{c}_j = \boldsymbol{C}_{\boldsymbol{B}}'\boldsymbol{Y}_j - c_{\boldsymbol{R}j} \geq 0, j = 1, \ldots, n-m$. Upon parametrizing $\boldsymbol{r}_k$, the revised optimality criterion becomes

$$\boldsymbol{C}_{\boldsymbol{B}}'\boldsymbol{B}^{-1}\boldsymbol{R}^* - \boldsymbol{C}_{\boldsymbol{R}}' = \left(-\bar{c}_1, \ldots, -\bar{c}_k^*, \ldots, -\bar{c}_{n-m}\right) \geq \boldsymbol{O}',$$

where

$$\begin{aligned} -\bar{c}_k^* &= C_B' B^{-1} r_k^* - c_{Rk} = C_B' B^{-1}(r_k + \tau V) - c_{Rk} \\ &= -\bar{c}_k + \tau C_B' B^{-1} V = -\bar{c}_k + \tau U' V \geq 0 \end{aligned} \tag{10.5}$$

and $U' = C_B' B^{-1}$ is a $(1 \times m)$ vector of nonnegative dual variables. Since $-\bar{c}_j \geq 0, j = 1,\ldots,k,\ldots,n-m$, primal optimality will be preserved if (10.5) holds. Now, as τ increases, primal optimality must be maintained. Given this restriction, what is the largest value of τ (its critical value, τ_c) for which (10.5) holds?

To find τ_c, we shall first assume that $U'V \geq 0$. Then τ can be increased indefinitely while still maintaining the revised optimality criterion. Next, if $U'V < 0$, then $-\bar{c}_k^* \geq 0$ for

$$\tau \leq \frac{\bar{c}_k}{U'V} = \tau_c. \tag{10.6}$$

In this regard, as τ increases to τ_c, it follows that $-\bar{c}_k^* = 0$ so that the case of multiple optimal solutions emerges. That is, the current basic feasible solution (call it $X_B^{(1)}$) remains optimal and an alternative optimal basic feasible solution ($X_B^{(2)}$) obtains when r_k is pivoted into the basis, with the primal simplex criterion determining the vector to be removed from the current basis. And as τ increases a bit beyond τ_c, it follows that $-\bar{c}_k^* < 0$ so that $X_B^{(2)}$ becomes the unique optimal basic feasible solution.

As indicated above, the choice of the direction vector V is completely arbitrary. However, if only a single component r_{ik} of the activity vector r_k is to change, then $V = e_i$(or $-e_i$). In this regard, if $r_k^* = r_k + \tau e_l$, then (10.5) becomes $-\bar{c}_k^* = -\bar{c}_k + \tau u_l \geq 0$. So for this choice of V, with $u_l \geq 0$, τ may be increased without bound while still preserving primal optimality. Next if $r_k^* = r_k + \tau e_l$, (10.5) simplifies to $-\bar{c}_k^* = -\bar{c}_k - \tau u_l \geq 0$. In this instance, (10.6) may be rewritten as

$$\tau \leq \frac{-\bar{c}_k}{u_l} = \tau_c. \tag{10.6.1}$$

If τ then increases above this critical value, $-\bar{c}_k^*$ turns negative so that r_k is pivoted into the basis to maintain primal optimality.

When τ is increased slightly above $\tau = \tau_c^{(1)}$, it may be the case that: τ may be increased without bound while still preserving primal optimality; or there exists a second critical value of τ beyond which the new basis ceases to yield an optimal feasible solution. Relative to this latter instance, since r_k has entered the optimal basis, we now have to determine a procedure for parametrizing a basis vector. Before doing so, however, let us examine Example 10.4.

Example 10.4 The optimal simplex matrix relative to the problem

$$\begin{aligned}
&max\, f = 2x_1 + 3x_2 + 4x_3 \text{ s.t.}\\
&2x_1 + x_2 + 4x_3 \le 100\\
&x_1 + 3x_2 + x_3 \le 40\\
&x_j, \ge 0, j = 1,2,3,
\end{aligned}$$

is

$$\begin{bmatrix} \frac{5}{11} & 0 & 1 & \frac{3}{11} & -\frac{1}{11} & 0 & 23.63 \\ \frac{2}{11} & 1 & 0 & -\frac{1}{11} & \frac{4}{11} & 0 & 5.46 \\ \frac{4}{11} & 0 & 0 & \frac{9}{11} & \frac{8}{11} & 1 & 110.91 \end{bmatrix},$$

with $x_2 = 5.46$, $x_3 = 23.63$, and $f = 110.91$. Let us parametrize $\boldsymbol{r}_1$ by choosing $\boldsymbol{V}' = (-2, 1)$. Then

$$\boldsymbol{r}_1^* = \boldsymbol{r}_1 + \tau \boldsymbol{V} = \begin{bmatrix} 2 \\ 1 \end{bmatrix} + \tau \begin{bmatrix} -2 \\ 1 \end{bmatrix} = \begin{bmatrix} 2-2\tau \\ 1+\tau \end{bmatrix}.$$

And since

$$\boldsymbol{U}'\boldsymbol{V} = \left(\frac{9}{11}, \frac{8}{11}\right) \begin{bmatrix} -2 \\ 1 \end{bmatrix} = -10/11,$$

it follows that $\tau_c = \tau_c^{(1)} = (-4/11)/(-10/11) = 2/5$. Hence

$$\left(\boldsymbol{r}_1^*\right)_{\tau_c^{(1)}} = \begin{bmatrix} 6/5 \\ 7/5 \end{bmatrix}.$$

Moreover, since $\boldsymbol{Y}_j = \boldsymbol{B}^{-1}\boldsymbol{r}_j, j = 1, \ldots, n-m$, it follows that, when $\boldsymbol{r}_j$ is replaced in $\boldsymbol{R}$ by $\left(\boldsymbol{r}_j^*\right)_{\tau_c^{(1)}}, \boldsymbol{Y}_j$ is replaced in $\boldsymbol{B}^{-1}\boldsymbol{R}$ by $\boldsymbol{Y}_j^* = \boldsymbol{B}^{-1}\left(\boldsymbol{r}_j^*\right)_{t_c^{(1)}}$. Then

$$\boldsymbol{Y}_1^* = \boldsymbol{B}^{-1}\left(\boldsymbol{r}_1^*\right)_{\tau_c^{(1)}} = \begin{bmatrix} \frac{3}{11} & -\frac{1}{11} \\ -\frac{1}{11} & \frac{4}{11} \end{bmatrix} \begin{bmatrix} \frac{6}{5} \\ \frac{7}{5} \end{bmatrix} = \begin{bmatrix} \frac{1}{5} \\ \frac{2}{5} \end{bmatrix}.$$

Finally, $-\bar{c}_1^* = -\bar{c}_1 + \tau_c^{(1)} U'V = \frac{4}{11} + \frac{2}{5}\left(-\frac{10}{11}\right) = 0$. Once $Y_1^*, -\bar{c}_1^*$ are substituted into the preceding optimal simplex matrix, we obtain

$$\begin{bmatrix} \frac{1}{5} & 0 & 1 & \frac{3}{11} & -\frac{1}{11} & 0 & 23.63 \\ \frac{2}{5} & 1 & 0 & -\frac{1}{11} & \frac{4}{11} & 0 & 5.46 \\ 0 & 0 & 0 & \frac{9}{11} & \frac{8}{11} & 0 & 110.91 \end{bmatrix} \rightarrow \begin{bmatrix} 0 & -\frac{1}{2} & 1 & \frac{7}{22} & -\frac{3}{11} & 0 & 20.90 \\ 1 & \frac{5}{2} & 0 & -\frac{5}{22} & \frac{10}{11} & 0 & 13.65 \\ 0 & 0 & 0 & \frac{9}{11} & \frac{8}{11} & 1 & 110.91 \end{bmatrix}$$

and thus the alternative optimal basic feasible solution consists of $x_1 = 13.65$, $x_3 = 20.90$, and $f = 110.91$. ∎

Next, to parametrize a vector within the optimal basis, let us replace the kth column $\boldsymbol{b}_k$ of the basis matrix $\boldsymbol{B} = [\boldsymbol{b}_1,\ldots,\boldsymbol{b}_k,\ldots,\boldsymbol{b}_m]$ by $\boldsymbol{b}_k^* = \boldsymbol{b}_k + \tau \boldsymbol{V}$ so as to obtain $\boldsymbol{B}^* = [\boldsymbol{b}_1,\ldots,\boldsymbol{b}_k^*,\ldots,\boldsymbol{b}_m]$. Since $\boldsymbol{b}_k^* = \sum_{i=1}^{m} y_{ik}\boldsymbol{b}_i = \boldsymbol{B}\boldsymbol{Y}_k^*$, it follows that

$$\boldsymbol{Y}_k^* = \boldsymbol{B}^{-1}\boldsymbol{b}_k^* = \begin{bmatrix} \boldsymbol{\beta}_1 \\ \vdots \\ \boldsymbol{\beta}_k \\ \vdots \\ \boldsymbol{\beta}_m \end{bmatrix} b_k^* = \begin{bmatrix} \boldsymbol{\beta}_1 \boldsymbol{b}_k + \tau \boldsymbol{\beta}_1 \boldsymbol{V} \\ \vdots \\ \boldsymbol{\beta}_k \boldsymbol{b}_k + \tau \boldsymbol{\beta}_k \boldsymbol{V} \\ \vdots \\ \boldsymbol{\beta}_m \boldsymbol{b}_k + \tau \boldsymbol{\beta}_m \boldsymbol{V} \end{bmatrix} = \begin{bmatrix} \tau \boldsymbol{\beta}_1 \boldsymbol{V} \\ \vdots \\ 1 + \tau \boldsymbol{\beta}_k \boldsymbol{V} \\ \vdots \\ \tau \boldsymbol{\beta}_m \boldsymbol{V} \end{bmatrix}.$$

From (10.A.1) of Appendix 10.A,

$$\boldsymbol{I}_m^* = \begin{bmatrix} 1 & \cdots & \frac{-\tau \boldsymbol{\beta}_1 \boldsymbol{V}}{1 + \tau \boldsymbol{\beta}_k \boldsymbol{V}} & \cdots & 0 \\ \vdots & & \vdots & & \vdots \\ 0 & \cdots & \frac{1}{1 + \tau \boldsymbol{\beta}_k \boldsymbol{V}} & \cdots & 0 \\ \vdots & & \vdots & & \vdots \\ 0 & \cdots & \frac{-\tau \boldsymbol{\beta}_m \boldsymbol{V}}{1 + \tau \boldsymbol{\beta}_k \boldsymbol{V}} & \cdots & 1 \end{bmatrix}.$$

And from (10.A.2), $(\boldsymbol{B}^*)^{-1} = \boldsymbol{I}_m^* \boldsymbol{B}^{-1}$ and thus

$$\boldsymbol{X}_B^* = (\boldsymbol{B}^*)^{-1}\boldsymbol{b} = \boldsymbol{I}_m^*\boldsymbol{B}^{-1}\boldsymbol{b} = \boldsymbol{I}_m^*\boldsymbol{X}_B$$

$$= \begin{bmatrix} x_{B1} - \dfrac{\tau\beta_1 V}{1+\tau\beta_k V}x_{Bk} \\ \vdots \\ \dfrac{1}{1+\tau\beta_k V}x_{Bk} \\ \vdots \\ x_{Bm} - \dfrac{\tau\beta_m V}{1+\tau\beta_k V}x_{Bk} \end{bmatrix}.$$

Since $\boldsymbol{X}_B^*$ must be primal feasible, we require that

$$1+\tau\beta_k V > 0,$$

$$x_{Bi} - \frac{\tau\beta_i V}{1+\tau\beta_k V}x_{Bk} \geq 0, i = 1,\ldots,m.$$

If $1 + \tau\beta_k V > 0$, then, upon solving the second inequality for τ and remembering that τ must be strictly positive,

$$\tau \leq \frac{x_{Bi}}{\beta_i V x_{Bk} - \beta_k V x_{Bi}}, \beta_i V x_{Bk} - \beta_k V x_{Bi} > 0, i = 1,\ldots,m. \tag{10.7}$$

Since (10.7) must hold for all $i \neq k$ when $1 + \tau\beta_k V > 0$, it follows that the upper limit or critical value to which τ can increase without violating primal feasibility is

$$\tau_c' = \min_{i\neq k}\left\{\frac{x_{Bi}}{\beta_i V x_{Bk} - \beta_k V x_{Bi}} \middle| \beta_i V x_{Bk} - \beta_k V x_{Bi} > 0\right\}. \tag{10.8}$$

To determine the effect of parametrizing the basis vector upon primal optimality, let us express the *revised optimality criterion* as

$$\begin{aligned}\boldsymbol{C}_{B*}'(\boldsymbol{B}^*)^{-1}\boldsymbol{R} - \boldsymbol{C}_R' &= \boldsymbol{C}_{B^*}'\boldsymbol{I}_m^*\boldsymbol{B}^{-1}\boldsymbol{R} - \boldsymbol{C}_R' \\ &= \boldsymbol{C}_{B^*}'\boldsymbol{I}_m^*[Y_1,\ldots,Y_{n-m}] - \boldsymbol{C}_R' \geq \boldsymbol{O}'\end{aligned}$$

or, in terms of its components

$$-\bar{c}_j = \sum_{i\neq k} c_{B^*i}y_{ij} + \frac{c_{B^*k}y_{kj}}{1+\tau\beta_k V} - \sum_{i\neq k}\frac{\tau c_{B^*i}\beta_i V}{1+\tau\beta_k V}y_{kj} - c_{Rj} \geq 0, j = 1,\ldots,n-m. \tag{10.9}$$

If we now add and subtract the term

$$c_{B^*k}y_{kj} + \frac{\tau c_{B^*k}\beta_k V}{1+\tau\beta_k V} y_{kj}$$

on the left-hand side of (10.9), the latter simplifies to

$$\begin{aligned} -\bar{c}_j^* &= \sum_i c_{B^*i}y_{ij} - \frac{\tau y_{kj}\sum_i c_{B^*i}\beta_i V}{1+\tau\beta_k V} - c_{Rj} \\ &= -\frac{\tau y_{kj}\boldsymbol{U}'V}{1+\tau\beta_k V} - \bar{c}_j \geq 0, j = 1,\ldots,n-m. \end{aligned} \tag{10.9.1}$$

If $1 + \tau\boldsymbol{\beta}_k V > 0$, and again remembering that τ must be strictly positive, we may solve (10.9.1) for τ as

$$\tau \leq \frac{-\bar{c}_j}{y_{kj}\boldsymbol{U}'V + \bar{c}_j\beta_k V}, y_{kj}\boldsymbol{U}'V + \bar{c}_j\beta_k V > 0, j = 1,\ldots,n-m. \tag{10.10}$$

Since (10.10) is required to hold for all $j = 1, \ldots, n - m$ when $1 + \tau\boldsymbol{\beta}_k V > 0$, we may write the critical value to which τ can increase while still preserving primal optimality as

$$\tau_c'' = \min_j \left\{ \frac{-\bar{c}_j}{y_{kj}\boldsymbol{U}'V + \bar{c}_j\beta_k V} \middle| y_{kj}\boldsymbol{U}'V + \bar{c}_j\beta_k V > 0 \right\}. \tag{10.11}$$

As we have just seen, the parametrization of a basis vector may affect both the feasibility and optimality of the current solution. This being the case, the critical value of τ that preserves both primal feasibility and optimality simultaneously is simply

$$\tau_c = \min\{\tau_c', \tau_c''\}. \tag{10.12}$$

Finally, the effect on the optimal value of f parametrizing $\boldsymbol{b}_k$ can be shown to be

$$f(\boldsymbol{X}_{B*}) = f(\boldsymbol{X}_B) - \frac{\tau x_{Bk}\boldsymbol{U}'V}{1+\tau\beta_k V}. \tag{10.13}$$

As far as the specification of the arbitrary direction vector V is concerned, it is sometimes the case that only a single component of $\boldsymbol{b}_i$ is to change, i.e. $V = \boldsymbol{e}_i$ (or $-\boldsymbol{e}_i$) so that $\boldsymbol{b}_k^* = \boldsymbol{b}_k + \tau\boldsymbol{e}_i$.

Example 10.5 We found previously that the optimal simplex matrix associated with the problem

$$\begin{aligned} max\, f &= x_1 + 4x_2 \text{ s.t.} \\ x_1 - x_2 &\le 10 \\ \frac{1}{2}x_1 + x_2 &\le 100 \\ x_2 &\le 40 \\ x_1, x_2 &\ge 0 \end{aligned}$$

is

$$\begin{bmatrix} 1 & 0 & 1 & 0 & 1 & 0 & 50 \\ 0 & 0 & -\frac{1}{2} & 1 & -\frac{3}{2} & 0 & 35 \\ 0 & 1 & 0 & 0 & 1 & 0 & 40 \\ 0 & 0 & 1 & 0 & 5 & 1 & 210 \end{bmatrix}$$

with $x_{B1} = x_1 = 50$, $x_{B2} = x_4 = 35$, $x_{B3} = x_2 = 40$, and $f = 210$. Currently

$$\boldsymbol{B} = [\boldsymbol{b}_1, \boldsymbol{b}_2, \boldsymbol{b}_3] = \begin{bmatrix} 1 & 0 & -1 \\ \frac{1}{2} & 1 & 1 \\ 0 & 0 & 1 \end{bmatrix}, \boldsymbol{B}_1 = \begin{bmatrix} \beta_1 \\ \beta_2 \\ \beta_3 \end{bmatrix} = \begin{bmatrix} 1 & 0 & 1 \\ -\frac{1}{2} & 1 & -\frac{3}{2} \\ 0 & 0 & 1 \end{bmatrix}.$$

To parametrize the second column of $\bar{A}$, which corresponds to the third column of $\boldsymbol{B}$ (here $k = 3$), we form

$$\boldsymbol{b}_3^* = \boldsymbol{b}_3 + \tau \boldsymbol{V} = \begin{bmatrix} -1 \\ 1 \\ 1 \end{bmatrix} + \tau \begin{bmatrix} 1 \\ 1 \\ 1 \end{bmatrix} = \begin{bmatrix} -1+\tau \\ 1+\tau \\ 1+\tau \end{bmatrix}.$$

Since

$$\frac{x_{B1}}{\beta_1 V x_{B3} - \beta_3 V x_{B1}} = \frac{50}{(2)(40)-(1)(50)} = \frac{5}{3}, \frac{x_{B2}}{\beta_2 V x_{B3} - \beta_3 V x_{B2}} = \frac{35}{(-1)(40)-(1)(35)} = -\frac{7}{15},$$

it follows from (10.8) that $\tau_c' = \frac{5}{3}$. Moreover, with

$$\frac{-\bar{c}_1}{y_{31}\boldsymbol{U}'V + \bar{c}_1\beta_3 V} = \frac{1}{(0)(6)+(-1)(1)} = -1, \frac{-\bar{c}_2}{y_{32}\boldsymbol{U}'V + \bar{c}_2\beta_3 V} = \frac{5}{(1)(6)+(-5)(1)} = 5,$$

(10.11) reveals $\tau_c'' = 5$. Then from (10.12), $\tau_c = \tau_c^{(1)} = 5/3$. In this regard we may now compute

$$(\boldsymbol{b}_3^*)_{\tau_c^{(1)}} = \begin{bmatrix} \frac{2}{3} \\ \frac{8}{3} \\ \frac{8}{3} \end{bmatrix}, (\boldsymbol{X}_B^*)_{\tau_c^{(1)}} = \begin{bmatrix} x_{B1} - \frac{\tau\beta_1 V}{1+\tau\beta_3 V} x_{B3} \\ x_{B2} - \frac{\tau\beta_2 V}{1+\tau\beta_3 V} x_{B3} \\ \frac{1}{1+\tau\beta_3 V} x_{B3} \end{bmatrix} = \begin{bmatrix} 50 - \frac{(5/3)(2)}{1+\left(\frac{5}{3}\right)(1)} 40 \\ 35 - \frac{(5/3)(-1)}{1+\left(\frac{5}{3}\right)(1)} 40 \\ \frac{1}{1+\left(\frac{5}{3}\right)(1)} 40 \end{bmatrix} = \begin{bmatrix} 0 \\ 60 \\ 15 \end{bmatrix},$$

$$(-\bar{c}_1^*)_{\tau_c^{(1)}} = -\frac{\tau y_{31} \boldsymbol{U}' V}{1+\tau\beta_3 V} - \bar{c}_1 = -\frac{\left(\frac{5}{3}\right)(0)(6)}{1+\left(\frac{5}{3}\right)(1)} - (-1) = 1,$$

$$(-\bar{c}_2^*)_{\tau_c^{(1)}} = -\frac{\tau y_{32} \boldsymbol{U}' V}{1+\tau\beta_3 V} - \bar{c}_2 = -\frac{\left(\frac{5}{3}\right)(1)(6)}{1+\left(\frac{5}{3}\right)(1)} - (-5) = \frac{5}{4}.$$

Additionally, since

$$(\boldsymbol{I}_3^*)_{\tau_c^{(1)}} = \begin{bmatrix} 1 & 0 & \frac{-\tau\beta_1 V}{1+\tau\beta_3 V} \\ 0 & 1 & \frac{-\tau\beta_2 V}{1+\tau\beta_3 V} \\ 0 & 0 & \frac{1}{1+\tau\beta_3 V} \end{bmatrix} = \begin{bmatrix} 1 & 0 & \frac{-\left(\frac{5}{3}\right)(2)}{1+\left(\frac{5}{3}\right)(1)} \\ 0 & 1 & \frac{-\left(\frac{5}{3}\right)(-1)}{1+\left(\frac{5}{3}\right)(1)} \\ 0 & 0 & \frac{1}{1+\left(\frac{5}{3}\right)(1)} \end{bmatrix} = \begin{bmatrix} 1 & 0 & -5/4 \\ 0 & 1 & 5/8 \\ 0 & 0 & 3/8 \end{bmatrix},$$

and remembering that $Y_j = B^{-1}r_j$, we may find $Y_j^* = (B^*)^{-1}r_j = I_m^* B^{-1} r_j = I_m^* Y_j, j = 1,2,$ as

$$\left(Y_1^*\right)_{\tau_c^{(1)}} = \begin{bmatrix} 1 & 0 & -5/4 \\ 0 & 1 & 5/8 \\ 0 & 0 & 3/8 \end{bmatrix} \begin{bmatrix} 1 \\ -\frac{1}{2} \\ 0 \end{bmatrix} = \begin{bmatrix} 1 \\ -\frac{1}{2} \\ 0 \end{bmatrix},$$

$$\left(Y_2^*\right)_{\tau_c^{(1)}} = \begin{bmatrix} 1 & 0 & -5/4 \\ 0 & 1 & 5/8 \\ 0 & 0 & 3/8 \end{bmatrix} \begin{bmatrix} 1 \\ -3/2 \\ 1 \end{bmatrix} = \begin{bmatrix} -1/4 \\ -7/8 \\ 3/8 \end{bmatrix}.$$

Finally, from (10.13), the adjusted value of f is

$$f\left(\left(X_B^*\right)_{\tau_c^{(1)}}\right) = f(X_B) - (\tau x_{B3} U' V)/(1 + \tau\beta_3 V)$$

$$= 210 - (5/3)(40)(6)/\left(1 + \left(\frac{5}{3}\right)(1)\right) = 60.$$

Once this information is substituted into the above optimal simplex matrix, we have

$$\begin{bmatrix} 1 & 0 & 1 & 0 & -\frac{1}{4} & 0 & 0 \\ 0 & 0 & -\frac{1}{2} & 1 & -\frac{7}{8} & 0 & 60 \\ 0 & 1 & 0 & 0 & \frac{3}{8} & 0 & 15 \\ 0 & 0 & 1 & 0 & \frac{5}{4} & 1 & 60 \end{bmatrix} \rightarrow \begin{bmatrix} -4 & 0 & -4 & 0 & 1 & 0 & 0 \\ -\frac{7}{2} & 0 & -4 & 1 & 0 & 0 & 60 \\ \frac{3}{2} & 1 & \frac{3}{2} & 0 & 0 & 0 & 15 \\ 5 & 0 & 6 & 0 & 0 & 1 & 60 \end{bmatrix},$$

where $x_{B1} = x_5 = 0$, $x_{B2} = x_4 = 660$, $x_{B3} = x_2 = 15$, and $f = 60$. Here the new simplex matrix has been obtained by a dual simplex pivot operation. (This latter pivot operation was necessary since, if τ assumes a value greater than $\tau_c^{(1)} = \frac{5}{3}, x_{B1} = x_1 = 50$ turns negative. Hence, to ensure primal feasibility as τ increases beyond 5/3, x_1 was pivoted out of the optimal basis.)

How much larger can τ become without violating primal feasibility or optimality? To answer this, let us first determine

$$\frac{x_{B1}}{\beta_1 V x_{B3} - \beta_3 V x_{B1}} = \frac{0}{(-3)(15) - \left(\frac{3}{2}\right)(0)} = 0,$$

$$\frac{x_{B2}}{\beta_2 V x_{B3} - \beta_3 V x_{B2}} = \frac{60}{(-3)(15) - \left(\frac{3}{2}\right)(60)} = -\frac{60}{135}.$$

Since each denominator is negative, τ_c' does not exist. Next,

$$\frac{-\bar{c}_1}{y_{31}\boldsymbol{U}'\boldsymbol{V}+\bar{c}_1\boldsymbol{\beta}_3\boldsymbol{V}}=\frac{5}{\left(\frac{3}{2}\right)(6)+(-5)\left(\frac{3}{2}\right)}=\frac{10}{3},\ \frac{-\bar{c}_2}{y_{32}\boldsymbol{U}'\boldsymbol{V}+\bar{c}_2\boldsymbol{\beta}_3\boldsymbol{V}}=\frac{6}{\left(\frac{3}{2}\right)(6)+(-6)\left(\frac{3}{2}\right)}=\frac{6}{0}$$

so that $\tau_c''=\tau_c^{(2)}=\frac{10}{3}$ by virtue (10.12). Then

$$(\boldsymbol{b}_3^*)_{\tau_c^{(2)}}=(\boldsymbol{b}_3^*)_{\tau_c^{(1)}}+\tau\boldsymbol{V}=\begin{bmatrix}4\\6\\6\end{bmatrix},(\boldsymbol{X}_B^*)_{\tau_c^{(2)}}=\begin{bmatrix}x_{B1}-\frac{\tau\boldsymbol{\beta}_1\boldsymbol{V}}{1+\tau\boldsymbol{\beta}_3\boldsymbol{V}}x_{B3}\\x_{B2}-\frac{\tau\boldsymbol{\beta}_2\boldsymbol{V}}{1+\tau\boldsymbol{\beta}_3\boldsymbol{V}}x_{B3}\\\frac{1}{1+\tau\boldsymbol{\beta}_3\boldsymbol{V}}x_{B3}\end{bmatrix}=\begin{bmatrix}0-\frac{\left(\frac{10}{3}\right)(-3)}{1+\left(\frac{10}{3}\right)\left(\frac{3}{2}\right)}15\\60-\frac{\left(\frac{10}{3}\right)(-3)}{1+\left(\frac{10}{3}\right)\left(\frac{3}{2}\right)}15\\\frac{1}{1+\left(\frac{10}{3}\right)\left(\frac{3}{2}\right)}15\end{bmatrix}=\begin{bmatrix}25\\85\\\frac{5}{2}\end{bmatrix},$$

$$(-\bar{c}_1^*)_{\tau_c^{(2)}}=-\frac{\tau y_{13}\boldsymbol{U}'\boldsymbol{V}}{1+\tau\boldsymbol{\beta}_3\boldsymbol{V}}-\bar{c}_1=-\frac{\left(\frac{10}{3}\right)\left(\frac{3}{2}\right)(6)}{1+\left(\frac{10}{3}\right)\left(\frac{3}{2}\right)}-(-5)=0,$$

$$(-\bar{c}_2^*)_{\tau_c^{(2)}}=-\frac{\tau y_{32}\boldsymbol{U}'\boldsymbol{V}}{1+\tau\boldsymbol{\beta}_3\boldsymbol{V}}-\bar{c}_2=-\frac{\left(\frac{10}{3}\right)\left(\frac{3}{2}\right)(6)}{1+\left(\frac{10}{3}\right)\left(\frac{3}{2}\right)}-(-6)=1.$$

Also, since

$$(\boldsymbol{I}_3^*)_{\tau_c^{(2)}}=\begin{bmatrix}1&0&\frac{-\tau\boldsymbol{\beta}_1\boldsymbol{V}}{1+\tau\boldsymbol{\beta}_3\boldsymbol{V}}\\0&1&\frac{-\tau\boldsymbol{\beta}_2\boldsymbol{V}}{1+\tau\boldsymbol{\beta}_3\boldsymbol{V}}\\0&0&\frac{1}{1+\tau\boldsymbol{\beta}_3\boldsymbol{V}}\end{bmatrix}=\begin{bmatrix}1&0&\frac{-\left(\frac{10}{3}\right)(-3)}{1+\left(\frac{10}{3}\right)\left(\frac{3}{2}\right)}\\0&1&\frac{-\left(\frac{10}{3}\right)(-3)}{1+\left(\frac{10}{3}\right)\left(\frac{3}{2}\right)}\\0&0&\frac{1}{1+\left(\frac{10}{3}\right)\left(\frac{3}{2}\right)}\end{bmatrix}=\begin{bmatrix}1&0&5/3\\0&1&5/3\\0&0&1/6\end{bmatrix},$$

it follows that

$$\left(\boldsymbol{Y}_1^*\right)_{\tau_c^{(2)}} = \begin{bmatrix} 1 & 0 & 5/3 \\ 0 & 1 & 5/3 \\ 0 & 0 & 1/6 \end{bmatrix} \begin{bmatrix} -4 \\ -\frac{7}{2} \\ \frac{3}{2} \end{bmatrix} = \begin{bmatrix} -\frac{3}{2} \\ -1 \\ \frac{1}{4} \end{bmatrix},$$

$$\left(\boldsymbol{Y}_2^*\right)_{\tau_c^{(2)}} = \begin{bmatrix} 1 & 0 & 5/3 \\ 0 & 1 & 5/3 \\ 0 & 0 & 1/6 \end{bmatrix} \begin{bmatrix} -4 \\ -4 \\ \frac{3}{2} \end{bmatrix} = \begin{bmatrix} -\frac{3}{2} \\ -\frac{3}{2} \\ \frac{1}{4} \end{bmatrix}.$$

Finally, from (10.13), the modified f value is

$$\begin{aligned} f\left(\left(\boldsymbol{X}_B^*\right)_{\tau_c^{(2)}}\right) &= f\left(\left(\boldsymbol{X}_B^*\right)_{\tau_c^{(1)}}\right) - \left(\tau x_{B3}\boldsymbol{U}'V\right)/\left(1+\tau\beta_3 V\right) \\ &= 60 - \left(\frac{10}{3}\right)(15)(6)\Big/\left(1+\left(\frac{10}{3}\right)\left(\frac{3}{2}\right)\right) = 10. \end{aligned}$$

Upon substituting these values into the latest optimal simplex matrix, we have

$$\begin{bmatrix} -\frac{3}{2} & 0 & -\frac{3}{2} & 0 & 1 & 0 & 25 \\ -1 & 0 & -\frac{3}{2} & 1 & 0 & 0 & 85 \\ \frac{1}{4} & 1 & \frac{1}{4} & 0 & 0 & 0 & \frac{5}{2} \\ 0 & 0 & 1 & 0 & 0 & 1 & 10 \end{bmatrix}$$

with $x_{B1} = x_5 = 25, x_{B2} = x_4 = 85, x_{B3} = x_2 = \frac{5}{2}$, and $f = 10$. Note that for this simplex matrix $-\overline{c}_1^*$ has decreased to zero, indicating that if τ increases above $\frac{10}{3}$, the optimality criterion is violated. To avoid this difficulty, let us pivot $\boldsymbol{b}_3$ out of the basis (or x_2 out of the set of basic variables) so that the preceding simplex matrix becomes

$$\begin{bmatrix} 0 & 6 & 0 & 0 & 1 & 0 & 40 \\ 0 & 4 & -\frac{1}{2} & 1 & 0 & 0 & 95 \\ 1 & 4 & 1 & 0 & 0 & 0 & 10 \\ 0 & 0 & 1 & 0 & 0 & 1 & 10 \end{bmatrix}$$

wherein $x_{B1} = x_5 = 40$, $x_{B2} = x_4 = 95$, $x_{B3} = x_1 = 10$, and $f = 10$. Since the coefficients on x_2 now constitute a nonbasic vector, we must return to the parametrization process outlined above and illustrated in the preceding example. In this regard, with $U'V = 1 > 0$, τ can be increased indefinitely while still preserving primal optimality so that the current parametrization process is terminated. ■

Now that we have seen how the various component parts of the primal program can be parametrized, we turn to Chapter 11 where the techniques developed in this chapter are applied in deriving output supply functions; input demand functions; marginal revenue productivity functions; marginal, total, and variable cost functions; and marginal and average productivity functions.

10.A Updating the Basis Inverse

We know that to obtain a new basic feasible solution to a linear programming problem via the simplex method, we need change only a single vector in the basis matrix at each iteration. We are thus faced with the problem of finding the inverse of a matrix $\hat{\boldsymbol{B}}$ that has only one column different from that of a matrix $\boldsymbol{B}$ whose inverse is known. So given $\boldsymbol{B} = [\boldsymbol{b}_1, \dots, \boldsymbol{b}_m]$ with $\boldsymbol{B}^{-1}$ known, we wish to replace the rth column $\boldsymbol{b}_r$ of $\boldsymbol{B}$ by $\boldsymbol{r}_j$ so as obtain the inverse of $\hat{\boldsymbol{B}} = \left[\boldsymbol{b}_1, \dots, \boldsymbol{b}_{r-1}, \boldsymbol{r}_j, \boldsymbol{b}_{r+1}, \dots, \boldsymbol{b}_m\right]$. Since the columns of $\boldsymbol{B}$ form a basis for $\boldsymbol{R}^m$, $\boldsymbol{r}_j = \boldsymbol{B}\boldsymbol{Y}_j$. If the columns of $\hat{\boldsymbol{B}}$ are also to yield a basis for $\boldsymbol{R}^m$, then it must be true that $y_{rj} \neq 0$. So if

$$\begin{aligned} \boldsymbol{r}_j &= \sum_{i=1}^{m} y_{ij}\boldsymbol{b}_i \\ &= y_{1j}\boldsymbol{b}_1 + \cdots + y_{r-1,j}\boldsymbol{b}_{r-1} + y_{rj}\boldsymbol{b}_r + y_{r+1,j}\boldsymbol{b}_{r+1} + \cdots + y_{mj}\boldsymbol{b}_m \end{aligned}$$

with $y_{rj} \neq 0$, then

$$\begin{aligned} \boldsymbol{b}_r &= -\frac{y_{1r}}{y_{rj}}\boldsymbol{b}_1 - \cdots - \frac{y_{r-1,j}}{y_{rj}}\boldsymbol{b}_{r-1} + \frac{1}{y_{rj}}\boldsymbol{r}_j - \frac{y_{r+1,j}}{y_{rj}}\boldsymbol{b}_{r+1} - \cdots - \frac{y_{mj}}{y_{rj}}\boldsymbol{b}_m \\ &= \hat{\boldsymbol{B}}\hat{\boldsymbol{Y}}_j, \end{aligned}$$

where $\hat{\boldsymbol{Y}}_j' = \left(-\frac{y_{1j}}{y_{rj}}, \dots, -\frac{y_{r-1,j}}{y_{rj}}, \frac{1}{y_{rj}}, -\frac{y_{r+1,j}}{y_{rj}}, \dots, -\frac{y_{mj}}{y_{rj}}\right)$. Hence $\boldsymbol{b}_r$ has been removed from the basis with $\boldsymbol{r}_j$ entered in its stead. If we next replace the rth column of the mth order identity matrix $\boldsymbol{I}_m$ by $\hat{\boldsymbol{Y}}_j$, we obtain

$$\hat{\boldsymbol{I}}_m = \left[\boldsymbol{e}_1, \dots, \boldsymbol{e}_{r-1}, \hat{\boldsymbol{Y}}_j, \boldsymbol{e}_{r+1}, \dots, \boldsymbol{e}_m\right], \tag{10.A.1}$$

where, as previously defined, $\boldsymbol{e}_j$ is the jth unit column vector. Now it is easily demonstrated that $\boldsymbol{B} = \hat{\boldsymbol{B}}\hat{\boldsymbol{I}}_m$ so that

$$\hat{\boldsymbol{B}}^{-1} = \hat{\boldsymbol{I}}_m\boldsymbol{B}^{-1}. \tag{10.A.2}$$

11

Parametric Programming and the Theory of the Firm

11.1 The Supply Function for the Output of an Activity (or for an Individual Product)

Let us determine how the firm reacts to a variation in the price of one of its outputs by generating a **supply function** for a particular product. Specifically, we want to determine the ceteris paribus amounts supplied at each possible price (the implication of the ceteris paribus assumption is that the firm's technology is taken as given or unchanging as are the prices of its other products and variable inputs and the quantities of its fixed inputs). It is further assumed that at each possible price the firm supplies an amount of the product that maximizes its gross profit so that, as required by our parametric analysis of the objective function, we move from one optimal basic feasible solution to the next as the price of the product under consideration varies continuously.

From (7.25.1) our problem is to initially

$$\begin{aligned} max\, f(\boldsymbol{X}) &= \left(\boldsymbol{P} - (\boldsymbol{A}^*)'\boldsymbol{Q}\right)'\boldsymbol{X} \quad \text{s.t.} \\ &\bar{\boldsymbol{A}}\boldsymbol{X} \leq \boldsymbol{b}, \boldsymbol{X} \geq \boldsymbol{0}. \end{aligned}$$

Once the price p_j (the jth component of $\boldsymbol{P}$) of one of the firm's output changes, it is replaced in the objective function by $p_j^* = p_j + \theta$ or $\boldsymbol{P}$ itself is replaced by $\boldsymbol{P}^* = \boldsymbol{P} + \theta \boldsymbol{e}_j$, where $\boldsymbol{e}_j$ denotes the jth unit column vector, $j = 1, \ldots, p$. Thus, the firm now desires to

$$\begin{aligned} max\, f(\boldsymbol{X}, \theta) &= \left(\boldsymbol{P} + \theta \boldsymbol{e}_j - (\boldsymbol{A}^*)'\boldsymbol{Q}\right)'\boldsymbol{X} \quad \text{s.t.} \\ &\bar{\boldsymbol{A}}\boldsymbol{X} \leq \boldsymbol{b}, \boldsymbol{X} \geq \boldsymbol{0}. \end{aligned}$$

If the problem presented in Example 7.3 is interpreted as an optimum product mix problem, with each activity corresponding to an individual product, then

Linear Programming and Resource Allocation Modeling, First Edition. Michael J. Panik.

we may, say, find the quantity of x_2 that the firm is willing to supply at a succession of prices. Our parametric problem is now to

$$\begin{aligned} max f &= 14x_1 + (12+\theta)x_2 + 11x_3 + 10x_4 \quad \text{s.t.} \\ 10x_1 + 8x_2 + 6x_3 + 4x_4 + x_5 \quad &= 40 \\ 2x_1 + 3x_2 + 5x_3 + 8x_4 \quad + x_6 &= 30 \\ x_1,\ldots,x_6 &\geq 0. \end{aligned}$$

The optimal simplex matrix associated with the original problem was found to be

$$\begin{bmatrix} 1 & \frac{11}{19} & 0 & -\frac{14}{19} & \frac{5}{38} & -\frac{3}{19} & 0 & \frac{10}{19} \\ 0 & \frac{7}{19} & 1 & \frac{36}{19} & -\frac{1}{19} & \frac{5}{19} & 0 & \frac{110}{19} \\ 0 & \frac{3}{19} & 0 & \frac{10}{19} & \frac{24}{19} & \frac{13}{19} & 1 & \frac{1350}{19} \end{bmatrix} \tag{11.1}$$

wherein the optimal basic feasible solution is $x_1 = 10/19$, $x_3 = 110/19$ and $f = 1350/19$.

To solve this parametric problem, let us adopt the following notation consistent with that used in Section 10.1. If the coefficient vector of the augmented objective function to the original problem is denoted as $C' = (c_1, c_2, c_3, c_4, c_5, c_6) = (14, 12, 11, 10, 0, 0)$ and for the parametrized problem the (arbitrary) direction vector is $S = e_2$, then, from the above optimal simplex matrix, $C'_B = (c_1, c_3) = (14,11), C'_R = (c_2, c_4, c_5, c_6) = (12,10,0,0), S'_B = (s_{B1}, s_{B2}) = (0,0)$, and $S'_R = (s_{R1}, s_{R2}, s_{R3}, s_{R4}) = (1,0,0,0)$. Then for $j = 1, 2, 3, 4$,

$$S'_B Y_1 - s_{R1} = (0,0)\begin{bmatrix} 11/19 \\ 7/19 \end{bmatrix} - 1 = -1,$$

$$S'_B Y_2 - s_{R2} = (0,0)\begin{bmatrix} -14/19 \\ 36/19 \end{bmatrix} - 0 = 0,$$

$$S'_B Y_3 - s_{R3} = (0,0)\begin{bmatrix} 5/38 \\ -1/19 \end{bmatrix} - 0 = 0,$$

$$S'_B Y_4 - s_{R4} = (0,0)\begin{bmatrix} -3/19 \\ 5/19 \end{bmatrix} - 0 = 0,$$

so that, from (10.2), $\theta_c = \theta_c^{(1)} = \bar{c}_1 / (S'_B Y_1 - s_{R1}) = 3/19$. (Since the minimum in (10.2) occurs for $j = 1$, r_1 will eventually be pivoted into the basis.) Thus, for $0 \leq \theta < 3/19$, the current basic feasible solution obtained from (11.1) remains optimal. Once $\theta = \theta_c^{(1)} = 3/19$, the adjusted problem has multiple optimal basic feasible solutions. To find this latter set of solutions, let us make the following adjustments in the above optimal simplex matrix. From (10.1.1),

$$\begin{aligned}
-\bar{c}_1^* &= -\bar{c}_1 + \theta_c\left(S_B' Y_1 - s_{R1}\right) = 3/19 + 3/19(-1) = 0, \\
-\bar{c}_2^* &= -\bar{c}_2 + \theta_c\left(S_B' Y_2 - s_{R2}\right) = 10/19 + 0 = 10/19, \\
-\bar{c}_3^* &= -\bar{c}_3 + \theta_c\left(S_B' Y_3 - s_{R3}\right) = 24/19 + 0 = 24/19, \\
-\bar{c}_4^* &= -\bar{c}_4 + \theta_c\left(S_B' Y_4 - s_{R4}\right) = 13/19 + 0 = 13/19.
\end{aligned}$$

(Since we are increasing the objective function coefficient of a variable that is currently nonbasic, no adjustment in the objective value is warranted.) Once these new optimality evaluators replace those now appearing in the last row of (11.1), we obtain

$$\begin{bmatrix}
1 & \frac{11}{19} & 0 & -\frac{14}{19} & \frac{5}{38} & -\frac{3}{19} & 0 & \frac{10}{19} \\
0 & \frac{7}{19} & 1 & \frac{36}{19} & -\frac{1}{19} & \frac{5}{19} & 0 & \frac{110}{19} \\
0 & 0 & 0 & \frac{10}{19} & \frac{24}{19} & \frac{13}{19} & 1 & \frac{1350}{19}
\end{bmatrix}. \tag{11.1.1}$$

From this new matrix, the first of the said multiple optimal basic feasible solutions is $x_1 = 10/19, x_3 = 110/19$ and $f = 1350/19$. To find the second, a pivot operation applied to this adjusted simplex matrix yields

$$\begin{bmatrix}
\frac{19}{11} & 1 & 0 & -\frac{14}{11} & \frac{5}{22} & -\frac{3}{11} & 0 & \frac{10}{11} \\
-\frac{7}{11} & 0 & 1 & \frac{26}{11} & -\frac{3}{22} & \frac{4}{11} & 0 & \frac{60}{11} \\
0 & 0 & 0 & \frac{10}{19} & \frac{24}{19} & \frac{13}{19} & 1 & \frac{1350}{19}
\end{bmatrix} \tag{11.2}$$

and thus $x_2 = 10/11$, $x_3 = 60/11$ and $f = 1350/19$.

How large of an increase in θ beyond $\theta_c^{(1)}$ is admissible before this basic feasible solution ceases to be optimal? With $C_B' = (c_2, c_3) = (12,11)$, $C_R' = (c_1, c_4, c_5, c_6) = (14,10,0,0)$, it follows that $S_B' = (1,0), S_R' = (0,0,0,0)$. Again, for $j = 1, 2, 3, 4$,

$$\begin{aligned}
S_B' Y_1 - s_{R1} &= 19/11, \\
S_B' Y_2 - s_{R2} &= 14/11, \\
S_B' Y_3 - s_{R3} &= 5/22, \\
S_B' Y_4 - s_{R4} &= 3/11.
\end{aligned}$$

Then, from (10.2), $\theta_c = \bar{c}_2/(c\mathbf{S}'_B\mathbf{Y}_2 - s_{R2}) = 55/133$ and thus $\theta_c^{(2)} = \theta_c^{(1)} + \theta_c = 4/7$. For $\theta_c^{(1)} = 3/19 < \theta < 4/7 = \theta_c^{(2)}$, the basic solution appearing in (11.2) remains optimal. However, once $\theta = \theta_c^{(2)} = 4/7$, the parametric problem again exhibits multiple optimal basic feasible solutions. To find the first of these solutions, let us compute $-\bar{c}_1^* = 5/7, -\bar{c}_2^* = 0, -\bar{c}_3^* = 19/14$, and $-\bar{c}_4^* = 4/7$. Moreover, $\Delta f = \theta_c x_2 = 50/133$ so that the new optimal f value is $f^* = f + \Delta f = 500/7$. (Note that since we are increasing the objective function coefficient of a variable that is currently basic, this adjustment in the optimal value of f is warranted.) Once these recomputed values are inserted into the last row of (11.2), the revised matrix appears as

$$\begin{bmatrix} \frac{19}{11} & 1 & 0 & -\frac{14}{11} & \frac{5}{22} & -\frac{3}{11} & 0 & \frac{10}{11} \\ -\frac{7}{11} & 0 & 1 & \frac{26}{11} & -\frac{3}{22} & \frac{4}{11} & 0 & \frac{60}{11} \\ \frac{5}{7} & 0 & 0 & 0 & \frac{19}{14} & \frac{4}{7} & 1 & \frac{500}{7} \end{bmatrix} \tag{11.2.1}$$

so that $x_2 = 10/11$, $x_3 = 60/11$, and $f = 500/7$. So for $\theta = \theta_c^{(2)} = 4/7$, (11.2.1) has rendered the first of the resulting pair of optimal solutions. To obtain the second, a pivot operation using (11.2.1) yields

$$\begin{bmatrix} \frac{18}{13} & 1 & \frac{7}{13} & 0 & \frac{2}{13} & -\frac{1}{13} & 0 & \frac{50}{13} \\ -\frac{7}{26} & 0 & \frac{11}{26} & 1 & -\frac{3}{52} & \frac{2}{13} & 0 & \frac{30}{13} \\ \frac{5}{7} & 0 & 0 & 0 & \frac{19}{14} & \frac{4}{7} & 1 & \frac{500}{7} \end{bmatrix} \tag{11.3}$$

with $x_2 = 50/13$, $x_4 = 30/13$, and $f = 500/7$.

We next look to the largest allowable increase in θ beyond $\theta_c^{(2)}$ that does not violate the optimality of the basic feasible solution found in (11.3). From $\mathbf{C}'_B = (12,10), \mathbf{C}'_R = (14,11,0,0)$ we have $\mathbf{S}'_B = (1,0), \mathbf{S}'_R = (0,0,0,0)$ and, for $j = 1, 2, 3, 4$,

$$\mathbf{S}'_B\mathbf{Y}_1 - s_{R1} = 18/13,$$
$$\mathbf{S}'_B\mathbf{Y}_2 - s_{R2} = 7/13,$$
$$\mathbf{S}'_B\mathbf{Y}_3 - s_{R3} = 2/13,$$
$$\mathbf{S}'_B\mathbf{Y}_4 - s_{R4} = 1/13,$$

so that $\theta_c = \bar{c}_4/(\mathbf{S}'_B\mathbf{Y}_4 - s_{R4}) = 52/7$. Then $\theta_c^{(3)} = \theta_c^{(2)} + \theta_c = 8$. For $\theta_c^{(2)} = 4/7 < \theta < \theta_c^{(3)} = 8$, the current basic feasible solution appearing in (11.3) is optimal while, for $\theta = \theta_c^{(3)} = 8$, the parametric problem again possesses two optimal basic feasible solutions. Once the adjusted values $-\bar{c}_1^* = 11, -\bar{c}_2^* = 4, -\bar{c}_3^* = 5/2, -\bar{c}_4^* = 0$, and $f^* = f + \Delta f = \frac{500}{7} + \frac{200}{7} = 100$ are inserted into the last row of (11.3), we obtain

$$\begin{bmatrix} \frac{18}{13} & 1 & \frac{7}{13} & 0 & \frac{2}{13} & -\frac{1}{13} & 0 & \frac{50}{13} \\ -\frac{7}{26} & 0 & \frac{11}{26} & 1 & -\frac{3}{52} & \frac{2}{13} & 0 & \frac{30}{13} \\ 11 & 0 & 4 & 0 & \frac{5}{2} & 0 & 1 & 100 \end{bmatrix} \tag{11.3.1}$$

with $x_2 = 50/13$, $x_4 = 30/13$, and $f = 100$ as the first optimal basic feasible solution; and upon pivoting in (11.3.1) to the following matrix, the second such solution is obtained, i.e. the new matrix

$$\begin{bmatrix} \frac{5}{4} & 1 & \frac{3}{4} & \frac{1}{2} & \frac{1}{8} & 0 & 0 & 5 \\ -\frac{7}{4} & 0 & \frac{11}{4} & \frac{13}{2} & -\frac{3}{8} & 1 & 0 & 15 \\ 11 & 0 & 4 & 0 & \frac{5}{2} & 0 & 1 & 100 \end{bmatrix} \tag{11.4}$$

yields $x_2 = 5$, $x_6 = 15$, and $f = 100$.

If θ increases slightly above $\theta_c^{(3)} = 8$, the optimal basic feasible solution contained within (11.4) is unique. Moreover, since now $\boldsymbol{C}_B' = (12,10)$, $\boldsymbol{C}_R' = (14,11,10,0)$, $\boldsymbol{S}_B' = (1,0)$, and $\boldsymbol{S}_R' = (0,0,0,0)$, we may form, for $j = 1, 2, 3, 4$,

$$\begin{aligned} \boldsymbol{S}_B'\boldsymbol{Y}_1 - s_{R1} &= 5/4, \\ \boldsymbol{S}_B'\boldsymbol{Y}_2 - s_{R2} &= 3/4, \\ \boldsymbol{S}_B'\boldsymbol{Y}_3 - s_{R3} &= 1/2, \\ \boldsymbol{S}_B'\boldsymbol{Y}_4 - s_{R4} &= 1/8. \end{aligned}$$

With each of these values positive, the current basic feasible solution remains optimal as θ increases without bound. In sum, the above sequence of optimal basic feasible solutions to the parametric problem is:

$$\begin{aligned} (x_1,x_3) &= (10/19,110/19) && \text{for } 0 \le \theta \le \frac{3}{19}; \\ (x_2,x_3) &= (10/11,60/11) && \text{for } \frac{3}{19} \le \theta \le \frac{4}{7}; \\ (x_2,x_4) &= (50/13,30/13) && \text{for } \frac{4}{7} \le \theta \le 8; \\ (x_2,x_6) &= (5,15) && \text{for } 8 \le \theta < +\infty. \end{aligned}$$

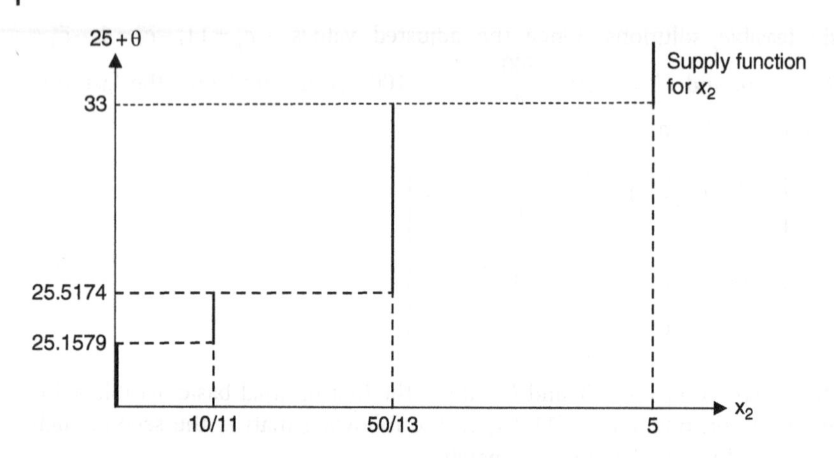

Figure 11.1 Supply function for x_2.

Figure 11.1 traces out the path taken by product two as θ, and thus the price of x_2, increases continuously through the derived set of critical values. (Note that the parametric price of x_2 is $p_2^* = p_2 + \theta = 25 + \theta$.)

As this figure reveals, the supply function for x_2 consists of a set of vertical line segments less the endpoints of each segment. This is because at each such point no unique level of quantity supplied emerges since, as exhibited above, multiple optimal basic feasible solutions exist at the critical values of θ. Hence, the supply function for x_2 shows the unique ceteris paribus quantities supplies at each level of price save for those price levels that correspond to critical values of θ or multiple optimal solutions. Clearly the supply function for x_2 is undefined along each horizontal gap depicted in Figure 11.1.

11.2 The Demand Function for a Variable Input

A second exercise involving the parametrization of the objective function is to see how the firm reacts to a change in the price of one of its variable inputs by deriving a **demand function** for the same. In this regard, let us determine the ceteris paribus amount of some variable input demanded at each possible price (we are again assuming that the firm's technology as well as its amounts of fixed inputs is unchanging and that the prices of the firm's output as well as the prices of all other variable inputs are constant). Additionally, it is also assumed that at each possible input price the firm employs an amount of the variable input that maximizes gross profit so that, as the price of the particular variable input under discussion continuously changes, we move from one optimal basic feasible solution to the next.

Again using (7.25.1) as our starting point, if the price q_i (the ith component of $\boldsymbol{Q}$) of one of the firm's variable inputs changes, it is replaced in the objective function by $q_i^* = q_i + \theta$ or $\boldsymbol{Q}$ itself is replaced by $\boldsymbol{Q}^* = \boldsymbol{Q} + \theta \boldsymbol{e}_i$, $i = 1, \ldots, m - l$. Hence, the firm now attempts to

$$max\, f(\boldsymbol{X},\theta) = \left(\boldsymbol{P} - (\boldsymbol{A}^*)'\boldsymbol{Q} - \theta(\boldsymbol{A}^*)'\boldsymbol{e}_i\right)'\boldsymbol{X} \text{ s.t.}$$
$$\bar{\boldsymbol{A}}\boldsymbol{X} \le \boldsymbol{b}, \boldsymbol{X} \ge \boldsymbol{0}.$$

And again looking to the problem presented in Example 7.3, let us determine the quantity demanded of the first variable input (factor three) when its price is allowed to change continuously. Thus, q_3 is replaced by $q_3 + \theta = 3 + \theta$ so that **total variable cost** (TVC) can be computed as follows:

$$TVC_1 = (3+\theta)\cdot 3 + 2 = 11 + 3\theta,$$
$$TVC_2 = (3+\theta)\cdot 1 + 10 = 13 + \theta,$$
$$TVC_3 = (3+\theta)\cdot 4 + 2 = 14 + 4\theta,$$
$$TVC_4 = (3+\theta)\cdot 3 + 6 = 15 + 3\theta,$$

and thus $TVC = \sum_{j=1}^{4}(TVC_j)\cdot x_j = (11+3\theta)x_1 + (13+\theta)x_2 + (14+4\theta)x_3 + (15+3\theta)x_4$. Upon subtracting TVC from total revenue to obtain gross profit, our parametrized problems appear as

$$\begin{aligned} max \quad & f = (14-3\theta)x_1 + (12-\theta)x_2 + (11-4\theta)x_3 + (10-3\theta)x_4 \quad \text{s.t.} \\ & 10x_1 + 8x_2 + 6x_3 + 4x_4 + x_5 = 40 \\ & 2x_1 + 3x_2 + 5x_3 + 8x_4 + x_6 = 30 \\ & x_1, \ldots, x_6 \ge 0. \end{aligned} \tag{11.5}$$

The optimal simplex matrix associated with the aforementioned problem is

$$\begin{bmatrix} 1 & \frac{11}{19} & 0 & -\frac{14}{19} & \frac{5}{38} & -\frac{3}{19} & 0 & \frac{10}{19} \\ 0 & \frac{7}{19} & 1 & \frac{36}{19} & -\frac{1}{19} & \frac{5}{19} & 0 & \frac{110}{19} \\ 0 & \frac{3}{19} & 0 & \frac{10}{19} & \frac{24}{19} & \frac{13}{19} & 1 & \frac{1350}{19} \end{bmatrix} \tag{11.6}$$

with $x_1 = 10/19$, $x_3 = 110/19$ and $f = 1350/19$. By virtue of the structure of (11.6) we have $\boldsymbol{C}' = (c_1, c_2, c_3, c_4, c_5, c_6) = (14,12,11,10,0,0)$, $\boldsymbol{C}_{\boldsymbol{B}}' = (c_1, c_3) = (14,11)$, and $\boldsymbol{C}_{\boldsymbol{R}}' = (c_2, c_4, c_5, c_6) = (12,10,0,0)$. From the parametric objective function appearing in (11.5) we have as our direction vector $\boldsymbol{S}' = (-3, -1, -4, -3, 0, 0)$ so that $\boldsymbol{S}_{\boldsymbol{B}}' = (s_{B1}, s_{B2}) = (-3,-4)$, $\boldsymbol{S}_{\boldsymbol{R}}' = (s_{R1}, s_{R2}, s_{R3}, s_{R4},) = (-1,-3,0,0)$ and, $j =$ 1, 2, 3, 4,

$$S_B'Y_1 - s_{R1} = (-3,-4)\begin{bmatrix} 11/19 \\ 7/19 \end{bmatrix} - (-1) = -42/19,$$

$$S_B'Y_2 - s_{R2} = (-3,-4)\begin{bmatrix} -14/19 \\ 36/19 \end{bmatrix} - (-3) = -45/19,$$

$$S_B'Y_3 - s_{R3} = (-3,-4)\begin{bmatrix} 5/38 \\ -1/19 \end{bmatrix} - 0 = -7/38,$$

$$S_B'Y_4 - s_{R4} = (-3,-4)\begin{bmatrix} -3/19 \\ 5/19 \end{bmatrix} - 0 = -11/19.$$

Employing (10.2) we find that $\theta_c = \theta_c^{(1)} = \bar{c}_1/(S_B'Y_1 - s_{R1}) = \frac{1}{14}$. (With the minimum in (10.2) occurring for $j = 1$, $\boldsymbol{r}_1$ will eventually be pivoted into the basis.) So for $0 \le \theta < \frac{1}{14} = \theta_c^{(1)}$, the current basic feasible solution given by (11.6) remains optimal. Once $\theta = \theta_c^{(1)} = 1/14$, the adjusted problem exhibits multiple optimal basic feasible solutions. To determine this latter set of solutions, let us make the following modifications in (11.6). Using (10.1.1),

$$-\bar{c}_1^* = -\bar{c}_1 + \theta_c(S_B'Y_1 - s_{R1}) = \frac{3}{19} + \frac{1}{14}\left(-\frac{42}{19}\right) = 0,$$

$$-\bar{c}_2^* = -\bar{c}_2 + \theta_c(S_B'Y_2 - s_{R2}) = \frac{10}{19} + \frac{1}{14}\left(-\frac{45}{19}\right) = \frac{5}{14},$$

$$-\bar{c}_3^* = -\bar{c}_3 + \theta_c(S_B'Y_3 - s_{R3}) = \frac{24}{19} + \frac{1}{14}\left(-\frac{7}{38}\right) = \frac{5}{4},$$

$$-\bar{c}_4^* = -\bar{c}_4 + \theta_c(S_B'Y_4 - s_{R4}) = \frac{13}{19} + \frac{1}{14}\left(-\frac{11}{19}\right) = \frac{9}{14}.$$

Once these new optimality evaluators replace these currently appearing in the last row of (11.6), we obtain

$$\begin{bmatrix} 1 & \frac{11}{19} & 0 & -\frac{14}{19} & \frac{5}{38} & -\frac{3}{19} & 0 & \frac{10}{19} \\ 0 & \frac{7}{19} & 1 & \frac{36}{19} & -\frac{1}{19} & \frac{5}{19} & 0 & \frac{110}{19} \\ 0 & 0 & 0 & \frac{5}{14} & \frac{5}{4} & \frac{9}{14} & 1 & \frac{485}{7} \end{bmatrix}. \tag{11.6.1}$$

(Note also that once $\theta = \theta_c = \frac{1}{14}$, f decreases to $f^* = f + \Delta f = f - 3\theta x_1 - 4\theta x_3 =$ 485/7.) From this new matrix we have the first of the two multiple optimal basic feasible solutions, namely $x_1 = 10/19$, $x_3 = 110/19$, and $f = 485/7$. To find the second, a pivot operation applied to (11.6.1) yields

$$\begin{bmatrix} \frac{19}{11} & 1 & 0 & -\frac{14}{11} & \frac{5}{22} & -\frac{3}{11} & 0 & \frac{10}{11} \\ -\frac{7}{11} & 0 & 1 & \frac{26}{11} & -\frac{3}{22} & \frac{4}{11} & 0 & \frac{60}{11} \\ 0 & 0 & 0 & \frac{5}{14} & \frac{5}{4} & \frac{9}{14} & 1 & \frac{485}{7} \end{bmatrix} \tag{11.7}$$

with $x_2 = 10/11$, $x_3 = 60/11$, and $f = 485/7$.

How large of an increase in θ beyond $\theta_c^{(1)}$ is admissible before this latter basic feasible solution ceases to be optimal? With $\boldsymbol{C}'_B = (12,11), \boldsymbol{C}'_R = (14,10,0,0)$, it follows that $\boldsymbol{S}'_B = (-1,-4), \boldsymbol{S}'_R = (-3,-3,0,0)$, and for $j = 1,2,3,4$,

$$\begin{aligned} \boldsymbol{S}'_B \boldsymbol{Y}_1 - s_{R1} &= 42/11, \\ \boldsymbol{S}'_B \boldsymbol{Y}_2 - s_{R2} &= -57/11, \\ \boldsymbol{S}'_B \boldsymbol{Y}_3 - s_{R3} &= 7/22, \\ \boldsymbol{S}'_B \boldsymbol{Y}_4 - s_{R4} &= -13/11. \end{aligned}$$

Then $\theta_c = \bar{c}_2 / (\boldsymbol{S}'_B \boldsymbol{Y}_2 - s_{R2}) = 55/798$ with $\theta_c^{(2)} = \theta_c^{(1)} + \theta_c = 8/57$. For $\theta_c^{(1)} = \frac{1}{14} < \theta < \frac{8}{57} = \theta_c^{(2)}$, the basic solution in (11.7) remains optimal. But once $\theta = \theta_c^{(2)} = 8/57$, the parametric problem again exhibits multiple optimal basic feasible solutions. To find the first of these solutions, let us compute $-\bar{c}_1^* = 5/19$, $-\bar{c}_2^* = 0$, $-\bar{c}_3^* = 145/114$, and $-\bar{c}_4^* = 32/57$. Once these recomputed values are inserted into the last row of (11.7) (with $f^* = f + \Delta f = f - \theta x_2 - 4\theta x_3 = 3860/57$), the revised matrix appears as

$$\begin{bmatrix} \frac{19}{11} & 1 & 0 & -\frac{14}{11} & \frac{5}{22} & -\frac{3}{11} & 0 & \frac{10}{11} \\ -\frac{7}{11} & 0 & 1 & \frac{26}{11} & -\frac{3}{22} & \frac{4}{11} & 0 & \frac{60}{11} \\ \frac{15}{57} & 0 & 0 & 0 & \frac{145}{114} & \frac{32}{57} & 1 & \frac{3860}{57} \end{bmatrix} \tag{11.7.1}$$

with $x_2 = 10/11$, $x_3 = 60/11$, and $f = 3860/57$. So for $\theta = \theta_c^{(2)} = 8/57$, (11.7.1) yields the first of the resulting pair of optimal solutions. To obtain the second, a pivot operation on (11.7.1) yields

$$\begin{bmatrix} \frac{18}{13} & 1 & \frac{7}{13} & 0 & \frac{2}{13} & -\frac{1}{13} & 0 & \frac{50}{13} \\ -\frac{7}{26} & 0 & \frac{11}{26} & 1 & -\frac{3}{52} & \frac{2}{13} & 0 & \frac{30}{13} \\ \frac{15}{57} & 0 & 0 & 0 & \frac{145}{114} & \frac{32}{57} & 1 & \frac{3860}{57} \end{bmatrix} \tag{11.8}$$

so that $x_2 = 50/13$, $x_4 = 30/13$, and $f = 3860/57$.

We next examine the largest allowable increment in θ beyond $\theta_c^{(2)}$ that does not violate the optimality of the basic feasible solution found in (11.8). From $\boldsymbol{C}_B' = (12,10)$, $\boldsymbol{C}_R' = (14,11,0,0)$, we have $\boldsymbol{S}_B' = (-1,-3)$, $\boldsymbol{S}_R' = (-3,-4,0,0)$, and, for $j = 1,2,3,4$,

$$\begin{aligned} \boldsymbol{S}_B' \boldsymbol{Y}_1 - s_{R1} &= 63/26, \\ \boldsymbol{S}_B' \boldsymbol{Y}_2 - s_{R2} &= 57/26, \\ \boldsymbol{S}_B' \boldsymbol{Y}_3 - s_{R3} &= 1/52, \\ \boldsymbol{S}_B' \boldsymbol{Y}_4 - s_{R4} &= -5/13, \end{aligned}$$

so that $\theta_c = 416/285$ and $\theta_c^{(3)} = \theta_c^{(2)} + \theta_c = 8/5$. For $\theta_c^{(2)} = \frac{8}{57} < \theta < \theta_c^{(3)} = 8/5$, the current basic feasible solution appearing in (11.8) is optimal while, for $\theta = \theta_c^{(3)} = \frac{8}{5}$, the parametric problem again possesses two optimal basic feasible solutions. Once the adjusted values $-\bar{c}_1^* = 19/5, -\bar{c}_2^* = 16/5, -\bar{c}_3^* = 13/10$, and $-\bar{c}_4^* = 0$ are inserted into the last row of (11.8) (here $f^* = f + \Delta f = f - \theta x_2 - 3\theta x_4 = 52$), we obtain

$$\begin{bmatrix} \frac{18}{13} & 1 & \frac{7}{13} & 0 & \frac{2}{13} & -\frac{1}{13} & 0 & \frac{50}{13} \\ -\frac{7}{26} & 0 & \frac{11}{26} & 1 & -\frac{3}{52} & \frac{2}{13} & 0 & \frac{30}{13} \\ \frac{19}{5} & 0 & \frac{16}{5} & 0 & \frac{13}{10} & 0 & 1 & 52 \end{bmatrix} \tag{11.8.1}$$

with $x_2 = 50/13$, $x_4 = 30/13$, and $f = 52$ as the first optimal basic feasible solution. And upon pivoting in (11.8.1) to the following matrix, the second such solution obtains, i.e. the new matrix

$$\begin{bmatrix} \frac{5}{4} & 1 & \frac{3}{4} & \frac{1}{2} & \frac{1}{8} & 0 & 0 & 5 \\ -\frac{7}{4} & 0 & \frac{11}{4} & \frac{13}{2} & -\frac{3}{8} & 1 & 0 & 15 \\ \frac{19}{5} & 0 & \frac{16}{5} & 0 & \frac{13}{10} & 0 & 1 & 52 \end{bmatrix} \tag{11.9}$$

yields $x_2 = 5$, $x_6 = 15$, and $f = 52$.

If θ increases slightly above $\theta_c^{(3)} = 8/5$, the optimal basic feasible solution contained in (11.9) is unique. Moreover, since now $\boldsymbol{C}_B' = (12,10)$, $\boldsymbol{C}_R' = (14,11,0,0)$, $\boldsymbol{S}_B' = (-1,0)$, and $\boldsymbol{S}_R' = (-3,-4,-3,0)$, we again form, for $j = 1, 2, 3, 4$,

$$S_B'Y_1 - s_{R1} = 7/4,$$
$$S_B'Y_2 - s_{R2} = 13/4,$$
$$S_B'Y_3 - s_{R3} = 5/2,$$
$$S_B'Y_4 - s_{R4} = -1/8.$$

Hence $\theta_c = 52/5$ and $\theta_c^{(4)} = \theta_c^{(3)} + \theta_c = 12$. For $\theta_c^{(3)} = \frac{8}{5} < \theta < 12 = \theta_c^{(4)}$, the basic feasible solution found in (11.9) is optimal while for $\theta = \theta_c^{(4)} = 12$, the parametric problem has two optimal basic feasible solutions. Inserting the adjusted optimality evaluators $-\bar{c}_1^* = 22, -\bar{c}_2^* = 37, -\bar{c}_3^* = 26$, and $-\bar{c}_4^* = 0$ into (11.9) (with $f^* = f + \Delta f = f - \theta x_2 = 0$) yields

$$\begin{bmatrix} \frac{5}{4} & 1 & \frac{3}{4} & \frac{1}{2} & \frac{1}{8} & 0 & 0 & 5 \\ -\frac{7}{4} & 0 & \frac{11}{4} & \frac{13}{2} & -\frac{3}{8} & 1 & 0 & 15 \\ 22 & 0 & 37 & 26 & 0 & 0 & 1 & 0 \end{bmatrix} \tag{11.9.1}$$

so that the first solution is $x_2 = 5$, $x_6 = 15$, and $f = 0$; and pivoting in (11.9.1) generates

$$\begin{bmatrix} 10 & 8 & 6 & 4 & 1 & 0 & 0 & 40 \\ 2 & 3 & 5 & 8 & 0 & 1 & 0 & 30 \\ 22 & 0 & 37 & 26 & 0 & 0 & 1 & 0 \end{bmatrix} \tag{11.10}$$

with $x_5 = 40$, $x_6 = 30$, and $f = 0$ as the second.

Given that now $C_B' = (0,0)$, $C_R' = (12,14,11,10)$, $S_B' = (0,0)$, and $S_R' = (-3,-1,-4,-3)$, it follows that $S_B'Y_j - s_{Rj} > 0, j = 1,2,3,4$. Thus, the optimal basic feasible solution obtained from (11.10) remains optimal as θ increases without limit. A summary of the preceding sequence of optimal basic feasible solutions to this parametric problem appears as:

$$(x_1,x_3) = (10/19,110/19) \quad \text{for } 0 \le \theta \le \frac{1}{14};$$
$$(x_2,x_3) = (10/11,60/11) \quad \text{for } \frac{1}{14} \le \theta \le \frac{8}{57};$$
$$(x_2,x_4) = (50/13,30/13) \quad \text{for } \frac{8}{57} \le \theta \le \frac{8}{5};$$
$$(x_2,x_6) = (5,15) \quad \text{for } \frac{8}{5} \le \theta \le 12;$$
$$(x_5,x_6) = (40,30) \quad \text{for } 12 \le \theta < +\infty.$$

We may now trace out the path taken by input three (i.e. the first variable input) as its price increases continuously through the above set of critical values, where the parametric price of this input is $q_3^* = q_3 + \theta = 3 + \theta$. To determine the quantities of factor three utilized as its price changes, let us examine the third component of each of the four activity vectors given in Example 7.3:

$$\begin{bmatrix} - \\ - \\ 3 \\ - \end{bmatrix}, \begin{bmatrix} - \\ - \\ 1 \\ - \end{bmatrix}, \begin{bmatrix} - \\ - \\ 4 \\ - \end{bmatrix}, \begin{bmatrix} - \\ - \\ 3 \\ - \end{bmatrix}.$$

Hence, the amounts of factor three used at each of the preceding basic feasible solutions are respectively:

$$3\left(\frac{10}{19}\right) + 4\left(\frac{110}{19}\right) = \frac{470}{19};$$
$$1\left(\frac{10}{11}\right) + 4\left(\frac{60}{11}\right) = \frac{250}{11};$$
$$1\left(\frac{50}{13}\right) + 3\left(\frac{30}{13}\right) = \frac{140}{13};$$
$$1(5) \;\; +0\,(15) \;=\; 5;$$
$$0\,(40) + 0\,(30) = 0.$$

The parametric price of factor three plotted against these employment levels appears in Figure 11.2.

As this diagram reveals, the demand function for the first variable input consists of a set of vertical line segments less their endpoints. At the endpoints no unique level of quantity demanded exists since multiple optimal basic feasible

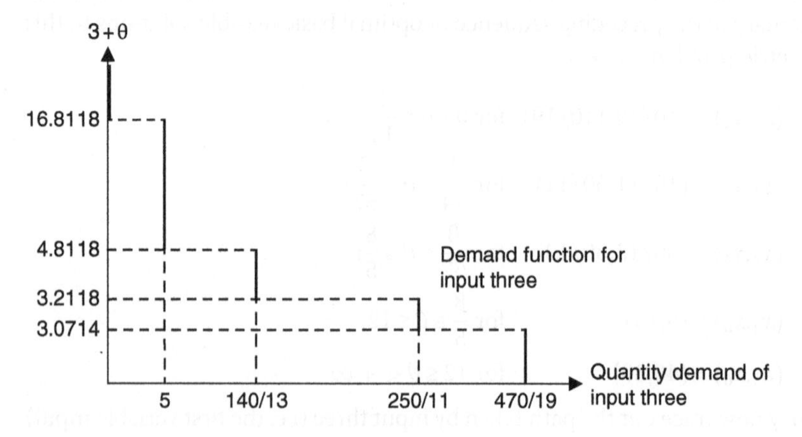

Figure 11.2 Demand function for input three.

solutions are found at the critical values of θ. Thus, the demand function for the first variable input shows the unique ceteris paribus quantities demanded at each price level except for those price levels that correspond to critical values of θ or multiple optimal basic feasible solutions.

One final point is in order. The parametric problem at hand has considered only an increase in the price of factor three. However, it may be the case that one or more alternative optimal basic solutions emerge when the price of the said input decreases. Clearly, such decreases must be examined if we are to generate the entire demand function for the variable input. In this instance, the parametric price is $q_3^* = q_3 - \theta = 3 - \theta$. Then $\boldsymbol{S}' = (3, 1, 4, 3, 0, 0)$ so that, as can be easily demonstrated, $\boldsymbol{S}_B'\boldsymbol{Y}_j - s_{Rj} > 0, j = 1,2,3,4$. Hence, no new basic feasible solution emerges when the price of factor three declines so that no modification of the above demand schedule is warranted.

11.3 The Marginal (Net) Revenue Productivity Function for an Input[1]

We shall now examine how the firm's net revenue changes given that the supply of one of its fixed inputs can be augmented. Specifically, we shall generate the **marginal (net) revenue productivity function** for a given input, i.e. we wish to find the ceteris paribus rate of change in net revenue with respect to a particular fixed factor. (Here, too, the firm's technology is taken as given, the output and variable factor prices are held constant, and the quantities of all other fixed inputs are invariant.) In addition, at each possible requirements level for the fixed input, the firm seeks to maximize its gross profit or net revenue. This being the case, the procedure underlying the parametrization of the requirements vector stipulates that the product or activity mix varies as we move from one optimal basic feasible solution to the next when some fixed input requirement is considered continuously variable.

Using (7.25.1), the problem is to initially

$$max\, f(\boldsymbol{X}) = \left(\boldsymbol{P} - (\boldsymbol{A}^*)'\boldsymbol{Q}\right)'\boldsymbol{X} \text{ s.t.}$$
$$\bar{\boldsymbol{A}}\boldsymbol{X} \le \boldsymbol{b}, \boldsymbol{X} \ge \boldsymbol{0}.$$

1 In some formulations of the theory of the firm, the objective function is interpreted as a net revenue function instead of a gross profit function as specified here. So when determining the objective function response to a change in the availability of some fixed input, the "convention" is to construct the marginal (net) revenue productivity function. To be consistent with this terminology, the objective function in this particular parametric experiment can alternatively be thought of as a net revenue function.

Once the ith requirement b_i (the ith component of $\boldsymbol{b}$) changes, it is replaced in the requirements vector by $b_i^* = b_i + \phi$ or $\boldsymbol{b}$ is replaced by $\boldsymbol{b}^* = \boldsymbol{b} + \phi \boldsymbol{e}_i$, where $\boldsymbol{e}_i$ is the ith unit column vector, $i = 1, \ldots, l$. Thus, the firm desires to

$$\begin{aligned} max f(\boldsymbol{X}) &= (\boldsymbol{P} - (\boldsymbol{A}^*)'\boldsymbol{Q})'\boldsymbol{X} \text{ s.t.} \\ \bar{\boldsymbol{A}}\boldsymbol{X} &\leq \boldsymbol{b} + \phi \boldsymbol{e}_i, \boldsymbol{X} \geq \boldsymbol{0}. \end{aligned}$$

Again interpreting the problem solved in Example 7.3 as an optimum product mix problem, we may, say, determine the marginal (net) revenue product of input b_1 as this fixed factor is allowed to increase (and then decrease) in value. The resulting parametric problem thus assumes the form

$$\begin{aligned} max f = 14x_1 + 12x_2 + 11x_3 + 10x_4 \text{ s.t.} \\ 10x_1 + 8x_2 + 6x_3 + 4x_4 + x_5 \qquad &= 40 + \phi \\ 2x_1 + 3x_2 + 5x_3 + 8x_4 \qquad + x_6 &= 30 \\ x_1, \ldots, x_6 \geq 0. \end{aligned}$$

To initiate the solution to this problem, let us start with the optimal simplex matrix presented at the end of Example 7.3, namely

$$\begin{bmatrix} 1 & \frac{11}{19} & 0 & -\frac{14}{19} & \frac{5}{38} & -\frac{3}{19} & 0 & \frac{10}{19} \\ 0 & \frac{7}{19} & 1 & \frac{36}{19} & -\frac{1}{19} & \frac{5}{19} & 0 & \frac{110}{19} \\ 0 & \frac{3}{19} & 0 & \frac{10}{19} & \frac{24}{19} & \frac{13}{19} & 1 & \frac{1350}{19} \end{bmatrix} \tag{11.11}$$

wherein $x_1 = 10/19$, $x_3 = 110/19$, and $f = 1350/19$. Since, for $\boldsymbol{t} = \boldsymbol{e}_1$, we have

$$\boldsymbol{b}^* = \boldsymbol{b} + \phi \boldsymbol{e}_1 = \begin{bmatrix} 40 + \phi \\ 30 \end{bmatrix},$$

it follows from (11.11) that

$$\boldsymbol{X}_{\boldsymbol{B}}^* = \boldsymbol{X}_{\boldsymbol{B}} + \phi \boldsymbol{B}^{-1}\boldsymbol{e}_1 = \begin{bmatrix} 10/19 \\ 110/19 \end{bmatrix} + \phi \begin{bmatrix} 5/38 & -3/19 \\ -1/19 & 5/19 \end{bmatrix} \begin{bmatrix} 1 \\ 0 \end{bmatrix} = \begin{bmatrix} \frac{10}{19} + \frac{5}{38}\phi \\ \frac{110}{19} - \frac{1}{19}\phi \end{bmatrix}. \tag{11.12}$$

With $\boldsymbol{\beta}_2 \boldsymbol{e}_1 = \beta_{21} = -1/19 < 0$, (10.4) yields $\phi_c = \phi_c^{(1)} = -x_{B2}/\beta_{21} = (-110/19)/(-1/19) = 110$. Then, from (11.12)

$$(\boldsymbol{X}_{\boldsymbol{B}}^*)_{\phi_c^{(1)}} = \begin{bmatrix} 15 \\ 0 \end{bmatrix}$$

and, with $\Delta f = u_1 \Delta b_1 = u_1 \phi_c = (24/19)110 = 2640/19$, $f^* = f + \Delta f = 210$. Once these adjusted values of the basic variable and f are substituted into (11.11) we have

$$\begin{bmatrix} 1 & \frac{11}{19} & 0 & -\frac{14}{19} & \frac{5}{38} & -\frac{3}{19} & 0 & 15 \\ 0 & \frac{7}{19} & 1 & \frac{36}{19} & -\frac{1}{19} & \frac{5}{19} & 0 & 0 \\ 0 & \frac{3}{19} & 0 & \frac{10}{19} & \frac{24}{19} & \frac{13}{19} & 1 & 210 \end{bmatrix}.$$

An appropriate dual simplex pivot operation thus yields

$$\begin{bmatrix} 1 & \frac{3}{2} & \frac{5}{2} & 4 & 0 & \frac{1}{2} & 0 & 15 \\ 0 & -7 & -19 & -36 & 1 & -5 & 0 & 0 \\ 0 & 9 & 24 & 46 & 0 & 7 & 1 & 210 \end{bmatrix}. \tag{11.13}$$

Hence, our new optimal basic feasible solution is $x_1 = 15$, $x_5 = 0$ and $f = 210$. Note that for this solution the new requirements vector is

$$\boldsymbol{b}^*_{\phi_c^{(1)}} = \boldsymbol{b} + \phi_c^{(1)} \boldsymbol{e}_1 = \begin{bmatrix} 150 \\ 30 \end{bmatrix}.$$

How large of an increase ϕ beyond $\phi_c^{(1)}$ preserves the feasibility of the current solution? Given that ϕ is to increase further, we now have

$$\boldsymbol{b}^* = \boldsymbol{b}^*_{\phi_c^{(1)}} + \phi \boldsymbol{e}_1 = \begin{bmatrix} 150 + \phi \\ 30 \end{bmatrix}$$

and, from (11.13),

$$\boldsymbol{X}^*_{\boldsymbol{B}} = \left(\boldsymbol{X}^*_{\boldsymbol{B}}\right)_{\phi_c^{(1)}} + \phi \boldsymbol{B}^{-1} \boldsymbol{e}_1 = \begin{bmatrix} 15 \\ \phi \end{bmatrix}.$$

Clearly $\boldsymbol{\beta}_i \boldsymbol{e}_1 \geq 0$ for $i = 1, 2$ so that ϕ may increase without bound yet still preserve the feasibility of the current optimal solution.

To complete the marginal (net) revenue productivity function for the fixed input b_1, let us consider a decrease in the latter by replacing $\boldsymbol{e}_1$ by $-\boldsymbol{e}_1$, i.e. now $\boldsymbol{t} = -\boldsymbol{e}_1$. In this instance

$$\boldsymbol{b}^* = \boldsymbol{b} + \phi \boldsymbol{e}_1 = \begin{bmatrix} 40 - \phi \\ 30 \end{bmatrix}$$

and, from (11.11),

$$X_B^* = X_B^* - \phi B^{-1}e_1 = \begin{bmatrix} \frac{10}{19} - \frac{5}{38}\phi \\ \frac{110}{19} + \frac{1}{19}\phi \end{bmatrix}. \tag{11.14}$$

With $-\beta_1 e_1 = -\beta_{11} = -5/38 < 0$, (10.4) renders $\phi_c = \phi_c^{(2)} = -x_{B1}/(-\beta_{11}) = 4$. Then, from (11.14),

$$\left(X_B^*\right)_{\phi_c^{(2)}} = \begin{bmatrix} 0 \\ 6 \end{bmatrix}$$

and $f^* = f + \Delta f = f - u_1 \Delta b_1 = f - u_1 \phi_c = 66$. If this new set of basic variables along with the revised f value is substituted into (11.11), we have

$$\begin{bmatrix} 1 & \frac{11}{19} & 0 & -\frac{14}{19} & \frac{5}{38} & -\frac{3}{19} & 0 & 0 \\ 0 & \frac{7}{19} & 1 & \frac{36}{19} & -\frac{1}{19} & \frac{5}{19} & 0 & 6 \\ 0 & \frac{3}{19} & 0 & \frac{10}{19} & \frac{24}{19} & \frac{13}{19} & 1 & 66 \end{bmatrix}$$

and, upon pivoting to a new basis, we obtain

$$\begin{bmatrix} -\frac{19}{14} & -\frac{11}{14} & 0 & 1 & -\frac{5}{28} & \frac{3}{14} & 0 & 0 \\ \frac{18}{7} & \frac{106}{19} & 1 & 0 & \frac{2}{7} & -\frac{1}{7} & 0 & 6 \\ \frac{5}{7} & \frac{4}{7} & 0 & 0 & \frac{19}{14} & \frac{4}{7} & 1 & 66 \end{bmatrix} \tag{11.15}$$

with $x_3 = 6$, $x_4 = 0$, and $f = 66$. At this new optimal solution the revised requirements vector has the form

$$b^*_{\phi_c^{(2)}} = b - \phi_c^{(2)} e_1 = \begin{bmatrix} 36 \\ 30 \end{bmatrix}.$$

To determine the largest decrease in ϕ that does not violate the feasibility of the preceding solution, we first form

$$b^* = b^*_{\phi_c^{(2)}} - \phi e_1 = \begin{bmatrix} 36 - \phi \\ 30 \end{bmatrix}.$$

In addition, (11.15) is used to determine

$$X_B = \left(X_B^*\right)_{\phi_c^{(2)}} - \phi B^1 e_1 = \begin{bmatrix} \frac{5}{28}\phi \\ 6 - \frac{2}{7}\phi \end{bmatrix}. \tag{11.16}$$

Given that $-\beta_2 e_1 = -\beta_{21} = -2/7 < 0$, (10.4) implies that $\phi_c = \phi_c^{(3)} = x_{B2}/(-\beta_{21}) = 21$. Then, from (11.16),

$$\left(X_B^*\right)_{\phi_c^{(3)}} = \begin{bmatrix} 15/4 \\ 0 \end{bmatrix}$$

and $f^* = f + \Delta f = f - u_1 \Delta b_1 = f - u_1 \phi_c = 75/2$. Using this latter set of values to update (11.15) we have

$$\begin{bmatrix} -\frac{19}{14} & -\frac{11}{14} & 0 & 1 & -\frac{5}{28} & \frac{3}{14} & 0 & \frac{15}{4} \\ \frac{18}{7} & \frac{106}{19} & 1 & 0 & \frac{2}{7} & -\frac{1}{7} & 0 & 0 \\ \frac{5}{7} & \frac{4}{7} & 0 & 0 & \frac{19}{14} & \frac{4}{7} & 1 & \frac{75}{2} \end{bmatrix}$$

and, upon pivoting to a new optimal basic feasible solution, we obtain

$$\begin{bmatrix} \frac{7}{2} & \frac{53}{7} & \frac{3}{2} & 1 & \frac{1}{4} & 0 & 0 & \frac{15}{4} \\ -18 & -39 & -7 & 0 & -2 & 1 & 0 & 0 \\ 11 & \frac{160}{7} & 4 & 0 & \frac{5}{2} & 0 & 1 & \frac{75}{2} \end{bmatrix} \tag{11.17}$$

and thus $x_4 = 15/4$, $x_6 = 0$, and $f = 75/2$. Here the associated requirements vector appears as

$$b^*_{\phi_c^{(3)}} = b_{\phi_c^{(2)}} - \phi_c^{(3)} e_1 = \begin{bmatrix} 15 \\ 30 \end{bmatrix}.$$

Let us decrease ϕ further still by forming

$$b^* = b^*_{\phi_c^{(3)}} - \phi e_1 = \begin{bmatrix} 15 - \phi \\ 30 \end{bmatrix}$$

and, from (11.17),

$$X_B^* = \left(X_B^*\right)_{\phi_c^{(3)}} - \phi B^{-1} e_1 = \begin{bmatrix} \frac{15}{4} - \frac{1}{4}\phi \\ 2\phi \end{bmatrix}. \tag{11.18}$$

Since $-\beta_1 e_1 = -\beta_{11} = -1/4 < 0$, (10.4) yields $\phi_c = \phi_c^{(4)} = -x_{B1}/(-\beta_{11}) = 15$. Then from (11.18),

$$(X_B^*)_{\phi_c^{(4)}} = \begin{bmatrix} 0 \\ 30 \end{bmatrix}$$

and $f^* = f - u_1\phi_c = 0$. Substituting these recomputed values for the basic variables and the revised objective value into (11.17) gives us

$$\begin{bmatrix} \frac{7}{2} & \frac{53}{7} & \frac{3}{2} & 1 & \frac{1}{4} & 0 & 0 & 0 \\ -18 & -39 & -7 & 0 & -2 & 1 & 0 & 30 \\ 11 & \frac{160}{7} & 4 & 0 & \frac{5}{2} & 0 & 1 & 0 \end{bmatrix}. \tag{11.19}$$

Since there are no negative entries in the first row of this matrix, we cannot employ a dual simplex pivot operation to obtain a new matrix (further increases in ϕ violates the feasibility of the current solution) so that (11.19) contains the final optimal basic feasible solution, i.e. $x_4 = 0$, $x_6 = 30$, and $f = 0$, and the associated requirements vector is

$$b^*_{\phi_c^{(4)}} = b^*_{\phi_c^{(3)}} - \phi_c^{(4)} e_1 = \begin{bmatrix} 0 \\ 30 \end{bmatrix}.$$

If we examine the above sequence of optimal basic feasible solutions and collect the results pertaining to the values of f, b_1, and u_1 and construct Table 11.1,

Table 11.1 Marginal (net) revenue productivity of b_1.

Net revenue f	Level of b_1	Change in f, Δf	Change in b_1, Δb_1	Marginal (net) revenue productivity of b_1, $\Delta f/\Delta b_1$	Dual variable, u_1
0	0	37.5	15	2.500	2.500
37.5	15	28.5	21	1.3571	1.3571
66	36	5.0526	4	1.2632	1.2632
71.0526	40	138.9474	110	1.2632	1.2632
210	150	0	ϕ	0	0
210	150+ϕ	.	.	.	
.	.	.	.	.	
.	.	.	.	.	
.	.	0		0	
210	$+\infty$				

then the data contained therein can be used to construct the net revenue function as well as the marginal (net) revenue productivity function (Figure 11.3). A glance at the last two columns of this table reveals a rather interesting result. The marginal (net) revenue productivity of the fixed input b_1 is actually the dual

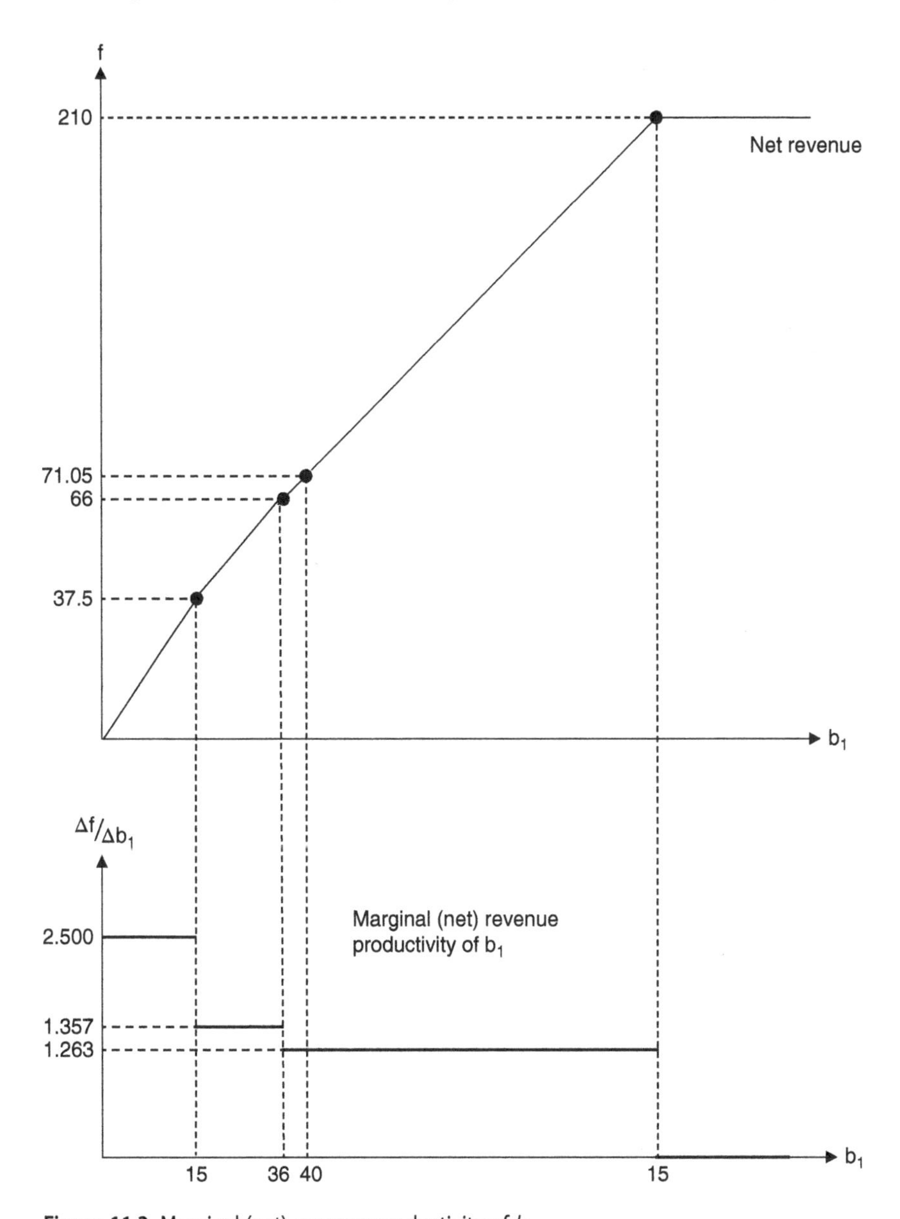

Figure 11.3 Marginal (net) revenue productivity of b_1.

variable u_1 since the latter can be interpreted as the (internal) rate at which the production process transforms units of the first fixed input into value of output or net revenue. Clearly net revenue is a piecewise-linear function of the level of b_1. That is, this function is linear only over specific subintervals with its constituent line segments meeting at vertices, the latter corresponding to degenerate optimal basic feasible solutions wherein only one product is produced. Moreover, its slope, the marginal (net) revenue productivity of $b_1$or the ceteris paribus rate of change in net revenue with respect to b_1, is discontinuous, i.e. $\Delta f/\Delta b_1$ is constant over each subinternal and nonincreasing.

11.4 The Marginal Cost Function for an Activity (or Individual Product)

A final experiment concerning the behavior of the multiproduct firm pertains to the derivation of the **marginal cost function** for a given product. In this regard, we wish to determine the ceteris paribus rate of change in the total cost of producing some product with respect to increases in its output level. (As required for this type of analysis, the firm's technology is assumed unchanging, as are the prices of its products and variable inputs. Additionally, its fixed inputs are available in set amounts for the production period under consideration.)

In this model, the total cost of a given product is the sum of two individual cost components. The first is a direct cost element that simply amounts to the variable cost of producing the product (as depicted by the outlay on the variable inputs) while the second, an indirect cost element, is the implicit opportunity or imputed cost connected with its production. To gain some insight into this latter cost component, we need only note that when a firm produces several different outputs, an increase in the production of one of them, in the presence of fixed inputs, must be accompanied by a decrease in the production of the others. So if a given product is produced, the firm must forgo the gross profit obtainable from the production of the other products.

To find the direct effect that the output of a particular product, say x_r, has on gross profit and ultimately on the rate of change of its total production cost, we must modify (7.25.1) slightly by setting its gross profit coefficient in the objective function equal to zero. In addition, we shall transfer all terms involving x_r in the structural constraints to the right-hand side of the same. Thus, the modified problem appears as

$$max\, f(x_1,\ldots,x_p) = \sum_{\substack{j=1 \\ j\neq r}}^{p} \left(p_j - \sum_{i=l+1}^{m} q_i a_{ij} \right) x_j \quad \text{s.t.}$$

$$\sum_{\substack{j=1 \\ j\neq r}}^{p} a_{ij} x_j \le b_i - a_{ir} x_r, i = 1,\ldots,l,$$

$$x_j \ge 0, j = 1,\ldots,p.$$

Clearly the variable x_r is combined with the requirements b_i, $i = 1, \ldots, l$, to yield the amounts of the fixed inputs available for the remaining products as x_r increases. Thus, $\boldsymbol{b}$ is transformed to $\bar{\boldsymbol{b}} = \boldsymbol{b} - x_r \bar{\boldsymbol{a}}_r$. If $\boldsymbol{t} = -\bar{\boldsymbol{a}}_2, \bar{\boldsymbol{b}}^* = \bar{\boldsymbol{b}} + \phi \boldsymbol{t} = \bar{\boldsymbol{b}} - \phi \bar{\boldsymbol{a}}_2 = \boldsymbol{b} - (x_r + \phi)\bar{\boldsymbol{a}}_2$. Therefore, ϕ may be interpreted as the change in x_r, Δx_r, so that we may determine the critical values of Δx_r simply by finding the critical values of ϕ. And as stipulated by our parametric analysis of the requirements vector, the firm adjusts its product mix at each possible x_r level in order to maximize its gross profit. Hence, we move from one optimal basic feasible solution to the next as ϕ, and thus x_r, varies continuously.

If we apply this procedure to the problem solved in Example 7.3, with $r = 2$, then we wish to

$$\begin{aligned} max \quad f &= 14x_1 + 11x_3 + 10x_4 \quad \text{s.t.} \\ 10x_1 + 6x_3 + 4x_4 + x_5 \quad\quad &= 40 - 8x_2 \\ 2x_1 + 5x_3 + 8x_4 \quad + x_6 &= 30 - 3x_2 \\ x_1,\ldots,x_6 &\ge 0. \end{aligned}$$

Since the optimal simplex matrix associated with the original version of this problem is

$$\begin{bmatrix} 1 & \frac{11}{19} & 0 & -\frac{14}{19} & \frac{5}{38} & -\frac{3}{19} & 0 & \frac{10}{19} \\ 0 & \frac{7}{19} & 1 & \frac{36}{19} & -\frac{1}{19} & \frac{5}{19} & 0 & \frac{110}{19} \\ 0 & \frac{3}{19} & 0 & \frac{10}{19} & \frac{24}{19} & \frac{13}{19} & 1 & \frac{1350}{19} \end{bmatrix}, \quad (11.1)$$

the optimal simplex matrix associated with this modified problem is

$$\begin{matrix} x_1 & x_3 & x_4 & x_5 & x_6 & & \\ \end{matrix}$$
$$\begin{bmatrix} 1 & 0 & -\frac{14}{19} & \frac{5}{38} & -\frac{3}{19} & 0 & \left(\frac{10}{19}-\frac{11}{19}x_2\right) \\ 0 & 1 & \frac{36}{19} & -\frac{1}{19} & \frac{5}{19} & 0 & \left(\frac{110}{19}-\frac{7}{19}x_2\right) \\ 0 & 0 & \frac{10}{19} & \frac{24}{19} & \frac{13}{19} & 1 & \left(\frac{1350}{19}-\frac{231}{19}x_2\right) \end{bmatrix}, \tag{11.20}$$

where the values of the basic variables appearing in the right-hand column of (11.20) are determined as

$$\boldsymbol{B}^{-1}\bar{\boldsymbol{b}} = \boldsymbol{B}^{-1}(\boldsymbol{b}-x_2\bar{\boldsymbol{a}}_2) = \begin{bmatrix} 5/38 & -3/19 \\ -1/19 & 5/19 \end{bmatrix}\left(\begin{bmatrix} 40 \\ 30 \end{bmatrix} - x_2\begin{bmatrix} 8 \\ 3 \end{bmatrix}\right) = \begin{bmatrix} \frac{10}{19}-\frac{11}{19}x_2 \\ \frac{110}{19}-\frac{7}{19}x_2 \end{bmatrix};$$

$$\boldsymbol{C}_B'\boldsymbol{B}^{-1}\bar{\boldsymbol{b}} = (14,11)\begin{bmatrix} \frac{10}{19}-\frac{11}{19}x_2 \\ \frac{110}{19}-\frac{7}{19}x_2 \end{bmatrix} = \frac{1350}{19}-\frac{231}{19}x_2.$$

(Note that $f = \frac{1350}{19} - \frac{231}{19}x_2$ depicts gross profit from the production of other products as a decreasing function of x_2. At each level of x_2 its implicit opportunity or imputed cost is the decrease in gross profit from the maximum attainable when $x_2 = 0$ or $\frac{1350}{19} - f$. Currently, $\frac{1350}{19} - f = \frac{231}{19}x_2$, where the imputed cost per unit of x_2 is $231/19 = \boldsymbol{U}'\bar{\boldsymbol{Q}}_2 = u_1\bar{a}_{12} + u_2\bar{a}_{22} = (24/19)8 + (13/19)3$.) For $x_2 = 0$, the optimal basic feasible solution to the original problem is $x_1 = 10/19$, $x_3 = 110/19$, and $f = 1350/19$. As x_2 increases in value, (11.20) reveals that the values of the basic variables decline. With each of the basic variables a function of the parameter $\Delta x_2 = \phi$, the problem is to determine the largest allowable increase in x_2 which does not violate the feasibility of the current basic feasible solution.

Keeping in mind that $\phi = \Delta x_2$ and $\boldsymbol{t}' = (-8, -3)$, we first form

$$\boldsymbol{b}^* = \boldsymbol{b} + \phi\boldsymbol{t} = \begin{bmatrix} 40-8\phi \\ 30-3\phi \end{bmatrix}.$$

Then, from (11.1),

$$X_B^* = X_B + \phi B^{-1}t = \begin{bmatrix} \frac{10}{19} \\ \frac{110}{19} \end{bmatrix} + \phi \begin{bmatrix} \frac{5}{38} & -\frac{3}{19} \\ -\frac{1}{19} & \frac{5}{19} \end{bmatrix} \begin{bmatrix} -8 \\ -3 \end{bmatrix} = \begin{bmatrix} \frac{10}{19} - \frac{11}{19}\phi \\ \frac{110}{19} - \frac{7}{19}\phi \end{bmatrix}. \tag{11.21}$$

With $\boldsymbol{\beta}_1 \boldsymbol{t} = -7/19$, $\boldsymbol{\beta}_1 \boldsymbol{t} = -7/19$, (10.4) yields $\phi_c^{(1)} = -x_{B1}/\boldsymbol{\beta}_1 \boldsymbol{t} = 10/11$. Then, from (11.21),

$$\left(X_B^*\right)_{\phi_c^{(1)}} = \begin{bmatrix} 0 \\ 60/11 \end{bmatrix}$$

and, with $\Delta f = -u_1 \Delta b_1 - u_2 \Delta b_2 = -(24/19)(8)(10/11) - (13/19)(3)(10/11) = -210/19, f^* = f + \Delta f = 60$. Substituting these adjusted values into (11.20) yields

$$\begin{bmatrix} 1 & 0 & -\frac{14}{19} & \frac{5}{38} & -\frac{3}{19} & 0 & 0 \\ 0 & 1 & \frac{36}{19} & -\frac{1}{19} & \frac{5}{19} & 0 & \frac{60}{11} \\ 0 & 0 & \frac{10}{19} & \frac{24}{19} & \frac{13}{19} & 1 & 60 \end{bmatrix} \rightarrow$$

$$\begin{bmatrix} -\frac{19}{14} & 0 & 1 & -\frac{5}{28} & \frac{3}{14} & 0 & 0 \\ \frac{18}{7} & 1 & 0 & \frac{2}{7} & -\frac{1}{7} & 0 & \frac{60}{11} \\ \frac{5}{7} & 0 & 0 & \frac{19}{14} & \frac{4}{7} & 1 & 60 \end{bmatrix}. \tag{11.22}$$

Thus the new basic feasible solution is $x_3 = 60/11$, $x_4 = 0$, and $f = 60$, with the new requirements vector appearing as

$$\boldsymbol{b}^*_{\phi_c^{(1)}} = \boldsymbol{b} + \phi_c^{(1)} \boldsymbol{t} = \begin{bmatrix} 360/11 \\ 300/11 \end{bmatrix}.$$

Given that ϕ may be increased further we now have

$$\boldsymbol{b}^* = \boldsymbol{b}^*_{\phi_c^{(1)}} + \phi \boldsymbol{t} = \begin{bmatrix} \frac{360}{11} - 8\phi \\ \frac{300}{11} - 3\phi \end{bmatrix}$$

and, from (11.22),

$$X_B^* = \left(X_B^*\right)_{\phi_c^{(1)}} + \phi B^{-1}t = \begin{bmatrix} \frac{11}{14}\phi \\ \frac{60}{11} - \frac{13}{7}\phi \end{bmatrix}. \tag{11.23}$$

Since $\beta_2 t = -13/7$, (10.4) renders $\phi_c^{(2)} = -x_{B2}/\beta_2 t = 420/143$. Then, from (11.23),

$$\left(X_B^*\right)_{\phi_c^{(2)}} = \begin{bmatrix} 30/13 \\ 0 \end{bmatrix}$$

and $f^* = f + \Delta f = f - U'\bar{a}_2\phi_c^{(2)} = 300/13$. Inserting this revised information into (11.22) yields

$$\begin{bmatrix} -\frac{19}{14} & 0 & 1 & -\frac{5}{28} & \frac{3}{14} & 0 & \frac{30}{13} \\ \frac{18}{7} & 1 & 0 & \frac{2}{7} & -\frac{1}{7} & 0 & 0 \\ \frac{5}{7} & 0 & 0 & \frac{19}{14} & \frac{4}{7} & 1 & \frac{300}{13} \end{bmatrix} \rightarrow$$
$$\begin{bmatrix} \frac{7}{2} & \frac{3}{2} & 1 & \frac{1}{4} & 0 & 0 & \frac{30}{13} \\ -18 & -7 & 0 & -2 & 1 & 0 & 0 \\ 11 & 4 & 0 & \frac{5}{2} & 0 & 1 & \frac{300}{13} \end{bmatrix}. \tag{11.24}$$

Here $x_4 = 30/13$, $x_6 = 0$, and $f = 300/13$, and the revised requirements vector is

$$b^*_{\phi_c^{(2)}} = b^*_{\phi_c^{(1)}} + \phi_c^{(2)} t = \begin{bmatrix} 120/13 \\ 240/13 \end{bmatrix}.$$

Given that ϕ is to be increased again while still preserving primal feasibility, we next form

$$b^* = b^*_{\phi_c^{(2)}} + \phi t = \begin{bmatrix} \frac{120}{13} - 8\phi \\ \frac{240}{13} - 3\phi \end{bmatrix}$$

and, from (11.24),

$$X_B^* = \left(X_B^*\right)_{\phi_c^{(2)}} + \phi B^{-1} t = \begin{bmatrix} \frac{30}{13} - 2\phi \\ 13\phi \end{bmatrix}. \tag{11.25}$$

With $\beta_1 t = -2$, (10.4) provides us with $\phi_c^{(3)} = -x_{B1}/\beta_1 t = 15/13$. Then, using (11.25),

$$\left(X_B^*\right)_{\phi_c^{(3)}} = \begin{bmatrix} 0 \\ 15 \end{bmatrix}$$

and $f^* = f + \Delta f = f - U'\bar{a}_2 \phi_c^{(3)} = 0$. Substituting these adjusted values of the basic variables into (11.24) gives us

$$\begin{bmatrix} \frac{95}{14} & \frac{3}{2} & 1 & \frac{1}{4} & 0 & 0 & 0 \\ -18 & -7 & 0 & -2 & 1 & 0 & 15 \\ 11 & 4 & 0 & \frac{5}{2} & 0 & 1 & 0 \end{bmatrix} \tag{11.26}$$

so that the final optimal basic feasible solution for this parametric problem amounts to $x_4 = 0$, $x_6 = 15$, and $f = 0$. The appropriate requirements vector for the current problem with $\phi = \phi_c^{(3)}$ is

$$b^*_{\phi_c^{(3)}} = \begin{bmatrix} 0 \\ 15 \end{bmatrix}.$$

If we summarize the information generated by the above sequence of optimal basic feasible solutions in Table 11.2 and then in Figure 11.4, we may determine, in the last two columns of the same, the total cost of producing each level of x_2 as well as the ceteris paribus rate of change in total cost with respect to x_2 (i.e. the marginal cost of increasing the output of x_2). Here the marginal cost of x_2 is $MC_2 = \Delta TC_2/\Delta x_2$ (= implicit opportunity cost/unit + variable cost/unit $= U'\bar{a}_2 + 13$). At each vertex of the total cost curve in Figure 11.4b there are two slopes or marginal cost values so that the marginal cost function is discontinuous at the implied x_2 values. Thus, the marginal cost function for the product x_2 corresponds to a set of horizontal line segments (Figure 11.4c) and is undefined along each vertical gap.

If we next introduce the price of x_2 into the picture, the total revenue curve for product x_2, $TR_2 = p_2 x_2 = 25x_2$, appears in Figure 11.4b. With gross profit $TR_2 - TC_2$ negative for each level of x_2, product x_2 is unprofitable and thus is not produced (as the optimal basic feasible solution to the problem in Example 7.3 reveals since x_2 is nonbasic). Looked at from another perspective, since in Figure 11.4c marginal revenue $MR_2 = dTR_2/dx_2 = p_2 = 25 < MC_2$ for all x_2

Table 11.2 The marginal cost function for output x_2.

Change in x_2, $\Delta x_2 = \phi_c$	Level of x_2	Gross profit, f	Implicit opportunity cost, $OC_2 = \frac{1350}{19} - f = U_2'\bar{a}_2 x_2$	Variable cost, $VC_2 = 13x_2$	Total cost, $TC_2 = OC_2 + VC_2$	Change in total cost, ΔTC_2	Marginal cost, $MC_2 = \Delta TC_2/\Delta x_2$
0.9091	0	71.0526	0	0	0	22.8709	25.1577
2.9730	0.9091	60	11.0526	11.8183	22.8709	75.1048	25.5719
1.1538	3.8461	23.0769	47.9757	50	97.9757	38.0769	33.0013
	5.0	0	71.0526	65	136.0526		

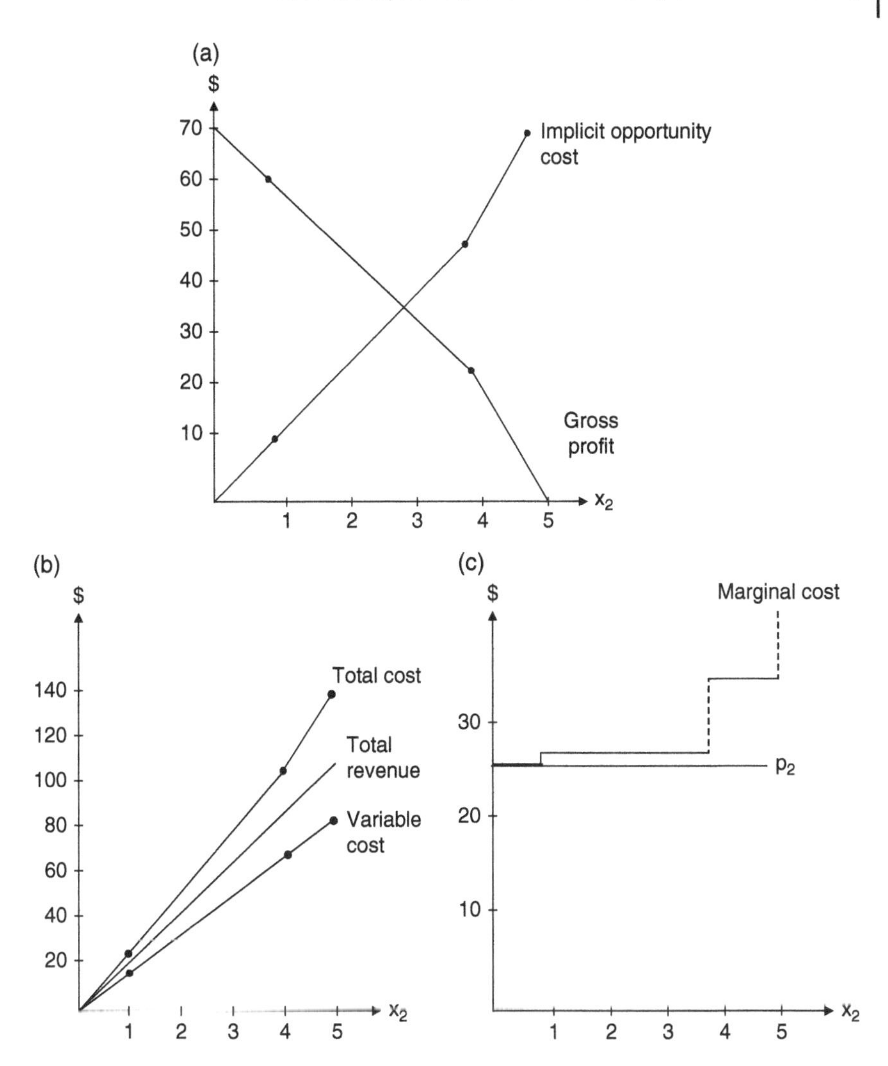

Figure 11.4 (a) Gross profit vs. implicit opportunity cost; (b) Total cost vs. total revenue; (c) Marginal cost function for output x_2.

values, the firm adds more to total cost than to total revenue as x_2 increases so that gross profit, while everywhere negative, becomes even more so as the excess of MC_2 over MR_2 increases.

If the price of x_2 were slightly higher so that the intersection between the price line and marginal cost occurs along a horizontal segment of the latter, then no unique output level of x_2 is discernible. However, if at some new higher price for x_2 the price line cuts through a vertical gap, then a unique output level for x_2

results since such an intersection point corresponds to a vertex of the total cost function and thus an optimal basic feasible solution. Thus, the firm will supply the quantity of x_2 corresponding to any price within the vertical gap between successive marginal cost line segments. Thus, each vertical gap represents a segment of the supply curve for x_2. In fact, if we compare Figures 11.1 and 11.4c we see that there exists a one-to-one correspondence between the horizontal segments of the marginal cost curve for x_2 and the horizontal gaps in the x_2 supply curve as well as between the vertical gaps of the x_2 marginal cost curve and the vertical segments (less the end points) of the supply curve for x_2.

11.5 Minimizing the Cost of Producing a Given Output

In our previous discussions concerning the optimizing behavior of the firm our analysis has been restricted to maximizing gross profit subject to the requirements that the amounts of the fixed inputs utilized cannot exceed their available amounts and that each activity level is nonnegative. Let us now pose a symmetrical problem. Specifically, given that a single product can be produced by a single activity or by a nonnegative linear combination of p separate activities, the firm now attempts to minimize the total variable input cost of producing x^0 units of output subject to the restrictions that the sum of the p nonnegative activity levels equals x^0 and that the amounts of the fixed inputs required to product x^0 cannot exceed the amounts available. In this regard, the implied optimization problem appears as

$$\begin{aligned} &min\, C(x_1,\ldots,x_p) = \sum_{j=1}^{p} TVC_j x_j = \sum_{j=1}^{p}\left(\sum_{i=l+1}^{m} q_i a_{ij}\right)x_j \quad \text{s.t.} \\ &\sum_{j=1}^{p} a_{ij}x_j \le b_i, i = 1,\ldots,l \\ &\sum_{j=1}^{p} x_j = x^0 \\ &x_j \ge 0, j = 1,\ldots,p, \end{aligned} \tag{11.27}$$

or, in matrix form,

$$\begin{aligned} &min \quad C(X) = Q'A^{*}X \quad \text{s.t.} \\ &\bar{A}X \le b \\ &1'X = x^0 \\ &X \ge O. \end{aligned} \tag{11.27.1}$$

From the information presented in Example 7.3, with $x^0 = 120/19$, (11.27) appears as

$$\begin{aligned} &\textit{min} \quad C = 11x_1 + 13x_2 + 14x_3 + 15x_4 \text{ or} \\ &\textit{max} \quad g = -C = -11x_1 - 13x_2 - 14x_3 - 15x_4 \quad \text{s.t.} \\ &\quad 10x_1 + 8x_2 + 6x_3 + 4x_4 \le 40 \\ &\quad 2x_1 + 3x_2 + 5x_3 + 8x_4 \le 30 \\ &\quad x_1 + x_2 + x_3 + x_4 = \frac{120}{19} \\ &\quad x_1, \ldots, x_4 \ge 0. \end{aligned}$$

If x_5, x_6 depict nonnegative slack variables and x_7 represents a nonnegative artificial variable, then, via the M-penalty method, this problem may be respecified as

$$\begin{aligned} &\textit{max} \quad g^* = -11x_1 - 13x_2 - 14x_3 - 15x_4 - Mx_7 \quad \text{s.t.} \\ &\quad 10x_1 + 8x_2 + 6x_3 + 4x_4 + x_5 = 40 \\ &\quad 2x_1 + 3x_2 + 5x_3 + 8x_4 + x_6 = 30 \\ &\quad x_1 + x_2 + x_3 + x_4 + x_7 = \frac{120}{19} \\ &\quad x_1, \ldots, x_7 \ge 0. \end{aligned}$$

Then the optimal simplex matrix (with the column corresponding to the artificial variable deleted) associated with this problem is

$$\begin{bmatrix} 1 & \frac{1}{3} & 0 & 0 & \frac{1}{2} & \frac{1}{3} & 0 & \frac{10}{19} \\ 0 & -\frac{1}{3} & 0 & 1 & \frac{1}{2} & \frac{2}{3} & 0 & 0 \\ 0 & 1 & 1 & 0 & -1 & -1 & 0 & \frac{110}{19} \\ 0 & \frac{1}{3} & 0 & 0 & 1 & \frac{1}{3} & 1 & -\frac{1650}{19} \end{bmatrix} \tag{11.28}$$

where in the optimal basic feasible solution to the above (original) problem is $x_1 = 10/19$, $x_3 = 110/19$, $x_4 = 0$, and $g^* = g = -C = -1650/19$ or $C = 1650/19$. (Note that if total variable cost = 1650/19 is subtracted from total revenue = 3000/19, the difference is the previously determined gross profit level of 1350/19.)

11.6 Determination of Marginal Productivity, Average Productivity, Marginal Cost, and Average Cost Functions

In Section 7.3 we derived the total productivity function for an input b_1 along with its associated marginal and average productivity functions. We shall now reexamine this derivation in the light of our parametric programming techniques so that the procedure can be formalized. Moreover, once the unit price of the variable input is introduced into the model, we may also obtain the total variable and average variable cost functions along with the marginal cost function. Additionally, if the price per unit of output is also introduced into the model, then we can determine the short-run (gross) profit maximizing level of output.

To this end, let us assume that the firm's initial optimization problem assumes the general form indicated by (7.13) and that the firm's technology can be summarized by the activities

$$\boldsymbol{a}_1 = \begin{bmatrix} 10 \\ 2 \end{bmatrix}, \boldsymbol{a}_2 = \begin{bmatrix} 8 \\ 3 \end{bmatrix}, \boldsymbol{a}_3 = \begin{bmatrix} 6 \\ 5 \end{bmatrix}, \boldsymbol{a}_4 = \begin{bmatrix} 4 \\ 8 \end{bmatrix}.$$

Moreover, upper limits on the availability of the first and second inputs b_1, b_2 are 40 and 30 units, respectively. In this regard, we seek to

$$\begin{aligned} max \quad & f = x_1 + x_2 + x_3 + x_4 \quad \text{s.t.} \\ & 10x_1 + 8x_2 + 6x_3 + 4x_4 \le 40 \\ & 2x_1 + 3x_2 + 5x_3 + 8x_4 \le 30 \\ & x_1, \dots, x_4 \ge 0. \end{aligned}$$

The optimal simplex matrix associated with this problem is

$$\begin{bmatrix} \frac{19}{11} & 1 & 0 & -\frac{14}{11} & \frac{5}{22} & -\frac{3}{11} & 0 & \frac{10}{11} \\ -\frac{7}{11} & 0 & 1 & \frac{26}{11} & -\frac{3}{22} & \frac{4}{11} & 0 & \frac{60}{11} \\ \frac{1}{11} & 0 & 0 & \frac{1}{11} & \frac{1}{11} & \frac{1}{11} & 1 & \frac{70}{11} \end{bmatrix} \tag{11.29}$$

with $x_2 = 10/11$, $x_3 = 60/11$. and $f = 70/11$.

Let us now look to the following parametric experiment. Holding b_2 fixed at 30 units, let us treat b_1 as continuously variable so that the supply of b_1 can be augmented. Thus, the parametrized requirements vector is, for direction vector $\boldsymbol{t} = \boldsymbol{e}_1$,

$$\boldsymbol{b}^* = \boldsymbol{b} + \phi \boldsymbol{e}_1 = \begin{bmatrix} 40 + \phi \\ 30 \end{bmatrix}.$$

At each possible requirements level for b_1 the firm seeks to maximize $f = \sum_{j=1}^{4} x_j$, i.e. as we parametrize the requirements vector the activity mix must vary in a fashion such that optimality is preserved when we move from one basic feasible solution to the next.

From (11.29),

$$X_B^* = X_B + \phi B^{-1} e_1 = \begin{bmatrix} \frac{10}{11} + \frac{5}{22}\phi \\ \frac{60}{11} - \frac{3}{22}\phi \end{bmatrix}. \tag{11.30}$$

Then, from (10.4), $\phi_c = \phi_c^{(1)} = 40$ so that (11.30) becomes

$$\left(X_B^*\right)_{\phi_c^{(1)}} = \begin{bmatrix} 10 \\ 0 \end{bmatrix}$$

and, with $\Delta f = u_1 \Delta b_1 = u_1 \phi_c = 40/11$, $f^* = f + \Delta f = 10$. Substituting the adjusted values of the basic variables and f into (11.29) yields, after a dual simplex pivot,

$$\begin{bmatrix} 0 & 1 & \frac{19}{7} & \frac{36}{7} & -\frac{1}{7} & \frac{5}{7} & 0 & 10 \\ 1 & 0 & -\frac{11}{7} & -\frac{26}{7} & \frac{3}{14} & -\frac{4}{7} & 0 & 0 \\ 0 & 0 & \frac{1}{7} & \frac{3}{7} & \frac{1}{14} & \frac{1}{7} & 1 & 10 \end{bmatrix}. \tag{11.31}$$

Hence, the new optimal basic feasible solution is $x_1 = 0$, $x_2 = 10$, and $f = 10$. (For this solution, the new requirements vector is $b^*_{\phi_c^{(1)}} = b + \phi_c^{(1)} e_1 = \begin{bmatrix} 80 \\ 30 \end{bmatrix}$.)

Continuing with the usual procedure of determining the extent of the increase in ϕ that does not violate the feasibility of the current solution we obtain, for

$$b^* = b^*_{\phi_c^{(1)}} + \phi e_1 = \begin{bmatrix} 80 + \phi \\ 30 \end{bmatrix}$$

and, from (11.31),

$$X_B = \left(X_B^*\right)_{\phi_c^{(1)}} + \phi B^{-1} e_1 = \begin{bmatrix} 10 + \frac{1}{7}\phi \\ \frac{3}{14}\phi \end{bmatrix},$$

$$\phi_c = \phi_c^{(2)} = 70 \left(\text{so that } \left(X_B^*\right)_{\phi_c^{(2)}} = \begin{bmatrix} 0 \\ 15 \end{bmatrix} \right),$$

$$f^* = f + \Delta f = f + u_1 \phi_c = 15.$$

Using this revised information in (11.31) enables us to determine

$$\begin{bmatrix} 0 & -7 & -19 & -36 & 1 & -5 & 0 & 0 \\ 1 & \frac{3}{2} & \frac{5}{2} & 4 & 0 & \frac{1}{2} & 0 & 15 \\ 0 & \frac{1}{2} & \frac{3}{2} & 3 & 0 & \frac{1}{2} & 1 & 15 \end{bmatrix}. \tag{11.32}$$

Here $x_1 = 15$, $x_5 = 0$, $f = 15$, and the corresponding requirements vector is

$$\boldsymbol{b}^*_{\phi_c^{(2)}} = \boldsymbol{b}^*_{\phi_c^{(1)}} + \phi_c^{(2)}\boldsymbol{e}_1 = \begin{bmatrix} 150 \\ 30 \end{bmatrix}.$$

Since for this current optimal basic feasible solution we have

$$\boldsymbol{X_B} = \left(\boldsymbol{X_B^*}\right)_{\phi_c^{(2)}} + \phi \boldsymbol{B}^{-1}\boldsymbol{e}_1 = \begin{bmatrix} \phi \\ 15 \end{bmatrix},$$

ϕ may be increased without bound and yet the feasibility of the current solution will be preserved.

To complete the total product of b_1 function, let us now consider a decrease in the said input by replacing $\boldsymbol{e}_1$ by $-\boldsymbol{e}_1$. In this instance

$$\boldsymbol{b}^* = \boldsymbol{b} - \phi\boldsymbol{e}_1 = \begin{bmatrix} 40-\phi \\ 30 \end{bmatrix},$$

and, using (11.29),

$$\boldsymbol{X_B^*} = \boldsymbol{X_B} - \phi\boldsymbol{B}^{-1}\boldsymbol{e}_1 = \begin{bmatrix} \frac{10}{11} - \frac{5}{22}\phi \\ \frac{60}{11} + \frac{3}{22}\phi \end{bmatrix},$$

$$\phi_c = \phi_c^{(3)} = 4 \left(\text{and thus } \left(\boldsymbol{X_B^*}\right)_{\phi_c^{(3)}} = \begin{bmatrix} 0 \\ 6 \end{bmatrix} \right),$$

$$f^* = f + \Delta f = f - u_1\phi_c = 6.$$

Then the updated version of (11.29) transforms to

$$\begin{bmatrix} -\frac{19}{14} & -\frac{11}{14} & 0 & 1 & -\frac{5}{28} & \frac{3}{14} & 0 & 0 \\ \frac{17}{7} & \frac{13}{7} & 1 & 0 & \frac{2}{7} & -\frac{1}{7} & 0 & 6 \\ \frac{3}{14} & \frac{1}{14} & 0 & 0 & \frac{3}{28} & \frac{1}{14} & 1 & 6 \end{bmatrix} \tag{11.33}$$

with $x_3 = 6$, $x_4 = 0$, and $f = 6$. For this new solution the requirements vector has the form

$$\boldsymbol{b}^*_{\phi_c^{(3)}} = \boldsymbol{b} - \phi_c^{(3)} \boldsymbol{e}_1 = \begin{bmatrix} 36 \\ 30 \end{bmatrix}.$$

Next, for

$$\boldsymbol{b}^* = \boldsymbol{b}^*_{\phi_c^{(3)}} - \phi \boldsymbol{e}_1 = \begin{bmatrix} 36 - \phi \\ 30 \end{bmatrix},$$

(11.33) is used to obtain

$$\boldsymbol{X}^*_{\boldsymbol{B}} = (\boldsymbol{X}^*{}_{\boldsymbol{B}})_{\phi_c^{(3)}} - \phi \boldsymbol{B}^{-1} \boldsymbol{e}_1 = \begin{bmatrix} \frac{5}{28}\phi \\ 6 - \frac{2}{7}\phi \end{bmatrix},$$

$$\phi_c = \phi_c^{(4)} = 21 \left(\text{here } (\boldsymbol{X}^*_{\boldsymbol{B}})_{\phi_c^{(4)}} = \begin{bmatrix} 15/4 \\ 0 \end{bmatrix} \right),$$

$$f^* = f + \Delta f = f - u_1 \phi_c = 15/14.$$

Using these values to update (11.33) we have, after pivoting,

$$\begin{bmatrix} \frac{5}{2} & 2 & \frac{3}{2} & 1 & \frac{1}{4} & 0 & 0 & \frac{15}{4} \\ -18 & -13 & -7 & 0 & -2 & 1 & 0 & 0 \\ \frac{3}{2} & 1 & \frac{1}{2} & 0 & \frac{1}{4} & 0 & 1 & \frac{15}{4} \end{bmatrix}. \tag{11.34}$$

For this solution $x_4 - 15/4$, $x_6 - 0$, $f - 15/4$ and the requirements vector for the current problem is

$$\boldsymbol{b}^*_{\phi_c^{(4)}} = \boldsymbol{b}^*_{\phi_c^{(3)}} - \phi_c^{(4)} \boldsymbol{e}_1 = \begin{bmatrix} 15 \\ 30 \end{bmatrix}.$$

One final round of calculations ultimately yields the simplex matrix derived from the updated version of (11.34), namely

$$\begin{bmatrix} \frac{5}{2} & 2 & \frac{3}{2} & 1 & \frac{1}{4} & 0 & 0 & 0 \\ -18 & -13 & -7 & 0 & -2 & 1 & 0 & 30 \\ \frac{3}{2} & 1 & \frac{1}{2} & 0 & \frac{1}{4} & 0 & 1 & 0 \end{bmatrix}. \tag{11.35}$$

Since there are no negative elements in the first row of this matrix, we cannot pivot to a new basic feasible solution using the dual simplex method (further increases in ϕ violate the feasibility of the current solution) so that (11.35) depicts the final optimal basic feasible solution. So for this matrix $x_4 = 0$, $x_6 = 30, f = 0$, and the associated requirements vector is

$$\boldsymbol{b}^*_{\phi_c^{(5)}} = \boldsymbol{b}^*_{\phi_c^{(5)}} - \phi_c^{(5)}\boldsymbol{e}_1 = \begin{bmatrix} 0 \\ 30 \end{bmatrix}.$$

If we collect the information on f, b_1, and u_1 provided by the above sequence of optimal basic feasible solutions and construct Table 11.3, then the summarized data contained therein can be used to construct the total product of b_1 function along with the associated marginal and average productivity functions of b_1 (Figure 11.5). Here the **marginal product** of b_1 is the ceteris paribus rate of change in output with respect to b_1 while the **average product** of b_1 is the ceteris paribus level of output per unit of b_1. (As required by the ceteris paribus assumption, the firm's technology is taken as given, factor prices are held constant, and the quantity of the fixed input b_2 is invariant.) (If we compare the MP_1, u_1 columns of the above table, we see that they are one in the same. This is to be expected, since the marginal product of b_1 is the dual variable u_1 – the latter representing the (internal) rate at which the firm's technology transforms

Table 11.3 Total, marginal, and average products of input b_1.

Total product, f	Level of b_1	Change in, f, Δf	Change in b_1, Δb_1	Marginal Product of b_1, $MP_1 = \Delta f/\Delta b_1$	Dual variable u_1	Average product of b_1, $AP_1 = f/b_1$
0	0	3.75	15	0.25	0.25	0.25
15/4	15	2.25	21	0.1071	0.1071	0.1666
6	36	0.3636	4	0.0909	0.0909	0.1591
70/11	40	3.6364	40	0.0909	0.0909	0.125
10	80	5	70	0.0714	0.0714	0.10
15	150	0	ϕ	0	0	15/(150+ϕ)
15	150+ϕ	.	.	.		.
.	.	.	.	.		.
.	.	.	.	.		.
.	.					
15	+∞	0		0	0	0

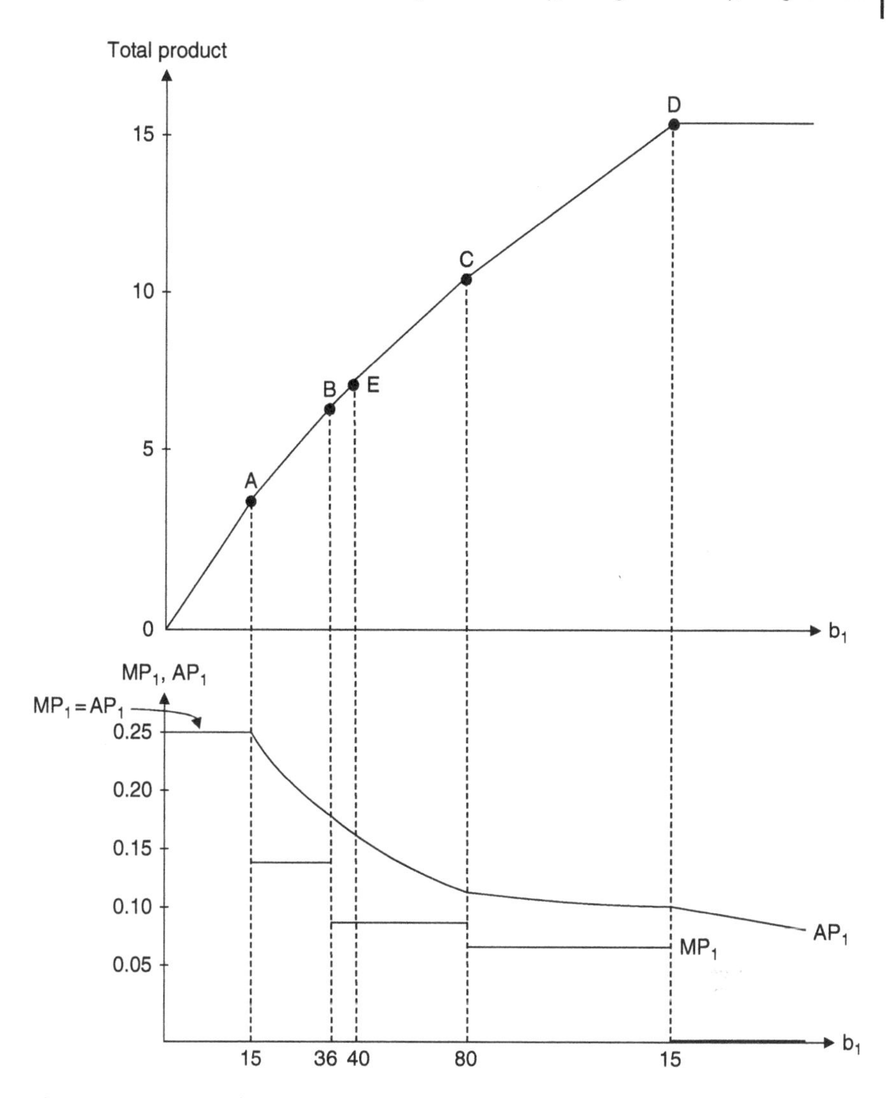

Figure 11.5 Total product, marginal product, and average product of input b_1.

units of b_1 into units of output.) As expressed in Figure 11.5, the total product of b_1 is a piecewise-linear function of the level of b_1, i.e. it is linear only over specific subintervals with its constituent line segments meeting at vertices, the latter corresponding to degenerate optimal basic feasible solutions wherein only one activity is used. The slope of this function, the marginal product of b_1, $\Delta f/\Delta b_1$, is discontinuous – it is constant over each subinterval and nonincreasing as b_1 increases. Next, the average product of b_1, f/b_1, can be measured by the slope of a ray from the origin to a point on the total product of b_1 function. Clearly the average product of b_1 is continuous and strictly decreasing as b_1 increases.

To derive the total variable cost function, let us assume that the price per unit of b_1 is \$4. Then from the preceding table we may determine the total variable cost function, $TVC = 4b_1$, along with the marginal cost and average variable cost functions (see the Table 11.4 along with Figure 11.6). Here total variable cost represents the ceteris paribus (*supra*) minimum cost of producing each of the maximal output levels obtained from the above parametric analysis and **marginal cost** is the ceteris paribus rate of change in total variable cost with respect to output while **average variable cost** is the ceteris paribus level of total variable cost per unit of output. As Figure 11.6 reveals, total variable cost is a piecewise-linear function of the level of output (its individual line segments meet at vertices that must also correspond to degenerate optimal basic feasible solutions). Moreover, its slope or marginal cost, $\Delta TVC/\Delta f$, is discontinuous – it is constant over specific subintervals and nondecreasing as output increases. Finally, average variable cost, TVC/f, can also be depicted by the slope of a ray from the origin to a point on the total variable cost function. Thus, average variable cost must be continuous and strictly increasing as output increases. When total output reaches its maximum value, the average variable cost function becomes vertical.

In the preceding discussion, we generated the total variable cost function in an indirect fashion in that we first determined from our parametric analysis of the

Table 11.4 Total variable cost, marginal cost, and average cost.

Total variable cost, $TVC = 4b_1$	Change in TVC, ΔTVC	Marginal cost, $MC = \Delta TVC/\Delta f$	Average variable cost, $AVC = TVC/f$
0	60	16	16
60	84	37.33	24
144	16	44.0	24.42
160	160	44.0	32
320	280	56	40
600	4ϕ	$+\infty$	$40 + 4/5\ \phi$
$600 + 4\phi$	.	.	.
.	.	.	.
.	.	.	.
.			$+\infty$
$+\infty$			

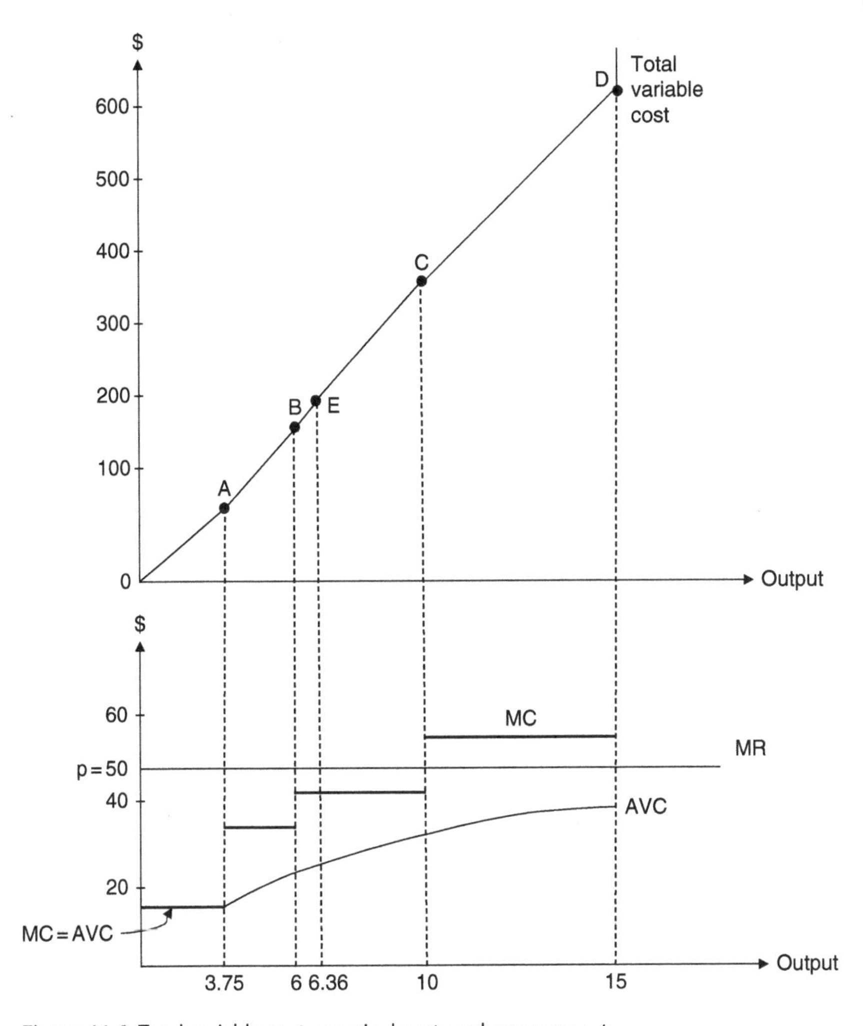

Figure 11.6 Total variable cost, marginal cost, and average cost.

requirements vector a sequence of optimal basic feasible solutions, each of which yielded a maximal output value and the appropriate b_1 level needed to support it (given $b_2 = 30$). Since total variable cost at each maximal output level represents the minimum expenditure on the variable input b_1, we may directly obtain these minimal cost levels by alternatively applying our parametric analysis to the problem

$$\begin{aligned} min\ \ TVC &= 4v_1 = 4(10x_1 + 8x_2 + 6x_3 + 4x_4) \quad \text{s.t.} \\ 2x_1 + 3x_2 + 5x_3 + 8x_4 &\leq 30 \\ x_1 + x_2 + x_3 + x_4 &= x^0 \\ x_1, \ldots, x_4 &\geq 0, \end{aligned} \tag{11.36}$$

where x^0 represents a given level of output.

For instance, if $x^0 = 10$, the optimal simplex matrix associated with this problem is (via the M-penalty method)

$$\begin{bmatrix} 0 & 1 & 3 & 6 & 1 & 0 & 10 \\ 1 & 0 & -2 & -5 & -1 & 0 & 0 \\ 0 & 0 & 8 & 24 & 8 & 1 & -320 \end{bmatrix}$$

So that $x_1 = 0$, $x_2 = 10$, and $TVC = 320$. (Note that these same x_1, x_2 values emerged earlier in (11.31) when we obtained a maximal output level of 10 units.) In terms of Figure 11.6, this solution corresponds to point C on the total variable cost function. And as x^0 is varied parametrically, the other vertices on this function (points O, A, B, and D) can be systematically generated, each corresponding to a degenerate optimal basic feasible solution. The marginal and average variable cost functions can then be directly obtained as byproducts of the resulting sequence of solutions.

If the setting of our discussion is changed to requirements space, Figure 11.7 depicts the simple activities $\boldsymbol{a}_j$ and their respective radial projections P_j, $j = 1,\ldots,4$. The solution to the above minimization problem corresponds to point C in this diagram, a point on the $f = 10$ isoquant (which also corresponds to point C on the total product curve in Figure 11.5 and also to point C on the total variable cost function in Figure 11.6). And as x^0 is varied parametrically, we obtain the remaining points in the sequence of optimal degenerate basic feasible solutions, namely points O, A, B, and D. Since these points, along with C, represent the minimum cost of producing successively larger output levels, OABCD represents the firm's short-run expansion path – the locus of economically efficient (least-cost) input combinations corresponding to successively higher outputs. So whether we perform a parametric experiment involving the maximization of output with a particular component of the requirements vector continuously variable or the minimization of total variable cost with the output requirement continuously variable, we ultimately obtain the same sets of information emerging from the resulting sequence of optimal solutions (e.g. the set of one-to-one correspondences between points O, A, B, C, and D in each of the Figures 11.5–11.7).

One final task remains, namely the determination of the optimal output point on the expansion path. To accomplish this, we must introduce the unit price of output into the model, i.e. we need only compare price (marginal revenue) and

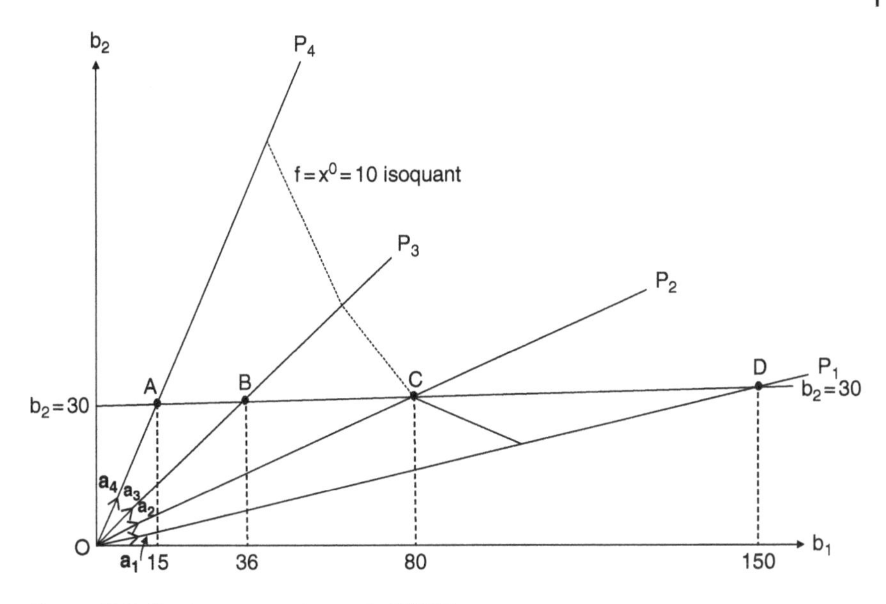

Figure 11.7 Short-run expansion path OABCD.

marginal cost. If, for example, product price is $p = \$50$ (Figure 11.6), then the optimal output level is 10 units. (This same result may alternatively be determined by Maximizing gross profit = Total revenue – Total variable cost = 50 $\sum_{j=1}^{4} x_j - 4v_1$ subject to the first structural constraint in (11.36), i.e.

$$\begin{aligned} max\ f &= 10x_1 + 18x_2 + 26x_3 + 34x_4 \quad \text{s.t.} \\ & 2x_1 + 3x_2 + 5x_3 + 8x_4 \le 30 \\ & x_1, \ldots, x_4 \ge 0.) \end{aligned} \tag{11.37}$$

Then total profit is 50 (10) – 4 (80) – 500 – 320 – \$180. (Note that this is the same profit figure that emerges upon directly solving (11.37).)

Figure 11.7 Short-run expansion path OABCD.

marginal cost. If, for example, product price is $p = \$50$ (Figure 11.6), then the optimal output level is 10 *units*. (This same result may alternatively be determined by Maximizing gross profit = Total revenue − Total variable cost = $50\sum_{j=1}^{4} x_j - 4x_1$ subject to the first structural constraint in (11.36), i.e.

$$\max f = 10x_1 + 18x_2 + 22x_3 + 31x_4 \quad \text{s.t.}$$

$$2x_1 + 3x_2 + 5x_3 + 8x_4 \leq 30 \tag{11.37}$$

$$x_1, \ldots, x_4 \geq 0.)$$

Then total profit is 50 (10) − 4 (80) = 500 − 320 = \$180. (Note that this is the same profit figure that emerges upon directly solving (11.37).)

12

Duality Revisited

12.1 Introduction

The material presented in this chapter is largely theoretical in nature and may be considered to be a more "refined" approach to the study of linear programming and duality. That is to say, the mathematical techniques employed herein are quite general in nature in that they encompass a significant portion of the foundations of linear (as well as nonlinear) programming. In particular, an assortment of mathematical concepts often encountered in the calculus, along with the standard matrix operations which normally underlie the theoretical development of linear programming, are utilized to derive the Karush-Kuhn-Tucker necessary conditions for a constrained extremum and to demonstrate the formal equivalence between a solution to the primal maximum problem and the associated saddle-point problem. Additionally, the duality and complementary slackness theorems of the preceding chapter are reexamined in the context of this "alternative" view of the primal and dual problems.

12.2 A Reformulation of the Primal and Dual Problems

Turning to the primal problem, let us maximize $f(X) = C'X$ subject to $AX \leq b$, $X \geq O$, $X \in \mathcal{E}^n$, where A is of order $(m \times n)$ with rows $a_1, \ldots, a_m$, and b is an $(m \times 1)$ vector with components $b_1, \ldots, b_m$. Alternatively, if

$$\bar{A} = \begin{bmatrix} A \\ \cdots \\ -I_n \end{bmatrix}, \bar{b} = \begin{bmatrix} b \\ \cdots \\ O \end{bmatrix}$$

are, respectively, of order $(m + n \times n)$ and $(m + n \times 1)$, then we may maximize $f(X) = C'X$ subject to $\bar{b} - \bar{A}X \geq O$, where $X \in \mathcal{E}^n$ is now unrestricted. In this

Linear Programming and Resource Allocation Modeling, First Edition. Michael J. Panik.

formulation $x_j \geq 0$, $j = 1, \ldots, n$, is treated as a structural constraint so that $\bar{\boldsymbol{b}} - \bar{A}X \geq \boldsymbol{O}$ defines a **region of feasible** or **admissible solutions** $\subseteq \mathcal{E}^n$.

If X_0 yields an optimal solution to the primal maximum problem, then no permissible change in X (i.e. one that does not violate any of the constraints specifying $\mathcal{K}$) can improve the optimal value of the objective function. How may we characterize such admissible changes? It is evident that, if starting at some feasible X, we can find a vector $\boldsymbol{h}$ such that a small movement along it violates no constraint, then $\boldsymbol{h}$ specifies a feasible direction at X. More formally, let $\delta(X_0)$ be a suitably restricted spherical δ – neighborhood about the point $X_0 \in \mathcal{E}^n$. Then the $(n \times 1)$ vector $\boldsymbol{h}$ is a **feasible direction** at X_0 if there exists a scalar $t \geq 0$, $0 \leq t < \delta$, such that $X_0 + t\boldsymbol{h}$ is feasible. In this light, for $\delta(X_0)$ again a spherical δ – neighborhood about the feasible point $X_0 \in \mathcal{E}^n$, the **set of feasible directions** at X_0, $\mathcal{D}(X_0)$, is the collection of all $(n \times 1)$ vectors $\boldsymbol{h}$ such that $X_0 + t\boldsymbol{h}$ is feasible for t sufficiently small, i.e.

$$\mathcal{D}(X_0) = \{\boldsymbol{h} | \text{if } X_0 \in \mathcal{K}, \text{ then } X_0 + t\boldsymbol{h} \in \mathcal{K}, 0 \leq t < \delta\}.$$

Here $\mathcal{D}(X_0)$ is the **tangent support cone** consisting of all feasible directions at X_0 and is generated by the tangent or supporting hyperplanes to $\mathcal{K}$ at X_0 (Figure 12.1). And since each such hyperplane

$$\begin{aligned} -\bar{\boldsymbol{a}}_i\boldsymbol{h} &= -\bar{\boldsymbol{a}}_i(X - X_0) \\ &= \bar{b}_i - \bar{\boldsymbol{a}}_i X = 0, i = 1, \ldots, m + n,^1 \end{aligned}$$

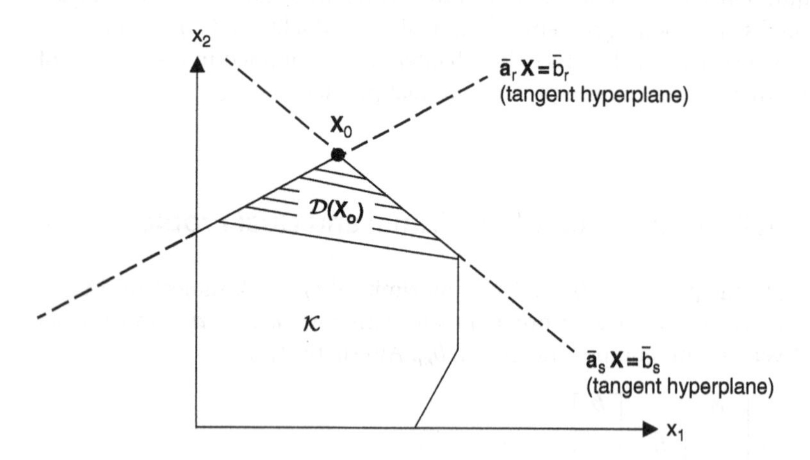

Figure 12.1 Set of feasible directions $\mathcal{D}(X_0)$.

1 Here $\bar{\boldsymbol{a}}_i$ is the ith row of $\bar{A}$ while $\bar{b}_i$ is the ith component of $\bar{\boldsymbol{b}}$. Specifically: $\bar{\boldsymbol{a}}_1 = \boldsymbol{a}_1, \ldots, \bar{\boldsymbol{a}}_m = \boldsymbol{a}_m$; $\bar{\boldsymbol{a}}_{m+j} = -\boldsymbol{e}_j', j = 1, \ldots, n$; $\bar{b}_1 = b_1, \ldots, \bar{b}_m = b_m$; and $\bar{b}_{m+j} = 0, j = 1, \ldots, n$.

specifies a closed half-space $\bar{a}_i X \le \bar{b}_i, i = 1, m+n,$ the tangent support cone $\mathcal{D}(X_0)$ represents that portion of the intersection of these closed half-planes in the immediate vicinity of (X_0).

To discern the relationship between $\mathcal{D}(X_0)(X_0$ optimal) and: (i) $f(X_0)$; (ii) the constraints $\bar{\boldsymbol{b}} - \bar{A}X_0 \ge \boldsymbol{O}$, let us consider the following two theorems, starting with Theorem 12.1.

Theorem 12.1 If the primal objective function attains its maximum at X_0, then any movement from X_0 along a feasible direction $\boldsymbol{h}$ cannot increase the value of f, i.e.

$$df^0 = C'\boldsymbol{h} = C'(X - X_0) \le 0 \text{ or } C'X \le C'X_0.$$

Proof. If $f(X_0) = C'X_0$ is the maximum of f over $\mathcal{K}$, then

$$C'(X_0 + t\boldsymbol{h}) - C'X_0 = tC'\boldsymbol{h} \le 0$$

and thus $C'\boldsymbol{h} \le 0$. Q.E.D.

Hence f must decrease along any feasible direction $\boldsymbol{h}$. Geometrically, no feasible direction may form an acute angle ($<\pi/2$) between itself and C.

To set the stage for our next theorem, we note that for any optimal $X \in \mathcal{K}$, it is usually the case that not all the inequality constraints are binding or hold as an equality. To incorporate this observation into our discussion, let us divide all of the constraints $\bar{\boldsymbol{b}} - \bar{A}X \ge \boldsymbol{O}$ into two mutually exclusive classes – those that are binding at $X, \bar{b}_i - \bar{a}_i X = 0,$ and those which are not, $\bar{b}_i - \bar{a}_i X > 0, i = 1, \ldots, m+n.$ So if

$$\mathcal{J} = \left\{ i \middle| \bar{b}_i - \boldsymbol{a}_i X = 0, X \in \mathcal{E}^n \right\}$$

depicts the index set of binding constraints, then $\bar{b}_i - \bar{a}_i X > 0$ for $i \notin \mathcal{I}$. In this regard, we have Theorem 12.5.

Theorem 12.2 The vector $\boldsymbol{h} \in \mathcal{D}(X_0)$ (X_0 optimal) if and only if $-\boldsymbol{a}_i\boldsymbol{h} \ge 0, i \in \mathcal{J}$.

Proof. (necessity) Let $X_0 \in \mathcal{K}$ with $\bar{a}_i X_0 = \bar{b}_i, i \in \mathcal{J}$. For $\boldsymbol{h} \in \mathcal{D}(X_0)$, $\bar{a}_i(X_0 + t\boldsymbol{h}) \le \bar{b}_i, 0 \le t < \delta,$ for some i. If $i \in \mathcal{I}$, $\bar{a}_i X_0 = \bar{b}_i$ and thus $t\bar{a}_i h \le 0$ or $-\bar{a}_i\boldsymbol{h} \ge 0$ for all $i \in \mathcal{J}$. (sufficiency). Let $-\bar{a}_i\boldsymbol{h} \ge 0$ for all $i \in \mathcal{J}$. Then for $i \notin \mathcal{J}$, $\bar{a}_i X_0 < \bar{\boldsymbol{b}}_i$ and thus

$$\bar{a}_i(X_0 + t\boldsymbol{h}) = \bar{a}_i X_0 + t\bar{a}_i\boldsymbol{h} < \bar{b}_i$$

for all $t \ge 0$. If $i \in \mathcal{J}$, $\bar{a}_i X_0 = \bar{b}_i$ so that

$$\bar{a}_i(X_0 + t\boldsymbol{h}) = \bar{b}_i + t\bar{a}_i\boldsymbol{h} \le \bar{b}_i$$

for each $t \ge 0$. Q.E.D.

To interpret this theorem, we note that $\bar{a}_i'$ may be characterized as an inward pointing or interior normal to $-\bar{a}_i h = 0$ $\left(\text{or } \bar{b} - \bar{a}_i X = 0\right)$ at X_0. So if h is a feasible direction, it makes a nonobtuse angle ($\leq \pi/2$) with all of the interior normals to the boundary of $\mathcal{K}$ at X_0. Geometrically, these interior normal or gradients of the binding constraints form a finite cone containing all feasible directions making nonobtuse angles with the supporting hyperplanes to $\mathcal{K}$ at X_0. Such a cone is polar to the tangent support cone $\mathcal{D}(X_0)$ and will be termed the **polar support cone** $\mathcal{D}(X_0)^+$(Figure 12.2a) Thus, $\mathcal{D}(X_0)^+$ is the cone spanned by the gradients

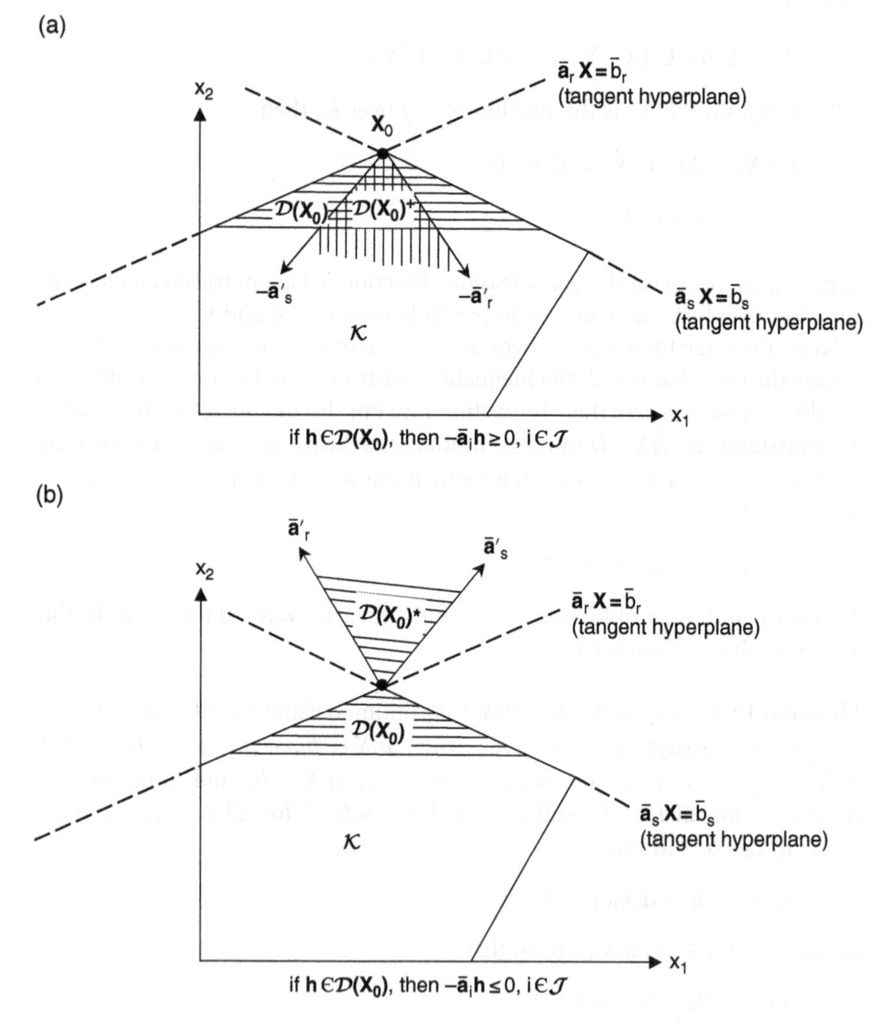

Figure 12.2 (a) Polar support cone $\mathcal{D}(X_0)^+$; (b) Dual support cone $\mathcal{D}(X_0)^*$.

$-\bar{a}_i'$ such that for $h \in \mathcal{D}(X_0)$, $-\bar{a}_i h \geq 0$, $i \in \mathcal{J}$. Looked at from another perspective, $\bar{a}_i$ may be considered an outward-pointing or exterior normal to the boundary of $\mathcal{K}$ at X_0. In this instance, if h is a feasible direction, it must now make a non-acute angle ($\geq \pi/2$) with all of the outward-pointing normals to the boundary of $\mathcal{K}$ at X_0. Again looking to geometry, the exterior normals or negative gradients of the binding constraints form a finite cone containing all feasible directions making nonacute angles with the hyperplanes tangent to $\mathcal{K}$ at X_0. This cone is the dual of the tangent support cone $\mathcal{D}(X_0)$ and is termed the **dual support cone** $\mathcal{D}(X_0)^*$(Figure 12.2b). Thus, $\mathcal{D}(X_0)^*$ is the cone spanned by the negative gradients $\bar{a}_i'$ such that, for all $h \in \mathcal{D}(X_0), \bar{a}_i h \leq 0$, $i \in \mathcal{J}$.

At this point, let us collect our major results. We found that if $f(X)$ subject to $\bar{b}_i - \bar{a}_i X = 0$, $i \in \mathcal{J}$, attains its maximum at X_0, then

$$C'h \leq 0 \text{ for all } h \text{ satisfying } \bar{a}_i h \leq 0, i \notin \mathcal{J}, h \in \mathcal{D}(X_0). \tag{12.1}$$

How may we interpret this condition? Given any $h \in \mathcal{D}(X_0)$, (12.1) holds if the gradient of f or C lies within the finite cone spanned by the exterior normals $\bar{a}_i'$, $i \in \mathcal{J}$, i.e. $C \in \mathcal{D}(X_0)^*$(or if $-C$ is contained within the finite cone generated by the interior normals $-\bar{a}_i'$, $i \in \mathcal{J}$, i.e. $-C \in \mathcal{D}(X_0)^+$). Hence (12.1) requires that the gradient of f be a nonnegative linear combination of the negative gradients of the binding constraints at X_0 (Figure 12.3). In this regard, there must exist real numbers $\bar{u}_i^0 \geq 0$ such that

$$C = \sum_i \bar{u}_i^0 \bar{a}_0', i \in \mathcal{J}. \tag{12.2}$$

Under what conditions will the numbers $\bar{u}_i^0 \geq 0$, $i \in \mathcal{J}$, exist? To answer this question, let us employ Theorem 12.3.

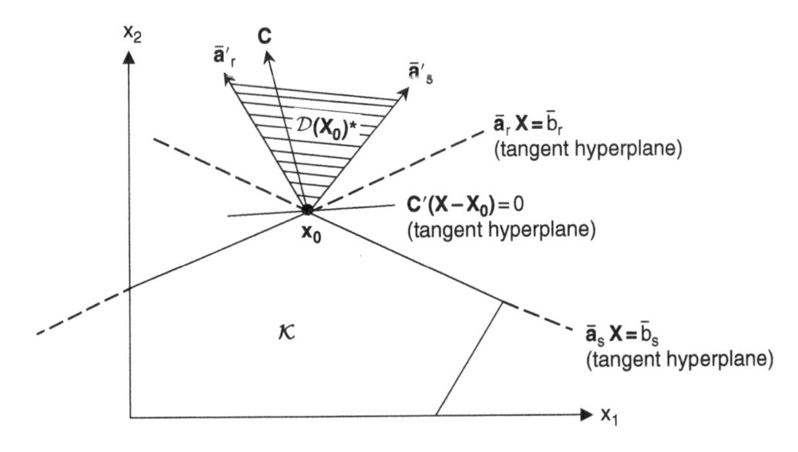

Figure 12.3 $C \in \mathcal{D}(X_0)^*$.

Theorem 12.3 Minkowski-Farkas Theorem. A necessary and sufficient condition for the n-component vector $\boldsymbol{V}$ to lie within the finite cone spanned by the columns of the $(m \times n)$ matrix $\boldsymbol{B}$ is that $\boldsymbol{V}'\boldsymbol{Y} \leq 0$ for all $\boldsymbol{Y}$ satisfying $\boldsymbol{B}'\boldsymbol{Y} \leq \boldsymbol{O}$, i.e. there exists an n-component vector $\boldsymbol{\lambda} \geq \boldsymbol{O}$ such that $\boldsymbol{B\lambda} = \boldsymbol{V}$ if and only if $\boldsymbol{V}'\boldsymbol{Y} \leq 0$ for all $\boldsymbol{Y}$ satisfying $\boldsymbol{B}'\boldsymbol{Y} \leq \boldsymbol{O}$.

Proof. (sufficiency) if $\boldsymbol{B\lambda} = \boldsymbol{V}$, $\boldsymbol{\lambda} \geq \boldsymbol{O}$, then $\boldsymbol{\lambda}'\boldsymbol{B}' = \boldsymbol{V}'$ and $\boldsymbol{\lambda}'\boldsymbol{B}'\boldsymbol{Y} = \boldsymbol{V}'\boldsymbol{Y} \leq 0$ for all $\boldsymbol{Y}$ for which $\boldsymbol{B}'\boldsymbol{Y} \leq \boldsymbol{O}$. (necessity) if $\boldsymbol{V}'\boldsymbol{Y} \leq 0$ for each $\boldsymbol{Y}$ satisfying $\boldsymbol{B}'\boldsymbol{Y} \leq \boldsymbol{O}$, then the (primal) linear programming problem

$$\max f(\boldsymbol{Y}) = \boldsymbol{V}'\boldsymbol{Y} \text{ subject to } \boldsymbol{B}'\boldsymbol{Y} \leq \boldsymbol{O}, \boldsymbol{Y} \text{ unrestricted}$$

has as its optimal solution $\boldsymbol{Y} = \boldsymbol{O}$. Hence, the dual problem

$$\min g(\boldsymbol{\lambda}) = \boldsymbol{O}'\boldsymbol{\lambda} \text{ subject to } \boldsymbol{B\lambda} = \boldsymbol{V}, \boldsymbol{\lambda} \geq \boldsymbol{O}$$

also has an optimal solution by Duality Theorem 6.4. Hence there exists at least one $\boldsymbol{\lambda} \geq \boldsymbol{O}$ satisfying $\boldsymbol{B\lambda} = \boldsymbol{V}$. Q.E.D.

Returning now to our previous question, in terms of the notation used above, if: $\boldsymbol{C} = \boldsymbol{V}$; the vectors $\bar{a}_i'$, $i \in \mathcal{J}$, are taken to be the columns of $\boldsymbol{B}$ (i.e. $\boldsymbol{B} = \bar{\boldsymbol{A}}'$); and the $\bar{u}_i^0$s, $i \in \mathcal{J}$, are the elements of $\boldsymbol{\lambda} \geq \boldsymbol{O}$, then a necessary and sufficient condition for $\boldsymbol{C}$ to lie within the finite cone generated by the vectors $\bar{a}_i'$, is that $\boldsymbol{C}'\boldsymbol{h} \leq 0$ for all $\boldsymbol{h}$ satisfying $\bar{\boldsymbol{a}}_i\boldsymbol{h} \leq 0$, $i \in \mathcal{J}$. Hence, there exist real numbers $\bar{u}_i^0 \geq 0$ such that (12.2) holds.

We may incorporate all of the constraints (active or otherwise) $\bar{b}_i - \bar{\boldsymbol{a}}_i\boldsymbol{X} = 0, i = 1,\ldots,m+n$, into our discussion by defining the scalar $\bar{u}_i^0$ as zero whenever $\bar{b}_i - \bar{\boldsymbol{a}}_i\boldsymbol{X} > 0$ or $i \notin \mathcal{J}$. Then (12.2) is equivalent to the system

$$\left.\begin{array}{l} \boldsymbol{C} - \sum_{i=1}^{m+n} \bar{u}_i^0 \bar{\boldsymbol{a}}_i' = \boldsymbol{O} \\ \bar{u}_i^0\left(\bar{b}_i - \bar{\boldsymbol{a}}_i\boldsymbol{X}_0\right) = 0 \\ \bar{b}_i - \bar{\boldsymbol{a}}_i\boldsymbol{X}_0 \geq 0 \\ \bar{u}_i^0 \geq 0 \\ \boldsymbol{X}_0 \text{ unrestricted} \end{array} \quad i = 1,\ldots,m+n \right\} \text{ or } \left\{\begin{array}{l} \boldsymbol{C} - \bar{\boldsymbol{A}}'\bar{\boldsymbol{U}}_0 = \boldsymbol{O} \\ \bar{\boldsymbol{U}}_0'\left(\bar{\boldsymbol{b}} - \bar{\boldsymbol{A}}\boldsymbol{X}_0\right) = 0 \\ \bar{\boldsymbol{b}} - \bar{\boldsymbol{A}}\boldsymbol{X}_0 \geq \boldsymbol{O} \\ \bar{\boldsymbol{U}}_o \geq \boldsymbol{O} \\ \boldsymbol{X}_0 \text{ unrestricted} \end{array}\right. \tag{12.3}$$

where $\bar{\boldsymbol{U}}_0' = \left(\bar{u}_1^0,\ldots,\bar{u}_{m+n}^0\right)$. Note that $\bar{u}_i^0\left(\bar{b}_i - \bar{\boldsymbol{a}}_i\boldsymbol{X}\right) = 0$ for all i values since, if $\bar{b}_i - \bar{\boldsymbol{a}}_i\boldsymbol{X} > 0$, then $\bar{u}_1^0 = 0$; while if $\bar{b}_i - \bar{\boldsymbol{a}}_i\boldsymbol{X} = 0$, then $\bar{u}_1^0 \geq 0$.

When we formed the structural constraint inequality $\bar{b}_i - \bar{\boldsymbol{A}}\boldsymbol{X} \geq \boldsymbol{O}$, the n nonnegativity conditions $\boldsymbol{X} \geq \boldsymbol{O}$ were treated as structural constraints. That is, $x_j \geq 0$ was converted to $\bar{b}_{m+j} - \bar{\boldsymbol{a}}_{m+j}\boldsymbol{X} = \boldsymbol{e}_j'\boldsymbol{X} \geq \boldsymbol{0}, j = 1,\ldots,n$. However, if the nonnegativity conditions are not written in this fashion, but appear explicitly as

$X \geq O$, then (12.3) may be rewritten in an equivalent form provided by Theorem 12.4.

Theorem 12.4 Karush-Kuhn-Tucker Theorem for Linear Programs (A Necessary and Sufficient Condition). The point $X_0 \in \mathcal{E}^n$ solves the primal problem

$$\max f(X) = C'X \text{ subject to } b - AX \geq O, X \geq O,$$

if and only if

$$\left.\begin{array}{l}(\text{a})\, C - A'U_0 \leq O\\ (\text{b})\, X_0'(C - A'U_0) = 0\\ (\text{c})\, U_0'(b - AX_0) = 0\\ (\text{d})\, b - AX_0 \geq O\\ (\text{e})\, U_o \geq O\\ (\text{f})\, X_0 \geq O\end{array}\right\} \text{or} \left\{\begin{array}{l}(\text{a.1})\, C - \sum_{i=1}^{m} u_i^0 a_i' \leq O \text{ or } c_j - \sum_{i=1}^{m} u_i^0 a_{ij} \leq 0, j = 1, \ldots, n\\ (\text{b.1}) \sum_{j=1}^{n} x_j^0 \left(c_j - \sum_{i=1}^{m} u_i^0 a_{ij}\right) = 0\\ (\text{c.1}) \sum_{i=1}^{m} u_i^0 (b_i - a_i X_0) = 0\\ \left.\begin{array}{l}(\text{d.1})\, b_i - a_i X_0 \geq 0\\ (\text{e.1})\, u_i^0 \geq 0\end{array}\right\} i = 1, \ldots, m\\ (\text{f.1})\, x_j^0 \geq 0, j = 1, \ldots, n,\end{array}\right. \tag{12.4}$$

where U_0 is an $(m \times 1)$ vector with components $u_1^0, \ldots u_m^0$.[2]

Proof. (necessity) to solve the primal problem, let us employ the technique of Lagrange. In this regard, we first convert each of the inequality constraints to an equality by subtracting, from its left-hand side, a nonnegative slack variable. (Actually, since we require that each slack variable be nonnegative, its square will be subtracted from the left-hand side to ensure its nonnegativity.) That is, $b_i - a_i X \geq 0$ is transformed to $b_i - a_i X - x_{n+i}^2 = 0, i = 1, \ldots, m$, while $x_j \geq 0$ is converted to $x_j - x_{n+m+j}^2 = 0, j = 1, \ldots n$, where $x_{n+i}^2, i = 1, \ldots, m; x_{n+m+j}^2, j = 1, \ldots, n$, are all squares of slack variables. Hence, the primal Lagrangian appears as

$$\begin{aligned} &L(x_1, \ldots, x_n, x_{n+1}, \ldots, x_{n+m}, x_{n+m+1}, \ldots, x_{2n+m}, u_1, \ldots, u_m, v_1, \ldots, v_n)\\ &\quad = \sum_{j=1}^{n} c_j x_j + \sum_{i=1}^{m} u_i \left(b_i - a_i X - x_{n+i}^2\right)\\ &\quad + \sum_{j=1}^{n} v_j \left(x_j - x_{n+m+j}^2\right), \end{aligned}$$

2 Here (12.4) represents the Karush-Kuhn-Tucker conditions for a maximum of a function subject to inequality constraints. On all this, see Panik (1976).

where $u_1, \ldots, u_m, v_1, \ldots, v_n$ are Lagrange multipliers. In matrix terms, the primal Lagrangian may be written as

$$L(\boldsymbol{X},\boldsymbol{X}_{s1},\boldsymbol{X}_{s2},\boldsymbol{U},\boldsymbol{V}) = \boldsymbol{C}'\boldsymbol{X} + \boldsymbol{U}'\left(\boldsymbol{b}-\boldsymbol{A}\boldsymbol{X}-\boldsymbol{X}_{s1}^2\right) + \boldsymbol{V}'\left(\boldsymbol{X}-\boldsymbol{X}_{s2}^2\right),$$

where

$$\boldsymbol{X}_{s1} = \begin{bmatrix} x_{n+1} \\ \vdots \\ x_{n+m} \end{bmatrix}, \boldsymbol{X}_{s1}^2 = \begin{bmatrix} x_{n+1}^2 \\ \vdots \\ x_{n+m}^2 \end{bmatrix}, \boldsymbol{X}_{s2} = \begin{bmatrix} x_{n+m+1} \\ \vdots \\ x_{2n+m} \end{bmatrix}, \boldsymbol{X}_{s2}^2 = \begin{bmatrix} x_{n+m+1}^2 \\ \vdots \\ x_{2n+m}^2 \end{bmatrix}, \boldsymbol{U} = \begin{bmatrix} u_1 \\ \vdots \\ u_m \end{bmatrix}, \boldsymbol{V} = \begin{bmatrix} v_1 \\ \vdots \\ v_n \end{bmatrix}.$$

Here $\boldsymbol{X}_{s1}, \boldsymbol{X}_{s2}^2$, and $\boldsymbol{U}$ are of order $(m \times 1)$ while $\boldsymbol{X}_{s2}, \boldsymbol{X}_{s2}^2$, and $\boldsymbol{V}$ have order $(n \times 1)$. Then

$$\begin{aligned}
&\text{(a)} \quad \nabla_{\boldsymbol{X}} L = \boldsymbol{C} - \boldsymbol{A}'\boldsymbol{U} + \boldsymbol{V} = \boldsymbol{O} \\
&\text{(b)} \quad \nabla_{\boldsymbol{X}_{s1}} L = -2\boldsymbol{U}'\boldsymbol{X}_{s1} = 0 \\
&\text{(c)} \quad \nabla_{\boldsymbol{X}_{s2}} L = -2\boldsymbol{V}'\boldsymbol{X}_{s2} = 0 \\
&\text{(d)} \quad \nabla_{\boldsymbol{U}} L = \boldsymbol{b} - \boldsymbol{A}\boldsymbol{X} - \boldsymbol{X}_{s1}^2 = \boldsymbol{O} \\
&\text{(e)} \quad \nabla_{\boldsymbol{V}} L = \boldsymbol{X} - \boldsymbol{X}_{s2}^2 = \boldsymbol{O}.
\end{aligned} \tag{12.5}$$

Let us first transform (12.5b, c) to $\boldsymbol{U}'\boldsymbol{X}_{s1}^2 = \boldsymbol{O}, \left(\boldsymbol{X}_{s2}^2\right)'\boldsymbol{V} = \boldsymbol{O}$, respectively. Then, from $\boldsymbol{U}'\boldsymbol{X}_{s1}^2 = \boldsymbol{O}$ and (12.5d) we obtain, at $\boldsymbol{X}_0$ and $\boldsymbol{U}_0, \boldsymbol{U}_0'(\boldsymbol{b}-\boldsymbol{A}\boldsymbol{X}_0) = 0$. That is, if the constraint $b_i - \boldsymbol{a}_i\boldsymbol{X} \geq 0$ is not binding at $\boldsymbol{X}_0$, then $x_{n+i}^0 > 0, u_i^0 = 0$; while if it is binding, $x_{n+i}^0 = 0, u_i^0 \geq 0, i = 1,\ldots,m$. Hence, at least one of each pair $\left(u_i^0, b_i - \boldsymbol{a}_i\boldsymbol{X}_0\right)$ vanishes, thus guaranteeing that $\sum_{i=1}^m u_i^0(b_i - \boldsymbol{a}_i\boldsymbol{X}_0) = 0$. Next, combining $\boldsymbol{X}_{s2}'\boldsymbol{V} = 0$ and (12.5e) yields $\boldsymbol{X}'\boldsymbol{V} = 0$. From (12.5a) we have $-\boldsymbol{V} = \boldsymbol{C} - \boldsymbol{A}'\boldsymbol{U}$. Then, at $\boldsymbol{X}_0$ and $\boldsymbol{U}_0$, these latter two expressions yield $\boldsymbol{X}_0'(\boldsymbol{C} - \boldsymbol{A}'\boldsymbol{U}_0) = 0$ or $\sum_{j=1}^n x_j^0\left(c_j - \sum_{i=1}^m u_i^0 a_{ij}\right) = 0$. In this regard, either $x_j^0 = 0$, allowing c_j to be less than $\sum_{i=1}^m u_i^0 a_{ij}$ or $c_j - \sum_{i=1}^m u_i^0 a_{ij} = 0$, in which case x_j^0 may be positive. To see this, let us assume, without loss of generality, that the first $k < m$ structural constraints are binding at $\boldsymbol{X}_0$. Then $b_i - \boldsymbol{a}_i\boldsymbol{X}_0 = 0\left(x_{n+i}^0 = 0, u_i^0 \geq 0\right), i = 1,\ldots,k$, while $b_i - \boldsymbol{a}_i\boldsymbol{X}_0 > 0\left(x_{n+i}^0 > 0, u_i^0 = 0\right), i = k+1,\ldots,m$. Now, if the primal problem possesses a solution at $\boldsymbol{X}_0$, $\boldsymbol{C}'\boldsymbol{h} \leq 0$ for those vectors $\boldsymbol{h}$ that do not violate $\boldsymbol{a}_i\boldsymbol{h} = 0$, $i = 1, \ldots, k$. If we multiply each equality $\boldsymbol{a}_i\boldsymbol{h} = 0$ by some constant u_i^0 and form the sum $\boldsymbol{C}'\boldsymbol{h} - \sum_{i=1}^k u_i^0\boldsymbol{a}_i\boldsymbol{h} \leq 0$, we ultimately obtain

$$\left(\boldsymbol{C}' - \sum_{i=1}^m u_i^0\boldsymbol{a}_i'\right)'\boldsymbol{h} \leq 0. \tag{12.6}$$

Let us further assume that at least one component of X_0, say x_l^0, is strictly positive (and thus $x_{n+m+l}^0 > 0, v_l^0 = 0$). Then there exists a set of sufficiently small positive and negative deviations $h_l = \delta, -\delta$ such that, for $h_l = \delta > 0$, (12.6) becomes

$$c_l - \sum_{i=1}^{k} u_i^0 a_{il} \leq 0 \tag{12.7}$$

while for $h_l = -\delta$, (12.6) may be written as

$$c_l - \sum_{i=1}^{k} u_i^0 a_{il} \geq 0 \tag{12.8}$$

whence, upon combining (12.7), (12.8),

$$c_l - \sum_{i=1}^{k} u_i^0 a_{il} = 0.$$

If $x_l^0 = 0$ (so that $x_{n+m+l}^0 = 0, v_l^0 > 0$), h_l can only be positive. Hence $h_l = \delta > 0$ in (12.6) yields

$$c_l - \sum_{i=1}^{k} u_i^0 \boldsymbol{a}_{il} \leq 0.$$

In general then,

$$\boldsymbol{C} - \sum_{i=1}^{m} u_i^0 \boldsymbol{a}_i' \leq \boldsymbol{O}.$$

(sufficiency) For $\boldsymbol{X} \in \mathcal{K}$, let us express the Lagrangian of f as

$$\boldsymbol{C}'\boldsymbol{X} \leq \boldsymbol{C}'\boldsymbol{X} + \boldsymbol{U}_0'(\boldsymbol{b} - \boldsymbol{A}\boldsymbol{X}) + \boldsymbol{V}_0'\boldsymbol{X} = L(\boldsymbol{X}, \boldsymbol{U}_0, \boldsymbol{V}_0).$$

With $L(\boldsymbol{X}, \boldsymbol{U}_0, \boldsymbol{V}_0)$ concave in $\boldsymbol{X}$ over $\mathcal{K}$, the linear approximation to L at $\boldsymbol{X}_0$ is

$$\begin{aligned} L(\boldsymbol{X}, \boldsymbol{U}_0, \boldsymbol{V}_0) &\leq L(\boldsymbol{X}, \boldsymbol{U}_0, \boldsymbol{V}_0) + \nabla_{\boldsymbol{X}} L(\boldsymbol{X}, \boldsymbol{U}_0, \boldsymbol{V}_0)'(\boldsymbol{X} - \boldsymbol{X}_0) \\ &= \boldsymbol{C}'\boldsymbol{X}_0 + (\boldsymbol{C} - \boldsymbol{A}'\boldsymbol{U}_0 + \boldsymbol{V}_0)'(\boldsymbol{X} - \boldsymbol{X}_0) \\ &= \boldsymbol{C}'\boldsymbol{X}_0.^3 \end{aligned} \tag{12.9}$$

Hence $\boldsymbol{C}'\boldsymbol{X}_0 \geq L(\boldsymbol{X}, \boldsymbol{U}_0, \boldsymbol{V}_0) \geq \boldsymbol{C}'\boldsymbol{X}$, $\boldsymbol{X} \in \mathcal{K}$, and thus $\boldsymbol{X}_0$ solves the primal problem. Q.E.D.

3 Since $L(\boldsymbol{X}, \boldsymbol{U}_0, \boldsymbol{V}_0)$ is linear in $\boldsymbol{X}$, it is concave over $\mathcal{K}$. Moreover, with L concave, it lies everywhere on or beneath its tangent hyperplane at $\boldsymbol{X}_0$ as (12.9) indicates.

We note briefly that if the primal problem appears as

$$\max f(X) = C'X \quad \text{s.t.}$$
$$AX \geq b, X \geq O,$$

Then (12.4) becomes

$$\begin{aligned} &(a)\ C - A'U_0 \geq O \\ &(b)\ X_0'(C - A'U_0) = 0 \\ &(c)\ U_0'(b - AX_0) = 0 \\ &(d)\ b - AX_0 \geq O \\ &(e)\ U_o \geq O \\ &(f)\ X_0 \geq O. \end{aligned} \tag{12.4.1}$$

We now turn to a set of important corollaries to the preceding theorem (Corollaries 12.1, 12.2, and 12.3).

Corollary 12.1 Let (12.4) hold so that the primal maximum problem possesses an optimal solution. Then the dual problem

$$\min g(U) = b'U \text{ subject to } A'U \geq C, U \geq O$$

also has an optimal solution U_0 with $C'X_0 = b'U_0$.

Proof. (From 12.4a, e) $A'U_0 \geq C$ and $U_0 \geq O$ so that U_0 is a feasible solution to the dual problem. With $AX_0 \leq b$, we have $X_0'A' \leq b'$ or $X_0'A'U_0 \leq b'U$. Since $A'U \geq C, X_0'C = C'X_0 \leq b'U$ for U feasible. Moreover, from (12.4c, d), $X_0'C - X_0'A'U_0 = U_0'b - U_0'AX_0$ or $C'X_0 = b'U_0$ since $X_0'A'U_0 = U_0'AX_0$. Hence U_0 must be an optimal solution to the dual problem. Q.E.D.

Next,

Corollary 12.2 If (12.4) holds so that the primal maximum and dual minimum problems have optimal solutions $(X_0', (X_s)_0'), (U_0', (U_s)_0')$, respectively, then the (weak) complementary slackness condition $(U_s)_0'X_0 + U_0'(X_s)_0 = 0$ holds.

Proof. If X_s represents an $(m \times 1)$ vector of nonnegative primal slack variables and U_s depicts an $(n \times 1)$ vector of nonnegative dual surplus variables, then (12.4b, c) become $X_0'(-U_s)_0 = U_0'X_s$or $(U_s)_0'X_0 + U_0'(X_s)_0 = 0$. Alternatively, (12.5a) reveals that $(U_s)_0 = -V_0$ so that the (weak) complementary slackness condition may be rewritten in terms of the primal variables and Lagrange multipliers as $U_0'(X_s)_0 - V_0'X_0 = 0$. Q.E.D.

Finally,

Corollary 12.3 The optimal value of the *i*th dual variable u_i^0 is generally a measure of the amount by which the optimal value of the primal objective function changes given a small (incremental) change in the amount of the *i*th requirement b_i, with all other requirements b_l, $l \neq i$, $i = 1, \ldots, m$, remain constant, i.e.

$$\frac{\partial f^0}{\partial b_i} = u_i^0. \tag{12.10}$$

Proof. Let $\boldsymbol{X}_0$, $\boldsymbol{U}_0$ be feasible solutions to the primal maximum and dual minimum problems, respectively, with $f^0 = \boldsymbol{C}'\boldsymbol{X}_0$, $g^0 = \boldsymbol{b}'\boldsymbol{U}_0$. If (12.4) holds, then (from Corollary 12.1) $f^0 = g^0$, i.e. $\boldsymbol{X}_0$ and $\boldsymbol{U}_0$ are optimal solutions and thus $f^0 = \sum_{i=1}^{m} b_i u_i^0$ so that $\frac{\partial f^0}{\partial b_i} = u_i^0$. Q. E. D.

So if the right-hand side b_i of the *i*th primal structural constraint were changed by a sufficiently small amount ε_i, then the corresponding change in the optimal value of *f*, f^0, would be $df^0 = u_i^0 \varepsilon_i$. Moreover, since $df^0 = \sum_{i=1}^{m} \frac{\partial f^0}{\partial b_i} db_i = \sum_{i=1}^{m} u_i^0 db_i$, we may determine the amount by which the optimal value of the primal objective function changes given small (marginal) changes in the right-hand sides of any specific subset of structural constraints. For instance, let us assume, without loss of generality, that the right-hand sides of the first $t < m$ primal structural constraints are increased by sufficiently small increments $\varepsilon_1, \ldots, \varepsilon_t$, respectively. Then the optimal value of the primal objective function would be increased by $df^0 = \sum_{i=1}^{t} u_i^0 \varepsilon_i$.

Throughout the preceding discussion the reader may have been wondering why the qualification "generally" was included in the statement of Corollary 12.3. The fact of the matter is that there exist values of, say, b_k, for which $\partial f^0 / \partial b_k$ is not defined, i.e. points where $\partial f^0 / \partial b_k$ is discontinuous so that (12.10) does not hold. To see this, let us express *f* in terms of b_k, as

$$f(b_k) = \max_{X} \left\{ f(X) \middle| \boldsymbol{a}_i \boldsymbol{X}^i \leq b_i^0, i \neq k, i = 1, \ldots, m;\ \boldsymbol{a}_k \boldsymbol{X} \leq b_k, \boldsymbol{X} \geq \boldsymbol{O} \right\},$$

where the b_i^0s depict fixed levels of the requirements $b_i^0, i \neq k, i = 1, \ldots, m$.

If in Figure 12.4 we take area OABCG to be the initial feasible region when $b_k = b_k'$ (here the line segment CG consists of all feasible points $\boldsymbol{X}$ that satisfy $\boldsymbol{a}_k \boldsymbol{X} = b_k'$, then increasing the *k*th requirement from b_k' to b_k'' changes the optimal extreme point from C to D. The result of this increase in b_k is a concomitant increase in f^0 proportionate to the distance $E_k = b_k'' - b_k'$, where the constant of proportionality is u_k^0. Thus, over this range of b_k values, $\partial f^0 / \partial b_k$ exists

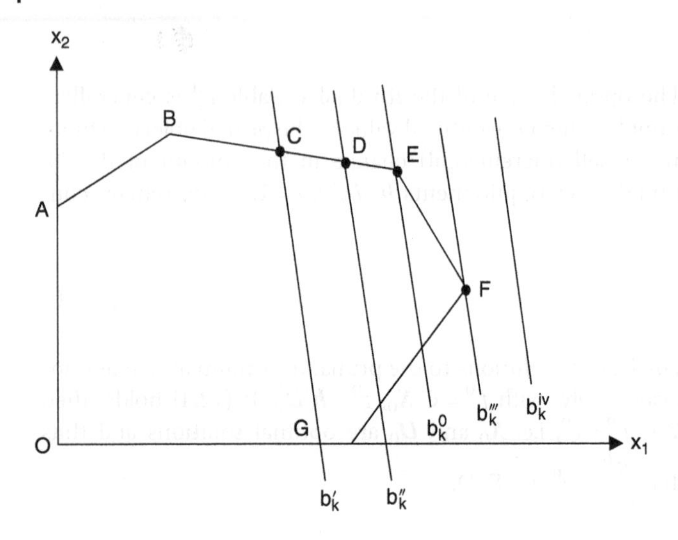

Figure 12.4 Increasing the requirements of b_k.

(is continuous). If we again augment the level of b_k by the same amount as before, this time shifting the kth structural constraint so that it passes through point E, $\partial f^0/\partial b_k$ becomes discontinuous since now the maximal point moves along a more steeply sloping portion (segment EF) of the boundary of the feasible region so that any further equal parallel shift in the kth structural constraint would produce a much larger increment in the optimal value of f than before. Thus, each time the constraint line $\boldsymbol{a}_k\boldsymbol{X} = b_k$ passes through an extreme point of the feasible region, there occurs a discontinuity in $\partial f^0/\partial b_k$.

To relate this argument directly to $f(b_k)$ above, let us describe the behavior of $f(b_k)$ as b_k assumes the values $b_k', b_k'', b_k^0, b_k'''$ and b_k^{IV}. In Figure 12.5 the piecewise continuous curve $f(b_k)$ indicates the effect on $f(b_k)$ precipitated by changes in b_k. (Note that each time the constraint $\boldsymbol{a}_k\boldsymbol{X} = b_k$ passes through an extreme point of the feasible region in Figure 12.4, the $f(b_k)$ curve assumes a kink, i.e. a point of discontinuity in its derivative. Note also that for a requirement level of $b_k > b_k'''$ [say b_k^{iv}], f cannot increase any further so that beyond F', $f(b_k)$ is horizontal) As far as the dual objective function $g(\boldsymbol{u}) = \boldsymbol{b}'\boldsymbol{u}$ is concerned, g may be expressed as $g(b_k) = b_k u_k +$ constant since $\sum_{i \neq k}^{m} b_i u_i$ is independent of b_k. For $b_k = b_k^0$ (b_k^0 optimal), $f(b_k^0) = g(b_k^0)$. Moreover, since the dual structural constraints are unaffected by changes in b_k, dual feasibility is preserved for variations in b_k away from b_k^0 so that, in general, for any pair of feasible solutions to the primal and dual problems, $f(b_k) \leq g(b_k)$, i.e. $g(b_k)$ is tangent to $f(b_k)$ for $b_k = b_k^0$ (point E' in Figure 12.5) while $f(b_k^0 + \varepsilon_k) < g(b_k^0 + \varepsilon_k)$ for $\varepsilon_k \left(\gtrless 0\right)$ sufficiently small.

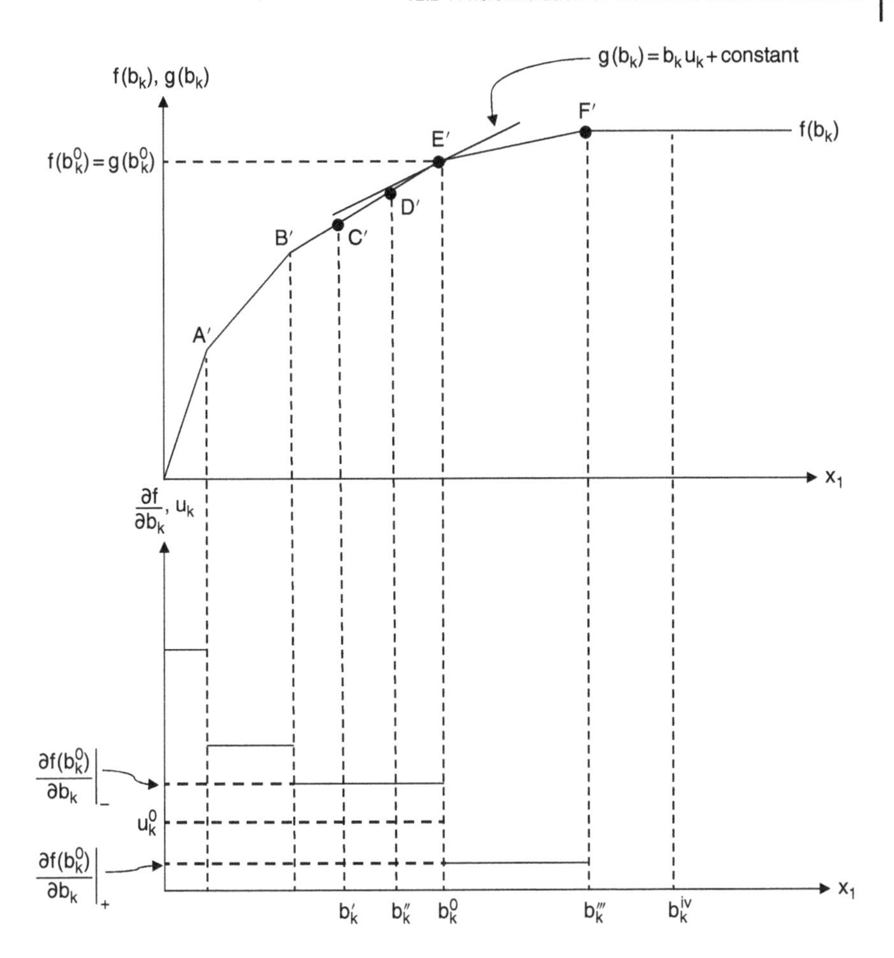

Figure 12.5 Tracking $f(b_k)$ as b_k increases.

In this regard, for $\varepsilon_k \to 0+, \partial g\left(b_k^0\right)/\partial b_k = u_k^0 > \partial f\left(b_k^0\right)/\partial b_k\big|_+$; while for $\varepsilon_k \to 0-$, $u_k^0 < \partial f\left(b_k^0\right)/\partial b_k\big|_-$, whence $\partial f\left(b_k^0\right)/\partial b_k\big|_+ < u_k^0 < \partial f\left(b_k^0\right)/\partial b_k\big|_-$. So while both the left- and right-hand derivatives of $f(b_k)$ exist at $b_k^0, \partial f b_k^0/\partial b_k$ itself does not exist since $\partial f\left(b_k^0\right)/\partial b_k\big|_+ \neq \partial f\left(b_k^0\right)/\partial b_k\big|_-$ i.e. $\partial f/\partial b_k$ possesses a finite discontinuity at b_k^0. In fact, $\partial f/\partial b_k$ is discontinuous at each of the points, A′, B′, E′ and F′ in Figure 12.5. For $b_k = b_k''$, it is evident that $\partial f\left(b_k''\right)/\partial b_k\big|_+ = u_k'' = \partial f\left(b_k''\right)/\partial b_k\big|_-$ since $f(b_k)$, $g(b_k)$ coincide all along the segment B′ E′ of Figure 12.5. In this instance, we may unequivocally write $\partial f\left(b_k''\right)/\partial b_k = u_k^0$. Moreover, for $b_k = b_k^{iv}$ (see Figures 12.4 and 12.5), $\partial f\left(b_k^{iv}\right)/\partial b_k = u_k^{iv} = 0$ since the constraint

$\boldsymbol{a}_k X = b_k^{iv}$ is **superfluous**, i.e. it does not form any part of the boundary of the feasible region. In general, then, $\partial f/\partial b_k|_+ \leq u_k \leq \partial f/\partial b_k|_-$.

To formalize the preceding discussion, we state Theorem 12.5.

Theorem 12.5 Let the requirements b_i be held fixed at $b_i^0, i = 1,\ldots,m$. If $f(b_k) = \max_X \{f(X)|\boldsymbol{a}_i X \leq b_i^0, i \neq k, i = 1,\ldots,m; \boldsymbol{a}_k X \leq b_k, X \geq O\}$ possesses finite right- and left-hand derivatives at an optimal point $b_k = b_k^0$, then u_k^0, the optimal value of the corresponding dual variable, lies between these derivatives, i.e.

$$\left.\frac{\partial f\left(b_k^0\right)}{\partial b_k}\right|_+ \leq u_k^0 \leq \left.\frac{\partial f\left(b_k^0\right)}{\partial b_k}\right|_- \quad \text{for all } k.$$

Moreover, if $\partial f\left(b_k^0\right)/\partial b_k\big|_+ = \partial f\left(b_k^0\right)/\partial b_k\big|_-$ so that $\partial f\left(b_k^0\right)/\partial b_k$ exists, then

$$\frac{\partial f\left(b_k^0\right)}{\partial b_k} = u_k^0.$$

Proof. Since $g(b_k) = b_k u_k + \text{constant}$ is linear and $g\left(b_k^0\right) = f\left(b_k^0\right) = \max_X \{f(X)|\boldsymbol{a}_i X \leq b_i^0, i \neq k, i = 1,\ldots,m; \boldsymbol{a}_k X \leq b_k^0, X \geq O\}$ for b_k^0 optimal (see, for instance, point E of Figure 12.4) and $f\left(b_k^0 + \varepsilon_k\right) < g\left(b_k^0 + \varepsilon_k\right), \varepsilon_k$ unrestricted in sign and sufficiently small, it follows that, with $\varepsilon_k > 0$,

$$\frac{g\left(b_k^0 + \varepsilon_k\right) - g\left(b_k^0\right)}{\varepsilon_k} = u_k^0 \geq \frac{f\left(b_k^0 + \varepsilon_k\right) - f\left(b_k^0\right)}{\varepsilon_k}$$

and thus

$$u_k^0 \geq \lim_{\varepsilon_k \to 0+} \frac{f\left(b_k^0 + \varepsilon_k\right) - f\left(b_k^0\right)}{\varepsilon_k} = \left.\frac{\partial f\left(b_k^0\right)}{\partial b_k}\right|_+;$$

while for $\varepsilon_k < 0$,

$$u_k^0 \leq \frac{f\left(b_k^0 + \varepsilon_k\right) - f\left(b_k^0\right)}{\varepsilon_k}$$

so that

$$u_k^0 \leq \lim_{\varepsilon_k \to 0+} \frac{f\left(b_k^0 + \varepsilon_k\right) - f\left(b_k^0\right)}{\varepsilon_k} = \left.\frac{\partial f\left(b_k^0\right)}{\partial b_k}\right|_-.$$

Hence $\partial f\left(b_k^0\right)/\partial b_k\big|_+ \leq u_k^0 \leq \partial f\left(b_k^0\right)/\partial b_k\big|_-$ for all k. The *moreover* portion of the theorem holds by definition. Q.E.D.

To summarize: if both the left- and right-hand derivatives $\partial f^o/\partial b_i|_-$, $\partial f^o/\partial b_i|_+$ exist and their common value is u_i^0, then $\partial f^o/\partial b_i = u_i^0$; otherwise $\partial f^o/\partial b_i|_+ \leq u_i^0 \leq \partial f^o/\partial b_i|_-, i = 1,\ldots,m$.

12.3 Lagrangian Saddle Points (Belinski and Baumol 1968; Williams 1963)

If we structure the primal-dual pair of problems as

PRIMAL PROBLEM	DUAL PROBLEM
$\max f(X) = C'X$ s.t.	$\min g(U) = b'U$ s.t.
$AX \le b, X \ge O$	$A'U \ge C, U \ge O$

then the Lagrangian associated with the primal problem is $L(X, U) = C'X + U'(b - AX)$ with $X \ge O$, $U \ge O$. And if we now express the dual problem as maximize $\{-b'U\}$ subject to $-A'U \le -C, U \ge O$, then the Lagrangian of the dual problem is $M(U, X) = -b'U + X'(-C + A'U)$, where again $U \ge O$, $X \ge O$. Moreover, since $M(U, X) = -b'U + X'C + X'A'U = -C'X - U'b + U'AX = -L(X, U)$, we see that the Lagrangian of the dual problem is actually the negative of the Lagrangian of the primal problem. Alternatively, if $x_j \ge 0$ is converted to $\bar{b}_{m+j} - \bar{a}_{m+j}X = e_j'X \ge 0, j = 1, \ldots, n$, then the revised primal problem appears as

$$\max f(X) = C'X \quad \text{s.t.}$$
$$\bar{A}X \le \bar{b}, X \quad \text{unrestricted,}$$

with associated Lagrangian $L(X, \bar{U}) = C'X + \bar{U}'(\bar{b} - \bar{A}X)$, where $\bar{U}' = (u_1, \ldots, u_{n+m}) \ge O'$. As far as the dual of this revised problem is concerned, we seek to

$$\min g(\bar{U}) = \bar{b}'\bar{U} \quad \text{s.t.}$$
$$\bar{A}'\bar{U} = C, \bar{U} \ge O.$$

In this instance $M(\bar{U}, X) = -\bar{b}'\bar{U} + X'\left(-C + \bar{A}'\bar{U}\right) = -L(X, \bar{U})$ also.

We next turn to the notion of a saddle point of the Lagrangian. Specifically, a point $(X_0, U_o) \varepsilon\, \mathcal{E}^{n+m}$ is termed a **saddle point** of the Lagrangian $L(X, U) = C'X + U'(b - AX)$ if

$$L(X, U_0) \le L(X_0, U_0) \le L(X_0, U) \tag{12.11}$$

for all $X \ge O$, $U \ge O$. Alternatively, $(\bar{X}_0, \bar{U}_o) \varepsilon\, \mathcal{E}^{2n+m}$ is a saddle point of $L(X, \bar{U}) = C'X + \bar{U}'(\bar{b} - \bar{A}X)$ if

$$L(X, \bar{U}_0) \le L(X_0, \bar{U}_0) \le L(X_0, \bar{U}) \tag{12.12}$$

for all X unrestricted and $U \ge O$. What these definitions imply is that $L(X, U)$ $(L(X, \bar{U}))$ simultaneously attains a maximum with respect to X and a minimum with respect to $U(\bar{U})$. Hence (12.11), (12.12) appear, respectively, as

$$L(X_0, U_0) = \max_{X \ge O}\left\{\min_{U \ge O} L(X, U)\right\} = \min_{U \ge O}\left\{\max_{X \ge O} L(X, U)\right\}, \tag{12.11.1}$$

$$L(X_0,\bar{U}_0) = \max_{X}\left\{\min_{\bar{U}\geq O} L(X,\bar{U})\right\} = \min_{\bar{U}\geq O}\left\{\max_{X} L(X,\bar{U})\right\}, X \text{ unrestricted.} \tag{12.12.1}$$

Under what conditions will $L(X,U)(L(X,\bar{U}))$ possess a saddle point at $(X_0,U_0)((X_0,\bar{U}_0))$? To answer these questions, we must look to the solution of what may be called the **saddle point problem**: To find a point (X_0, U_0) such that (12.11) holds for all $(X, U)\varepsilon\ \mathcal{E}^{n+m}, X \geq O, U \geq O$; or, to find a point $(X_0,\bar{U}_0)$ such that (12.12) holds for all $(X,\bar{U})\varepsilon\ \mathcal{E}^{2n+m}$, X unrestricted, $\bar{U} \geq O$.

In the light of (12.11.1), (12.12.1) we shall ultimately see that determining a saddle point of a Lagrangian corresponds to maxi-minimizing or mini-maximizing it. To this end, we look to Theorem 12.6.

Theorem 12.6 (A Necessary and Sufficient Condition). Let $L(X,\bar{U})$ be defined for all $(X,\bar{U})\varepsilon\ \mathcal{E}^{2n+m}$, where X is unrestricted and $\bar{U} \geq O$. Then $L(X,\bar{U})$ has a saddle point at $(X_0,\bar{U}_0)$ if and only if

(a) $L(X,\bar{U}_0)$ attains a maximum of X_0,
(b) $\bar{b}-\bar{A}X_0 \geq O$, (12.13)
(c) $\bar{U}'_0(\bar{b}-\bar{A}X_0) = 0$.

Proof. (necessity) The left-hand inequality of (12.12) is equivalent to (12.13a) while its right-hand inequality implies that $C'X_0 + \bar{U}(\bar{b}-\bar{A}X_0) \geq C'X_0 + \bar{U}_0'(\bar{b}-\bar{A}X_0)$ or $(\bar{U}-\bar{U}_0)'(\bar{b}-\bar{A}X_0) \geq 0$. If $\bar{b}-\bar{A}X_0 \leq O$, a $\bar{U}$ may be chosen sufficiently large so that $\bar{U}-\bar{U}_0 \geq O$ or $(\bar{U}-\bar{U}_0)'(\bar{b}-\bar{A}X_0) \leq 0$, whence $L(X_0,\bar{U}) \geq L(X_0,\bar{U}_0)$ is violated. Hence, (12.13b) must hold. If $\bar{U} = O$, then $(\bar{U}-\bar{U}_0)(\bar{b}-\bar{A}X_0) = \bar{U}'_0(\bar{b}-\bar{A}X_0) \leq 0$. But since $\bar{U}_0 \geq O, \bar{b}-\bar{A}X_0 \geq O$ together imply $\bar{U}'_0(\bar{b}-\bar{A}X_0) \geq 0$, we see that (12.13c) must hold as well. So if $L(X, U)$ assumes a saddle point at $(X_0,\bar{U}_0)$, (12.13) holds.

(sufficiency) If X_0 maximizes $L(X,\bar{U}_0)$, then $L(X_0,\bar{U}_0) \geq L(X,\bar{U}_0)$. If $\bar{U}'_0(\bar{b}-\bar{A}X_0) = 0$, then $L(X_0,\bar{U}_0) = C'X_0$. Hence $L(X_0,\bar{U}) = C'X_0 + \bar{U}'(\bar{b}-\bar{A}X_0) = L(X_0,\bar{U}_0) + \bar{U}'(\bar{b}-\bar{A}X_0)$ or $L(X_0,\bar{U}) \geq L(X_0,\bar{U}_0)$ since $\bar{U}'(\bar{b}-\bar{A}X_0) \geq 0$ if $\bar{b}-\bar{A}X_0 \geq O$. So if (12.13) is satisfied, $L(X,\bar{U})$ has a saddle point at $(X_0,\bar{U}_0)$. Q.E.D.

Our discussion in this chapter has centered around the solution of two important types of problems – the primal linear programming problem and the saddle point problem. As we shall now see, there exists an important connection between them. Specifically, what does the attainment of a saddle point of the

Lagrangian of the primal problem imply about the existence of a solution to the primal problem? To answer this, we cite Theorem 12.7.

Theorem 12.7 (A Sufficient Condition). Let $L(X,\bar{U})$ be defined for all $(X,\bar{U})\varepsilon$ $\mathcal{E}^{2n+m}$, where X is unrestricted and $\bar{U} \geq O$. If $(X_0,\bar{U}_0)$ is a saddle point for $L(X,\bar{U})$, then X_0 solves the primal problem.

Proof. Let $(X_0,\bar{U}_0)$ be a saddle point for $L(X,\bar{U})$.Then by the previous theorem, (12.13) holds and thus $L(X_0,\bar{U}_0) \geq L(X,\bar{U}_0)$or $C'X_0 + \bar{U}_0'(\bar{b}-\bar{A}X_0)$ $\geq C'X + \bar{U}_0'(\bar{b}-\bar{A}X_0)$. Since $\bar{U}_0'(\bar{b}-\bar{A}X_0) = 0$ and $\bar{U}_0'(\bar{b}-\bar{A}X) \geq 0, C'X_0 \geq$ $C'X$ for all X. Q.E.D.

The importance of the developments in this section is that they set the stage for an analysis of the Karush-Kuhn-Tucker equivalence theorem. This theorem (Theorem 12.5) establishes the notion that the existence of an optimal solution to the primal problem is equivalent to the existence of a saddle point of its associated Lagrangian, i.e. solving the primal problem is equivalent to maxi-minimizing (mini-maximizing) its Lagrangian.

Theorem 12.8 Karush-Kuhn-Tucker Equivalence Theorem (A Necessary and Sufficient Condition). A vector X_0 is an optimal solution to the primal linear programming problem

$$\begin{aligned} &\max f(X) = C'X \quad \text{s.t.} \\ &\quad \bar{b}-\bar{A}X \geq O, X \text{ unrestricted,} \end{aligned}$$

if and only if there exists a vector $\bar{U} \geq O$ such that $(X_0,\bar{U}_0)$ is a saddle point of the Lagrangian $L(X,\bar{U})$, in which case

$$\begin{aligned} &C-\bar{A}'\bar{U}_0 = O \\ &\bar{U}_0'(\bar{b}-\bar{A}X_0) = 0 \\ &\bar{b}-\bar{A}X_0 \geq O. \end{aligned}$$

Proof. (sufficiency) Let $L(X_0,\bar{U}_0)$ solve the saddle point problem. Then by Theorem 12.6, system (12.13) obtains. In this regard, if X_0 maximizes $L(X,\bar{U}_0) = C'X + \bar{U}_0'(\bar{b}-\bar{A}X)$, then $\nabla_X L(X_0,\bar{U}_0) = C - A'U_0 = O$. Hence, this latter expression replaces (12.13a) so that (12.13) is equivalent to (12.3), i.e. if $L(X,\bar{U})$ has a saddle point at $(X_0,\bar{U}_0)$, then (12.3) holds. Thus X_0 solves the primal problem.

(necessity) Let X_0 solve the primal problem.

Then, by Corollary 12.1, the optimality of X_0 implies the existence of a $\bar{U}_0 \geq O$ such that $C'X_0 = \bar{b}'\bar{U}_0$. And since $\bar{U}_0$ represents an optimal solution to the dual problem, $\bar{U}_0'(\bar{b}-\bar{A}X_0)=0$. Since X_0 satisfies $\bar{b}-\bar{A}X_0 \geq O$, then, for any $\bar{U} \geq O$, it follows that $\bar{U}'(\bar{b}-\bar{A}X_0) \geq 0$. Thus

$$L(X_0,\bar{U}_0)=C'X_0+\bar{U}_0'(\bar{b}-\bar{A}X_0) \leq C'X_0+\bar{U}'(\bar{b}-\bar{A}X_0).$$

Hence the right-hand side of (12.12) is established. To verify its left-hand side, we note that since $C-\bar{A}'U_0=O$ is equivalent to (12.13a), $L(X_0,\bar{U}_0) \geq L(X,\bar{U}_0)$ by the sufficiency portion of Theorem 12.6. Q.E.D.

It is instructive to analyze this equivalence from an alternative viewpoint. We noted above that solving the primal linear programming problem amounts to maxi-minimizing (mini-maximizing) its associated Lagrangian. In this regard, if we express the primal problem as

$$\begin{aligned} &\max f(X)=C'X \quad \text{s.t.} \\ &\quad b-AX \geq O, X \geq O, \end{aligned}$$

then, if $L(X, U_0)$ has a maximum in the X-direction at X_0, it follows from (12.4a, b, f) that

$$\left.\begin{array}{l} \nabla_X L^0 \leq O \\ (\nabla_X L^0)'X_0=0 \\ X_0 \geq O \end{array}\right\} \text{ or } \left\{\begin{array}{l} C-A'U_0 \leq O \\ X_0'(C-A'U_0)=0 \\ X_0 \geq O, \end{array}\right.$$

while, if $L(X_0, U)$ attains a minimum in the U-direction at U_0, then, from (12.4c, d, e),

$$\left.\begin{array}{l} -\nabla_U L^0 \leq O \text{ or } \nabla_U L^0 \geq O \\ (\nabla_U L^0)'U_0=0 \\ U_0 \geq O \end{array}\right\} \text{ or } \left\{\begin{array}{l} b-A'X_0 \geq O \\ U_0'(b-A'X_0)=0 \\ U_0 \geq O. \end{array}\right.$$

Thus, X_0 is an optimal solution to the preceding primal problem if and only if there exists a vector $U_0 \geq O$ such that (X_0, U_0) is a saddle point of $L(X, U)$, in which case system (12.4) is satisfied.

One final point is in order. As a result of the discussion underlying the Karush-Kuhn-Tucker equivalence theorem, we have Corollary 12.4.

Corollary 12.4 Let (12.3) hold so that the primal maximum problem has an optimal solution X_0, Then the dual minimum problem

$$\begin{aligned} &\min g(\bar{U})=\bar{b}'\bar{U} \quad \text{s.t.} \\ &\quad \bar{A}'\bar{U}=C, \bar{U} \geq O \end{aligned}$$

has an optimal solution $\bar{U}_0 \geq O$ with $C'X_0=\bar{b}'\bar{U}_0$.

Proof. If X_0 solves the primal problem, then, by the preceding theorem, there exists a vector $\bar{U}_0 \geq O$ such that $(X_0, \bar{U}_0)$ is a saddle point of $L(X,\bar{U}) = -M(\bar{U},X)$, whence $M(\bar{U}_0,X) \geq M(\bar{U}_0,X_0) \geq M(\bar{U},X_0)$, i.e. $M(\bar{U},X)$ attains a maximum with respect to $\bar{U}$ and a minimum with respect to X and thus possesses a saddle point at $(\bar{U}_0,X_0)$. And since $\bar{U}_0'(\bar{b}-\bar{A}X_0) = X_0'\left(C-\bar{A}'\bar{U}_0\right) = 0$, it follows that $\bar{U}_0'\bar{b}-\bar{U}_0'\bar{A}X_0 = X_0'C - X_0'\bar{A}'\bar{U}_0 = 0$ or $C'X_0 = \bar{b}'\bar{U}_0$. Q.E.D.

12.4 Duality and Complementary Slackness Theorems

(Dreyfus and Freimer 1962)

We now turn to an alternative exposition of two of the aforementioned fundamental theorems of linear programming (Chapter 6), namely the duality and complementary slackness theorems. First, Theorem 12.5 explains the duality theorem.

Theorem 12.9 Duality Theorem.
A feasible solution to the primal maximum problem

$$\begin{aligned} &\max f(X) = C'X \quad \text{s.t.} \\ &\bar{b}-\bar{A}X \geq O, \quad X \text{ unrestricted} \end{aligned}$$

is optimal if and only if the dual minimum problem

$$\begin{aligned} &\min g(\bar{U}) = \bar{b}'\bar{U} \quad \text{s.t.} \\ &\bar{A}'\bar{U} = C, \bar{U} \geq O \end{aligned}$$

has a feasible solution $\bar{U}_0$ with $\bar{b}'\bar{U}'_0 = C'X_0$.

Proof. (sufficiency) Let $\bar{U}_0$ be a feasible solution to the dual problem with $C'X_0 = \bar{b}'\bar{U}_0$. Since X_0 is, by hypothesis, a feasible solution to the primal problem, Corollary 12.4 informs us that X_0 is optimal.

(necessity) For *any* particular requirements vector β, let the differentiable function $f(\beta)$ depict the maximum possible value of f when $\bar{A}X \leq \beta$. Then at the optimal solution X_0 to the primal problem, $f^0 = f(\bar{b}) = C'X_0$. Moreover, it is evident that if any component of $\bar{b}$, say $\bar{b}_i$, were increased, additional feasible solutions may be admitted to our discussion without excluding any solutions that were originally feasible so that f^0 would not decrease, i.e.

$$\frac{\partial f^0}{\partial \bar{b}_i} = \bar{u}_i^0 \geq 0 \text{ for all } i. \tag{12.14}$$

And if $\sum_{j=1}^{n}\bar{a}_{ij}x_j^0 < \bar{b}_i$ for some i at X_0, then increasing $\bar{\boldsymbol{b}}_i$ clearly does not affect the optimal value of f so that

$$\frac{\partial f^0}{\partial \bar{b}_i} = \bar{u}_i^0 = 0 \text{ for any } \bar{\boldsymbol{a}}_i X_0 < \bar{b}_i, \tag{12.15}$$

where $\bar{\boldsymbol{a}}_i X_0$ is the ith row of $\bar{A}X_0$.

Let us now consider a suitably restricted positive increase h_j in the jth component of X_0 from x_j^0 to $\hat{x}_j = x_j^0 + h_j$, where now

$$\hat{X} = X_0 + h_j \boldsymbol{e}_j \text{ unrestricted.}$$

Here $\hat{X}$ represents a solution to the modified primal problem

$$\max f(X) = C'X \quad \text{s.t.}$$
$$\bar{A}X \le \bar{\boldsymbol{b}} + h_j \bar{\boldsymbol{\alpha}}_j, \quad X \text{ unrestricted,}$$

where $\bar{\boldsymbol{\alpha}}_j$ is the jth column of $\bar{A}$. Clearly $\hat{X}$ is feasible since

$$\bar{A}(X_0 + h_j \boldsymbol{e}_j) \le \bar{\boldsymbol{b}} + h_j \bar{\boldsymbol{\alpha}}_j \text{ or}$$
$$\bar{A}X_0 + h_j \bar{\boldsymbol{\alpha}}_j \le \bar{\boldsymbol{b}} + h_j \bar{\boldsymbol{\alpha}}_j.$$

In addition, at $\hat{X}$, the objective value of the modified problem,

$$C'(X_0 + h_j \boldsymbol{e}_j) = C'X_0 + h_j c_j$$
$$= f(\bar{\boldsymbol{b}}) + h_j c_j,$$

must be less than or equal to the maximal value of the modified problem, i.e.

$$f(\bar{\boldsymbol{b}} + h_j \bar{\boldsymbol{\alpha}}_j) \ge f(\bar{\boldsymbol{b}}) + h_j \boldsymbol{c}_j \text{ or} \tag{12.16}$$

$$\frac{f(\bar{\boldsymbol{b}} + h_j \bar{\boldsymbol{\alpha}}_j) - f(\bar{\boldsymbol{b}})}{h_j} \ge c_j, h_j > 0. \tag{12.16.1}$$

Applying a first-order Taylor expansion to $f(\bar{\boldsymbol{b}} + h_j \bar{\boldsymbol{\alpha}}_j)$ at $h_j = 0$ yields

$$f(\bar{\boldsymbol{b}} + h_j \bar{\boldsymbol{\alpha}}_j) = f(\bar{\boldsymbol{b}}) + h_j \nabla f(\bar{\boldsymbol{b}} + \tau h_j \bar{\boldsymbol{\alpha}}_j)' \bar{\boldsymbol{\alpha}}_j, 0 < \tau < 1,$$

or, utilizing (12.16.1),

$$\nabla f(\bar{\boldsymbol{b}} + \tau h_j \bar{\boldsymbol{\alpha}}_j)' \bar{\boldsymbol{\alpha}}_j \ge c_j, 0 < \tau < 1.$$

Then

$$\lim_{h_j \to 0} \nabla f(\bar{\boldsymbol{b}} + \tau h_j \bar{\boldsymbol{\alpha}}_j)' \bar{\boldsymbol{\alpha}}_j = \nabla f(\bar{\boldsymbol{b}})' \bar{\boldsymbol{\alpha}}_j$$
$$= \sum_{i=1}^{m+n} \frac{\partial f^0}{\partial b_i} \bar{\alpha}_{ij} \ge c_j, h_j > 0. \tag{12.17}$$

With x_j^0 unrestricted, a sufficiently small negative increment h_j in x_j^0 is admissible so that $\hat{X} = X_0 + h_j e_j, h_j < 0$, is also feasible for the modified primal problem. In this instance (12.16) becomes

$$\frac{f(\bar{b} + h_j \bar{\alpha}_j) - f(\bar{b})}{h_j} \le c_j, h_j < 0. \tag{12.16.2}$$

A second application of Taylor's formula now yields

$$\nabla f(\bar{b} + \tau h_j \bar{\alpha}_j)' \bar{\alpha}_j \le c_j, 0 < \tau < 1,$$

so that

$$\begin{aligned} \lim_{h_j \to 0} \nabla f(\bar{b} + \tau h_j \bar{\alpha}_j)' \bar{\alpha}_j &= \nabla f(\bar{b})' \bar{\alpha}_j \\ &= \sum_{i=1}^{m+n} \frac{\partial f^0}{\partial \bar{b}_i} \bar{\alpha}_{ij} \le c_j, h_j < 0. \end{aligned} \tag{12.18}$$

Since infinitesimally small positive and negative changes h_j in x_j^0 preserve feasibility when x_j^0 is unrestricted, (12.17), (12.18) together imply that

$$\sum_{i=1}^{m+n} \frac{\partial f^0}{\partial \bar{b}_i} \bar{\alpha}_{ij} = c_j, \quad x_j^0 \text{ unrestricted.} \tag{12.19}$$

From (12.11) we know that $\frac{\partial f^0}{\partial \bar{b}_i} = \bar{u}_i^0 \ge 0$. Hence, $\bar{U}_0 \ge O$ satisfies the dual structural constraints(12.19) and thus represents a feasible solution to the dual minimum problem. (We note briefly that if the primal problem appears as

$$\begin{aligned} &\max f(X) = C'X \quad \text{s.t.} \\ &\quad AX \le b, X \ge O \end{aligned}$$

then (12.19) replaces

$$\begin{aligned} &\sum_{i=1}^{m} \frac{\partial f^0}{\partial b_i} \alpha_{ij} = c_j, \quad x_j^0 > 0; \\ &\sum_{i=1}^{m} \frac{\partial f^0}{\partial b_i} \alpha_{ij} \ge c_j, \quad x_j^0 = 0, \end{aligned}$$

where $\frac{\partial f^0}{\partial b_i} = u_i^0 \ge 0$ is the ith component of U_0 is a feasible solution to the associated dual minimum problem.)

It now remains to demonstrate that the primal and dual objective values are the same. First, for x_j^0 unrestricted, (12.19) implies that

$$c_j = \sum_{i=1}^{m+n} \bar{u}_i^0 \bar{a}_{ij} = \bar{\alpha}_j' \bar{U}_0,$$

where $\bar{\alpha}_j' \bar{U}_0$ is the jth component of the $(n \times 1)$ vector $\bar{A}' \bar{U}_0$. Then $c_j x_j^0 = \left(\bar{\alpha}_j' \bar{U}_o\right) x_j^0$. And for some $x_j^0 = 0$, $c_j x_j^0 = \left(\bar{\alpha}_j' \bar{U}_o\right) x_j^0 = 0$ so that, in either instance,

$$c_j x_j^0 = \left(\bar{\alpha}_j' \bar{U}_o\right) x_j^0 \text{ for any } j.$$

Summing over all j thus yields

$$\sum_{j=1}^{n} c_j x_j^0 = \sum_{j=1}^{n} (\bar{\alpha}'_j \bar{U}_0) x_j^0 \text{ or}$$
$$C' X_0 = \bar{U}'_0 \bar{A} X_0. \tag{12.20}$$

Next, if for some specific i value $\bar{a}_i X_0 = \bar{b}_i$, then $\bar{u}_i^0 (\bar{a}_i X_0) = \bar{u}_i^0 \bar{b}_i$; while if $\bar{a}_i X_0 < \bar{b}_i$, (12.15) indicates that $\bar{u}_i^0 = 0$ so that $\bar{u}_i^0 (\bar{a}_i X_0) = \bar{u}_i^0 \bar{b}_i = 0$. In each case, then,

$$\bar{u}_i^0 (\bar{a}_i X_0) = \bar{u}_i^0 \bar{b}_i \text{ for any } i.$$

Summing over all i thus yields

$$\sum_{i=1}^{m+n} \bar{u}_i^0 (\bar{a}_i X_0) = \sum_{i=1}^{m+n} \bar{u}_i^0 \bar{b}_i \text{ or}$$
$$\bar{U}'_0 \bar{A} X_0 = \bar{U}'_0 \bar{b} \tag{12.21}$$

so that, from (12.20) and (12.21), $C' X_0 = \bar{b}' \bar{U}_0$. Q.E.D.

Next, if, as above, we express the primal-dual pair of problems as

PRIMAL PROBLEM	**DUAL PROBLEM**
$\max f(X) = C'X$ s.t.	$\min g(\bar{U}) = \bar{b}' \bar{U}$ s.t.
$\bar{A} X \leq \bar{b}$, X unrestricted,	$\bar{A}' \bar{U} \geq C, \bar{U} \geq O,$

then their associated Lagrangians may be written, respectively, as

$$L(X, \bar{U}, \bar{X}_s) = C'X + \bar{U}' (\bar{b} - \bar{A} X - \bar{X}_s),$$
$$M(\bar{U}, X) = -\bar{b}' \bar{U} + X' \left(C - \bar{A}' \bar{U}\right),$$

where $\bar{X}_s$ is an $(m + n \times 1)$ vector of nonnegative primal slack variables, i.e.

$$\bar{X}_s = \begin{bmatrix} \bar{x}_{n+1} \\ \vdots \\ \bar{x}_{n+m} \\ \bar{x}_{n+m+1} \\ \vdots \\ \bar{x}_{2n+m} \end{bmatrix} \geq O \text{ with } \begin{cases} \bar{x}_{n+i} = b_i - \bar{a}_i X \geq 0, i = 1, \ldots, m; \\ \bar{x}_{n+m+j} = \bar{b}_{m+j} - \bar{a}_{m+j} X \\ \quad = e_j' X \\ \quad = x_j \text{ unrestricted}, j = 1, \ldots, n. \end{cases}$$

In this light, we now look to Theorem 12.5.

Theorem 12.10 Complementary Slackness Theorem. The vectors $X_0, \bar{U}_0$ solve the primal and dual problems

$$\begin{aligned} &\max f(X) = C'X \quad \text{s.t.} && \min g(\bar{U}) = \bar{b}'\bar{U} \quad \text{s.t.} \\ &\bar{A}X \leq \bar{b}, \quad X \text{ unrestricted}, && \bar{A}'\bar{U} \geq C, \bar{U} \geq O, \end{aligned}$$

respectively, if and only if $X_0, \bar{U}_0$ satisfy the (weak) complementary slackness conditions

$$U_0'(\bar{X}_s)_0 = 0 \text{ or } \begin{cases} \bar{u}_i^0 \bar{x}_{n+i}^0 = 0, i = 1, \ldots, m; \\ \bar{u}_{m+j}^0 \bar{x}_{n+m+j}^0 = 0, j = 1, \ldots, n. \end{cases} \tag{12.22}$$

Proof. (necessity) If X_0 solves the primal problem, then there exists a vector $\bar{U}_0 \geq O$ such that $(X_0, \bar{U}_0)$ is a saddle point of $L(X, \bar{U}) = -M(\bar{U}, X)$, whence

$$\begin{aligned} L\left(X_0, \bar{U}_0, (\bar{X}_s)_0\right) + M(\bar{U}_0, X_0) &= C'X_0 + \bar{U}_0'\left(\bar{b} - \bar{A}X_0 - (\bar{X}_s)_0\right) \\ &\quad - \bar{b}'\bar{U}_0 - X_0'\left(C - \bar{A}'\bar{U}_0\right) \\ &= -\bar{U}_0'(\bar{X}_s)_0 = 0 \end{aligned}$$

or $\bar{U}_0'(\bar{X}_s)_0 = 0$. With $\bar{U}_0' = \left(\bar{u}_1^0, \ldots, \bar{u}_m^0, \bar{u}_{m+1}^0, \ldots, \bar{u}_{m+n}^0\right)$, it follows that

$$\begin{aligned} \bar{U}_0'(\bar{X}_s)_0 &= \sum_{i=1}^{m} \bar{u}_i^0 \bar{x}_{n+i}^0 + \sum_{j=1}^{n} \bar{u}_{m+j}^0 \bar{x}_{n+m+j}^0 \\ &= \sum_{i=1}^{m} \bar{u}_i^0 \bar{x}_{n+i}^0 + \sum_{j=1}^{n} \bar{u}_{m+j}^0 x_j^0 = 0. \end{aligned}$$

(sufficiency) If $\bar{U}_0(\bar{X}_s)_0 = \bar{U}_0'\left(\bar{b} - \bar{A}X\right) = 0$, then $L(X_0, \bar{U}_0) = C'X_0$, $M(\bar{U}_0, X_0) = -\bar{b}'\bar{U}_0 = -L(X_0, \bar{U}_0)$ and thus $C'X_0 = \bar{b}'\bar{U}_0$. Hence, $X_0, \bar{U}_0$ yield optimal solutions to the primal and dual problems, respectively. Q.E.D.

By way of interpretation: (i) if the primal slack variable $\bar{x}_{n+i}^0, i = 1, \ldots, m$, corresponding to the ith primal structural constraint is positive, then the associated dual variable $u_i^0, i = 1, \ldots, m$, must equal zero. Conversely, if $u_i^0, i = 1, \ldots, m$, is positive, then the ith primal structural constraint must hold as an equality with $\bar{x}_{n+i}^0 = 0, i = 1, \ldots, m$. Before turning to point (ii), let us examine the form of the jth dual structural constraint. Since $A = [\boldsymbol{\alpha}_1, \ldots, \boldsymbol{\alpha}_n]$,

$$\bar{A}'\bar{U} = \left[A' \vdots -I_n\right] \begin{bmatrix} U \\ \bar{u}_{m+1} \\ \vdots \\ \bar{u}_{m+n} \end{bmatrix}$$

$$= \left[\begin{array}{c|c} \boldsymbol{\alpha}_1' & \\ \vdots & -I_n \\ \boldsymbol{\alpha}_n' & \end{array}\right] \begin{bmatrix} U \\ \bar{u}_{m+1} \\ \vdots \\ \bar{u}_{m+n} \end{bmatrix} = C$$

and thus the jth dual structural constraint appears as $\boldsymbol{\alpha}_j'U - \bar{u}_{m+j} = c_j, j = 1, \ldots, n$. In this light, (ii) if the jth primal variable $\bar{x}_j^0, j = 1, \ldots, n$, is different from zero, then the dual surplus variable $\bar{u}_{m+j}^0, j = 1, \ldots, n$, associated with the jth dual structural constraint $\boldsymbol{\alpha}_j'U - \bar{u}_{m+j} = c_j, j = 1, \ldots, n$, equals zero. Conversely, if $\bar{u}_{m+j}^0 > 0, j = 1, \ldots, n$, then the jth dual structural constraint is not binding and thus the jth primal variable $\bar{x}_j^0, j = 1, \ldots, n$, must be zero.

13

Simplex-Based Methods of Optimization

13.1 Introduction

In this chapter the standard simplex method, or a slight modification thereof, will be used to solve an assortment of specialized linear as well as essentially or structurally nonlinear decision problems. In this latter instance, a set of transformations and/or optimality conditions will be introduced that linearizes the problem so that the simplex method is only indirectly applicable. An example of the first type of problem is game theory, while the second set of problems includes quadratic and fractional functional programming.

13.2 Quadratic Programming

The quadratic programming problem under consideration has the general form

$$\begin{aligned} &\max f(\boldsymbol{X}) = \boldsymbol{C}'\boldsymbol{X} + \boldsymbol{X}'\boldsymbol{Q}\boldsymbol{X} \quad \text{s.t.} \\ &\boldsymbol{A}\boldsymbol{X} \leq \boldsymbol{b}, \boldsymbol{X} \geq \boldsymbol{O}. \end{aligned} \tag{13.1}$$

Here, the objective function is the sum of a **linear form** and a **quadratic form** and is to be optimized in the presence of a set of linear inequalities. Additionally, both $\boldsymbol{C}$ and $\boldsymbol{X}$ are of order $(p \times 1)$, $\boldsymbol{Q}$ is a $(p \times p)$ symmetric coefficient matrix, $\boldsymbol{A}$ is the $(m \times p)$ coefficient matrix of the linear structural constraint system, and $\boldsymbol{b}$ (assumed $\geq \boldsymbol{O}$) is the $(m \times 1)$ requirements vector. If $\boldsymbol{Q}$ is not symmetric, then it can be transformed, by a suitable redefinition of coefficients, into a symmetric matrix without changing the value of $\boldsymbol{X}'\boldsymbol{Q}\boldsymbol{X}$. For instance, if a matrix $\boldsymbol{A}$ from $\boldsymbol{X}'\boldsymbol{A}\boldsymbol{X}$ is not symmetric, then it can be replaced by a symmetric matrix $\boldsymbol{B}$ if $b_{ij} = b_{ji} = \frac{a_{ij} + a_{ji}}{2}$ for all i, j or $\boldsymbol{B} = (\boldsymbol{A} + \boldsymbol{A}')/2$. Thus, $b_{ij} + b_{ji} = 2b_{ij} = a_{ij} + a_{ji}$ is the coefficient of $x_i x_j$, $i \neq j$, in

$$\begin{aligned} X'BX &= \sum_{i=1}^{n}\sum_{j=1}^{n}\frac{a_{ij}+a_{ji}}{2}x_ix_j \\ &= X'[(A+A')/2]X = \frac{1}{2}X'AX + \frac{1}{2}X'AX = X'AX. \end{aligned}$$

So if

$$A = \begin{bmatrix} 3 & 1 & 2 \\ 4 & 2 & 1 \\ 3 & 5 & -1 \end{bmatrix},$$

then, under this transformation,

$$B = \begin{bmatrix} 3 & 5/2 & 5/2 \\ 5/2 & 2 & 3 \\ 5/2 & 3 & -1 \end{bmatrix}.$$

In this regard, for the quadratic function $X'AX = 2x_1^2 + 4x_1x_2 + 3x_2^2$, the coefficient matrix may simply be assumed symmetric and written

$$Q = \begin{bmatrix} 2 & 2 \\ 2 & 3 \end{bmatrix};$$

or if

$$Q = \begin{bmatrix} 1 & 3 \\ 5 & 1 \end{bmatrix},$$

the associated quadratic function may be expressed as $X'QX = x_1^2 + 8x_1x_2 + x_2^2$. For additional details on quadratic forms, see Appendix 13.A.

We mentioned in earlier chapters that an optimal solution to a linear programming problem always occurs at a vertex (extreme point) of the feasible region $\mathcal{K} = \{X | AX \leq b, X \geq O\}$ or at a convex combination of a finite number of extreme points of $\mathcal{K}$ (i.e. along an edge of $\mathcal{K}$). This is not the case for a quadratic programming problem. An optimal solution to a quadratic program may occur at an extreme point or along an edge or at an interior point of $\mathcal{K}$ (see Figure 13.1a–c, respectively, wherein $\mathcal{K}$ and the contours of f are presented).

To further characterize the attainment of an optimal solution to (13.1), let us note that with $\mathcal{K}$ a convex set, any local maximum of f will also be the global maximum if f is concave. Clearly $C'X$ is concave since it is linear, while $X'QX$ is concave if it is negative semi-definite or negative definite. Thus, $f = C'X + X'QX$ is concave (since the sum of a finite number of concave functions is itself a concave function) on $\mathcal{K}$ so that the global maximum of f is attained. In what

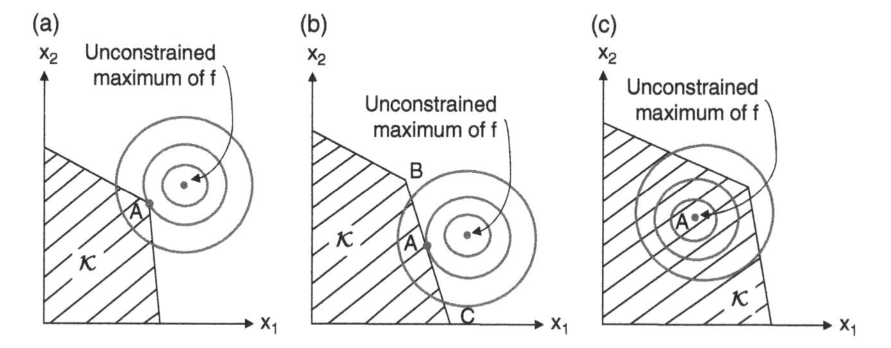

Figure 13.1 (a) Constrained maximum of *f* at extreme point A. (b) Constrained maximum of *f* occurs along the edge BC. (c) Constraints not binding at A.

follows, we shall restrict $\boldsymbol{Q}$ to the negative definite case. Then f may be characterized as strictly concave so that it takes on a unique or strong global maximum on $\mathcal{K}$.[1] If $\boldsymbol{X'QX}$ is negative semi-definite, then it may be transformed to a negative definite quadratic form by making an infinitesimally small change in the diagonal elements of $\boldsymbol{Q}$, i.e. $\boldsymbol{X'QX}$ is replaced by $\boldsymbol{X'}(\boldsymbol{Q} - \varepsilon\, \boldsymbol{I}_p)\boldsymbol{X} = \boldsymbol{X'QX} - \varepsilon\, \boldsymbol{X'I}_p\boldsymbol{X} < 0$ since $\boldsymbol{X'QX} \leq \boldsymbol{0}$ for any $\boldsymbol{X}$ and $-\varepsilon\, \boldsymbol{X'I}_p\boldsymbol{X} < 0$ for any $\boldsymbol{X} \neq \boldsymbol{O}$.

From (13.1) we may form the Lagrangian function

$$L(\boldsymbol{X},\boldsymbol{X}_{s1},\boldsymbol{X}_{s2},\boldsymbol{U}_1,\boldsymbol{U}_2) = \boldsymbol{C'X} + \boldsymbol{X'QX} + \boldsymbol{U}_1'(\boldsymbol{b} - \boldsymbol{AX} - \boldsymbol{X}_{s1}^2) + \boldsymbol{U}_2'(\boldsymbol{X} - \boldsymbol{X}_{s2}^2).$$

Then from the Karush-Kuhn-Tucker necessary and sufficient conditions for a constrained maximum (12.4),

$$\left.\begin{array}{l} \text{(a)}\ \boldsymbol{C} + 2\boldsymbol{QX} - \boldsymbol{A'U}_1 \leq \boldsymbol{O} \\ \text{(b)}\ \boldsymbol{X'}(\boldsymbol{C} + 2\boldsymbol{QX} - \boldsymbol{A'U}_1) = 0 \\ \text{(c)}\ \boldsymbol{U}_1'(\boldsymbol{b} - \boldsymbol{AX}) = 0 \\ \text{(d)}\ \boldsymbol{b} - \boldsymbol{AX} \geq \boldsymbol{O} \\ \text{(e)}\ \boldsymbol{X} \geq \boldsymbol{O}, \boldsymbol{U}_1 \geq \boldsymbol{O} \end{array}\right\} \text{ or } \left\{\begin{array}{l} \boldsymbol{C} + 2\boldsymbol{QX} - \boldsymbol{A'U}_1 + \boldsymbol{U}_2 = \boldsymbol{O} \\ \boldsymbol{X'U}_2 = 0 \\ \boldsymbol{U}_1'\boldsymbol{Y} = 0 \\ \boldsymbol{b} - \boldsymbol{AX} - \boldsymbol{Y} = \boldsymbol{O} \\ \boldsymbol{X} \geq \boldsymbol{O}, \boldsymbol{U}_1 \geq \boldsymbol{O}, \boldsymbol{U}_2 \geq \boldsymbol{O}, \boldsymbol{Y} \geq \boldsymbol{O}. \end{array}\right. \tag{13.2}$$

Before proceeding to the development of a solution algorithm for the quadratic programming problem, let us examine an important interpretation of (13.2).

1 If the problem were

$$\min f(\boldsymbol{X}) = \boldsymbol{C'X} + \boldsymbol{X'QX} \quad \text{s.t.}$$
$$\boldsymbol{AX} \geq \boldsymbol{b}, \boldsymbol{X} \geq \boldsymbol{O},$$

then $\boldsymbol{Q}$ would be restricted to the positive definite case. In this instance, f is strictly convex on $\mathcal{K} = \{\boldsymbol{X} | \boldsymbol{AX} \geq \boldsymbol{b}, \boldsymbol{X} \geq \boldsymbol{O}\}$ so that it assumes a unique strong global constrained minimum.

Specifically, if $\boldsymbol{X}_o$ solves (13.1), then there exists a vector $\boldsymbol{U}_1^o$ such that $(\boldsymbol{X}_o, \boldsymbol{U}_1^o)$ is a saddle point (see Chapter 12 for a review of this concept) of the Lagrangian $L(\boldsymbol{X}, \boldsymbol{U}_1) = \boldsymbol{C}'\boldsymbol{X} + \boldsymbol{X}'\boldsymbol{Q}\boldsymbol{X} + \boldsymbol{U}_1'(\boldsymbol{b} - \boldsymbol{A}\boldsymbol{X})$, in which case (13.2) holds. To see this, we need only note that if $L(\boldsymbol{X}, \boldsymbol{U}_1)$ has a local maximum in the $\boldsymbol{X}$ direction at $\boldsymbol{X}_o$, then it follows from (13.2a, b, e) that

$$\left.\begin{aligned} &\nabla_X L^o \leq \boldsymbol{O} \\ &(\nabla_X L^o)'\boldsymbol{X}_o = 0 \\ &\boldsymbol{X}_o \geq \boldsymbol{O} \end{aligned}\right\} \text{ or } \left\{\begin{aligned} &\boldsymbol{C} + 2\boldsymbol{Q}\boldsymbol{X}_o - \boldsymbol{A}'\boldsymbol{U}_1^o \leq \boldsymbol{O} \\ &\boldsymbol{X}_o'(\boldsymbol{C} + 2\boldsymbol{Q}\boldsymbol{X}_o - \boldsymbol{A}'\boldsymbol{U}_1^o) = 0 \\ &\boldsymbol{X}_o \geq \boldsymbol{O}; \end{aligned}\right.$$

while if $L(\boldsymbol{X}_o, \boldsymbol{U}_1)$ attains a minimum in the $\boldsymbol{U}_1$ direction at $\boldsymbol{U}_1^o$, then, from (13.2d, c, e),

$$\left.\begin{aligned} &\nabla_{U_1} L^o \geq \boldsymbol{O} \\ &(\nabla_{U_1} L^o)'\boldsymbol{U}_1 = 0 \\ &\boldsymbol{U}_1^0 \geq \boldsymbol{O} \end{aligned}\right\} \text{ or } \left\{\begin{aligned} &\boldsymbol{b} - \boldsymbol{A}\boldsymbol{X}_o \geq \boldsymbol{O} \\ &(\boldsymbol{U}_1^o)'(\boldsymbol{b} - \boldsymbol{A}\boldsymbol{X}_o) = 0 \\ &\boldsymbol{U}_1^o \geq \boldsymbol{O}. \end{aligned}\right.$$

Upon examining (13.2) we see that solving the quadratic programming problem amounts to finding a solution in $2(m + p)$ nonnegative variables to the linear system

$$\begin{aligned} &(\text{a})\ 2\boldsymbol{Q}\boldsymbol{X} - \boldsymbol{A}'\boldsymbol{U}_1 + \boldsymbol{U}_2 = -\boldsymbol{C} \\ &(\text{b})\ \boldsymbol{A}\boldsymbol{X} \qquad\qquad + \boldsymbol{Y} = \boldsymbol{b} \end{aligned}$$

$$\text{or} \quad \begin{bmatrix} 2\boldsymbol{Q} & -\boldsymbol{A}' & \boldsymbol{I}_p & \boldsymbol{O} \\ \boldsymbol{A} & \boldsymbol{O} & \boldsymbol{O} & \boldsymbol{I}_m \end{bmatrix} \begin{bmatrix} \boldsymbol{X} \\ \boldsymbol{U}_1 \\ \boldsymbol{U}_2 \\ \boldsymbol{Y} \end{bmatrix} = \begin{bmatrix} -\boldsymbol{C} \\ \boldsymbol{b} \end{bmatrix} \tag{13.2.1}$$

that satisfies the $m + p$ (nonlinear) complementary slackness conditions $\boldsymbol{X}'\boldsymbol{U}_2 = \boldsymbol{U}_1'\boldsymbol{Y} = 0$. Under this latter set of restrictions, at least p of the $2p$ x_js, u_{2j}s must be zero while at least m of the $2m$ u_{1i}s, y_is must equal zero. Thus, at least $m + p$ of the $2(m + \text{p})$ variables x_j, u_{2j}, u_{1j}, y_i must vanish at the optimal solution. In this regard, an optimal solution to (13.2.1) must be a basic feasible solution (Barankin and Dorfman 1958). So, if the (partitioned) vector $(\boldsymbol{X}_o, \boldsymbol{U}_1^o, \boldsymbol{U}_2^o, \boldsymbol{Y}^o)$ is a feasible solution to (13.2.1) and the complementary slackness conditions are satisfied (no more than $m + p$ of the mutually complementary variables are positive), then the resulting solution is also a basic feasible solution to (13.2.1) with the $\boldsymbol{X}_o$ component solving the quadratic programming problem.

To actually obtain an optimal basic feasible solution to (13.2.1) we shall use the complementary pivot method presented in Section 13.4. But first we look to dual quadratic programs.

13.3 Dual Quadratic Programs

A well-known result from duality theory in generalized nonlinear programming that is applicable for our purposes here is that if the Lagrangian for the **primal quadratic program** (13.1) is written as $L(X,U_1)=C'X+X'QX+U_1'(b-AX)$, then the **dual quadratic program** appears as

$$\min\{g(X,U_1)=L(X,U)-X'\nabla_X L\} \text{ s.t.}^2$$
$$\nabla_X L\le O, U_1\ge O, X \text{ unrestricted}$$

or

$$\begin{aligned}\min g(X,U_1)&=b'U_1-X'QX \text{ s.t.}\\ A'U_1-2QX&\ge C, U_1\ge O, X \text{ unrestricted.}\end{aligned} \tag{13.3}$$

In what follows, a few important theorems for quadratic programming problems will be presented. The first two demonstrate the similarity between linear and quadratic programming as far as the relationship of the primal and dual objective functions is concerned. To this end, we state Theorems 13.1 as follows.

Theorem 13.1 Let the $(p\times p)$ symmetric matrix Q be negative semi-definite. If X_o is a feasible solution to the primal problem (13.1) and $(\hat{X},\hat{U})$ is a feasible solution to the dual problem (13.3), then

$$C'X_o+X_o'QX_o\le b'\hat{U}_1-\hat{X}'Q\hat{X},$$

i.e. the dual objective value provides an upper bound to the primal objective value.

Proof. If $(\hat{X},\hat{U}_1)$ is a feasible solution to the dual problem, $C+2Q\hat{X}-A'\hat{U}_1\le O$. Since $X_o\ge O$,

$$X_o'C+2X_o'Q\hat{X}-X_o'A'\hat{U}_1\le 0. \tag{13.4}$$

Adding $b'\hat{U}_1$ to both sides of this inequality yields

$$C'X_o+2X_o'Q\hat{X}+(b'-X_o'A')\hat{U}_1\le b'\hat{U}_1. \tag{13.5}$$

Since $(X_o-\hat{X})'Q(X_o-\hat{X})=X_o'QX_o+\hat{X}'Q\hat{X}-2X_o'Q\hat{X}\le 0$ for Q negative semi-definite, it follows from (13.5) that

$$C'X_o+X_o'QX_o+(b'-X_o'A')\hat{U}_1\le b'\hat{U}_1-\hat{X}'Q\hat{X}$$

2 Panik (1976).

and thus, for $b' - X_o'A' \geq O'$,

$$C'X_o + X_o'QX_o \leq b'\hat{U}_1 - \hat{X}'Q\hat{X}. \text{ Q.E.D.}$$

Next,

Theorem 13.2 If X_o is a feasible solution to the primal problem (13.1) and $(\hat{X}, \hat{U})$ is a feasible solution to the dual problem (13.3) with $f(X_o) = g(\hat{X}, \hat{U}_1)$, then $X_o, (\hat{X}, \hat{U}_1)$ are optimal solutions to the primal and dual problems, respectively.

Proof. Let $\bar{X}$ be any other feasible solution to the primal problem. Then $C'\bar{X} + \bar{X}'Q\bar{X} \leq b'\hat{U}_1 - \hat{X}'Q\hat{X} = C'X_o + X_o'QX_o$ and thus X_o is a maximal solution to the primal problem. Similarly, if $(\bar{X}, \bar{U}_1)$ is any other feasible solution to the dual problem, then $b'\bar{U}_1 - \bar{X}'Q\bar{X} \geq C'X_o + X_o'QX_o = b'\hat{U}_1 - \hat{X}'Q\hat{X}$ and thus $(\hat{X}, \hat{U}_1)$ is a minimal solution to the dual problem. Q.E.D.

Finally, the principal duality result of this section is presented by

Theorem 13.3 (Dorn 1960; Saaty and Bram 1964). Let the $(p \times p)$ symmetric matrix Q be negative semi-definite. If X_o solves the primal problem (13.1), then there exists a vector U_1^o such that (X_o, U_1^o) solves the dual problem (13.3) with $f(X_o) = g(X_o, U_1^o)$. Conversely, if (X_o, U_1^o) solves the dual problem, then there exists a vector $\hat{X}$ such that $\hat{X}$ solves the primal problem with $f(\hat{X}) = g(X_o, U_1^o)$.

Proof. Let X_o be an optimal solution to the primal problem. Then from (13.2a, e), (X_o, U_1^o) is a feasible solution to the dual problem, i.e. $C + 2QX_o - A'U_1^o \leq O, U_1^o \geq O$. If (X, U_1) is any other feasible solution to the dual, then

$$\begin{aligned} g(X_o, U_1^0) - g(X, U_1^0) &= b'U_1^0 - X_o'QX_o - b'U_1 + X'QX \\ &= b'U_1^0 - b'U_1 + (X - X_o)'Q(X - X_o) + 2X_o'Q(X - X_o). \end{aligned}$$

With Q negative semi-definite,

$$g(X_o, U_1^0) - g(X, U_1) \leq b'U_1^0 - b'U_1 + 2X_o'Q(X - X_o).$$

From (13.2b, c), $b'U_1^0 = C'X_o + 2X_o'QX_o$ so that

$$g(X_o, U_1^0) - g(X, U_1) \leq C'X_o - b'U_1 + 2X_o'QX.$$

Employing (13.4) (with $\hat{X} = X, \hat{U}_1 = U_1$),

$$g(X_o, U_1^0) - g(X, U_1) \leq -b'U_1 + X_o'AU_1 = (X_o'A - b')U_1.$$

Since $X_o'A - b' \leq O', g(X_o, U_1^0) - g(X, U_1) \leq 0$ so that (X_o, U_1^0) minimizes g. And since $b'U_1^o = C'X_o + 2X_o'QX_o$,

$$g(X_o, U_1^0) = b'U_1^o - X_o'QX_o = C'X_o + X_o'QX_o = f(X_o).$$

Thus, the first half of this theorem is proved.

To verify the converse portion, let (X_o, U_1^0) be an optional solution to the dual program. Upon forming the Lagrangian of the equivalent dual problem

$$\max\{-g(X, U_1) = -bU_1 + X'QX\} \text{ s.t.}$$
$$-A'U_1 + 2QX \leq -C, U_1 \geq O, X \text{ unrestricted}$$

we have

$$M(X, \hat{X}, \bar{X}, U_1, U_{s1}^2, U_{s2}^2) = -b'U_1 + X'QX + \hat{X}'(-C - 2QX + A'U_1 - U_{s1}^2)$$
$$+ \bar{X}'(U_1 - U_{s2}^2).$$

Then from the first-order conditions

$$\frac{\partial M}{\partial X} = 2QX - 2Q\hat{X} = O$$
$$\frac{\partial M}{\partial \hat{X}} = -C - 2QX + A'U_1 - U_{s1}^2 = O$$
$$\frac{\partial M}{\partial \bar{X}} = U_1 - U_{s1}^2 = O$$
$$\frac{\partial M}{\partial U_1} = -b + A\hat{X} + \bar{X} = O$$
$$\frac{\partial M}{\partial U_{s1}} = -2\hat{X}'U_{s1} = 0$$
$$\frac{\partial M}{\partial U_{s2}} = -2\bar{X}'U_{s2} = 0$$

we obtain the Karush-Kuhn-Tucker necessary and sufficient conditions

$$\begin{aligned} &\text{(a) } b - A\hat{X} \geq O \\ &\text{(b) } (U_1^o)'(b - A\hat{X}) = 0 \\ &\text{(c) } \hat{X}'(-C - 2QX_o + A'U_1^o) = 0 \\ &\text{(d) } QX_o - Q\hat{X} = O \\ &\text{(e) } \hat{X} \geq O, U_1 \geq O. \end{aligned} \tag{13.6}$$

From (13.6a,e), $\hat{X}$ is a feasible solution to the primal problem. And since (X_o, U_1^0) is a feasible solution to the dual, (13.6d) implies that $(\hat{X}, U_1^0)$ satisfies

(13.2a, e). Moreover, (13.6b) is (13.2c) while both (13.6c, d) imply (13.2b). Hence (13.2) is satisfied so that $\hat{X}$ is an optimal solution to the primal problem. From (3.6d),

$$\begin{aligned} g(X_o, U_1^0) - f(\hat{X}) &= b'U_1^o - X_o'QX_o - C'\hat{X} + \hat{X}'Q\hat{X} \\ &= b'U_1^o - C'\hat{X} - 2\hat{X}'QX_o. \end{aligned}$$

Then using (13.6c) we have $-C'\hat{X} - 2\hat{X}'QX_o = \hat{X}'A'U_1^o$ so that

$$g(X_o, U_1^0) - f(\hat{X}) = b'U_1^o - \hat{X}'A'U_1^o = (U_1^0)'(b - A\hat{X}) = 0$$

by virtue of (13.6b). Q.E.D.

Example 13.1 Determine the dual solution to the primal quadratic program

$$\begin{aligned} &\max f = x_1 + 2x_2 - \frac{1}{2}x_1^2 - \frac{1}{2}x_2^2 \text{ s.t.} \\ &2x_1 + 3x_2 \leq 6 \\ &x_1 + 4x_2 \leq 5 \\ &x_1, x_2 \geq 0. \end{aligned}$$

Using (13.3), the dual problem appears as

$$\begin{aligned} &\min g = 6u_1 + 5u_2 + \frac{1}{2}x_1^2 + \frac{1}{2}x_2^2 \text{ s.t.} \\ &2u_1 + u_2 + x_1 \geq 1 \\ &3u_1 + 4u_2 + x_2 \geq 2 \\ &u_1, u_2 \geq 0; x_1, x_2 \text{ unrestricted.} \end{aligned}$$

Once the primal problem is solved, the dual becomes, for $X = X_o$, a simple linear programming problem involving u_1 alone, i.e. for $X_o' = (13/17, 18/17)$,

$$\begin{aligned} &\min g - 0.8529 = 6u_1 + 5u_2 \quad \text{s.t.} \\ &2u_1 + u_2 \geq \frac{4}{17} \\ &3u_1 + 4u_2 \geq \frac{16}{17} \\ &u_1, u_2 \geq 0. \end{aligned}$$

The optimal solution to this problem is $u_1^o = 0, u_2^o = {}^4/_7$, and $g^o = 2.03$. (A technique for solving the primal quadratic program will be provided in the next section.) ■

13.4 Complementary Pivot Method (Lemke 1965)

A computational routine that explicitly requires that a complementarity condition holds at each iteration is Lemke's **complementary pivot method.** This technique is designed to generate a solution to the **linear complementarity problem** (LCP): given a vector $\boldsymbol{q} \in \mathcal{R}^n$ and an nth order matrix $\boldsymbol{M}$, find a vector $\boldsymbol{Z} \in \mathcal{R}^n$ such that

$$\begin{aligned} &\text{a. } \boldsymbol{q} + \boldsymbol{MZ} \geq \boldsymbol{O} \\ &\text{b. } \boldsymbol{Z} \geq \boldsymbol{O} \\ &\text{c. } \boldsymbol{Z}'(\boldsymbol{q} + \boldsymbol{MZ}) = 0. \end{aligned} \tag{13.7}$$

If $\boldsymbol{W} = \boldsymbol{q} + \boldsymbol{MZ}$, then (13.7) becomes: find vectors $\boldsymbol{W}, \boldsymbol{Z} \in \mathcal{R}^n$, such that

$$\begin{aligned} &\text{a. } \boldsymbol{W} - \boldsymbol{MZ} \geq \boldsymbol{q} \\ &\text{b. } \boldsymbol{W} \geq \boldsymbol{O}, \boldsymbol{Z} \geq \boldsymbol{O} \\ &\text{c. } \boldsymbol{W}'\boldsymbol{Z} = 0. \end{aligned} \tag{13.8}$$

Here $\boldsymbol{W}'\boldsymbol{Z} = \sum_{i=1}^{n} w_i z_i = 0 \text{ or } w_i z_i = 0, i = 1, \ldots, n,$ is called the **complementarity condition,** where the variables w_i and z_i, called a **complementary pair,** are said to be complements of each other, i.e. at least one variable in each pair $(w_i, z_i,)$ must be zero.

A nonnegative solution $(\boldsymbol{W}, \boldsymbol{Z})$ to the linear system (13.8a) is called a **feasible solution** to the LCP. Moreover, a feasible solution $(\boldsymbol{W}, \boldsymbol{Z})$ to the LCP that also satisfies the complementarity condition on $\boldsymbol{W}'\boldsymbol{Z} = 0$ is termed a **complementary solution.** (Note that (13.7) and (13.8) contain no objective function; and there is only a single structural constraint in the variables $\boldsymbol{W}, \boldsymbol{Z}$, namely the complementary restriction $\boldsymbol{W}'\boldsymbol{Z} = 0$.)

If $\boldsymbol{q} \geq \boldsymbol{O}$, then there exists a trivial **complementary basic solution** given by $\boldsymbol{W} = \boldsymbol{q}$, $\boldsymbol{Z} = \boldsymbol{O}$. Hence (13.8) has a nontrivial solution only when $\boldsymbol{q} \ngeq \boldsymbol{O}$. In this circumstance, the initial basic solution given by $\boldsymbol{W} = \boldsymbol{q}$, $\boldsymbol{Z} = \boldsymbol{O}$ is infeasible even if the complementarity condition $\boldsymbol{W}'\boldsymbol{Z} = 0$ holds.

For the general quadratic program

$$\begin{aligned} &\max f(\boldsymbol{X}) = \boldsymbol{C}'\boldsymbol{X} + \boldsymbol{X}'\boldsymbol{QX} \text{ s.t.} \\ &\quad \boldsymbol{AX} \leq \boldsymbol{b}, \boldsymbol{X} \geq \boldsymbol{O}, \end{aligned} \tag{13.1}$$

let us repeat the Karush-Kuhn-Tucker conditions for (13.1) as

$$\begin{aligned} &\boldsymbol{C} + 2\boldsymbol{QX} - \boldsymbol{A}'\boldsymbol{U}_1 + \boldsymbol{U}_2 = \boldsymbol{O} \\ &\quad \boldsymbol{X}'\boldsymbol{U}_2 = 0 \\ &\quad \boldsymbol{U}_1'\boldsymbol{Y} = 0 \\ &\quad \boldsymbol{b} - \boldsymbol{AX} - \boldsymbol{Y} = \boldsymbol{O} \\ &\quad \boldsymbol{X} \geq \boldsymbol{O}, \boldsymbol{U}_1 \geq \boldsymbol{O}, \boldsymbol{U}_2 \geq \boldsymbol{O}, \boldsymbol{Y} \geq \boldsymbol{O} \end{aligned} \tag{13.2}$$

or

$$\begin{aligned} 2QX - A'U_1 + U_2 &= -C \\ AX + Y \qquad &= b \\ X'U_2 + U_1'Y \quad &= 0 \\ X \geq O, U_1 \geq O, U_2 &\geq O, Y \geq O. \end{aligned} \tag{13.9}$$

Let us define

$$W = \begin{bmatrix} U_2 \\ Y \end{bmatrix}, Z = \begin{bmatrix} X \\ U_1 \end{bmatrix}, q = \begin{bmatrix} -C \\ b \end{bmatrix}, \text{and } M = \begin{bmatrix} -2Q & A' \\ -A & O \end{bmatrix}.$$

Then (13.9) can be rewritten in *condensed form* as

$$\begin{aligned} &W - MZ = q \\ &W'Z = 0 \\ &W \geq O, Z \geq O. \end{aligned} \tag{13.10}$$

As we shall now see, these conditions are solved by a pivoting process. This will be accomplished by introducing an *auxiliary (artificial) variable* $z_o \in \mathcal{R}$ and a sum vector $\mathbf{1} \in \mathcal{R}^{n+m}$ into (13.10) so as to obtain

$$\begin{aligned} &W - MZ - z_o\mathbf{1} = q \\ &W \geq O, Z \geq O, z_o \geq 0 \\ &W'Z = 0 \text{ or } w_i z_i = 0, i = 1, \ldots, n, \end{aligned} \tag{13.11}$$

for at least $n-1$ of the i values given $z_o \geq 0$.

The algorithm starts with an initial basic feasible solution to (13.8) of the form $W = q, Z \geq O$, where $q \ngeq O$. To avoid having at least one $w_i < 0$, the nonnegative artificial variable z_o will be introduced at a sufficiently positive level (take $z_o = -\min_i \{q_i, i = 1, \ldots, n\}$) into the left-hand side of $W - MZ = q$. Hence our objective is to find vectors W, Z and a variable z_o such that (13.11) holds.

With each new right-hand side value $q_i + z_o$ nonnegative in (13.11), a basic solution to the same amounts to $W + z_o\mathbf{1} \geq O, Z = O$. While this basic solution is nonnegative and satisfies all the relationships in (13.11) (including complementarity), it is infeasible for the original LCP (13.8) since $z_o > 0$. A solution such as this will be termed an **almost complementary basic solution**. So while a complementary basic solution to (13.8) is one that contains exactly one basic variable from each complementary pair of variables (w_i, z_i), $i = 1, \ldots, n$, an almost complementary basic solution of (13.8) is a basic solution for (13.11) in which z_o is a basic variable and there is exactly one basic variable from each of only $n - 1$ complementary pairs of variables.

The complementary pivot method has us move from one *almost* complementary basic solution of (13.8) to another until we reach a complementary basic solution to (13.8), in which case we have w_i, $z_i = 0$ for all $i = 1, \ldots, n$. At this point, the algorithm is terminated. To see this, let us rewrite (13.11) as

$$\begin{array}{l} w_1 \qquad - m_{11}z_1 - \cdots - m_{1r}z_r - \cdots - m_{1n}z_n - z_o = q_1 \\ \quad w_2 \quad - m_{21}z_1 - \cdots - m_{2r}z_r - \cdots - m_{2n}z_n - z_o = q_2 \\ \cdots\cdots\cdots\cdots\cdots\cdots\cdots\cdots\cdots\cdots\cdots\cdots \\ \qquad w_n - m_{n1}z_1 - \cdots - m_{nr}z_r - \cdots - m_{nn}z_n - z_o = q_n \end{array}$$

with associated simplex matrix

$$\begin{array}{c} \begin{array}{cccccccccccc} w_1 & w_2 & & w_r & & w_n & z_1 & & z_r & & z_n & z_0 \end{array} \\ \left[\begin{array}{ccccccccccccc} 1 & 0 & \cdots & 0 & \cdots & 0 & -m_{11} & \cdots & -m_{1r} & \cdots & -m_{1n} & -1 & q_1 \\ 0 & 1 & \cdots & 0 & \cdots & 0 & -m_{21} & \cdots & -m_{2r} & \cdots & -m_{2n} & -1 & q_2 \\ \cdots & & & & & & & & & & & & \cdots \\ 0 & 0 & \cdots & 1 & \cdots & 0 & -m_{r1} & \cdots & -m_{rr} & \cdots & -m_{rn} & \boxed{-1} & q_r \\ \cdots & & & & & & & & & & & & \cdots \\ 0 & 0 & \cdots & 0 & \cdots & 1 & -m_{n1} & \cdots & -m_{nr} & \cdots & -m_{nn} & -1 & q_n \end{array}\right]. \end{array} \tag{13.12}$$

To find an initial almost complementary basic solution, the artificial variable z_o is made basic by having it replace the current basic variable with the most negative q_i value. Suppose $q_r = \min_i\{q_i, i = 1, \ldots, n\}$. Hence, z_o replaces w_r in the set of basic variables. To accomplish this we pivot on the (circled) element -1 in (13.12). This yields

$$\begin{array}{c} \begin{array}{ccccccccccc} w_1 & w_2 & & w_r & & w_n & z_1 & & z_r & & z_n \end{array} \\ \left[\begin{array}{ccccccccccccc} 1 & 0 & \cdots & -1 & \cdots & 0 & \overline{m}_{11} & \cdots & \overline{m}_{1r} & \cdots & \overline{m}_{1n} & 0 & \overline{q}_1 \\ 0 & 1 & \cdots & -1 & \cdots & 0 & \overline{m}_{21} & \cdots & \overline{m}_{2r} & \cdots & \overline{m}_{2n} & 0 & \overline{q}_2 \\ \cdots & & & & & & & & & & & & \cdots \\ 0 & 0 & \cdots & -1 & \cdots & 0 & \overline{m}_{r1} & \cdots & \overline{m}_{rr} & \cdots & \overline{m}_{rn} & 1 & \overline{q}_r \\ \cdots & & & & & & & & & & & & \cdots \\ 0 & 0 & \cdots & -1 & \cdots & 0 & \overline{m}_{n1} & \cdots & \overline{m}_{nr} & \cdots & \overline{m}_{nn} & 0 & \overline{q}_n \end{array}\right], \end{array} \tag{13.13}$$

where

$$\bar{q}_r = -q_r \bar{q}_i = q_i - q_r, i \neq r, i = 1, \ldots, n,$$

$$\bar{m}_{rj} = m_{rj}, j = 1, \ldots, n,$$

$$\bar{m}_{ij} = m_{rj} - m_{ij}, i \neq r, j = 1, \ldots, n.$$

As (13.13) reveals, an almost complementary basic solution to (13.8) is $w_i = \bar{q}_i, i \neq r, z_o = \bar{q}_r$, and $z_i = 0, i = 1, \ldots, n$.

The complementary pivot algorithm generates a sequence of almost complementary basic solutions until z_o becomes nonbasic or zero. Moreover, pivoting must be done in a fashion such that: (i) complementarity between the variables w_i, z_i is maintained at each iteration; and (ii) each successive basic solution is nonnegative.

Now, at the moment, both w_r and z_r are nonbasic ($w_r z_r = 0$ is satisfied). Since w_r turned nonbasic in (13.13), the appropriate variable to choose for entry into the set of basic variables is its complement z_r. In fact, this selection criterion for determining the nonbasic variable to become basic is referred to as the **complementary pivot rule**: choose as the incoming basic variable the one complementary to the basic variable, which just turned nonbasic.

Once the entering variable is selected, the outgoing variable can readily be determined from the standard simplex exit criterion, i.e. w_k is replaced by z_k in the set of basic variables if

$$\hat{\theta} = \frac{\bar{q}_k}{\bar{m}_{kr}} = \min_i \left\{ \frac{\bar{q}_i}{\bar{m}_{ir}}, \bar{m}_{ir} > 0, i = 1, \ldots, n \right\}. \tag{13.14}$$

A pivot operation is then performed using $\bar{m}_{kr}$ as the pivot element. Once w_k turns nonbasic, the complementary pivot rule next selects z_k for entry into the set of basic variables. Complementary pivoting continues until one of two possible outcomes obtains, at which point the complementary pivot algorithm terminates:

1) The simplex exit criterion selects row r as the pivot row and z_o turns nonbasic. The resulting solution is a complementary basic solution to (13.8).
2) No $\bar{m}_{ir} > 0$ in (13.14). (In this latter instance, the problem either has no feasible solution or, if a primal (dual) feasible solution exists, the primal (dual) objective function is unbounded.)

Example 13.2 Let us use the complementary pivot routine to solve the quadratic program:

$$\max f = -2x_1 + x_2 - 3x_1^2 - 4x_1x_2 - 2x_2^2 \text{ s.t.}$$

$$3x_1 + x_2 \leq 6$$

$$-\frac{3}{2}x_1 - 2x_2 \leq -6$$

$$x_1, x_2 \geq 0.$$

For this problem:

$$W=\begin{bmatrix}U_2\\ Y\end{bmatrix}=\begin{bmatrix}w_1\\ w_2\\ w_3\\ w_4\end{bmatrix}, Z=\begin{bmatrix}X\\ U_1\end{bmatrix}=\begin{bmatrix}z_1\\ z_2\\ z_3\\ z_4\end{bmatrix}, q=\begin{bmatrix}-C\\ b\end{bmatrix}=\begin{bmatrix}2\\ -1\\ 6\\ -6\end{bmatrix}, Q=\begin{bmatrix}-3 & -2\\ -2 & -2\end{bmatrix}$$

$$A=\begin{bmatrix}3 & 1\\ -\frac{3}{2} & -2\end{bmatrix}, \text{and } M=\begin{bmatrix}-2Q & A'\\ -A & O\end{bmatrix}=\begin{bmatrix}6 & 4 & 3 & -\frac{3}{2}\\ 4 & 4 & 1 & -2\\ -3 & -1 & 0 & 0\\ \frac{3}{2} & 2 & 0 & 0\end{bmatrix}.$$

(Note that, as required, Q is negative definite since $M_1 = -3 < 0$, $M_2 = 2 > 0$.)
Going directly to (13.12) we have, via (13.11),

$$\begin{array}{c} \begin{matrix} w_1 & w_2 & w_3 & w_4 & z_1 & z_2 & z_3 & z_4 & z_o & \end{matrix}\\ \left[\begin{array}{ccccccccc|c} 1 & 0 & 0 & 0 & -6 & -4 & -3 & \frac{3}{2} & -1 & -2\\ 0 & 1 & 0 & 0 & -4 & -4 & -1 & 2 & -1 & -1\\ 0 & 0 & 1 & 0 & 3 & 1 & 0 & 0 & -1 & 6\\ 0 & 0 & 0 & 1 & -\frac{3}{2} & -2 & 0 & 0 & \textcircled{-1} & -6 \end{array}\right] \end{array} \begin{array}{l} w_1z_1 = 2(0) = 0\\ w_2z_2 = (-1)0 = 0\\ w_3z_3 = 6(0) = 0\\ w_4z_4 = (-6)0 = 0 \end{array} \tag{13.15}$$

Clearly, the requisite complementarity conditions hold in (13.15). And since $6 = -\min\{q_i, i = 1, \ldots, 4\}$, z_o displaces w_4 in the set of basic variables. Upon pivoting on the circled element in (13.15) we obtain an almost complementary basic solution (see (13.16)) consisting of $w_1 = 8$, $w_2 = 5$, $w_3 = 12$, and $z_o = 6$.

$$\begin{array}{c} \begin{matrix} & & & w_4 & & & & z_4 & z_o & \end{matrix}\\ \left[\begin{array}{cccccccccc} 1 & 0 & 0 & -1 & -\frac{9}{2} & -2 & -3 & \frac{3}{2} & 0 & 8\\ 0 & 1 & 0 & -1 & -\frac{5}{2} & -2 & -1 & \textcircled{2} & 0 & 5\\ 0 & 0 & 1 & -1 & \frac{9}{2} & 3 & 0 & 0 & 0 & 12\\ 0 & 0 & 0 & -1 & \frac{3}{2} & 2 & 0 & 0 & 1 & 6 \end{array}\right] \end{array} \begin{array}{l} w_1z_1 = 8(0) = 0\\ w_2z_2 = 5(0) = 0\\ w_3z_3 = 2(0) = 0\\ w_4z_4 = 0(0) = 0 \end{array} \tag{13.16}$$

Since w_4 turned nonbasic, its complementary variable z_4 may be introduced into the set of basic variables. With $\frac{5}{2} = \min\left\{\frac{16}{3}, \frac{5}{2}\right\}$, we see that the circled element in (13.16) serves as the pivot. Upon pivoting, we obtain a second, almost complementary basic solution consisting of $w_1 = \frac{17}{4}$, $w_3 = 12$, $z_4 = \frac{5}{2}$, and $z_o = 6$ (see (13.17)).

$$\begin{array}{c} \begin{matrix} & w_2 & & & & z_2 & & z_o & & \end{matrix} \\ \begin{bmatrix} 1 & -\frac{3}{4} & 0 & -\frac{1}{4} & -\frac{21}{8} & -\frac{1}{2} & -\frac{9}{4} & 0 & 0 & \frac{17}{4} \\ 0 & \frac{1}{2} & 0 & -\frac{1}{2} & -\frac{5}{4} & -1 & -\frac{1}{2} & 1 & 0 & \frac{5}{2} \\ 0 & 0 & 1 & -1 & \frac{9}{2} & 3 & 0 & 0 & 0 & 12 \\ 0 & 0 & 0 & -1 & \frac{3}{2} & \textcircled{2} & 0 & 0 & 1 & 6 \end{bmatrix} \end{array} \quad \begin{aligned} w_1z_1 &= \frac{17}{4}(0) = 0 \\ w_2z_2 &= 0(0) = 0 \\ w_3z_3 &= 12(0) = 0 \\ w_4z_4 &= 0\left(\frac{5}{2}\right) = 0 \end{aligned} \tag{13.17}$$

Given that w_2 has turned nonbasic, its complementary variable z_2 can be made basic. Since $3 = \min\{3, 4\}$, the circled element in (13.17) becomes the pivot. A pivot operation now yields a complementary basic solution (since z_o turns nonbasic) with values $w_1 = \frac{23}{4}, w_3 = 3, z_2 = 3,$ and $z_4 = \frac{11}{2}$ (see (13.18)).

$$\begin{array}{c} \begin{matrix} & & & & & & & & z_o & \end{matrix} \\ \begin{bmatrix} 1 & -\frac{3}{4} & 0 & 0 & -\frac{9}{4} & 0 & -\frac{9}{4} & 0 & \frac{1}{4} & \frac{23}{4} \\ 0 & \frac{1}{2} & 0 & -1 & -\frac{1}{2} & 0 & -\frac{1}{2} & 1 & \frac{1}{2} & \frac{11}{2} \\ 0 & 0 & 1 & \frac{1}{2} & \frac{9}{4} & 0 & 0 & 0 & -\frac{3}{2} & 3 \\ 0 & 0 & 0 & -\frac{1}{2} & \frac{3}{4} & 1 & 0 & 0 & \frac{1}{2} & 3 \end{bmatrix} \end{array} \quad \begin{aligned} w_1z_1 &= \frac{23}{4}(0) = 0 \\ w_2z_2 &= 0(3) = 0 \\ w_3z_3 &= 3(0) = 0 \\ w_4z_4 &= 0\left(\frac{11}{2}\right) = 0 \end{aligned} \tag{13.18}$$

Since we ultimately have

$w_1 = u_{21} = 23/4$

$w_2 = u_{22} = 0$

$w_3 = x_3 = 3$ (the first structural constraint is not binding)

$w_4 = x_4 = 0$ (the second structural constraint is binding)

$$z_1 = x_1 = 0$$
$$z_2 = x_2 = 3$$
$$z_3 = u_{11} = 0$$
$$z_4 = u_{12} = 11/2$$

it follows that $\max f = -15$. ∎

13.5 Quadratic Programming and Activity Analysis

In Chapter 7 we considered a generalized linear activity analysis model in which a perfectly competitive firm produces p different products, each corresponding to a separate activity operated at the level x_j, $j = 1, \ldots, p$. The exact profit maximization model under consideration was given by (7.27), i.e.

$$\max f(x_1,\ldots,x_p) = \sum_{j=1}^{p}\left(p_j - \sum_{i=l+1}^{m} q_i a_{ij} - h_j\right) x_j \quad \text{s.t.}$$

$$\sum_{j=1}^{p} a_{ij}x_j \le b_i, i = 1,\ldots,l,$$

$$x_j \ge 0, j = 1,\ldots,p,$$

or, in matrix form,

$$\max f(\boldsymbol{X}) = (\boldsymbol{P} - (\boldsymbol{A}^*)'\boldsymbol{Q} - \boldsymbol{H})'\boldsymbol{X} \quad \text{s.t.}$$
$$\bar{\boldsymbol{A}}\boldsymbol{X} \le \boldsymbol{b}, \boldsymbol{X} \ge \boldsymbol{O}.$$

In this formulation, total revenue is $TR = \sum_{j=1}^{p} p_j x_j = \boldsymbol{P}'\boldsymbol{X}$, total variable input cost is $TVC = \sum_{j=1}^{p}\left(\sum_{i=l+1}^{m} q_i a_{ij}\right) x_j = \boldsymbol{Q}'\boldsymbol{A}^*\boldsymbol{X}$, and total conversion cost is $TCC = \sum_{j=1}^{p} h_j x_j = \boldsymbol{H}'\boldsymbol{X}$. Here p_j (the jth component of $\boldsymbol{P}$) is the constant price per unit of output of activity j, q_i (the ith component of $\boldsymbol{Q}$) is the constant unit price of the ith variable input, and $h_j = \sum_{i=l+1}^{m} r_{ij} a_{ij} = \boldsymbol{r}'\bar{\boldsymbol{a}}_j$ is the cost of converting the $m - l$ fixed inputs to the operation of the jth activity at the unit level, where r_{ij} (the ith component of $\boldsymbol{r}_j$) is the cost of converting one unit of ith fixed factor to the jth activity.

Let us now relax the assumption of given product prices p_j and assume, instead, that the firm is a monopolist in the markets for the p outputs. In this regard, there exists a downward sloping demand curve for each of the p products so that an inverse relationship exists between p_j and x_j, e. g., $p_j = a_j - b_j x_j$, $dp_j/dx_j = -b_j < 0$. Then total revenue from sales of the output of the jth activity

$TR_j = p_j x_j = a_j x_j - b_j x_j^2$. Hence total revenue is now $TR = \sum_{j=1}^{p} TR_j = \sum_{j=1}^{p} a_j x_j - \sum_{j=1}^{p} b_j x_j^2 = \boldsymbol{a}'\boldsymbol{X} - \boldsymbol{X}'\boldsymbol{B}\boldsymbol{X}$, where $\boldsymbol{a}' = (a_1, \ldots, a_p)$ and $\boldsymbol{B}$ is a $(p \times p)$ symmetric positive definite diagonal matrix of the form

$$\boldsymbol{B} = \begin{pmatrix} b_1 & & & \boldsymbol{O} \\ & b_2 & & \\ & & \ddots & \\ \boldsymbol{O} & & & b_p \end{pmatrix}.$$

Clearly total revenue under this new specification of price behavior is the sum of a linear and a quadratic form in $\boldsymbol{X}$. Moreover, the derivative of total revenue TR with respect to $\boldsymbol{X}$ is marginal revenue MR or

$$MR = \nabla TR = \boldsymbol{a} - 2\boldsymbol{B}\boldsymbol{X}$$
$$= \begin{bmatrix} a_1 - 2b_1 x_1 \\ a_2 - 2b_2 x_2 \\ \vdots \\ a_p - 2b_p x_p \end{bmatrix} = \begin{bmatrix} MR_1 \\ MR_2 \\ \vdots \\ MR_p \end{bmatrix},$$

i.e. the jth component of this vector is the marginal revenue from the sale of x_j or $MR_j = dTR_j/dx_j = a_j - 2b_j x_j$, $j = 1, \ldots, p$.

In view of this respecification of total revenue, the above profit maximization model appears as

$$\max f(x_1, \ldots, x_p) = \sum_{j=1}^{p} \left(a_j - b_j x_j - \sum_{i=l+1}^{m} q_i a_{ij} - h_j \right) x_j \quad \text{s.t.}$$

$$\sum_{j=1}^{p} a_{ij} x_j \le b_i, i = 1, \ldots, l,$$

$$x_j \ge 0, j = 1, \ldots, p,$$

or

$$\max f(\boldsymbol{X}) = \left(\boldsymbol{a} - \boldsymbol{B}\boldsymbol{X} - (\boldsymbol{A}^*)'\boldsymbol{Q} - \boldsymbol{H}\right)'\boldsymbol{X}$$
$$= \left(\boldsymbol{a} - (\boldsymbol{A}^*)'\boldsymbol{Q} - \boldsymbol{H}\right)'\boldsymbol{X} - \boldsymbol{X}'\boldsymbol{B}\boldsymbol{X} \quad \text{s.t.} \tag{13.19}$$

$$\bar{\boldsymbol{A}}\boldsymbol{X} \le \boldsymbol{b}, \boldsymbol{X} \ge \boldsymbol{O}.$$

Additionally, the Karush-Kuhn-Tucker necessary and sufficient conditions for an optimum appear (using (13.2)) as:

$$\begin{aligned}
&\text{(a) } \boldsymbol{a}-(\boldsymbol{A}^{*})'\boldsymbol{Q}-\boldsymbol{H}-2\boldsymbol{B}\boldsymbol{X}-\bar{\boldsymbol{A}}'\boldsymbol{U}_o \le \boldsymbol{O}\\
&\text{(b) } \boldsymbol{X}_o'\left[\boldsymbol{a}-(\boldsymbol{A}^{*})'\boldsymbol{Q}-\boldsymbol{H}-2\boldsymbol{B}\boldsymbol{X}-\bar{\boldsymbol{A}}'\boldsymbol{U}_o\right]=0\\
&\text{(c) } \boldsymbol{U}_o'(\boldsymbol{b}-\bar{\boldsymbol{A}}\boldsymbol{X}_o)=0\\
&\text{(d) } \boldsymbol{b}-\bar{\boldsymbol{A}}\boldsymbol{X}_o \ge \boldsymbol{O}\\
&\text{(e) } \boldsymbol{X}_o \ge \boldsymbol{O}, \boldsymbol{U}_o \ge \boldsymbol{O}.
\end{aligned} \tag{13.20}$$

Upon examining (13.20) we see that $\bar{\boldsymbol{A}}'\boldsymbol{U}_o \ge \boldsymbol{a}-2\boldsymbol{B}\boldsymbol{X}-(\boldsymbol{A}^{*})'\boldsymbol{Q}-\boldsymbol{H}$ or, in terms of the individual components of this inequality, the imputed cost of operating activity j at the unit level is at least as great as the gross profit margin for activity j, the latter being expressed as the marginal revenue of activity j less the total cost of operating activity j at the unit level less the cost of converting the $m - l$ fixed factors to the operation of the jth activity at the unit level. (The interpretation of (13.20b, c, d, e) is similar to the one advanced for (12.28) and will not be duplicated here.)

Turning next to the dual program associated with (13.19) we have

$$\begin{aligned}
&\min g(\boldsymbol{X},\boldsymbol{U})=\boldsymbol{b}'\boldsymbol{U}+\boldsymbol{X}'\boldsymbol{B}\boldsymbol{X} \quad \text{s.t.}\\
&\bar{\boldsymbol{A}}'\boldsymbol{U}-2\boldsymbol{B}\boldsymbol{X} \ge \boldsymbol{a}-(\boldsymbol{A}^{*})'\boldsymbol{Q}-\boldsymbol{H}\\
&\boldsymbol{U} \ge \boldsymbol{O}, \boldsymbol{X} \text{ unrestricted}
\end{aligned} \tag{13.21}$$

or, for $\boldsymbol{X}$ fixed at $\boldsymbol{X}_o$ (the optimal solution vector to the primal quadratic program),

$$\begin{aligned}
&\min g(\boldsymbol{X}_o,\boldsymbol{U})-\boldsymbol{X}_o'\boldsymbol{B}\boldsymbol{X}_o=\boldsymbol{b}'\boldsymbol{U} \quad \text{s.t.}\\
&\bar{\boldsymbol{A}}'\boldsymbol{U} \ge \boldsymbol{a}-2\boldsymbol{B}\boldsymbol{X}_o-(\boldsymbol{A}^{*})'\boldsymbol{Q}-\boldsymbol{H}, \boldsymbol{U} \ge \boldsymbol{O}.
\end{aligned} \tag{13.21.1}$$

Either (13.21) or (13.21.1) is appropriate for considering the dual program from a computational viewpoint. However, for purposes of interpreting the dual program in an activity analysis context, let us first examine the generalized nonlinear dual objective from which $g(\boldsymbol{X}, \boldsymbol{U})$ in (13.21) was derived, namely

$$\begin{aligned}
g(\boldsymbol{X},\boldsymbol{U}) &= L-\boldsymbol{X}'\nabla_{\boldsymbol{X}}L\\
&= \left(\boldsymbol{a}-(\boldsymbol{A}^{*})'\boldsymbol{Q}-\boldsymbol{H}\right)'\boldsymbol{X}-\boldsymbol{X}'\boldsymbol{B}\boldsymbol{X}+\boldsymbol{U}'(\boldsymbol{b}-\bar{\boldsymbol{A}}\boldsymbol{X})\\
&\quad -\boldsymbol{X}'\left(\boldsymbol{a}-(\boldsymbol{A}^{*})'\boldsymbol{Q}-\boldsymbol{H}-2\boldsymbol{B}\boldsymbol{X}-\bar{\boldsymbol{A}}'\boldsymbol{U}\right)\\
&= \boldsymbol{b}'\boldsymbol{U}+\left[\left(\boldsymbol{a}-(\boldsymbol{A}^{*})'\boldsymbol{Q}-\boldsymbol{H}\right)'\boldsymbol{X}-\boldsymbol{X}'\boldsymbol{B}\boldsymbol{X}-\boldsymbol{U}'\bar{\boldsymbol{A}}\boldsymbol{X}\right]\\
&\quad -\boldsymbol{X}'\left[\boldsymbol{a}-(\boldsymbol{A}^{*})'\boldsymbol{Q}-\boldsymbol{H}-2\boldsymbol{B}\boldsymbol{X}-\bar{\boldsymbol{A}}'\boldsymbol{U}\right].
\end{aligned} \tag{13.22}$$

The first term on the right-hand side of (13.22) is simply the total imputed value of the firm's supply of fixed or scarce resources (the dual objective in a linear activity analysis model). The second term can be interpreted as **economic rent** (the difference between total profit and the total imputed cost of all inputs *used*).[3] Finally, if the square-bracketed portion of the third term (which is nonpositive by virtue of (13.20a)) is thought of as a set of accounting or opportunity losses generated by a marginal increase in the level of operation of activity j, then the entire third term is a weighted sum of these losses (the weights being the various activity levels) and thus amounts to the marginal opportunity loss of all outputs. At an optimal solution, however, (13.20b) reveals that this third term is zero (Balinski and Baumol 1968). Next, upon examining the dual structural constraint in (13.21) we see that $\bar{A}'U \geq a - 2BX - (A^*)'Q - H$ so that, as in the linear activity model, the imputed cost of operating each activity at the unit level must equal or exceed its gross profit margin. In sum, the dual problem seeks a constrained minimum to the total imputed value of all scarce resources plus payments to economic rent plus losses due to unprofitable activities.

13.6 Linear Fractional Functional Programming

(Charnes and Cooper 1962; Lasdon 1970; Martos 1964; Craven 1978)

In what follows we shall employ the simplex algorithm to solve an optimization problem, known as a **linear fractional programming problem**, in which the objective function is nonlinear in that it is expressed as the ratio of two linear functions and the variables must satisfy a system of linear inequalities and nonnegativity conditions. Specifically, let us

$$\begin{aligned} &\text{maximize } f(X) = \frac{c_o + C'X}{d_o + D'X} \text{ s.t.} \\ &AX \leq b, X \geq O, \end{aligned} \tag{13.23}$$

where c_o, d_o are scalars, C and D are $(p \times 1)$ vectors with components c_j, d_j respectively, $j = 1, \ldots, p$, and A is of order $(m \times p)$. Although f is neither convex

3 In general, economic rent may be thought of as the difference between the actual earnings of a unit of a factor of production and its supply price. Actual earnings is simply the price that a factor receives for selling its services for a given period of time while its supply price is the minimum amount of money required to retain it in its current use. (If the costs of transfer of the factor from one use to another are zero, the supply price equals the maximum amount it could earn in any alternative endeavor and thus depicts an opportunity cost calculation.) In the short run, such returns are referred to as **quasi-rents** – the reward to a factor whose supply is fixed in the short run but not in the long run. For a further discussion on these concepts, see Ryan (1962).

nor concave, its contours $(c_o + C'X)/(d_o + D'X) =$ constant are hyperplanes. Moreover, any finite constrained local maximum of f is also global in character and occurs at an extreme point of the feasible region $\mathcal{K}$. That is,

Theorem 13.4 If $\mathcal{K}$ is a convex polyhedron and $d_o + D'X > 0$ for all $X \varepsilon \mathcal{K}$, then f assumes a finite maximum at an extreme point or at a convex combination of extreme points of $\mathcal{K}$.

Proof. Let $\bar{X}_1,\ldots,\bar{X}_l$ denote the extreme points of $\mathcal{K}$. If X is not an extreme point of $\mathcal{K}$, then it can be expressed as a convex combination of these extreme points as

$$X = \sum_{i=1}^{l} \theta_i \bar{X}_i, \theta_i \geq 0, \sum_{i=1}^{l} \theta_i = 1.$$

Assume that the maximum of f occurs at the extreme point $\bar{X}_o \varepsilon \mathcal{K}$ so that $f(\bar{X}_o) \geq f(\bar{X}_i), i = 1,\ldots,l,$ i.e.

$$\frac{c_o + C'X_o}{d_o + D'X_o} \geq \frac{c_o + C'\bar{X}_i}{d_o + D'\bar{X}_i}, i = 1,\ldots,l.$$

Then $(c_o + C'\bar{X}_o)(d_o + D'\bar{X}_i) \geq (d_o + D'\bar{X}_0)(c_o + C'\bar{X}_i)$ and thus, upon multiplying each side of this inequality by θ_i and summing over all values of i,

$$(c_o + C'\bar{X}_o)\left(d_o + \sum_{i=1}^{l} \theta_i D'\bar{X}_i\right) \geq (d_o + D'\bar{X}_0)\left(c_o + \sum_{i=1}^{l} \theta_i C'\bar{X}_i\right) \text{ or}$$

$$(c_o + C'\bar{X}_o)\left(d_o + \sum_{j=1}^{p} d_j \sum_{i=1}^{l} \theta_i \bar{x}_{ij}\right) \geq (d_o + D'\bar{X}_0)\left(c_o + \sum_{j=1}^{p} c_j \sum_{i=1}^{l} \theta_i \bar{x}_{ij}\right). \tag{13.24}$$

Using the expression for X in terms of the extreme points of $\mathcal{K}$, (13.24) becomes

$$(c_o + C'\bar{X}_o)\left(d_o + \sum_{j=1}^{p} d_j x_j\right) \geq (d_o + D'\bar{X}_0)\left(c_o + \sum_{j=1}^{p} c_j x_j\right) \text{ or}$$

$$(c_o + C'\bar{X}_o)(d_o + D'\bar{X}) \geq (d_o + D'\bar{X}_0)(c_o + C'\bar{X})$$

and thus

$$\frac{c_o + C'\bar{X}_o}{d_o + D'\bar{X}_o} \geq \frac{c_o + C'X}{d_o + D'X}, \text{i.e.,} f(\bar{X}_o) \geq f(X).$$

Hence, there exists an extreme point at which f attains its maximum value.

Next, let f assume its maximum at more than one extreme point, e.g. for $\bar{X}_1,\ldots,\bar{X}_k, k<l$, let $f(\bar{X}_o)=f(\bar{X}_1)=\cdots=f(\bar{X}_k)$. Let $\hat{X}$ be any convex combination of these extreme points, i.e.

$$\hat{X}=\sum_{i=1}^{k}\theta_i\bar{X}_i, \theta_i\geq 0, \sum_{i=1}^{k}\theta_i=1.$$

Then,

$$\frac{c_o+C'\hat{X}}{d_o+D'\hat{X}}=\frac{c_o+\sum_{i=1}^{k}\theta_i C'\bar{X}_i}{d_o+\sum_{i=1}^{k}\theta_i D'\bar{X}_i}=\frac{c_o+C'\bar{X}_o}{d_o+D'\bar{X}_o}$$

since $C'\bar{X}_o=C'\bar{X}_i, D'\bar{X}_o=D'\bar{X}_i, i=1,\ldots,k$. Q.E.D.

Let us next examine a couple of direct (in the sense that no variable transformation or additional structural constraints or variables are introduced) approaches to solving the above fractional program, which mirror the standard simplex method. For the first technique (Swarup 1965), let us start from an initial basic feasible solution and, under the assumption that $d_o+D'X>0$ for all $X \in \mathcal{K}$, demonstrate the conditions under which the objective value can be improved. Doing so will ultimately provide us with a set of conditions that an optimal basic feasible solution must satisfy.

Upon introducing the slack variables $x_{p+1},\ldots,x_n$ into the structural constraint system, we may write the $(m\times n)$ coefficient matrix of the same as $A=[B\vdots R]$, where the $(m\times m)$ matrix $B=[b_1,\ldots,b_m]$ has as its columns the columns of A corresponding to basic variables and the $(m\times n-m)$ matrix R contains all remaining columns of A. Then solving for the basic variables in terms of the nonbasic variables we have the familiar equality $X_B=B^{-1}b-B^{-1}RX_R$, where $X_B'=(x_{B1},\ldots,x_{Bm}), X_R'=(x_{R1},\ldots,x_{Rn-m})$. For $X_R=O$, we have the basic solution $X_B=B^{-1}b$. If $X_B\geq O$, the solution is deemed feasible. Let us partition the vectors C, D as

$$C=\begin{bmatrix}C_B\\ \cdots\\ C_R\end{bmatrix}, D=\begin{bmatrix}D_B\\ \cdots\\ D_R\end{bmatrix},$$

respectively, where the $(m\times 1)$ vectors C_B, D_B contain as their elements the objective function coefficients corresponding to the basic variables while the components of the $(n - m \times 1)$ vectors $\boldsymbol{C_R}$, $\boldsymbol{D_R}$ correspond to the objective function coefficients associated with nonbasic variables.

Let

$$z_{B1} = c_o + \boldsymbol{C}'_{\boldsymbol{B}}\boldsymbol{X}_{\boldsymbol{B}},$$

$$z_{B2} = d_o + \boldsymbol{D}'_{\boldsymbol{B}}\boldsymbol{X}_{\boldsymbol{B}}$$

and, for those nonbasic columns $\boldsymbol{r}_j$, $j = 1, \ldots, n - m$, of $\boldsymbol{R}$, let $\boldsymbol{r}_j = \boldsymbol{B}\boldsymbol{Y}_j$ so that $\boldsymbol{Y}_j = \boldsymbol{B}^{-1}\boldsymbol{r}_j$, i.e. $\boldsymbol{Y}_j$ is the jth column of $\boldsymbol{B}^{-1}\boldsymbol{R}$. If $z_{1j} = \boldsymbol{C}'_{\boldsymbol{B}}\boldsymbol{Y}_j$, $z_{2j} = \boldsymbol{D}'_{\boldsymbol{B}}\boldsymbol{Y}_j$, then the optimality evaluators may be expressed as

$$\bar{c}_{1j} = c_{Rj} - z_{1j} = c_{Rj} - \boldsymbol{C}'_{\boldsymbol{B}}\boldsymbol{Y}_j,$$

$$\bar{c}_{2j} = d_{Rj} - z_{2j} = d_{Rj} - \boldsymbol{D}'_{\boldsymbol{B}}\boldsymbol{Y}_j,$$

where c_{Rj}, d_{Rj} are, respectively, the jth component of $\boldsymbol{C_R}$, $\boldsymbol{D_R}$, $j = 1, \ldots, n - m$.

Our goal is now to find an alternative basic feasible solution that exhibits an improved value of $f = z_{B1}/z_{B2}$. If we change the basis one vector at a time by replacing $\boldsymbol{b}_j$ by $\boldsymbol{r}_j$, we obtain a new basis matrix $\hat{\boldsymbol{B}} = [\hat{\boldsymbol{b}}_1, \ldots, \hat{\boldsymbol{b}}_m]$, where $\hat{\boldsymbol{b}}_i = \boldsymbol{b}_i, i \neq r; \hat{\boldsymbol{b}}_r = \boldsymbol{r}_j$. Then the values of the new basic variables are, from (4.11.1),

$$x_{\hat{B}i} = x_{Bi} - \hat{\theta} y_{ij} = x_{Bi} - \frac{x_{Br}}{y_{rj}}, i \neq r;$$

$$x_{\hat{B}r} = \hat{\theta} = x_{Br}/y_{rj}.$$

And from (4.12), the new value of the objective function is

$$\frac{z_{\hat{B}1}}{z_{\hat{B}2}} = \frac{z_{B1} + \hat{\theta}\,\bar{c}_{1j}}{z_{B2} + \hat{\theta}\,\bar{c}_{2j}}.$$

Clearly the value of f will improve if

$$\frac{z_{B1} + \hat{\theta}\,\bar{c}_{1j}}{z_{B2} + \hat{\theta}\,\bar{c}_{2j}} > \frac{z_{B1}}{z_{B2}} \quad \text{or}$$

$$z_{B2}\left(z_{B1} + \hat{\theta}\,\bar{c}_{1j}\right) - z_{B1}\left(z_{B2} + \hat{\theta}\,\bar{c}_{2j}\right) > 0$$

since $z_{B2}, z_{B2} + \hat{\theta}\,\bar{c}_{2j} > 0$ given that $d_o + \boldsymbol{D}'\boldsymbol{X} > 0$. Simplifying the preceding inequality yields

$$\Delta_j = z_{B2}\,\bar{c}_{1j} - z_{B1}\,\bar{c}_{2j} > 0, j = 1, \ldots, n-m. \tag{13.25}$$

(If $\hat{\theta} - 0$, the value of f is unchanged and the degenerate case emerges.)

If for any $\boldsymbol{r}_j$ we find that $\Delta_j > 0$ and if at least one component y_{ij} of $\boldsymbol{Y}_j$ is positive (here $\boldsymbol{Y}_j$ corresponds to the first m elements of the jth nonbasic column of the simplex matrix), then it is possible to obtain a new basic feasible solution from the old one by replacing one of the columns of $\boldsymbol{B}$ by $\boldsymbol{r}_j$ with the result that $z_{\hat{B}1}/z_{\hat{B}2} > z_{B1}/z_{B2}$. For the entry criterion, let us adopt the convention that the incoming basic variable is chosen according to

$$\Delta_r = \max\{\Delta_j, j = 1, \ldots, n-m\},$$

i.e. x_{Br} enters the set of basic variables. (Note that this choice criterion gives us the largest increase in f.) In addition, the exit criterion is the same as the one utilized by the standard simplex routine. This procedure may be repeated until, in the absence of degeneracy, the process converges to an optimal basic feasible solution. Termination occurs when $\Delta_j \leq 0$ for all nonbasic columns $\boldsymbol{r}_j$, $j = 1, \ldots, n - m$. It is imperative to note that this method cannot be used if both c_o, d_o are zero, i.e. if we start from an initial basic feasible solution with $\boldsymbol{C_B} = \boldsymbol{C_D} = \boldsymbol{O}$ and $c_o = d_o = 0$, then $z_{B1} = z_{B2} = 0$ also and thus $\Delta_j = 0$ for all j, thus indicating that the initial basic feasible solution is optimal. Clearly, this violates the requirement that $d_o + \boldsymbol{D}'\boldsymbol{X} > 0$.

Example 13.3 (Swarup 1965, pp. 1034–1036) Let us

$$\max f = \frac{5x_1 + 3x_2}{1 + 5x_1 + 2x_2} \quad \text{s.t.}$$

$$3x_1 + 5x_2 \leq 15$$

$$5x_1 + 2x_2 \leq 10$$

$$x_1, x_2 \geq 0.$$

For x_3, x_4 nonnegative slack variables, the problem may be rewritten as

$$\max f = \frac{5x_1 + 3x_2 + 0x_3 + 0x_4}{1 + 5x_1 + 2x_2 + 0x_3 + 0x_4} \quad \text{s.t.}$$

$$3x_1 + 5x_2 + x_3 \quad = 15$$

$$5x_1 + 2x_2 \quad x_4 = 10$$

$$x_1, \ldots, x_4 \geq 0,$$

with $c_o = 0$, $d_o = 1$.

If we let x_3, x_4 serve as basic variables, then the initial basic feasible solution is $x_3 = 15$, $x_4 = 10$, and $f = 0$. In this instance $\boldsymbol{C_B} = \boldsymbol{D_B} = \boldsymbol{O}$, $\boldsymbol{B} = \boldsymbol{I}_2$, and

$$z_{B1} = c_o + \boldsymbol{C}_B'\boldsymbol{X}_B = 0$$

$$z_{B2} = d_o + \boldsymbol{D}_B'\boldsymbol{X}_B = 1.$$

Since the associated simplex matrix (with the usual objective function row deleted) assumes the form

$$\begin{bmatrix} 3 & 5 & 1 & 0 & 15 \\ \textcircled{5} & 2 & 0 & 1 & 10 \end{bmatrix}, \tag{13.26}$$

it follows that

$$\boldsymbol{Y}_1 = \begin{bmatrix} 3 \\ 5 \end{bmatrix}, \boldsymbol{Y}_2 = \begin{bmatrix} 5 \\ 2 \end{bmatrix}$$

with $z_{1j} = \boldsymbol{C}_B'\boldsymbol{Y}_j = 0, z_{2j} = \boldsymbol{D}_B'\boldsymbol{Y}_j = 0, j = 1,2$. Then

$$\bar{c}_{11} = c_{R1} - z_{11} = 5,$$

$$\bar{c}_{12} = c_{R2} - z_{12} = 3,$$

$$\bar{c}_{21} = d_{R1} - z_{21} = 5,$$

$$\bar{c}_{22} = d_{R2} - z_{22} = 2.$$

Using (13.25),

$$\Delta_1 = z_{B2}\,\bar{c}_{11} - z_{B1}\,\bar{c}_{21} = 1{\cdot}5 - 0{\cdot}5 = 5,$$

$$\Delta_2 = z_{B2}\,\bar{c}_{12} - z_{B1}\,\bar{c}_{22} = 1{\cdot}3 - 0{\cdot}2 = 3.$$

Since the maximum of these Δ values is 5, it follows that x_1 is introduced into the set of basic variables. Pivoting on the circled element in (13.26) yields the new simplex matrix

$$\begin{bmatrix} 0 & 19/5 & 1 & -3/5 & 9 \\ 1 & 2/5 & 0 & 1/5 & 2 \end{bmatrix} \tag{13.27}$$

with $x_{B1} = x_3 = 9$, $x_{B2} = x_1 = 2$, and $f = 1$. Since we do not have $\Delta_j \le 0$ for all j, let us undertake an additional round of calculations.

Since $x_{B1} = x_3$, $x_{B2} = x_1$ are now basic variables, it follows that

$$\boldsymbol{C_B} = \begin{bmatrix} 0 \\ 5 \end{bmatrix}, \boldsymbol{D_B} = \begin{bmatrix} 0 \\ 5 \end{bmatrix}, \text{and } \boldsymbol{B} = \begin{bmatrix} 1 & 3 \\ 0 & 5 \end{bmatrix}.$$

Then

$$z_{B1} = c_o + C'_B X_B = 0 + (0,5)\begin{bmatrix} 9 \\ 2 \end{bmatrix} = 10,$$

$$z_{B2} = d_o + D'_B X_B = 1 + (0,5)\begin{bmatrix} 9 \\ 2 \end{bmatrix} = 11,$$

$$z_{11} = C'_B Y_1 = (0,5)\begin{bmatrix} 19/5 \\ 2/5 \end{bmatrix} = 2,$$

$$z_{12} = C'_B Y_2 = (0,5)\begin{bmatrix} -3/5 \\ 1/5 \end{bmatrix} = 1,$$

$$z_{21} = D'_B Y_1 = (0,5)\begin{bmatrix} 19/5 \\ 2/5 \end{bmatrix} = 2,$$

$$z_{22} = D'_B Y_2 = (0,5)\begin{bmatrix} -3/5 \\ 1/5 \end{bmatrix} = 1,$$

$$\bar{c}_{11} = c_{R1} - z_{11} = 3 - 2 = 1,$$

$$\bar{c}_{12} = c_{R2} - z_{12} = 0 - 1 = -1,$$

$$\bar{c}_{21} = d_{R1} - z_{21} = 2 - 2 = 0,$$

$$\bar{c}_{22} = d_{R2} - z_{22} = 0 - 1 = -1.$$

By virtue of these calculations, we obtain

$$\Delta_1 = z_{B2}\,\bar{c}_{11} - z_{B1}\,\bar{c}_{21} = 11\cdot 1 - 10\cdot 0 = 11,$$

$$\Delta_2 = z_{B2}\,\bar{c}_{12} - z_{B1}\,\bar{c}_{22} = 11(-1) - 10(-1) = -1.$$

Upon introducing x_2 into the set of basic variables, (13.27) becomes

$$\begin{bmatrix} 0 & 1 & 5/19 & -3/19 & 45/19 \\ 1 & 0 & -2/19 & 5/19 & 20/19 \end{bmatrix} \tag{13.28}$$

with $x_{B1} = x_2 = 45/19$, $x_{B2} = x_1 = 20/19$, and $f = 235/209$. We still do not have all $\Delta_j \leq 0$ so that another iteration is warranted.

For $x_{B1} = x_2$, $x_{B2} = x_1$ basic variables, we have

$$C_B = \begin{bmatrix} 3 \\ 5 \end{bmatrix}, D_B = \begin{bmatrix} 2 \\ 5 \end{bmatrix}, \text{and } B = \begin{bmatrix} 5 & 3 \\ 2 & 5 \end{bmatrix}.$$

In addition,

$$\begin{array}{llll} z_{B1} = 235/19, & z_{11} = -5/19, & \bar{c}_{11} = -5/19, & \Delta_1 = -55/19, \\ z_{B2} = 11, & z_{12} = -16/19, & \bar{c}_{12} = -16/19, & \Delta_2 = 59/19. \\ & z_{21} = 0, & \bar{c}_{21} = 0, & \\ & z_{22} = -1, & \bar{c}_{22} = -1, & \end{array}$$

Thus x_4 turns basic and, from (13.28), we obtain

$$\begin{bmatrix} 3/5 & 1 & 1/5 & 0 & 3 \\ 19/5 & 0 & -2/5 & 1 & 4 \end{bmatrix}$$

with $x_{B1} = x_2 = 3$, $x_{B2} = x_4 = 4$, and $f = 9/7$. If the Δ_j's associated with this basic feasible solution are now determined, it is found that both are negative, thus indicating that the current solution is optimal. ■

Relative to the second direct technique (Bitran and Novaes 1973) for solving the fractional functional program, let us again assume that $d_o + \boldsymbol{D}'\boldsymbol{X} > 0$ for all $\boldsymbol{X} \varepsilon \mathcal{K}$ and use the standard simplex method to solve a related problem that is constructed from the former and appears as

$$\begin{aligned} &\max F(\boldsymbol{X}) = \boldsymbol{\lambda}'\boldsymbol{X} \quad \text{s.t.} \\ &\boldsymbol{AX} \leq \boldsymbol{b}, \boldsymbol{X} \geq \boldsymbol{O}, \text{where} \\ &\boldsymbol{\lambda} = \boldsymbol{C} - \left(\frac{\boldsymbol{C}'\boldsymbol{D}}{\boldsymbol{D}'\boldsymbol{D}}\right)\boldsymbol{D}. \end{aligned} \tag{13.29}$$

Here $\boldsymbol{\lambda}$ equals $\boldsymbol{C}$ minus the *vector projection of* $\boldsymbol{C}$ *onto* $\boldsymbol{D}$ (this latter notion is defined as a vector in the direction of $\boldsymbol{D}$ obtained by projecting $\boldsymbol{C}$ perpendicularly onto $\boldsymbol{D}$). Upon solving (13.29), we obtain the suboptimal solution point $\boldsymbol{X}^*$.

The next step in the algorithm involves utilizing $\boldsymbol{X}^*$ to construct a new objective function to be optimized subject to the same constraints, i.e. we now

$$\begin{aligned} &\max \bar{F} = [\boldsymbol{C} - f(\boldsymbol{X}^*)\boldsymbol{D}]'\boldsymbol{X} \quad \text{s.t.} \\ &\qquad \boldsymbol{AX} \leq \boldsymbol{b}, \boldsymbol{X} \geq \boldsymbol{O}. \end{aligned} \tag{13.30}$$

The starting point for solving (13.30) is the optimal simplex matrix for problem (13.42); all that needs to be done to initiate (13.30) is to replace the objective function row of the optimal simplex matrix associated with (13.29) by the objective in (13.30). The resulting basic feasible solution will be denoted as $\boldsymbol{X}^{**}$.

If $\boldsymbol{X}^{**} = \boldsymbol{X}^*$, then $\boldsymbol{X}^*$ represents the global maximal solution to the original fractional program; otherwise return to (13.30) with $\boldsymbol{X}^* = \boldsymbol{X}^{**}$ and repeat the process until the solution vector remains unchanged. (For a discussion on the geometry underlying this technique, see Bitran and Novaes 1973, pp. 25–26.)

Example 13.4 Let us resolve the problem appearing in Example 13.3 using this alternative computational method. Since

$$\boldsymbol{C} = \begin{bmatrix} 5 \\ 3 \end{bmatrix}, \boldsymbol{D} = \begin{bmatrix} 5 \\ 2 \end{bmatrix}, \boldsymbol{C}'\boldsymbol{D} = 31, \text{and } \boldsymbol{D}'\boldsymbol{D} = 29,$$

it follows that

$$\boldsymbol{\lambda} = \boldsymbol{C} - \left(\frac{\boldsymbol{C}'\boldsymbol{D}}{\boldsymbol{D}'\boldsymbol{D}}\right)\boldsymbol{D} = \begin{bmatrix} -10/29 \\ 25/29 \end{bmatrix}.$$

Hence, the problem at hand is to

$$\begin{aligned} &\max F = -10/29x_1 + 25/29x_2 \quad \text{s.t.} \\ &3x_1 + 5x_2 \le 15 \\ &5x_1 + 2x_2 \le 10 \\ &x_1, x_2 \ge 0. \end{aligned}$$

The optimal simplex matrix associated with this problem is

$$\begin{bmatrix} 3/5 & 1 & 1/5 & 0 & 0 & 3 \\ 19/5 & 0 & -2/5 & 1 & 0 & 4 \\ 25/29 & 0 & 5/29 & 0 & 1 & 75/29 \end{bmatrix} \tag{13.31}$$

so that $(\boldsymbol{X}^*)' = (0, 3)$ and

$$f(\boldsymbol{X}^*) = \frac{\boldsymbol{C}'\boldsymbol{X}^*}{1 + \boldsymbol{D}'\boldsymbol{X}^*} = \frac{9}{7}.$$

Then the objective function for the (13.30) problem is $\bar{F} = \left[\boldsymbol{C} - \frac{9}{7}\boldsymbol{D}\right]'\boldsymbol{X} = -10/7x_1 + 3/7x_2$. Upon inserting this revised objective into (13.31), we ultimately obtain

$$\begin{bmatrix} 3/5 & 1 & 1/5 & 0 & 0 & 3 \\ 19/5 & 0 & -2/5 & 1 & 0 & 4 \\ 10/7 & -3/7 & 0 & 0 & 1 & 0 \end{bmatrix} \to \begin{bmatrix} 3/5 & 1 & 1/5 & 0 & 0 & 3 \\ 19/5 & 0 & -2/5 & 1 & 0 & 4 \\ 59/35 & 0 & 3/35 & 0 & 1 & 9/7 \end{bmatrix}.$$

Here $(\boldsymbol{X}^{**})' = (0, 3) = (\boldsymbol{X}^*)'$ so that the optimal solution to the original problem is $x_2^* = 3$, $x_4^* = 4$, and $f^* = 9/7$. ∎

13.7 Duality in Linear Fractional Functional Programming (Craven and Mond 1973, 1976; Schnaible 1976; Wagner and Yuan 1968; Chadha 1971; Kydland 1962; and Kornbluth and Salkin 1972)

To obtain the dual of a linear fractional functional program, let us dualize its *equivalent linear program*. That is, under the variable transformation $\boldsymbol{Y} = t\boldsymbol{X}$, $t = (d_o + \boldsymbol{D}'\boldsymbol{X})^{-1}$, the linear fractional programming problem (13.23) or

$$\max f(\boldsymbol{X}) = \frac{c_o + \boldsymbol{C}'\boldsymbol{X}}{d_o + \boldsymbol{D}'\boldsymbol{X}} \quad \text{s.t.}$$
$$\boldsymbol{AX} \le \boldsymbol{b}, \boldsymbol{X} \ge \boldsymbol{O}$$

is equivalent to the linear program

$$\max \bar{f}(t, \boldsymbol{Y}) = c_o t + \boldsymbol{C}'\boldsymbol{Y} \quad \text{s.t.}$$
$$\text{(a)}\ d_o t + \boldsymbol{D}'\boldsymbol{Y} = \boldsymbol{1}$$
$$\text{(b)}\ -\boldsymbol{b}t + \boldsymbol{AY} \le \boldsymbol{O}$$
$$\text{(c)}\ \boldsymbol{Y} > \boldsymbol{O}, t \ge 0 \tag{13.32}$$

if $d_o + \boldsymbol{D}'\boldsymbol{X} > 0$ for all $\boldsymbol{X} \varepsilon \mathcal{K}$ and for each vector $\boldsymbol{Y} \ge \boldsymbol{O}$, the point $(t, \boldsymbol{Y}) = (0, \boldsymbol{Y})$ is not feasible, i.e. every feasible point $(t, \boldsymbol{Y})$ has $t > 0$ (Charnes and Cooper 1962). (This latter restriction holds if $\mathcal{K}$ is bounded. For if $(0, \hat{\boldsymbol{Y}})$ is a feasible solution to the equivalent linear program (13.32), then, under the above variable transformation, $\boldsymbol{A}\hat{\boldsymbol{Y}} \le t\boldsymbol{b} = \boldsymbol{O}, \hat{\boldsymbol{Y}} \ge \boldsymbol{O}$. If $\boldsymbol{X} \varepsilon \mathcal{K}$, then $\boldsymbol{X} + \theta\hat{\boldsymbol{Y}} \varepsilon \mathcal{K}$, $\theta \ge 0$, thus contradicting the boundedness of $\mathcal{K}$.) In this regard, if $\left(\tilde{t}, \tilde{\boldsymbol{Y}}\right)$ represents an optimal solution to the equivalent linear program, then $\tilde{\boldsymbol{X}} = \tilde{\boldsymbol{Y}}/\tilde{t}$ is an optimal solution to the original fractional problem.

To demonstrate this equivalence let

$$f = \frac{c_o + \boldsymbol{C}'(\boldsymbol{Y}/t)}{d_o + \boldsymbol{D}'(\boldsymbol{Y}/t)} = \frac{c_o t + \boldsymbol{C}'\boldsymbol{Y}}{d_o t + \boldsymbol{D}'\boldsymbol{Y}}.$$

Moreover, $\boldsymbol{AX} = \boldsymbol{A}(\boldsymbol{Y}/t) \le \boldsymbol{b}$ or $\boldsymbol{AY} - t\boldsymbol{b} \le \boldsymbol{O}$. If $d_o t + \boldsymbol{D}'\boldsymbol{Y} = 1$, then (13.32) immediately follows.

It now remains to dualize (13.32). Let

$$\bar{\boldsymbol{C}} = \begin{bmatrix} c_o \\ \boldsymbol{C} \end{bmatrix}, \bar{\boldsymbol{D}} = \begin{bmatrix} d_o \\ \boldsymbol{D} \end{bmatrix}, \bar{\boldsymbol{Y}} = \begin{bmatrix} t \\ \boldsymbol{Y} \end{bmatrix}, \text{and}\ \bar{\boldsymbol{A}} = \left[-\boldsymbol{b} \vdots \boldsymbol{A}\right].$$

Then (13.32) may be rewritten as

$$\begin{aligned}\max \bar{f}(\bar{Y}) &= \bar{C}'\bar{Y} \text{ s.t.}\\ \bar{D}'\bar{Y} &= 1\\ A\bar{Y} &\le O\\ \bar{Y} &\ge O.\end{aligned}$$

Upon replacing $\bar{D}'Y = 1$ by $\bar{D}'\bar{Y} \le 1, -\bar{D}'Y \le -1,$ the preceding problem becomes

$$\max \bar{f}(\bar{Y}) = \bar{C}'\bar{Y} \text{ s.t.}$$

$$\begin{bmatrix} \bar{D}' \\ -\bar{D}' \\ \bar{A} \end{bmatrix} \bar{Y} \le \begin{bmatrix} 1 \\ -1 \\ O \end{bmatrix}, \bar{Y} \ge O.$$

The symmetric dual to this problem is then

$$\min g(w,v) = w - v \text{ s.t.}$$

$$\left(\bar{D} \vdots -\bar{D} \vdots \bar{A}'\right) \begin{bmatrix} w \\ v \\ \bar{u} \end{bmatrix} \ge \bar{C}$$

$$w, v \ge 0, \bar{u} \ge O$$

or, for $\lambda = w - v$,

$$\begin{aligned}&\min g(\lambda) = \lambda \text{ s.t.}\\ &\text{(a) } d_o\lambda - b'\bar{u} \ge c_o\\ &\text{(b) } D\lambda + A'\bar{u} \ge C\\ &\text{(c) } \bar{u} \ge O, \lambda \text{ unrestricted.}\end{aligned} \tag{13.33}$$

As was the case with the primal and dual problems under linear programming, we may offer a set of duality theorems (Theorems 13.5 and 13.6) that indicate the relationship between (13.23), (13.33). To this end, we have

Theorem 13.5 If $X_o \in \mathcal{E}^p$ represents a feasible solution to the primal problem (13.23) and $(\lambda_o, \bar{u}_o) \in \mathcal{E}^{m+1}$ is a feasible solution to the dual problem (13.33), then $f(X_o) \le g(\lambda_o)$.

Proof. For X_o a feasible solution to the primal problem, $AX_o \le b$. With $\bar{u}_o \ge O$, it follows that $\bar{u}_o' AX_o \le \bar{u}_o' b$ or $X_o' A'\bar{u}_o \le b'\bar{u}_o$. For $(\lambda_o, \bar{u}_o)$ a feasible solution to the dual, (13.33a, b) yield $d_o\lambda_o - c_o \ge b'\bar{u}_o, D\lambda_o + A'\bar{u}_o \ge C,$ respectively.

Given $X_o \geq O, X'_o D\lambda_o + X'_o A'\bar{u}_o \geq X'_o C$ or $X'_o A'\bar{u}_o \geq X'_o C - X'_o D\lambda_o$. Hence $d_o\lambda_o - c_o \geq X'_o C - X'_o D\lambda_o$ or $\lambda_o \geq (c_o + C'X_o)/(d_o + D'X_o)$. Q. E. D.

We next have

Theorem 13.6 If $X_o \in \mathcal{E}^p$ represents an optimal solution to the primal problem (13.23), then there exists a vector $(\lambda_o, \bar{u}_o) \in \mathcal{E}^{m+1}$ that solves the dual problem (13.33) with $f(X_o) = g(\lambda_o)$. Conversely, if $(\lambda_o, \bar{u}_o)$ solves the dual problem, then there exists a vector $\hat{X}$ such that $\hat{X}$ solves the primal problem and $f(\hat{X}) = g(\lambda_o, \bar{u}_o)$.

Proof. From (12.4) we know that a necessary condition for X_o to solve (13.23) is the existence of a vector $Y_o \in \mathcal{E}^m$ such that

$$\begin{aligned}
&\nabla f(X_o) - A'Y_o \leq O \\
&X_o(\nabla f(X_o) - A'Y_o) = 0 \\
&Y'_o(b - AX_o) = 0 \\
&b - A'X_o \geq O \\
&X_o, Y_o \geq O
\end{aligned}$$

or, from (13.23),

$$\begin{aligned}
&(a)\ \frac{C(d_o + D'X_o) - D(c_o + C'X_o)}{(d_o + D'X_o)^2} - A'Y_o \leq O \\
&(b)\ \frac{X'_o C(d_o + D'X_o) - X'_o D(c_o + C'X_o)}{(d_o + D'X_o)^2} - X'_o A'Y_o = 0 \\
&(c)\ Y'_o(b - AX_o) = 0 \\
&(d)\ b - AX_o \geq O \\
&(e)\ X_o, Y_o \geq O.
\end{aligned} \tag{13.34}$$

For $\lambda_o = (c_o + C'X_o)/(d_o + D'X_o)$, (13.34a) becomes

$$\frac{C}{d_o + D'X_o} - \frac{D\lambda_o}{d_o + D'X_o} - A'Y \leq O.$$

Upon setting $Y_o(d_o + D'X_o) = \bar{u}_o$, the previous inequality simplifies to $D\lambda_o + A'\bar{u}_o \geq C$. From (13.34b),

$$\frac{X'_o C d_o - X'_o D c_o}{(d_o + D'X_o)^2} = X'_o A'Y_o. \tag{13.35}$$

And from (13.34c), (13.35),

$$b'Y_o = \frac{X_0'Cd_0 - X_o'Dc_o}{(d_o + D'X_o')^2},$$

$$b'\bar{u}_o = \frac{X_0'Cd_0 - X_o'Dc_o}{d_o + D'X_o'} = d_o\lambda_o - c_o.$$

Clearly, the vector $(\lambda_o, \bar{u}_o)$, where $\lambda_o = (c_o + C'X_o)/(d_o + D'X_o)$ and $\bar{u}_o = Y_o (d_o + D'X_o)$, provides us with a feasible solution to the dual problem. And since $\lambda_o \geq (c_o + C'X_o)/(d_o + D'X_o)$(by Theorem 13.5), the first half of the theorem is verified.

To prove the converse portion, let us transform (13.33) to the equivalent dual program

$$\begin{aligned} &\max\ -g = -\lambda \ \text{ s.t.} \\ &-d_o\lambda + b'\bar{u} \leq -c_o \\ &-D\lambda - A'\bar{u} \leq -C \\ &\bar{u} \geq O, \lambda \text{ unrestricted.} \end{aligned} \tag{13.33.1}$$

The Lagrangian associated with this problem is

$$M(\lambda, \bar{u}, \hat{x}, \hat{X}, \bar{X}, u_s, u_{s1}, u_{s2}) = -\lambda + \hat{x}(-c_o + d_o\lambda - b'\bar{u} - u_s^2)$$
$$+ \hat{X}'(-C + D\lambda + A'\bar{u} - u_{s1}^2) + \bar{X}'(\bar{u} - u_{s2}^2).$$

Then a necessary condition for $(\lambda_o, \bar{u}_o)$ to solve (13.33.1) is the existence of vectors $\bar{X}_o$, $(\hat{x}_o, \hat{X}_o)$ such that

$$\frac{\partial M}{\partial \lambda} = -1 + \hat{x}_o d_o + \hat{X}_o' D = 0$$

$$\frac{\partial M}{\partial u} = -\hat{x}_o b + A\hat{X}_o + \bar{X}_o = O$$

$$\frac{\partial M}{\partial \hat{x}} = -c_o + d_o\lambda_o - b'\bar{u}_o - u_s^2 = 0$$

$$\frac{\partial M}{\partial \hat{X}} = -C + D\lambda_o + A'\bar{u}_o - u_{s1}^2 = O$$

$$\frac{\partial M}{\partial \bar{X}} = \bar{u}_o - u_{s2}^2 = O$$

$$\frac{\partial M}{\partial u_{s1}} = -2\hat{X}_o{}' u_{s1} = 0$$

$$\frac{\partial M}{\partial u_{s2}} = -2\bar{X}_o{}' u_{s2} = 0$$

$$\frac{\partial M}{\partial u_s} = -2\hat{x}_o u_s = 0$$

or, for $\bar{X}_o \geq O$,

$$\begin{aligned}
&(\text{a})\, \hat{x}_o d_o + \hat{X}_o' D = 1 \\
&(\text{b})\, -\hat{x}_o b + A\hat{X}_o \leq O \\
&(\text{c})\, \hat{x}_o(-c_o + d_o\lambda_o - b'\bar{u}_o) = 0 \\
&(\text{d})\, \hat{X}_o'(-C + D\lambda_o + A'\bar{u}_o) = 0 \\
&(\text{e})\, \bar{u}_o'(-\hat{x}_o b + A\hat{X}_o) = 0 \\
&(\text{f})\, -c_o + d_o\lambda_o - b'\bar{u}_o \geq 0 \\
&(\text{g})\, -C + D\lambda_o + A'\bar{u}_o \geq O \\
&(\text{h})\, \bar{u}_o > O, \lambda_o \text{ unrestricted}, \hat{x}_o \geq 0, \hat{X}_o \geq O.
\end{aligned} \tag{13.36}$$

From (13.36b) we see that $\hat{x}_0 > 0$ since otherwise $A\hat{X}_o \leq O$, thus implying that $\hat{X}_o = O$. Clearly this violates (13.36a). Thus (13.36a, b) together indicate that $(\hat{x}_o \hat{X}_{o},)$ is a feasible solution to the equivalent linear program. If $X_o = \hat{X}_o/\hat{x}_o$, then (13.36b) reveals that we also obtain a feasible solution to the original linear program. From (13.36c, e), we have $\bar{u}_o' A\hat{X}_o = -\hat{x}_o c_o + \hat{x}_o d_o \lambda_o$. Combining this equation with (13.36d) yields $-\hat{X}_o' C + \hat{X}_o' D\lambda_o - \hat{x}_o c_o + \hat{x}_o d_o \lambda_o = 0$. If $X_o = \hat{X}_o/\hat{x}_o$ is substituted into the preceding equation, then we obtain $\lambda_o = (c_o + C'X_o)/(d_o + D'X_o)$. And since min $\lambda = -$ max $(-\lambda)$, the values of the objective functions for problems (13.23), (13.33) are thus equal. Q.E.D.

While (13.33) represents the dual of the equivalent linear program (13.32), what does the general nonlinear dual (Panik 1976) to (13.23) look like? Moreover, what is the connection between the dual variables in (13.33) and those found in the general nonlinear dual to (13.23)? To answer these questions, let us write the said dual to (13.33) as

$$\begin{aligned}
&\min g(X,U) = L(X,U) - X'\nabla_X L \quad \text{s.t.} \\
&\quad \nabla_X L \leq O, U \geq O, \quad X \text{ unrestricted},
\end{aligned}$$

where $L(X, U) = (c_o + C'X)/(d_o + D'X) + U'(b - AX)$ is the Lagrangian associated with the primal problem (13.23). In this regard, this dual becomes

$$\begin{aligned}
&\min g(X,U) = \frac{c_o + C'X}{d_o + D'X} + U'b - \frac{X'(Cd_o - Dc_o)}{(d_o + D'X)^2} \quad \text{s.t.} \\
&\frac{C(d_o + D'X) - D(c_o + C'X)}{(d_o + D'X)^2} - A'U \leq O, U \geq O, \ X \text{ unrestricted}.
\end{aligned} \tag{13.37}$$

Since $t = (d_o + \boldsymbol{D}'\boldsymbol{X})^{-1}$, the structural constraint in (13.37) becomes

$$t\boldsymbol{C} \leq t\boldsymbol{D}\left(\frac{c_o + \boldsymbol{C}'\boldsymbol{X}}{d_o + \boldsymbol{D}'\boldsymbol{X}}\right) - \boldsymbol{A}'\boldsymbol{U}. \tag{13.37.1}$$

Since at an optimal solution to the primal-dual pair of problems we must have $\lambda = (c_o + \boldsymbol{C}'\boldsymbol{X})/(d_o + \boldsymbol{D}'\boldsymbol{X})$ (see Theorem 13.6), (13.37.1) becomes

$$t\boldsymbol{C} \leq t\boldsymbol{D}\lambda + \boldsymbol{A}'\boldsymbol{U} \text{ or } \boldsymbol{C} \leq \boldsymbol{D}\lambda + \boldsymbol{A}'(\boldsymbol{U}/t). \tag{13.37.2}$$

Then the structural constraint in (13.37) is equivalent to (13.33b) if $\bar{\boldsymbol{u}} = \boldsymbol{u}/t$ or $\boldsymbol{u} = t\bar{\boldsymbol{u}}$. Moreover, when $\bar{\boldsymbol{u}} = \boldsymbol{u}/t$, the objective in (13.37) simplifies to λ, the same objective as found in (13.33). To summarize, solving (13.32) gives us $m + 1$ dual variables, one corresponding to each structural constraint. The dual variable for (13.32a) (namely λ) has the value attained by $\bar{f}$ at its optimum. The dual variables for (13.23) are then found by multiplying the dual variables for the constraints (13.32b) by t.

Example 13.5 Let us solve the problem presented in Example 13.3 above by generating a solution to its equivalent linear programming problem. From (13.32), we desire to

$$\begin{aligned} &\max \bar{f} = 5y_1 + 3y_2 \text{ s.t.} \\ &\quad t + 5y_1 + 2y_2 = 1 \\ &-15t + 3y_1 + 5y_2 \leq 0 \\ &-10t + 5y_1 + 2y_2 \leq 0 \\ &\quad t, y_1, y_2 \geq 0. \end{aligned} \tag{13.38}$$

For y_3 a nonnegative artificial variable and y_4, y_5 nonnegative slack variables, the optimal simplex matrix (using the two-phase method) is

$$\begin{bmatrix} 1 & \frac{19}{35} & 0 & -\frac{2}{35} & 0 & 0 & \frac{1}{7} \\ 0 & \frac{78}{35} & 1 & \frac{1}{35} & 0 & 0 & \frac{3}{7} \\ 0 & \frac{209}{35} & 0 & -\frac{22}{35} & 1 & 0 & \frac{4}{7} \\ 0 & \frac{59}{35} & 0 & \frac{3}{35} & 0 & 1 & \frac{9}{7} \end{bmatrix}.$$

Thus the optimal basic feasible solution for this problem is $t = \frac{1}{7}$, $y_2 = \frac{3}{7}$, $y_4 = \frac{4}{7}$, and $\bar{f} = \frac{9}{7}$. Then with $\boldsymbol{X} = {}^{\boldsymbol{Y}}/_t$, it follows that $x_2 = 3, x_5 = 4$, and $f = {}^9/_7$. Moreover, the dual variable corresponding to (13.32a) is $\lambda = 9/7$ while the dual variables corresponding to (13.32b) are respectively $\bar{u}_1 = 3/35, \bar{u}_2 = 0$. The accounting loss figure associated with y_1 is $\bar{u}_{s1} = 59/35$. (See also the last row of the optimal simplex matrix for Example 13.4.) Upon transforming these last three variables to those corresponding to the original fractional problem yields, upon multiplying by t, $u_1 = 3/245$, $u_2 = 0$, and $u_{s1} = 59/245$.

The dual problem associated with (13.51) is

$$\begin{aligned}
\min g = \ & \lambda \text{ s.t.} \\
& \lambda - 15\bar{u}_1 - 10\bar{u}_2 \geq 0 \\
& 5\lambda + 3\bar{u}_1 + 5\bar{u}_2 \geq 5 \\
& 2\lambda + 5\bar{u}_1 + 2\bar{u}_2 \geq 3 \\
& \lambda \text{ unrestricted}, \bar{u}_1, \bar{u}_2 \geq 0.
\end{aligned}$$

If $\lambda = w - v$, with both w, v nonnegative, $\bar{u}_3$ is a nonnegative slack variable, $\bar{u}_4$ and $\bar{u}_5$ are nonnegative surplus variables, and $\bar{u}_6$, $\bar{u}_7$ are nonnegative artificial variables, then the above dual problem may be rewritten as

$$\begin{aligned}
\max -g = & -w + v \text{ s.t.} \\
& -w + v + 15\bar{u}_1 + 10\bar{u}_2 + \bar{u}_3 = 0 \\
& 5w - 5v + 3\bar{u}_1 + 5\bar{u}_2 - \bar{u}_4 + \bar{u}_6 = 5 \\
& 2w - 2v + 5\bar{u}_1 + 2\bar{u}_2 + \bar{u}_5 + \bar{u}_7 = 3 \\
& w, v, \bar{u}_1, \ldots, \bar{u}_7 \geq 0.
\end{aligned}$$

The optimal simplex matrix (via the two-phase method) is

$$\begin{bmatrix}
0 & 0 & 1 & \frac{22}{35} & \frac{2}{35} & 0 & -\frac{1}{35} & 0 & \frac{3}{35} \\
1 & -1 & 0 & -\frac{4}{7} & -\frac{1}{7} & 0 & -\frac{3}{7} & 0 & \frac{9}{7} \\
0 & 0 & 0 & -\frac{209}{35} & -\frac{19}{35} & 1 & -\frac{78}{35} & 0 & \frac{59}{35} \\
0 & 0 & 0 & \frac{4}{7} & \frac{1}{7} & 0 & \frac{3}{7} & 1 & -\frac{9}{7}
\end{bmatrix}$$

with $w = 9/7, \bar{u}_1 = 3/35, \bar{u}_4 = 59/35$, and $-g = -w = -\lambda = -9/7$ or $g = 9/7$. ■

13.8 Resource Allocation with a Fractional Objective

We previously specified a generalized multiactivity profit-maximization model (7.27.1) as

$$\begin{aligned}
\text{maximize} f(\boldsymbol{X}) = & \left(\boldsymbol{P} - (\boldsymbol{A}^*)'\boldsymbol{Q} - \boldsymbol{H}\right)'\boldsymbol{X} \text{ s.t.} \\
& \bar{\boldsymbol{A}}\boldsymbol{X} \leq \boldsymbol{b}, \boldsymbol{X} \geq \boldsymbol{O}.
\end{aligned}$$

An assumption that is implicit in the formulation of this model is that all activities can be operated at the same rate per unit of time. However, some activities may be operated at a slower pace than others so that the time needed to produce a unit of output by one activity may vary substantially between activities. Since production time greatly influences the optimum activity mix, the model should explicitly reflect the importance of this attribute. Let us assume, therefore, that activity j is operated at an average rate of t_j units per minute (hour, etc.). Then total production time is $T(X) = \sum_{j=1}^{p} x_j / t_j = \boldsymbol{T}'\boldsymbol{X}$, where $\boldsymbol{T}' = \left({}^{1}/_{t_1},\ldots,{}^{1}/_{t_p}\right)$ and $\boldsymbol{T}$ is of order $(p \times 1)$. Note that $dT/dt_j < 0$ for each j, i.e. total production time is inversely related to the speed of operation of each activity.

If C denotes the (constant) rate per unit of time of operating the production facility and K represents fixed overhead cost, then the total overhead cost of operating the facility is $h(X) = CT'X + K$. In view of this discussion, total profit is $\hat{f}(X) = f(X) - h(X) = (\boldsymbol{P} - (A^*)'\boldsymbol{Q} - \boldsymbol{H} - C\boldsymbol{T})'\boldsymbol{X} - K$. We cannot directly maximize this function subject to the above constraints since two different activity mixes do not necessarily yield the same production time so that we must divide $\hat{f}$ by h to get average profit per unit of time

$$\boldsymbol{P}(\boldsymbol{X}) = \frac{\hat{f}(\boldsymbol{X})}{h(\boldsymbol{X})} = \frac{(\boldsymbol{P} - (A^*)'\boldsymbol{Q} - \boldsymbol{H} - C\boldsymbol{T})'\boldsymbol{X} - K}{C\boldsymbol{T}'\boldsymbol{X} + K}. \tag{13.39}$$

Hence, the adjusted multiactivity profit-maximization model involves maximizing (13.52) subject to $\bar{\boldsymbol{A}}\boldsymbol{X} \leq \boldsymbol{b}, \boldsymbol{X} \geq \boldsymbol{O}$.

For instance, from Example 7.3 the problem

$$\begin{aligned} &\text{maximize} f = 14x_1 + 12x_2 + 11x_3 + 10x_4 \quad \text{s.t.} \\ &\quad 10x_1 + 8x_2 + 6x_3 + 4x_4 \leq 40 \\ &\quad 2x_1 + 3x_2 + 5x_3 + 8x_4 \leq 30 \\ &\quad x_1,\ldots,x_4 \geq 0 \end{aligned}$$

has as its optimal basic feasible solution $x_1 = 10/19, x_3 = 110/19$, and $f = 1350/19$. Moreover, the shadow prices are $u_1 = 24/19$, $u_2 = 13/19$ while the accounting loss figure for products x_2, x_4 are, respectively, $u_{s2} = 3/19$, $u_{s4} = 10/19$. If we now explicitly introduce the assumption that some activities are operated at a faster rate than others, say $t_1 = 2$, $t_2 = 10$, $t_3 = 8$, and $t_4 = 2$, then the preceding problem becomes, for $C = 10$, $K = 100$,

$$\begin{aligned} &\text{maximize}\, P = \frac{9x_1 + 11x_2 + 9.75x_3 + 5x_4 - 100}{0.5x_1 + 0.1x_2 + 0.125x_3 + 0.5x_4 + 100} \quad \text{s.t.} \\ &10x_1 + 8x_2 + 6x_3 + 4x_4 \leq 40 \\ &2x_1 + 3x_2 + 5x_3 + 8x_4 \leq 30 \\ &x_1,\ldots,x_4 \geq 0. \end{aligned}$$

Upon converting this problem to the form provided by (13.32), i.e. to

$$\begin{aligned}
&\text{maximize} f = -100t + 9y_1 + 11y_2 + 9.75y_3 + 5y_4 \text{ s.t.} \\
&-40t + 10y_1 + 8y_2 + 6y_3 + 4y_4 \le 0 \\
&-30t + 2y_1 + 3y_2 + 5y_3 + 8y_4 \le 0 \\
&100t + 0.5y_1 + 0.1y_2 + 0.125y_3 + 0.5y_4 = 1 \\
&y_1, \ldots, y_4 \ge 0,
\end{aligned}$$

and solving (via the two-phase method) yields the optimal simplex matrix (with the artificial column deleted)

$$\begin{bmatrix} 0 & 1.7616 & 1 & 0 & -1.2447 & 0.2268 & -0.2742 & 0 & 0.0843 \\ 0 & -0.4303 & 0 & 1 & 2.5317 & -0.1392 & 0.3544 & 0 & 0.5063 \\ 1 & 0.0377 & 0 & 0 & 0.0308 & -0.00053 & -0.00169 & 0 & 0.0928 \\ 0 & 5.8038 & 0 & 0 & 5.6835 & 1.1424 & 0.4557 & 1 & 4.9367 \end{bmatrix}.$$

Thus $t_o = 0.0928$, $y_2 = 0.0843$, $y_3 = 0.5063$, and $\bar{f} = 4.9367$. With $X_o = Y_o/t_o$, the optimal solution, in terms of the original variables, is $x_2^o = 0.9084$ and $x_3^o = 5.4558$. In addition, the dual variables obtained from the above matrix are $\bar{u}_1^o = 1.424$ and $\bar{u}_2^o = 0.4557$. Then the original dual variables are $\bar{u}_1^o = t_o\bar{u}_1^o = 0.1060$, $\bar{u}_2^o = t_o\bar{u}_2^o = 0.0422$. Moreover, the computed accounting loss figures for activities one and four are, respectively, $\bar{u}_{s1}^o = 5.8038$, $\bar{u}_{s4}^o = 5.6835$. Upon transforming these values to the original accounting loss values yields $\bar{u}_{s1}^o = 0.5386$, $\bar{u}_{s4}^o = 0.5274$.

We noted earlier that for the standard linear programming resource allocation problem the dual variable u_i represents the change in the objective (gross profit) function "per unit change" in the ith scarce resource b_i. However, this is not the case for linear fractional programs, the reason being that the objective function in (13.23) is nonlinear, i.e. the dual variables evaluate the effect on f precipitated by infinitesimal changes in the components of the requirements vector $\boldsymbol{b}$ and not unit changes. To translate per unit changes in the b_i (which do not change the optimal basis) into changes in f for fractional objectives so that the "usual" shadow price interpretation of dual variables can be offered, let us write the change in f as

$$\Delta f = \frac{(d_o + \boldsymbol{D}'\boldsymbol{X}_o)\, \boldsymbol{U}_o'\boldsymbol{e}_i}{(d_o + \boldsymbol{D}'\boldsymbol{X}_o) + \boldsymbol{D}_B'\boldsymbol{e}_i}, i = i, \ldots, m, \tag{13.40}$$

where the $(m \times 1)$ vector $\boldsymbol{D_B}$ contains the coefficients of $\boldsymbol{D}$ corresponding to basic variables and $\boldsymbol{e}_i$ is the ith unit column vector (Martos 1964; Kydland 1962; Bitran and Magnanti 1976).

In view of this discussion, the change in f per unit change in b_1 is

$$\Delta f = \frac{(d_o + \boldsymbol{D}'\boldsymbol{X}_o)\,\boldsymbol{U}'_o\boldsymbol{e}_1}{(d_o + \boldsymbol{D}'\boldsymbol{X}_o) + \boldsymbol{D}'_B\boldsymbol{e}_1} = 0.1059;$$

while the change in f per unit change in b_2 is

$$\Delta f = \frac{(d_o + \boldsymbol{D}'\boldsymbol{X}_o)\,\boldsymbol{U}'_o\boldsymbol{e}_2}{(d_o + \boldsymbol{D}'\boldsymbol{X}_o) + \boldsymbol{D}'_B\boldsymbol{e}_2} = 0.0421.$$

13.9 Game Theory and Linear Programming

13.9.1 Introduction

We may view the notion of a *game of strategy* as involving a situation in which there is a contest or conflict situation between two or more players, where it is assumed that the players can influence the final outcome of the game, and the said outcome is not governed entirely by chance factors. The players can be individuals (two people engaging in a chess match, or multiple individuals playing a game of bridge) or, in a much broader sense, adversarial situations can emerge in a social, political, economic, or military context. Simply stated, a **game** is a set of rules for executing plays or moves. For example, the rules state what moves can be made, when they are to be made and by whom; what information is available to each participant; what are the termination criteria for a play or move, and so on. Moreover, after each play ends, we need to specify the reward or payoff to each player.

If the game or contest involves two individuals, organizations, or countries, it is called a **two-person game**. And if the sum of the payoffs to all players at the termination of the game is zero, then the game is said to be a **zero-sum game**, i.e. what one player wins, the other player loses. Hence a zero-sum, two-person game is one for which, at the end of each play, one player gains what the other player loses.

Suppose we denote the two players of a game as P_1 and P_2, respectively. Each player posits certain courses of action called **strategies**, i.e. these strategies indicate what one player will do for each possible move that his opponent might make. In addition, each player is assumed to have a finite number of strategies, where the number of strategies formulated by one player need not be the same for the other player. Hence the **set of strategies** for, say P_1 covers all possible alternative ways of executing a play of the game given all possible moves that P_2 might make.

It must be noted that for two-person games, one of the opponents can be *nature* so that *chance* influences certain moves (provided, of course, that the players themselves ultimately control the outcome). However, if nature does not influence any moves and both parties select a strategy, then we say that the outcome of the game is **strictly dominated**. But if *nature* has a hand in affecting pays and the outcome of a game is not strictly determined, then it makes sense to consider the notion of **expected outcome** (more on this point later on).

13.9.2 Matrix Games

Let us specify a two-person game G as consisting of the nonempty sets $\mathcal{R}$ and $\mathcal{C}$ and a real-valued function ϕ defined on a pair (r, c), where $r \in \mathcal{R}$ and $c \in \mathcal{C}$. Here the elements r, c of sets $\mathcal{R}, \mathcal{C}$ are the strategies for player P_1 and P_2, respectively, and the function ϕ is termed the **payoff function**. The number $\phi\ (r, c)$ is the amount that P_2 pays P_1 when P_1 plays strategy r and P_2 plays strategy c. If the sets $\mathcal{R}$ are finite, then the payoff ϕ can be represented by a matrix so that the game is called a **matrix game**.

For matrix games, if $\mathcal{R}, \mathcal{C}$ contain strategies $r_1, r_2, \ldots, r_m$ and $c_1, c_2, \ldots, c_n$, respectively, then the payoff to P_1 if he chooses strategy i and P_2 selects strategy j is $\phi\ (r_i, c_j) = a_{ij}, i = 1, \ldots, m; j = 1, \ldots, n$. If some moves are determined by chance, then the **expected payoff** to P_1 is also denoted by a_{ij}. So when P_1 chooses strategy i and P_2 selects strategy j, a_{ij} depicts the payoff to P_1 if the game is strictly determined or if nature makes some of the plays. In general, a_{ij} may be positive, negative, or zero. Given that P_1 has m strategies and P_2 has n strategies, the payoffs a_{ij} can be arranged into an $(m \times n)$ **payoff matrix:**

$$\underset{(m \times n)}{A} = \begin{bmatrix} a_{11} & a_{12} & \cdots & a_{1n} \\ a_{21} & a_{22} & \cdots & a_{2n} \\ \vdots & \vdots & \vdots & \vdots \\ a_{m1} & a_{m2} & \cdots & a_{mn} \end{bmatrix}.$$

Clearly the rows represent the m strategies of P_1 and the columns depict the n strategies of P_2, where it is assumed that each player knows the strategies of his opponent. Thus, row i of A gives the payoff (or expected payoff) to P_1 if he uses strategy i, with the actual payoff to P_1 determined by the strategy selected by P_2. A game is said to be in **normal form** if the strategies for each player are specified and the elements within A are given.

We can think of the a_{ij} elements of A as representing the payoffs to P_1 while the payoffs to P_2 are the negatives of these. The "conflict of interest" aspect of a game is easily understood by noting that P_1 is trying to win as much as possible, with P_2 trying to preclude P_1 from winning more than is practicable. In this case, P_1 will be regarded as the maximizing player while P_2 will try to minimize the winnings of P_1.

Given that P_1 selects strategy i, he is sure of winning $\min_j a_{ij}$. (If "nature" impacts some of the moves, then P_1's expected winnings are at least $\min_j a_{ij}$.) Hence P_1 should select the strategy that yields the maximum of these minima, i.e.

$$\max_i \min_j a_{ij}. \tag{13.41}$$

Since it is reasonable for P_2 to try to prevent P_1 from getting any more than is necessary, selecting strategy j ensures that P_1 will not get more than $\max_i a_{ij}$ no matter what P_1 does. Hence, P_2 attempts to minimize his maximum loss, i.e. P_2 selects a strategy for which

$$\min_j \max_i a_{ij}. \tag{13.42}$$

The strategies r_i, $i = 1, \ldots, m$, and c_j, $j = 1, \ldots, n$, are called **pure strategies**. Suppose P_1 selects the pure strategy r_l and P_2 selects the pure strategy c_k. If

$$a_{lk} = \max_i \min_j a_{ij} = \min_j \max_i a_{ij}, \tag{13.43}$$

then the game is said to possess a **saddle point solution**. Here r_l turns out to be the optimal strategy for P_1 while c_k is the optimal strategy for P_2 – with P_2 selecting c_k, P_1 cannot get more than a_{lk}, and with P_1 choosing r_l, he is sure of winning at least a_{lk}.

Suppose (13.43) does not hold and that, say,

$$a_{rv} = \max_i \min_j < \min_j \max_i a_{uk}.$$

In this circumstance, P_1 might be able to do better than a_{rv}, and P_2 might be able to decrease the payoff to P_1 below the a_{uk} level. For each player to pursue these adjustments, we must abandon the execution of pure strategies and look to the implementation of **mixed strategies** via a chance device. For P_1, pure strategy i with probability $u_i \geq 0$ and $\sum_{i=1}^{m} u_i = 1$ is randomly selected; and for P_2, pure strategy j with probability $v_j > 0, \sum_{j=1}^{n} v_j = 1$, is randomly chosen. (More formally, a mixed strategy for P_1 is a real-valued function f on $\mathcal{R}$ such that $f(r_i) = u_i \geq 0$, while a mixed strategy for P_2 is a real-valued function h on $\mathcal{C}$ such that

$h(c_j) = v_j \geq 0$.) Thus random selection determines the strategy each player will use, and neither is cognizant of the other's strategy or even of his own strategy until it is determined by chance. In sum:

$$u_i \geq 0, i = 1, \ldots, m, \sum_{i=1}^{m} u_i = 1 \qquad v_j \geq 0, j = 1, \ldots, n, \sum_{j=1}^{n} v_j = 1$$

or or

$$\underset{(m \times 1)}{\boldsymbol{U}} \geq 0, \boldsymbol{1}'\boldsymbol{U} = 1 \qquad \underset{(n \times 1)}{\boldsymbol{V}} \geq 0, \boldsymbol{1}'\boldsymbol{V} = 1,$$

Under a regime of mixed strategies, we can no longer be sure what the outcome of the game will be. However, we can determine the **expected payoff** to P_1. So if P_1 uses mixed strategy $\boldsymbol{U}$ and P_2 employs mixed strategy V, then the expected payoff to P_1 is

$$E(\boldsymbol{U}, \boldsymbol{V}) = \boldsymbol{U}'\boldsymbol{A}\boldsymbol{V} = \sum_{i,j} u_i a_{ij} v_j. \tag{13.44}$$

Using an argument similar to that used to rationalize (13.41) and (13.42), P_1 seeks a $\boldsymbol{U}$ that maximizes his expected winnings. For any $\boldsymbol{U}$ chosen, he is sure that his expected winnings will be at least $\min_V E(\boldsymbol{U}, \boldsymbol{V})$. P_1 then maximizes his payoff relative to $\boldsymbol{U}$ so that his expected winnings are at least

$$W_1^* = \max_U \min_V E(\boldsymbol{U}, \boldsymbol{V}). \tag{13.45}$$

Likewise, P_2 endeavors to find a V such that the expected winnings of P_1 do not exceed

$$W_2^* = \min_V \max_U E(\boldsymbol{U}, \boldsymbol{V}). \tag{13.46}$$

Now, if there exist mixed strategies $\boldsymbol{U}^*$, $\boldsymbol{V}^*$ such that $W_1^* = W_2^*$, then there exists a **generalized saddle point** of $\phi\ (\boldsymbol{U},\boldsymbol{V})$. P_1 should use the mixed strategy $\boldsymbol{U}^*$ and P_2 the mixed strategy $\boldsymbol{V}^*$ so that the expected payoff to P_1 is exactly W_1^*, the **value of the game**. How do we know that mixed strategies $\boldsymbol{U}$, V exist such that the value of the game is $W^* = W_1^* = W_2^*$? The answer is provided by Theorem 13.7.

Theorem 13.7 Fundamental Theorem of Game Theory.
$\boldsymbol{U}^*$ and $\boldsymbol{V}^*$ values always exist such that

$$W_1^* = \max_U \min_V E(\boldsymbol{U}, \boldsymbol{V}) = \min_V \max_U E(\boldsymbol{U}, \boldsymbol{V}) = W_2^*,$$

with

$$\boldsymbol{U}^* \geq \boldsymbol{O}, \boldsymbol{1}'\boldsymbol{U}^* = 1; \boldsymbol{V}^* \geq \boldsymbol{O}, \boldsymbol{1}'\boldsymbol{V}^* = 1.$$

Thus, according to this fundamental theorem, every matrix game has a solution in mixed strategies. Moreover, it has at most a single value $W^* = \phi\,(\boldsymbol{U}^*, \boldsymbol{V}^*)$, which is the **generalized saddle value** associated with the generalized saddle point $(\boldsymbol{U}^*, \boldsymbol{V}^*)$.

Example 13.6

a) Suppose $\boldsymbol{A}$ represents a (3×3) matrix game with the pure strategies of P_1 and P_2 bordering $\boldsymbol{A}$:

$$\begin{array}{cc|ccc} & & & \boldsymbol{P_2} & \\ & & c_1 & c_2 & c_3 \\ & r_1 & \textcircled{2} & 5 & 3 \\ \boldsymbol{P_1} & r_2 & 2 & \textcircled{0} & 3 \\ & r_3 & \boxed{\textcircled{3}} & \boxed{6} & \boxed{4} \end{array} = \boldsymbol{A}.$$

Then P_1's minimum winnings for each fixed i taken over all j's are: $\min_j a_{1j} = 2$, $\min_j a_{2j} = 0$, and $\min_j a_{3j} = 3$ (the circled elements in $\boldsymbol{A}$). Similarly, for each fixed j, P_1's maximum winnings taken over all i's are: $\max_i a_{i1} = 3$, $\max_i a_{i2} = 6$, and $\max_i a_{i3} = 4$ (the blocked items in $\boldsymbol{A}$). Then P_1 utilizes the **maximin strategy** $\max_i \min_j a_{ij} = 3$; and P_2 follows the **minimax strategy** $\min_j \max_i a_{ij} = 3$. Hence this matrix game has a saddle point at $a_{31} = 3$ since $a_{31} = 3 = \max_i \min_j a_{ij} = \min_j \max_i a_{ij} = 3$.

b) Let $\boldsymbol{A}$ represent a (3×4) matrix game depicting the pure strategies of P_1 and P_2:

$$\begin{array}{cc|cccc} & & & \boldsymbol{P_2} & & \\ & & c_1 & c_2 & c_3 & c_4 \\ & r_1 & 3 & \textcircled{1} & \boxed{4} & 2 \\ \boldsymbol{P_1} & r_2 & \boxed{8} & 2 & \textcircled{1} & 4 \\ & r_3 & 2 & \textcircled{0} & 1 & \boxed{6} \end{array} = \boldsymbol{A}.$$

Then, for fixed i values: $\min_j a_{1j} = 3$, $\min_j a_{2j} = 1$, and $\min_j a_{3j} = 0$. Also, for each fixed j value: $\max_i a_{i1} = 8$, $\max_i a_{i2} = 2$, $\max_i a_{i3} = 4$, and $\max_i a_{i4} = 6$. If again P_1 employs the maximin strategy $\max_i \min_j a_{ij} = 1$ and P_2 utilizes the minimax strategy $\min_j \max_i a_{ij} = 2$, then $1 = \max_i \min_j a_{ij} < \min_j \max_i a_{ij} = 2$.

Clearly this matrix game does not have a saddle point in pure strategies. ∎

13.9.3 Transformation of a Matrix Game to a Linear Program

Let us express a matrix game for players P_1 and P_2 as

$$\underset{(m \times n)}{A} = \begin{bmatrix} a_{11} & a_{12} & \cdots & a_{1n} \\ a_{21} & a_{22} & \cdots & a_{2n} \\ \vdots & \vdots & \vdots & \vdots \\ a_{m1} & a_{m2} & \cdots & a_{mn} \end{bmatrix} = [\boldsymbol{a}_1, \boldsymbol{a}_2, \ldots, \boldsymbol{a}_n],$$

where $\boldsymbol{a}_j$, $j = 1, \ldots, n$, is the jth column of $\boldsymbol{A}$. Suppose P_2 chooses to play the pure strategy j. If P_1 uses the mixed strategy $\boldsymbol{U} \geq \boldsymbol{O}$, $\boldsymbol{1}'\boldsymbol{U} = 1$, then the expected winnings of P_1 is

$$E(\boldsymbol{U}'\boldsymbol{a}_j) = \sum_{i=1}^{m} a_{ij} u_i. \tag{13.47}$$

Then P_1's expected payoff will be at least W_1 if there exists a mixed strategy $\boldsymbol{U}$ such that

$$\sum_{i=1}^{m} a_{ij} u_i \geq W_1, j = 1, \ldots, n, \tag{13.48}$$

i.e. P_1 can never expect to win more than the largest value of W_1 for which there exists a $\boldsymbol{U} > \boldsymbol{O}$, $\boldsymbol{1}'\boldsymbol{U} = 1$, and (13.48) holds. Considering each j, (13.47) becomes

$$\boldsymbol{A}'\boldsymbol{U} \geq W_1 \boldsymbol{1}, \boldsymbol{U} \geq \boldsymbol{O}, \boldsymbol{1}'\boldsymbol{U} = 1. \tag{13.49}$$

With W_1 assumed to be positive, we can set

$$y_i = \frac{u_i}{W_1}, i = 1, \ldots, n. \tag{13.50}$$

Clearly $y_i > 0$ since $W_1 > 0$. Under this variable, transformation (13.49) can be rewritten as

$$\boldsymbol{A}'\boldsymbol{Y} \geq \boldsymbol{1}, \boldsymbol{Y} \geq \boldsymbol{O}, \boldsymbol{1}'\boldsymbol{Y} = \frac{1}{W_1} = z. \tag{13.51}$$

Here P_1's *decision problem* involves solving the linear program

$$\begin{aligned} &\min z = \boldsymbol{1}'\boldsymbol{Y} \quad \text{s.t.} \\ &\boldsymbol{A}'\boldsymbol{Y} \geq \boldsymbol{1}, \boldsymbol{Y} > \boldsymbol{O}. \end{aligned} \tag{13.52}$$

Thus, minimizing $z = \dfrac{1}{W_1}$ will yield the maximum value of W_1, the expected winnings for P_1.

Next, suppose P_1 decides to play pure strategy i. From the viewpoint of P_2, he attempts to find a mixed strategy $\boldsymbol{V} \geq \boldsymbol{O}$ $\boldsymbol{1}'\boldsymbol{V} = 1$, which will give the smallest W_2 such that, when P_1 plays strategy i,

$$E(\boldsymbol{\alpha}_i \boldsymbol{V}) = \sum_{j=1}^{n} a_{ij} v_j \leq W_2, i = 1, \ldots, m, \tag{13.53}$$

where $\boldsymbol{\alpha}_i$ is the ith row of the payoff matrix $\boldsymbol{A}$ and $E(\boldsymbol{\alpha}_i \boldsymbol{V})$ is the expected payout of P_2. Considering each i, (13.53) becomes

$$\boldsymbol{A}\boldsymbol{V} \leq W_2 \boldsymbol{1}, \boldsymbol{V} \geq \boldsymbol{O}, \boldsymbol{1}'\boldsymbol{V} = 1. \tag{13.54}$$

With W_2 taken to be positive, let us define

$$x_j = \frac{v_j}{W_2} \geq 0, j = 1, \ldots, n. \tag{13.55}$$

Hence (13.54) becomes

$$\boldsymbol{A}\boldsymbol{X} \leq \boldsymbol{1}, \boldsymbol{X} \geq \boldsymbol{O}, \boldsymbol{1}'\boldsymbol{X} = \frac{1}{W_2} = z'. \tag{13.56}$$

Thus P_2's *decision problem* is a linear program of the form

$$\begin{aligned} &\max z' = \boldsymbol{1}'\boldsymbol{X} \quad \text{s.t.} \\ &\boldsymbol{A}\boldsymbol{X} \leq \boldsymbol{1}, \boldsymbol{X} \geq \boldsymbol{O}, \end{aligned} \tag{13.57}$$

i.e. maximizing $z' = \frac{1}{W_2}$ will yield the minimum value of W_2, the upper bound on the expected payout of P_2.

A moment's reflection reveals that (13.57) can be called the primal problem and (13.52) is its symmetric dual. Moreover, from our previous discussion of duality theory, there are always feasible solutions $\boldsymbol{U}$, $\boldsymbol{V}$ since P_1, P_2 can execute pure strategies. And with W_1, W_2 each positive, there exist feasible vectors $\boldsymbol{X}$, $\boldsymbol{Y}$ for the primal and dual problems, respectively, and thus the primal and dual problems have optimal solutions with max z' = min z or

$$W_2^* = \min W_2 = \max W_1 = W_1^*.$$

These observations thus reveal that we have actually verified the fundamental theorem of game theory by constructing the primal-dual pair of linear programs for P_1 and, P_2.

Example 13.7 For the matrix game presented in Example 13.6b, find the optimal strategies for P_1 and P_2. Here

$$\boldsymbol{A} = \begin{bmatrix} 3 & 1 & 4 & 2 \\ 8 & 2 & 1 & 4 \\ 2 & 0 & 1 & 6 \end{bmatrix}$$

and thus P_2's primal linear program (13.57) appears as

$$\begin{aligned}
\max z' &= x_1 + x_2 + x_3 + x_4 \text{ s.t.} \\
&3x_1 + x_2 + 4x_3 + x_4 \leq 1 \\
&8x_1 + 2x_2 + x_3 + 4x_4 \leq 1 \\
&2x_1 \qquad\qquad x_3 + 6x_4 \leq 1 \\
&x_1, x_2, x_3, x_4 \geq 0.
\end{aligned}$$

The reader can readily verify that at the optimal solution to this problem we have $x_1 = x_4 = 0$, $x_2 = 0.4286$, and $x_3 = 0.1429$ with $z' = 0.5714 = {}^1/_{W_2}$ or $W_2 = 1.75$. Then, since $x_j = v_j / W_2$, $j = 1, 2, 3, 4$, we can obtain $v_1 = v_4 = 0$, $v_2 = 0.75$, and $v_3 = 0.25$. Clearly $\sum_{j=1}^{4} v_j = v_2 + v_3 = 1$ as required.

In a similar vein, the dual to the preceding primal problem is P_1's linear program

$$\begin{aligned}
\min z &= y_1 + y_2 + y_3 + y_4 \text{ s.t.} \\
&3y_1 + 8y_2 + 2y_3 \geq 1 \\
&y_1 + 2y_2 \qquad \geq 1 \\
&4y_1 + y_2 + y_3 \geq 1 \\
&2y_1 + 4y_2 + 6y_3 \geq 1 \\
&y_1, y_2, y_3 \geq 0.
\end{aligned}$$

The optimal solution here is $y_1 = 0.1429$, $y_2 = 0.4286$, $y_3 = 0$, and $z = {}^1/_{W_1} = 0.5714$ so that $W_1 = 1.75$. Since $y_i = u_i / W_1$, $i = 1, 2, 3$, it follows that $u_1 = 0.25$, $u_2 = 0.75$, $u_3 = 0$, and $\sum_{i=1}^{3} u_i = u_1 + u_2 = 1$. Thus the generalized saddle value of the game in the mixed strategies

$$\boldsymbol{U}^* = \begin{bmatrix} 0.25 \\ 0.75 \\ 0 \end{bmatrix}, \boldsymbol{V}^* = \begin{bmatrix} 0 \\ 0.75 \\ 0.25 \\ 0 \end{bmatrix}$$

amounts to $\phi\ (\boldsymbol{U}^*, \boldsymbol{V}^*) = W_1^* = W_2^* = 1.75$. ■

13.A Quadratic Forms

13.A.1 General Structure

Suppose Q is a function of the n variables $x_1, \ldots, x_n$. Then Q is termed a quadratic form in $x_1, \ldots, x_n$ if

$$Q(x_1, \ldots, x_n) = \sum_{i=1}^{n} \sum_{j=1}^{n} a_{ij} x_i x_j, \tag{13.A.1}$$

where at least one of the constant coefficients $a_{ij} \neq 0$. More explicitly,

$$\begin{aligned}\sum_{i=1}^{n}\sum_{j=1}^{n} a_{ij}x_i x_j &= \sum_{j=1}^{n} a_{1j}x_1x_j + \sum_{j=1}^{n} a_{2j}x_2x_j + \cdots + \sum_{j=1}^{n} a_{nj}x_nx_j \\ &= a_{11}x_1^2 + a_{12}x_1x_2 + \cdots + a_{1n}x_1x_n \\ &\quad + a_{21}x_2x_1 + a_{22}x_2^2 + \cdots + a_{2n}x_2x_n \\ &\quad + a_{n1}x_nx_1 + a_{n2}x_nx_2 + \cdots + a_{nn}x_n^2.\end{aligned} \tag{13.A.2}$$

It is readily seen that Q is a homogeneous[4] polynomial of the second degree since each term involves either the square of a variable or the product of two different variables. Moreover, Q contains n^2 distinct terms and is continuous for all values of the variables x_i, $i = 1, \ldots, n$, and equals zero when $x_i = 0$ for all $i = 1, \ldots, n$.

In matrix form Q equals, for $X \in \mathcal{E}^n$, the scalar quantity

$$Q(x_1, \ldots, x_n) = \boldsymbol{X}'\boldsymbol{A}\boldsymbol{X}, \tag{13.A.3}$$

where

$$\underset{(n\times 1)}{\boldsymbol{X}} = \begin{bmatrix} x_1 \\ x_2 \\ \vdots \\ x_n \end{bmatrix}, \quad \underset{(n\times n)}{\boldsymbol{A}} = \begin{bmatrix} a_{11} & a_{12} & \cdots & a_{1n} \\ a_{21} & a_{22} & \cdots & a_{2n} \\ \vdots & \vdots & & \vdots \\ a_{na} & a_{n2} & \cdots & a_{nn} \end{bmatrix}.$$

To see this, let us first write

$$\boldsymbol{AX} = \begin{bmatrix} \sum_{j=1}^{n} a_{1j}x_j \\ \sum_{j=1}^{n} a_{2j}x_j \\ \vdots \\ \sum_{j=1}^{n} a_{nj}x_j \end{bmatrix}.$$

4 A form is **homogeneous of degree t** in the variables $x_1, \ldots, x_n$ if, when each variable in the form is multiplied by a scalar λ, the whole form is multiplied by λ^t, i.e. $Q(\lambda x_1, \lambda x_2, \ldots, \lambda x_n) = \lambda^t Q(x_1, x_2, \ldots, x_n)$.

Then

$$X'AX = (x_1, \ldots, x_n) \begin{bmatrix} \sum_{j=1}^{n} a_{1j}x_j \\ \sum_{j=1}^{n} a_{2j}x_j \\ \vdots \\ \sum_{j=1}^{n} a_{nj}x_j \end{bmatrix}$$

$$= x_1 \sum_{j=1}^{n} a_{1j}x_j + x_2 \sum_{j=1}^{n} a_{2j}x_j + \cdots + x_n \sum_{j=1}^{n} a_{nj}x_j$$

$$= \sum_{i=1}^{n} \sum_{j=1}^{n} a_{ij}x_i x_j.$$

From (13.A.2) it can be seen that $a_{ij} + a_{ji}$ is the coefficient of $x_i x_j$ since a_{ij}, a_{ji} are both coefficients of $x_i x_j = x_j x_i$, $i \neq j$.

Example 13.A.1

a) Find the quadratic form $Q(x_1, x_2, x_3)$ associated with the matrix

$$A = \begin{bmatrix} 1 & 2 & 1 \\ 3 & 5 & 2 \\ 1 & 1 & 2 \end{bmatrix}.$$

From (13.A.3) we have

$$X'AX = (x_1, x_2, x_3) \begin{bmatrix} 1 & 2 & 1 \\ 3 & 5 & 2 \\ 1 & 1 & 2 \end{bmatrix} \begin{bmatrix} x_1 \\ x_2 \\ x_3 \end{bmatrix} = (x_1, x_2, x_3) \begin{bmatrix} x_1 + 2x_2 + x_3 \\ 3x_1 + 5x_2 + 2x_3 \\ x_1 + x_2 + 2x_3 \end{bmatrix}$$

$$= x_1^2 + 5x_1x_2 + 2x_1x_3 + 5x_2^2 + 3x_2x_3 + 2x_3^2.$$

b) Find $X'AX$ when

$$A = \begin{bmatrix} 1 & 0 & 1 \\ 3 & 0 & 0 \\ 1 & 1 & 2 \end{bmatrix}.$$

Again using (13.A.3) we have

$$X'AX = x_1^2 + 3x_1x_2 + 2x_1x_3 + x_2x_3 + 2x_3^2.$$

■

13.A.2 Symmetric Quadratic Forms

Remember that if a matrix A is symmetric so that $A = A'$, then $a_{ij} = a_{ji}$, $i \neq j$. In this regard, a quadratic form $X'AX$ is **symmetric** if A is symmetric or $a_{ij} = a_{ji}$, $i \neq j$. Hence $a_{ij} + a_{ji} = 2a_{ij}$ is the coefficient on $x_i x_j$ since $a_{ij} = a_{ji}$ and a_{ij}, a_{ji} are both coefficients of $x_{ij} = x_{ji}$, $i \neq j$.

If A is not a symmetric matrix $(a_{ij} \neq a_{ji})$, we can always transform it into a symmetric matrix B by defining new coefficients

$$b_{ij} = b_{ji} = \frac{a_{ij} + a_{ji}}{2} \text{ for all } i,j. \tag{13.A.4}$$

Then $b_{ij} + b_{ji} = 2b_{ij}$ is the coefficient of $x_i x_j$, $i \neq j$, in

$$X'BX = \sum_{i=1}^{n}\sum_{j=1}^{n} \frac{a_{ij} + a_{ji}}{2} x_i x_j. \tag{13.A.5}$$

(Since $b_{ij} + b_{ji} = a_i + a_{ij}$, this redefinition of coefficients clearly leaves the value of Q unchanged and thus, under (13.A.4), $X'AX = X'BX$, $X \in \mathcal{E}^n$.) So given any quadratic form $X'AX$, the matrix A may be assumed to be symmetric. If it is not, it can be readily transformed into a symmetric matrix.

Example 13.A.2

a) Transform the matrix

$$A = \begin{bmatrix} 1 & 3 & 2 \\ 1 & -1 & 6 \\ 3 & 5 & 4 \end{bmatrix}$$

into a symmetric matrix B. From (13.A.4),

$$b_{11} = a_{11} = 1$$

$$b_{12} = b_{21} = \frac{a_{12} + a_{21}}{2} = \frac{3+1}{2} = 2$$

$$b_{13} = b_{31} = \frac{a_{13} + a_{31}}{2} = \frac{2+3}{2} = \frac{5}{2}$$

$$b_{22} = a_{22} = -1$$

$$b_{23} = b_{32} = \frac{a_{23} + a_{32}}{2} = \frac{6+5}{2} = \frac{11}{2}$$

$$b_{33} = a_{33} = 4.$$

Then

$$B = \begin{bmatrix} 1 & 2 & 5/2 \\ 2 & -1 & 11/2 \\ 5/2 & 11/2 & 4 \end{bmatrix}.$$

b) Find the matrix $\boldsymbol{A}$ associated with the quadratic form

$$\boldsymbol{A}'\boldsymbol{A}\boldsymbol{X} = 2x_1^2 - 3x_1x_2 + \frac{7}{2}x_1x_3 + x_1x_4 + x_2^2 + 6x_2x_3 - 8x_3x_4 + 2x_4^2.$$

Since $\boldsymbol{A}$ may be taken to be symmetric, we have

$$\boldsymbol{A} = \begin{bmatrix} 2 & -3/2 & 7/4 & 1/2 \\ -3/2 & 1 & 3 & 0 \\ 7/4 & 3 & 0 & -4 \\ 1/2 & 0 & -4 & 2 \end{bmatrix}.$$

■

13.A.3 Classification of Quadratic Forms

As we shall now see, there are, in all, five mutually exclusive and collectively exhaustive varieties of quadratic forms. Specifically:

A) **Definite Quadratic Form**

A quadratic form is said to be **positive definite** (**negative definite**) if it is positive (negative) at every point $\boldsymbol{X} \in \mathcal{E}^n$ except $\boldsymbol{X} = \boldsymbol{O}$. That is

a) $\boldsymbol{X}'\boldsymbol{A}\boldsymbol{X}$ is positive definite if $\boldsymbol{X}'\boldsymbol{A}\boldsymbol{X} > 0$ for every $\boldsymbol{X} \neq \boldsymbol{O}$;
b) $\boldsymbol{X}'\boldsymbol{A}\boldsymbol{X}$ is negative definite if $\boldsymbol{X}'\boldsymbol{A}\boldsymbol{X} < 0$ for every $\boldsymbol{X} \neq \boldsymbol{O}$.

(Clearly a form that is either positive or negative definite cannot assume both positive and negative values.)

B) **Semi-Definite Quadratic Form**

A quadratic form is said to be **positive semi-definite** (**negative semi-definite**) if it is nonnegative (nonpositive) at every point $\boldsymbol{X} \in \mathcal{E}^n$, and there exist points $\boldsymbol{X} \neq \boldsymbol{O}$ for which it is zero. That is:

a) $\boldsymbol{X}'\boldsymbol{A}\boldsymbol{X}$ is positive semi-definite if $\boldsymbol{X}'\boldsymbol{A}\boldsymbol{X} \geq 0$ for every $\boldsymbol{X}$ and $\boldsymbol{X}'\boldsymbol{A}\boldsymbol{X} = 0$ for some points $\boldsymbol{X} \neq \boldsymbol{O}$;
b) $\boldsymbol{X}'\boldsymbol{A}\boldsymbol{X}$ is negative semi-definite if $\boldsymbol{X}'\boldsymbol{A}\boldsymbol{X} \leq 0$ for every $\boldsymbol{X}$ and $\boldsymbol{X}'\boldsymbol{A}\boldsymbol{X} = 0$ for some points $\boldsymbol{X} \neq \boldsymbol{O}$.

C) **Indefinite Quadratic Forms**

A quadratic form is said to be **indefinite** if it is positive for some points $\boldsymbol{X} \in \mathcal{E}^n$ and negative for others.

Example 13.A.3

a) The quadratic form $\boldsymbol{X}'\boldsymbol{A}\boldsymbol{X} = x_1^2 + x_2^2$ is positive definite while $\boldsymbol{X}'(-\boldsymbol{A})\boldsymbol{X} = -x_1^2 - x_2^2$ is negative definite since both equal zero only at the point $\boldsymbol{X} = \boldsymbol{O}$.

b) The quadratic form $X'AX = x_1^2 - 2x_1x_2 + x_2^2 = (x_1 - x_2)^2$ is positive semi-definite and $X'(-A)X = -x_1^2 + x_1x_2 - x_2^2 = -(x_1 - x_2)^2$ is negative semi-definite since the former is never negative while the latter is never positive, yet both equal zero for $x_1 = x_2 \neq 0$.
c) The quadratic form $X'AX = x_1x_2 + x_2^2$ is indefinite since, on the one hand, $X'AX < 0$ for $x_1 = -2$, $x_2 = 1$ and, on the other, $X'AX > 0$ for $x_1 = -2$, $x_2 = 1$.

13.A.4 Necessary Conditions for the Definiteness and Semi-Definiteness of Quadratic Forms

We now turn to an assortment of theorems (13.A.1–13.A.3) that will enable us to identify the various types of quadratic forms.

Theorem 13.A.1 If a quadratic form $X'AX$, $X \in \mathcal{E}^n$, is positive (negative) definite, all of the terms involving second powers of the variables must have positive (negative) coefficients.

Note that the converse of this theorem does not hold since a quadratic form may have positive (negative) coefficients on all its terms involving second powers yet not be definite, e.g. $X'AX = x_1^2 - 2x_1x_2 + x_2^2 = (x_1 - x_2)^2$. Similarly,

Theorem 13.A.2 If a quadratic form $X'AX$, $X \in \mathcal{E}^n$, is positive (negative) semi-definite, all of the terms involving second powers of the variables must have nonnegative (nonpositive) coefficients.

(Here, too, the converse of the theorem is not true, e.g. the quadratic form $X'AX = x_1^2 - x_1x_2$ has nonnegative coefficients associated with its second-degree terms, yet is indefinite.)

Before stating our next theorem, let us define the ***k*th naturally ordered principal minor of *A*** as

$$M_k = \begin{vmatrix} a_{11} & \cdots & a_{1k} \\ \vdots & & \vdots \\ a_{k1} & \cdots & a_{kk} \end{vmatrix}, k = 1, \ldots, n,$$

or

$$M_{11} = a_{11}, M_2 = \begin{vmatrix} a_{11} & a_{12} \\ a_{21} & a_{22} \end{vmatrix}, M_3 = \begin{vmatrix} a_{11} & a_{12} & a_{13} \\ a_{21} & a_{22} & a_{23} \\ a_{31} & a_{32} & a_{33} \end{vmatrix}, \ldots, M_n = |A|.$$

In the light of this definition, we now look to

Theorem 13.A.3 If the quadratic form $X'AX$, $X \in \mathcal{R}^n$, is definite, the naturally ordered principal minors of A are *all* different from zero.

So, if any $M_k = 0$, $k = 1, \ldots, n$, the form is not definite – but it may be semi-definite or indefinite.

Example 13.A.4

Determine if the following quadratic forms are definite:

a) $X'AX = x_1^2 - x_2^2$. Here

$$A = \begin{bmatrix} 1 & 0 \\ 0 & -1 \end{bmatrix}$$

with $M_1 = 1$ and $M_2 = -1$. In fact, this quadratic form is indefinite since it is positive for some values of X and negative for others.

b) $X'AX = x_1^2 - 2x_1x_2 + x_2^2$. Here

$$A = \begin{bmatrix} 1 & -1 \\ -1 & 1 \end{bmatrix}$$

with $M_1 = 1$ and $M_2 = 0$. Clearly this quadratic form is not definite. But it is semi-definite.

c) $X'AX = x_1x_2 + x_2^2$. Then

$$A = \begin{bmatrix} 0 & \frac{1}{2} \\ \frac{1}{2} & 1 \end{bmatrix}$$

with $M_1 = 0$ and $M_2 = -\frac{1}{4}$. This quadratic form happens to be indefinite. ■

13.A.5 Necessary and Sufficient Conditions for the Definiteness and Semi-Definiteness of Quadratic Forms

Let us modify Theorem 13.A.3 so as to obtain Theorem 13.A.4.

Theorem 13.A.4 The quadratic form $X'AX$, $X \in \mathcal{E}^n$ is positive definite if and only if the naturally ordered principal minors of A are all positive, i.e.

$$M_k > 0, k = 1, \ldots, n.$$

We next have Theorem 13.A.5.

Theorem 13.A.5 The quadratic form $X'AX$, $X \in \mathcal{E}^n$, is negative definite if and only if the naturally ordered principal minors of A alternate in sign, the first being negative, i.e.

$$(-1)^k M_k = (-1)^k \begin{vmatrix} a_{11} & \cdots & a_{1k} \\ \vdots & & \vdots \\ a_{k1} & \cdots & a_{kk} \end{vmatrix} > 0, k = 1, \ldots, n,$$

or

$$M_1 = a_{11} < 0, M_2 = \begin{vmatrix} a_{11} & a_{12} \\ a_{21} & a_{22} \end{vmatrix} > 0, M_3 = \begin{vmatrix} a_{11} & a_{12} & a_{13} \\ a_{21} & a_{22} & a_{23} \\ a_{31} & a_{32} & a_{33} \end{vmatrix} < 0, \ldots,$$

$$M_n = (-1)^n |A| > 0.$$

A similar set of theorems (13.A.6 and 13.A.7) holds for semi-definite quadratic forms. Specifically,

Theorem 13.A.6 The quadratic form $X'AX$, $X \in \mathcal{E}^n$ is positive semi-definite if and only if *all* of the principal minors of A (not only the naturally ordered ones) are nonnegative.

Similarly,

Theorem 13.A.7 The quadratic form $X'AX$, $X \in \mathcal{E}^n$ is negative semi-definite if and only if *all* of the principal minors of A (not only the naturally ordered ones) alternate in sign, the first being nonpositive.

Example 13.A.5 Determine if the following quadratic forms are positive (or negative) definite.

a) $X'AX = 2x_1^2 + 2x_1x_2 + 6x_2^2 + 4x_2x_3 + x_3^2$. Here

$$A = \begin{bmatrix} 2 & 1 & 0 \\ 1 & 6 & 2 \\ 0 & 2 & 1 \end{bmatrix}$$

with $M_1 = 2$, $M_2 = 11$, and $M_3 = 3$. Hence $X'AX$ is positive definite via Theorem 13.A.4.

b) $\boldsymbol{X}'\boldsymbol{A}\boldsymbol{X} = -x_1^2 + x_1x_2 - x_2^2 - x_3^2$. With

$$\boldsymbol{A} = \begin{bmatrix} -1 & \frac{1}{2} & 0 \\ \frac{1}{2} & -1 & 0 \\ 0 & 0 & -1 \end{bmatrix},$$

it is readily verified that $M_1 = 1$ and $M_2 = \frac{3}{4}$, and $M_3 = -\frac{3}{4}$. Hence, by Theorem 13.A.5, this quadratic form is negative definite. ■

14

Data Envelopment Analysis (DEA)

14.1 Introduction (Charnes et al. 1978; Banker et al. 1984; Seiford and Thrall 1990; and Banker and Thrall 1992)

In this chapter we shall examine an important applications area of linear programming, namely the technique of **data envelopment analysis** (DEA) that was first introduced by Charnes, Cooper, and Rhodes (1978). Generally speaking, DEA is a computational procedure used to estimate multiple-input, multiple-output production correspondences so that the productive efficiency of decision making units (DMUs) can be scrutinized. It accomplishes this by using linear programming modeling to measure the productive efficiency of a DMU relative to that of a set of baseline DMUs. In particular, linear programming is used to estimate what is called a **best-practice extremal frontier**. It does so by constructing a piecewise linear surface, which essentially "rests on top" of the observations. This frontier "envelops" the data and enables us to determine a set of efficient projections to the envelopment surface. The nature of the **projection path** to the said surface depends on whether the linear program is input-oriented or output-oriented:

Input-oriented DEA – the linear program is formulated so as to determine the amount by which input usage of a DMU could be contracted in order to produce the same output levels as best practice DMUs.

Output-oriented DEA – the linear program is formulated so as to determine a DMU's potential output given its inputs if it operated as efficiently as the best-practice DMUs.

Moreover, the envelopment surface obtained depends on the scale assumptions made, e.g. we can posit either **constant returns to scale** (CRS) (output changes by the same proportion as the inputs change) or **variable returns to scale** (VRS) (the production technology can exhibit increasing, constant, or decreasing returns to scale).

Additional features of DEA models include the following:

a) They are **nonparametric** in that no specific functional form is imposed on the model, thus mitigating the chance of confounding the effects of misspecification of the functional form with those of inefficiency.
b) They are **nonstochastic** in that they make no accommodation for the effects of statistical noise on measuring the relative efficiency of a DMU.
c) They build on the individual firm efficiency measures of Farrell (1957) by extending the basic (engineering) ratio approach to efficiency measurement from a single-output, single-input efficiency index to multiple-output, multiple-input instances without requiring preassigned weights. This is accomplished by the extremal objective incorporated in the model, thus enabling the model to envelop the observations as tightly as possible.
d) They measure the efficiency of a DMU relative to all other DMUs under the restriction that all DMUs lie on or below the efficient or envelopment frontier.

Before we look to the construction of DEA models proper, some additional conceptual apparatus needs to be developed, namely: the set theoretic representation of a production technology, output, and input distance functions (IDFs), and various efficiency measures.

14.2 Set Theoretic Representation of a Production Technology (Färe et al. 2008; Lovell 1994; Färe and Lovell 1978; and Fried et al. 2008)

Suppose $\boldsymbol{x}$ is an $(n \times 1)$ variable input vector in R_{+}^{N} and $\boldsymbol{q}$ is an $(m \times 1)$ variable output vector in R_{+}^{M}. Then a multi-input, multi-output production technology can be described by the **technology set**

$$T(\boldsymbol{x},\boldsymbol{q}) = \{(\boldsymbol{x},\boldsymbol{q}) | \boldsymbol{x} \text{ can produce } \boldsymbol{q}\}. \tag{14.1}$$

(T is also termed the **production possibility set** or the **graph** of the technology.) Clearly T contains all feasible correspondences of inputs $\boldsymbol{x}$ capable of producing output levels $\boldsymbol{q}$. Moreover, the DMUs operate in T; and once T is determined, the location of a DMU within it enables us to gauge its relative efficiency and (local) returns to scale.

Looking to the properties of the technology set $T(\boldsymbol{x}, \boldsymbol{q})$ we note that: T is closed (it contains all of its boundary points); bounded (it has a finite diameter); and satisfies **strong disposability**, i.e. if $(\boldsymbol{x}, \boldsymbol{q}) \in T$, then $(\boldsymbol{x}_1, \boldsymbol{q}_1) \in T$ for all $(\boldsymbol{x}_1, -\boldsymbol{q}_1) \geq (\boldsymbol{x}, -\boldsymbol{q})$ or $\boldsymbol{x}_1 \geq \boldsymbol{x}$ and $\boldsymbol{q}_1 \leq \boldsymbol{q}$.

The technology set $T(\boldsymbol{x}, \boldsymbol{q})$ contains its **isoquants**

$$Isoq(T) = \{(\boldsymbol{x},\boldsymbol{q}) | (\boldsymbol{x},\boldsymbol{q}) \in T, (\delta\boldsymbol{x},\delta^{-1}\boldsymbol{q}) \notin T, 0 < \delta < 1\} \tag{14.2}$$

(for $(\boldsymbol{x}, \boldsymbol{q}) \in T$, *decreasing inputs* by a fraction δ, $0 < \delta < 1$, does not allow us to *increase outputs*), which, in turn, contain **efficient subsets**

$$Eff(T) = \{(\boldsymbol{x},\boldsymbol{q}) | (\boldsymbol{x},\boldsymbol{q}) \in T, (\boldsymbol{x}_1,\boldsymbol{q}_1) \notin T, 0 \le \boldsymbol{x}_1 \le \boldsymbol{x}, \boldsymbol{q}_1 \ge \boldsymbol{q}\}. \tag{14.3}$$

A multi-input, multi-output production technology can equivalently be represented by output and input correspondences derived from the technology set T. The **output correspondence** or **output set** $P(\boldsymbol{x})$ is the set of all output vectors $\boldsymbol{q}$ that can be produced using the input vector $\boldsymbol{x}$. Thus

$$P(\boldsymbol{x}) = \{\boldsymbol{q} | (\boldsymbol{x},\boldsymbol{q}) \in T\}. \tag{14.4}$$

Here P maps inputs $\boldsymbol{x}$ into subsets $P(\boldsymbol{x})$ of outputs. In terms of the properties of $P(\boldsymbol{x})$, for each $\boldsymbol{x}$, $P(\boldsymbol{x})$ satisfies:

a) $\boldsymbol{O} \in P(\boldsymbol{x})$.
b) Nonzero output levels cannot be produced from $\boldsymbol{x} = \boldsymbol{O}$.
c) $P(\boldsymbol{x})$ is closed ($P(\boldsymbol{x})$ contains all of its boundary points).
d) $P(\boldsymbol{x})$ is bounded above (unlimited outputs cannot be produced from a given $\boldsymbol{x}$).
e) $P(\boldsymbol{x})$ is convex (i.e. if $\boldsymbol{q}_1$ and $\boldsymbol{q}_2$ lie within $P(\boldsymbol{x})$, then $\boldsymbol{q}^* = \theta\boldsymbol{q}_1 + (1-\theta)\boldsymbol{q}_2 \in P(\boldsymbol{x})$, $0 \le \theta \le 1$. So if $\boldsymbol{q}_1$ and $\boldsymbol{q}_2$ can each be produced from a given $\boldsymbol{x}$, then any internal average of these output vectors can also be produced).
f) $P(\boldsymbol{x})$ satisfies **strong (free) disposability of outputs** (if $\boldsymbol{q} \in P(\boldsymbol{x})$ and $\boldsymbol{q}^* \le \boldsymbol{q}$, then $\boldsymbol{q}^* \in P(\boldsymbol{x})$). Looked at in an alternative fashion, we can say that outputs are strongly disposable if they can be disposed of without resource use or cost.[1]
g) $P(\boldsymbol{x})$ satisfies **strong disposability of inputs** (if $\boldsymbol{q}$ can be produced from $\boldsymbol{x}$, then $\boldsymbol{q}$ can be produced from any $\boldsymbol{x}^* \ge \boldsymbol{x}$).

Output sets $P(\boldsymbol{x})$ contain their **output isoquants**

$$Isoq\,P(\boldsymbol{x}) = \{\boldsymbol{q} | \boldsymbol{q} \in P(\boldsymbol{x}), \phi\boldsymbol{q} \notin P(\boldsymbol{x}), \phi > 1\}, \tag{14.5}$$

which, in turn, contain their **output efficient subsets**

$$Eff\,P(\boldsymbol{x}) = \{\boldsymbol{q} | \boldsymbol{q} \in P(\boldsymbol{x}), \boldsymbol{q}_1 \ge \boldsymbol{q} \text{ and } \boldsymbol{q}_1 \ne \boldsymbol{q} \Rightarrow \boldsymbol{q}_1 \notin P(\boldsymbol{x})\}. \tag{14.6}$$

A moment's reflection reveals that $Eff\,P(\boldsymbol{x}) \subseteq Isoq\,P(\boldsymbol{x}) \subseteq P(\boldsymbol{x})$ (see Figure 14.1a). The output set $P(\boldsymbol{x})$ is the area with the piecewise linear boundary OABCDEF,

1 $P(\boldsymbol{x})$ satisfies **weak disposability of outputs** (if a $\boldsymbol{q}$ can be produced from $\boldsymbol{x}$, then any contraction of $\boldsymbol{q}$, $\lambda\boldsymbol{q}$, $0 \le \lambda < 1$, can also be produced by $\boldsymbol{x}$). Clearly
strong disposability $\Rightarrow$ weak disposability;
weak disposability $\nRightarrow$ strong disposability.

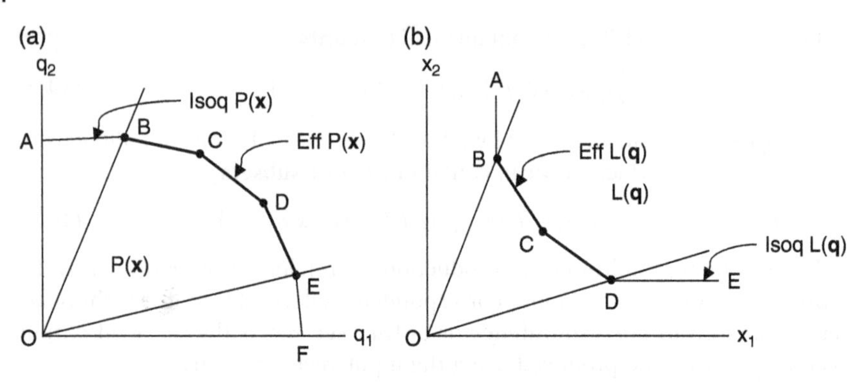

Figure 14.1 (a) Output-space representation of a technology; (b) Input-space representation of a technology.

the output isoquant *Isoq P*($\boldsymbol{x}$) is ABCDEF, and the output efficient subset of the output isoquant is BCDE (it has no horizontal or vertical components).

The **input correspondence** or **input set** $L(\boldsymbol{q})$ is the set of all input vectors $\boldsymbol{x}$ that can produce the output vector $\boldsymbol{q}$. Hence

$$L(\boldsymbol{q}) = \{\boldsymbol{x} | (\boldsymbol{x}, \boldsymbol{q}) \in T\}. \tag{14.7}$$

Looking to the properties of $L(\boldsymbol{q})$, for each $\boldsymbol{q}$, $L(\boldsymbol{q})$ satisfies:

a) $L(\boldsymbol{q})$ is closed ($L(\boldsymbol{q})$ contains all of its boundary points).
b) $L(\boldsymbol{q})$ is bounded below.
c) $L(\boldsymbol{q})$ is convex (i.e. if $\boldsymbol{x}_1, \boldsymbol{x}_2 \in L(\boldsymbol{q})$, then $\boldsymbol{x}^* = \theta\boldsymbol{x}_1 + (1-\theta)\boldsymbol{x}_2 \in L(\boldsymbol{q})$, $0 \le \theta \le 1$).
d) $L(\boldsymbol{q})$ satisfies **strong disposability of inputs** (if $\boldsymbol{x} \in L(\boldsymbol{q})$ and if $\boldsymbol{x}^* \ge \boldsymbol{x}$, then $\boldsymbol{x}^* \in L(\boldsymbol{q})$).

   (Inputs are **weakly disposable** if $\boldsymbol{x} \in L(\boldsymbol{q})$ implies that, for all $\lambda \ge 1$, $\lambda\boldsymbol{x} \in L(\boldsymbol{q})$.)

Input sets $L(\boldsymbol{q})$ contain their **input isoquants**

$$Isoq\, L(\boldsymbol{q}) = \{\boldsymbol{x} | \boldsymbol{x} \in L(\boldsymbol{q}), \theta\boldsymbol{x} \notin L(\boldsymbol{q}), 0 \le \theta < 1\}, \tag{14.8}$$

which, in turn, contain their **input efficient subsets**

$$Eff\, L(\boldsymbol{q}) = \{\boldsymbol{x} | \boldsymbol{x} \in L(\boldsymbol{q}), \boldsymbol{x}_1 \le \boldsymbol{x} \text{ and } \boldsymbol{x}_1 \ne \boldsymbol{x} \Rightarrow \boldsymbol{x}_1 \notin L(\boldsymbol{q})\}. \tag{14.9}$$

Given these constructs, it is readily seen that *Eff* $L(\boldsymbol{q}) \subseteq$ *Isoq* $L(\boldsymbol{q}) \subseteq L(\boldsymbol{q})$ (Figure 14.1b). The input set $L(\boldsymbol{q})$ is the area to the right and above the piecewise linear boundary ABCDE, the input isoquant *Isoq* $L(\boldsymbol{q})$ is ABCDE, and the input efficient subset of the input isoquant is BCD (it has no horizontal or vertical elements).

On the basis of the preceding discussion, we can redefine the output set $P(\boldsymbol{x})$ (Eq. (14.4)) and input set $L(\boldsymbol{q})$ (Eq. (14.7)) as

$$P(\boldsymbol{x}) = \{\boldsymbol{q} | \boldsymbol{x} \in L(\boldsymbol{q})\} \tag{14.4.1}$$

and

$$L(\boldsymbol{q}) = \{\boldsymbol{x} | \boldsymbol{q} \in P(\boldsymbol{x})\}, \tag{14.7.1}$$

respectively. This enables us to obtain the input and output correspondences from one another since if $\boldsymbol{q} \in P(\boldsymbol{x})$, then $\boldsymbol{x} \in L(\boldsymbol{q})$. Hence $P(\boldsymbol{x})$ and $L(\boldsymbol{q})$, being inversely related, contain the same information.

14.3 Output and Input Distance Functions (Färe and Lovell 1978; Fried et al. 2008; Coelli et al. 2005; and Shephard 1953, 1970)

Output and input distance functions serve as an alternative way of describing a multi-output, multi-input technology. Moreover, as we shall see in the next section, such functions will be utilized to measure the efficiency of a DMU in production.

We start with the **output distance function** (ODF) – considers a maximal proportional expansion of the output vector $\boldsymbol{q}$, given an input vector $\boldsymbol{x}$. The ODF is defined on the output set $P(\boldsymbol{x})$ as

$$d_O(\boldsymbol{x},\boldsymbol{q}) = min\{\delta | (\boldsymbol{q}/\delta) \in P(\boldsymbol{x})\} \tag{14.10}$$

(assuming that the minimum is actually attained).

To evaluate (14.10), let us see how δ is determined. Looking to Figure 14.2a, suppose $\delta^* = \min\{\delta | (\boldsymbol{q}/\delta) \in P(\boldsymbol{x})\}$, $\boldsymbol{q}$ is located at point A, with $(\boldsymbol{q}/\delta^*)$ found at point B. Then the ODF at Point A is $\delta^* = OA/OB < 1$ so that $(\boldsymbol{q}/\delta^*)$ lies on *Isoq* $P(\boldsymbol{x})$. By definition, δ^* is the smallest δ that puts $(\boldsymbol{q}/\delta^*)$ on the boundary of $P(\boldsymbol{x})$. (Clearly the value of the ODF at points B and C is 1.)

Key properties of the ODF are:

a) $d_O(\boldsymbol{x},\boldsymbol{O}) = 0$ for all $\boldsymbol{x} \geq \boldsymbol{O}$.
b) $d_O(\boldsymbol{x},\boldsymbol{q})$ is nonincreasing in $\boldsymbol{x}$ and nondecreasing in $\boldsymbol{q}$.
c) $d_O(\boldsymbol{x},\boldsymbol{q})$ is linearly homogeneous (or homogeneous of degree 1) in $\boldsymbol{q}$.[2]
d) if $\boldsymbol{q} \in P(\boldsymbol{x})$, then $d_O(\boldsymbol{x},\boldsymbol{q}) \leq 1$.
e) if $\boldsymbol{q} \in Isoq\ P(\boldsymbol{x})$, then $d_O(\boldsymbol{x},\boldsymbol{q}) = 1$.
f) $d_O(\boldsymbol{x},\boldsymbol{q})$ is quasi-convex[3] in $\boldsymbol{x}$ and convex[4] in $\boldsymbol{q}$.

We next examine the **input distance function** (IDF). Specifically, the IDF characterizes a production technology by specifying the minimal proportional

2 A real-valued function on $y = f(x_1, \ldots, x_n)$ is homogeneous of degree k in $x_1, \ldots, x_n$ if $f(tx_1, \ldots, tx_n) = t^k f(x_1, \ldots, x_n)$.

3 A real-valued function $y = f(\boldsymbol{x})$ defined on a convex set $S \subset R^n$ is **quasi-convex** on S if and only if for any pair of points $\boldsymbol{x}_1, \boldsymbol{x}_2 \in S$, and $0 \leq \theta \leq 1$, $f(\boldsymbol{x}_2) \leq f(\boldsymbol{x}_1)$implies that $f(\boldsymbol{x}_1) \geq f(\theta\boldsymbol{x}_2 + (1-\theta)\boldsymbol{x}_1)$.

4 A real-valued function $y = f(\boldsymbol{x})$ defined on a convex set $S \subset R^n$ is **convex** on S if and only if for any pair of points $\boldsymbol{x}_1, \boldsymbol{x}_2 \in S$, and $0 \leq \theta \leq 1$, $f(\theta\boldsymbol{x}_1 + (1-\theta)\boldsymbol{x}_2) \leq \theta f(\boldsymbol{x}_2) + (1-\theta)f(\boldsymbol{x}_1)$.

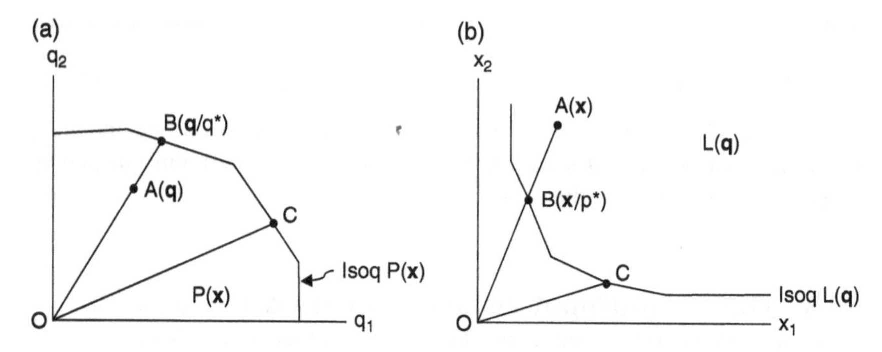

Figure 14.2 (a) Output distance function (ODF); (b) Input distance function (IDF).

contraction of the input vector $\boldsymbol{x}$, given the output vector $\boldsymbol{q}$. In this regard, the IDF can be expressed as

$$d_I(\boldsymbol{x},\boldsymbol{q}) = max\{\rho|(\boldsymbol{x}/\rho) \in L(\boldsymbol{q})\}. \tag{14.11}$$

(It is assumed that the maximum is realized.)

How is ρ to be determined? In Figure 14.2b, let $\rho^* = max\ \{\rho|(\boldsymbol{x}/\rho) \in L(\boldsymbol{q})\}$ with $\boldsymbol{x}$ found at point A and $(\boldsymbol{x}/\rho^*)$ located at point B. Then the IDF at point A is $\rho^* = OA/OB > 1$ so that $(\boldsymbol{x}, \rho^*)$ lies on $Isoq\, L(\boldsymbol{q})$. By definition, ρ^* is the largest ρ which puts $(\boldsymbol{x}/\rho^*)$ on the boundary of $L(\boldsymbol{q})$. (Note that the IDF at points B and C is 1.)

Properties of the IDF are:

a) $d_I(\boldsymbol{x}, \boldsymbol{q})$ is nondecreasing in $\boldsymbol{x}$ and nonincreasing in $\boldsymbol{q}$.
b) $d_I(\boldsymbol{x}, \boldsymbol{q})$ is linearly homogeneous (or homogeneous of degree 1) in $\boldsymbol{x}$.
c) If $\boldsymbol{x} \in L(\boldsymbol{q})$, then $d_I(\boldsymbol{x}, \boldsymbol{q}) \geq 1$.
d) If $\boldsymbol{x} \in Isoq\ L(\boldsymbol{q})$, then $d_I(\boldsymbol{x}, \boldsymbol{q}) = 1$.
e) $d_I(\boldsymbol{x}, \boldsymbol{q})$ is concave[5] in $\boldsymbol{x}$ and quasi-concave[6] in $\boldsymbol{q}$.

What is the connection between the output and IDFs? It should be evident that if $\boldsymbol{q} \in P(\boldsymbol{x})$, then $\boldsymbol{x} \in L(\boldsymbol{q})$. Moreover, if inputs and outputs are weakly disposable, then it can be shown that $d_O(\boldsymbol{x}, \boldsymbol{q}) \leq 1$ if and only if $d_I(\boldsymbol{x}, \boldsymbol{q}) \geq 1$. And if the technology is subject to CRS, then $d_O(\boldsymbol{x}, \boldsymbol{q}) = 1/d_I(\boldsymbol{x}, \boldsymbol{q})$ for all $\boldsymbol{x}$, $\boldsymbol{q}$.

---

5 A real-valued function $y = f(\boldsymbol{x})$ defined on a convex set $S \subset R^n$ is **concave** on S if and only if for any pair of points $\boldsymbol{x}_1, \boldsymbol{x}_2 \in S$, and $0 \leq \theta \leq 1$, $f(\theta\boldsymbol{x}_1 + (1-\theta)\boldsymbol{x}_2) \geq \theta f(\boldsymbol{x}_2) + (1-\theta)f(\boldsymbol{x}_1)$. (Note: f is concave if and only if $-f$ is convex.)

6 A real-valued function $y = f(\boldsymbol{x})$ defined on a convex set $S \subset R^n$ is **quasi-concave** on S if and only if for any pair of points $\boldsymbol{x}_1, \boldsymbol{x}_2 \in S$, and $0 \leq \theta \leq 1$, $f(\boldsymbol{x}_1) \leq f(\boldsymbol{x}_2)$implies that $f(\boldsymbol{x}_1) \leq f(\theta\boldsymbol{x}_2 + (1-\theta)\boldsymbol{x}_1)$. (Note: f is quasi-concave if and only if $-f$ is quasi-convex.)

14.4 Technical and Allocative Efficiency

As indicated above, the principal focus of DEA is the assessment of the *relative efficiency* of a DMU. Generally speaking, the efficiency of a DMU consists of two components: *technical* and *allocative*. Specifically, **technical efficiency** (TE) can have an input-conserving orientation or an output-augmenting orientation, i.e.

a) **Input-conserving orientation** – given the technology and output vector q, the DMU utilizes as little of the inputs x as possible;
b) **Output-augmenting orientation** – given the technology and input vector x, the DMU maximizes the outputs q produced.[7]

(14.12)

Next, **allocative efficiency** (AE) refers to the use of inputs in optimal proportions given their market prices and the production technology. In what follows, we shall develop in greater detail the notion of technical and AE (see Sections 14.4.1 and 14.4.2, respectively).

14.4.1 Measuring Technical Efficiency (Debreu 1951; Farrell 1957)

Let us assume that the production process exhibits CRS. In two dimensions, TE will be measured along a ray from the origin to the observed production point. (This obviously holds the relative proportions of inputs constant.) In this regard, the measures or indexes of TE that follow are termed **radial measures of TE**. Given this convention, we may modify our definitions of input-oriented and output-oriented TE (14.12a, b) to read:

a′) **Input-conserving orientation** – we are interested in determining the amount by which input quantities can be *proportionately reduced* without changing the output quantities produced.
b′) **Output-augmenting orientation** – we are interested in determining the amount by which output quantities can be *proportionately increased* without changing the input quantities used.

(14.13)

7 This description of technical efficiency is consistent with those given by Koopmans (1951) – a DMU is technically efficient if an increase in any output requires a reduction in at least one other output, or an increase in at least one input, and if a reduction in any input requires an increase in at least one other input or a reduction in at least one output; and by Charnes, Cooper, and Rhodes (1978):

i) Input orientation – a DMU is *not efficient* if it is possible to decrease any input without increasing any other input and without decreasing any output.
ii) Output-oriented – a DMU is *not efficient* if it is possible to increase any output without increasing any input and without decreasing any other output.

Thus, a DMU is *efficient* if neither (i) nor (ii) obtains.

Let us first examine the input-oriented case. Suppose that a DMU uses two inputs to produce a single output. If the production function appears as $q = f(x_1, x_2)$, then, under CRS, $k = 1$ in footnote 1. Set $t = {}^1/_q$ so that $f(x_1/q, x_2/q) = \frac{1}{q} f(x_1, x_2) = 1$. Hence, the technology can be represented by the unit isoquant. In this regard, let the **unit isoquant of fully efficient DMUs** be depicted as II' in Figure 14.3. Suppose P represents the observed input combination $\boldsymbol{x}$ that produces a unit of output. Then the *(absolute) technical inefficiency* of the DMU is represented by the distance QP – the amount by which all inputs could be proportionately reduced without a reduction in output so that QP/OP is the *percentage* by which all inputs need to be reduced to attain production efficiency. Since $QP = OP - OQ$, set

$$\frac{QP}{OP} = \frac{OP - OQ}{OP} = 1 - \frac{OQ}{OP} \tag{14.14}$$

so that we can measure the **degree of input-oriented TE** of a DMU as

$$TE_I = \frac{OQ}{OP} = 1 - \frac{QP}{OP}, 0 \le TE_I \le 1. \tag{14.15}$$

Here, TE_I depicts the ratio of inputs needed to produce q to the inputs actually used to produce q, given $\boldsymbol{x}$. (Note: if $TE_I = 1$, then a DMU is fully efficient since its input combination lies on the unit isoquant II', e.g. point Q in Figure 14.3.) Point Q is therefore point P's **technically efficient projection point** on the unit isoquant. More formally, we can state the **input-oriented measure of TE** (Debreu 1951; Farrell 1957). A **radial measure of the TE** of input vector $\boldsymbol{x}$ in the production of output vector $\boldsymbol{q}$ is

$$TE_I(\boldsymbol{x}, \boldsymbol{q}) = min\{\theta | \theta \boldsymbol{x} \in L(\boldsymbol{q})\}. \tag{14.16}$$

So for $\boldsymbol{x} \in L(\boldsymbol{q})$, $TE_I(\boldsymbol{x}, \boldsymbol{q}) \le 1$. Hence $\theta = 1$ indicates **radial TE** ($\boldsymbol{x} \in Isoq\, L(\boldsymbol{q})$) and $\theta < 1$ indicates the **degree of radial technical inefficiency**.

(In Figure 14.3, let $\theta = 1 - QP/OP$. Then θP puts us at point Q on the unit isoquant.) A moment's reflection reveals that: (i) $Eff\, L(\boldsymbol{q}) \subseteq Isoq\, L(\boldsymbol{q}) = \{\boldsymbol{x} | TE_I(\boldsymbol{x}, \boldsymbol{q}) = 1\}$; and (ii) $TE_I(\boldsymbol{x}, \boldsymbol{q}) = 1/d_I(\boldsymbol{x}, \boldsymbol{q})$. (For a fully efficient DMU, $TE_I(\boldsymbol{x}, \boldsymbol{q}) = 1 = d_I(\boldsymbol{x}, \boldsymbol{q})$.)

Second, we turn our attention to the *output-oriented case*. Let us assume that a DMU produces two outputs (q_1 and q_2) from a single input x. If the production possibility curve can be written as $x = g(q_1, q_2)$, then, under CRS, we can obtain $g(q_1/x, q_2/x,) = 1$. Hence, the technology can be represented by the unit

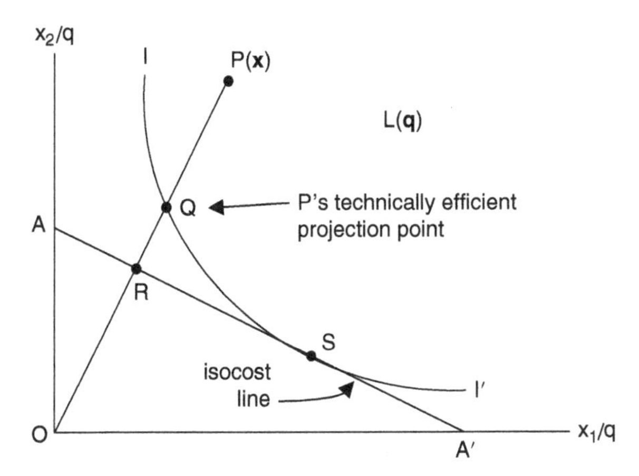

Figure 14.3 The unit isoquant II'.

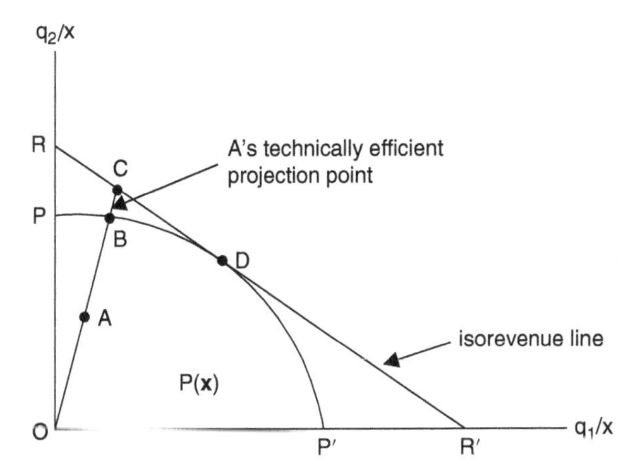

Figure 14.4 The unit production possibility curve PP'.

production possibility curve and thus the **unit production possibility curve of fully efficient DMUs** can be denoted PP' in Figure 14.4. Let point A represent the observed output combination that utilizes 1 unit of x. Then the (***absolute***) *technical inefficiency* of the DMU is depicted by the distance AB – the amount by which output could be proportionately increased without requiring more input so that AB/OA is the *percent* by which all outputs could be increased to attain production efficiency. Given that AB = OB – OA, set

$$\frac{AB}{OA} = \frac{OB - OA}{OA} = \frac{OB}{OA} - 1 \tag{14.17}$$

so that we can measure the **degree of output-oriented TE** of a DMU as

$$TE_0 = \frac{OB}{OA} = 1 + \frac{AB}{OA}, TE_0 \geq 1. \tag{14.18}$$

It is instructive to note that, whether measuring input-oriented or output-oriented TE, we are always forming the ratio between the optimal production point and the actual production point, e.g. see $TE_I = OQ/OP$ (Figure 14.3) and $TE_0 = OB/OA$(Figure 14.4). We may view TE_0 as indicating the ratio of outputs efficiently produced by x to the outputs actually produced by x, given $\boldsymbol{q}$. (Note: if $TE_0 = 1$, then a DMU is **fully efficient** since its output combination lies on the unit production possibility curve PP', i.e. point B in Figure 14.4.) Point B is thus A's **technically efficient projection point** on the unit production possibility curve. In this regard we can now state that **output-oriented measure of TE** (Debreu 1951; Farrell 1957). A **radial measure of the TE** of output vector $\boldsymbol{q}$ produced by input vector $\boldsymbol{x}$ is

$$TE_0(\boldsymbol{x}, \boldsymbol{q}) = max\{\phi | \phi\boldsymbol{q} \in P(\boldsymbol{x})\}. \tag{14.19}$$

So for $\boldsymbol{q} \in P(\boldsymbol{x})$, $TE_0(\boldsymbol{x}, \boldsymbol{q}) \geq 1$. That is, $\phi = 1$ indicates **radial TE** ($\boldsymbol{q} \in Isoq\ P(\boldsymbol{x})$) and $\phi > 1$ indicates the **degree of radial technical inefficiency**.

(In Figure 14.4, let $\phi = 1 + AB/OA$. Then ϕA puts us at point B on the unit production possibility curve.) It is also true that: (i) $Eff\ P(\boldsymbol{x}) \subseteq Isoq\ P(\boldsymbol{x}) = \{\boldsymbol{q} | TE_0(\boldsymbol{x}, \boldsymbol{q}) = 1\}$; and (ii) $TE_0(\boldsymbol{x}, \boldsymbol{q}) = 1/d_0(\boldsymbol{x}, \boldsymbol{q})$. (For a fully efficient DMU, $TE_0(\boldsymbol{x}, \boldsymbol{q}) = 1 = d_0(\boldsymbol{x}, \boldsymbol{q})$.)

14.4.2 Allocative, Cost, and Revenue Efficiency

(Lovell 1994; Coelli et al. 2005)

Suppose a DMU uses the inputs (x_1 and x_2) to produce a single output (q) under CRS. (The unit isoquant of fully efficient firms is again denoted II' in Figure 14.5.) Moreover, if the input prices (w_1 and w_2) are known and assumed fixed, then the *isocost line* AA' with slope $-w_1/w_2$ can also be plotted in Figure 14.5. Here P depicts the actual production point with input vector $\boldsymbol{x}$; Q is P's technically efficient projection point with input vector $\boldsymbol{x}$; and S denotes the cost-minimizing point with input vector $\boldsymbol{x}^*$. Given the isocost line AA', we know that **AE** involves the selection of an input mix that assigns factors of production to their optimal or best alternative uses based upon their opportunity costs. Clearly point S minimizes the cost of producing the single unit of output depicted by II'. Since the cost at R is the same as that of the allocatively efficient

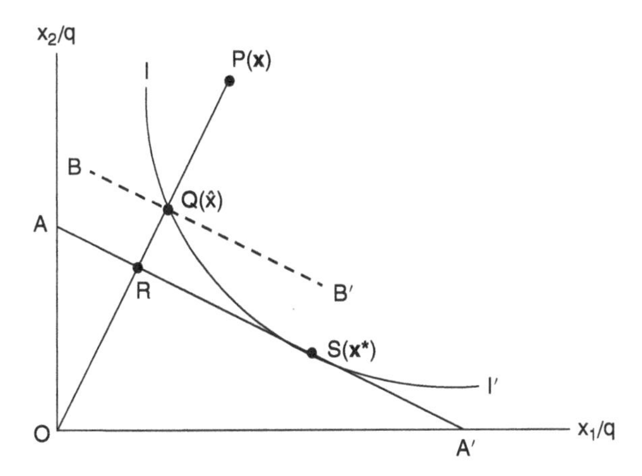

Figure 14.5 The unit isoquant II' and isocost line AA'.

Table 14.1 Technical vs. allocative efficiency input combinations

		P	Q	S
Efficiency	Technically efficient	No	Yes	Yes
Status	Allocatively efficient	No	No	Yes

point S, and is less than that incurred at the technically efficient point Q, we can measure AE as $AE = OR/OQ$.[8] In addition, since RQ represents the reduction in production cost that would occur if production were carried out at S instead of Q, then, under a parallel shift of the isocost line AA' to BB', RQ/OQ depicts the proportionate reduction in cost associated with a movement from Q to S.

Since TE and AE are measured along the same ray from the origin to the inefficient point P, the inefficient input proportions are held constant so that the efficiency measures TE and AE must be independent. In this regard, given $TE = OQ/OP$, the overall **cost efficiency** (CE) of the DMU can be expressed as the composite index

$$CE = TE \times AE = (OQ/OP)(OR/OQ) = OR/OP, 0 \le CE \le 1. \tag{14.20}$$

(Note: $CE \le TE$; and $CE \le 1/d_I(\boldsymbol{x}, \boldsymbol{q})$.)

It should be evident that AE and CE are input-oriented concepts. We now turn to an output-oriented measure of efficiency, namely revenue efficiency

8 Table 14.1 summarizes the efficiency status of input points P, Q, and S.

(RE). To this end, let the outputs q_1 and q_2 be produced by the single fixed input x under CRS. Then the **unit production possibility curve of fully efficient DMUs** can be represented as PP' in Figure 14.6. Additionally, if the output prices (p_1 and p_2) are observed and assumed fixed, then the *isorevenue line* EE' with slope $-p_1/p_2$ can also be plotted in Figure 14.6. Here A depicts the actual output point with output vector $\boldsymbol{q}$; B represents A's technically efficient projection point with output vector $\hat{\boldsymbol{q}}$; and D denotes the revenue maximizing point with output vector $\boldsymbol{q}^*$.[9]

Let us express the TE of output point A as $TE = OA/OB$ while the AE of the DMU can be determined as $AE = OB/OC$. Then the overall **RE** of the DMU is

$$RE = TE \times AE = (OA/OB)(OB/OC) = OA/OC, 0 \le RE \le 1. \tag{14.21}$$

(Note: $RE \le TE$; and $RE \ge 1/d_O(\boldsymbol{x}, \boldsymbol{q})$.)

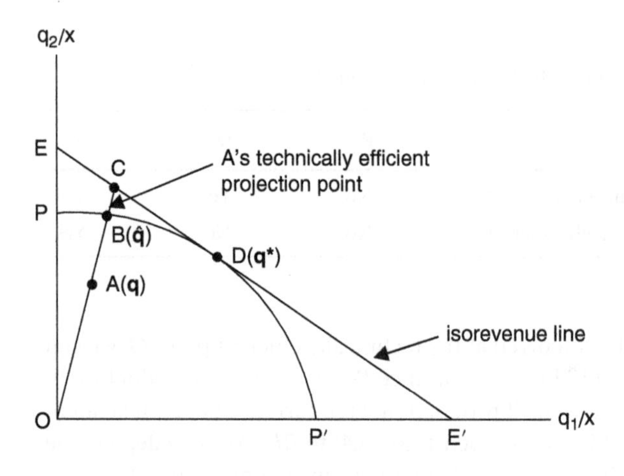

Figure 14.6 The unit production possibility curve PP' and isorevenue line EE'.

Table 14.2 Technical vs. allocative efficiency output combinations

		A	B	D
Efficiency	Technically efficient	No	Yes	Yes
Status	Allocatively efficient	No	No	Yes

9 The efficiency status of output points *A*, *B*, and *D* in Figure 14.6 appears in Table 14.2.

14.5 Data Envelopment Analysis (DEA) Modeling

(Charnes et al. 1978; Cooper et al. 2007; Lovell 1994; Banker et al. 1984; Banker and Thrall 1992)

In the models that follow, we shall employ linear programming techniques to construct a nonparametric, piecewise linear surface (or frontier) to cover or envelop the data points. Efficiency measures are then determined relative to this surface. Suppose we have N inputs, M outputs, and I DMUs. Then to calculate these efficiency measures, a set of input weights w_r, $r = 1, \ldots, N$, and output weights u_s, $s = 1, \ldots, M$, will be derived from the data and not taken in advance to be fixed. As will be noted below, a group of DMUs is used to evaluate each other, with each DMU displaying various degrees of managerial freedom in pursuing its individual goals.

For the jth DMU, the input and output vectors are, respectively,

$$\underset{(N\times 1)}{\boldsymbol{x}_j} = \begin{bmatrix} x_{1j} \\ x_{2j} \\ \vdots \\ x_{Nj} \end{bmatrix}, \quad \underset{(M\times 1)}{\boldsymbol{q}_j} = \begin{bmatrix} q_{1j} \\ q_{2j} \\ \vdots \\ q_{Mj} \end{bmatrix}, j = 1, \ldots, I,$$

where x_{rj} is the amount of the rth input used by DMU_j and q_{sj} is the amount of the sth output produced by the DMU_j. (All components of these input and output vectors are assumed to be nonnegative, with at least one component within each taken to be positive. Thus $\boldsymbol{x}_j$, $\boldsymbol{q}_j$ are **semi-positive**, i.e. $\boldsymbol{x}_j \geq 0$ and $\boldsymbol{x}_j \neq 0$; $\boldsymbol{q}_j \geq 0$ and $\boldsymbol{q}_j \neq 0$.) Then the $(N \times I)$ input matrix $\boldsymbol{X}$ can be written as

$$\underset{(N\times I)}{\boldsymbol{X}} = [\boldsymbol{x}_1, \boldsymbol{x}_2, \ldots, \boldsymbol{x}_I] = \begin{bmatrix} x_{11} & x_{12} & \cdots & x_{1I} \\ x_{21} & x_{22} & \cdots & x_{2I} \\ \vdots & \vdots & & \vdots \\ x_{N1} & x_{N2} & \cdots & x_{NI} \end{bmatrix};$$

and the $(M \times I)$ output matrix $\boldsymbol{Q}$ can be expressed as

$$\underset{(M\times I)}{\boldsymbol{Q}} = [\boldsymbol{q}_1, \boldsymbol{q}_2, \ldots, \boldsymbol{q}_I] = \begin{bmatrix} q_{11} & q_{12} & \cdots & q_{1I} \\ q_{21} & q_{22} & \cdots & q_{2I} \\ \vdots & \vdots & & \vdots \\ q_{M1} & q_{M2} & \cdots & q_{MI} \end{bmatrix}.$$

Next, let the $(N \times 1)$ vector of input weights and the $(M \times 1)$ vector of output weights be denoted, respectively, as

$$\underset{(N\times 1)}{\boldsymbol{W}} = \begin{bmatrix} w_1 \\ w_2 \\ \vdots \\ w_N \end{bmatrix}, \quad \underset{(M\times 1)}{\boldsymbol{U}} = \begin{bmatrix} u_1 \\ u_2 \\ \vdots \\ u_M \end{bmatrix}.$$

14.6 The Production Correspondence

We noted earlier that DMU_j has at least one positive value in both the input and output vectors $\boldsymbol{x}_j \in R^N$ and $\boldsymbol{q}_j \in R^M$, respectively. Let us refer to each pair of semi-positive input and output vectors $(\boldsymbol{x}_j, \boldsymbol{q}_j)$ as an **activity**. The set of feasible activities is termed the **production possibility set** R with these properties:

1) Observed activities $(\boldsymbol{x}_j, \boldsymbol{q}_j)$, $j = 1, \ldots, I$, belong to R.
2) If an activity $(\boldsymbol{x}, \boldsymbol{q}) \in R$, then $(t\boldsymbol{x}, t\boldsymbol{q}) \in R$ for t, a positive real scalar (the CRS assumption).
3) If activity $(\boldsymbol{x}, \boldsymbol{q}) \in R$, then any semi-positive activity $(\bar{\boldsymbol{x}}, \bar{\boldsymbol{q}})$, with $\bar{\boldsymbol{x}} \geq \boldsymbol{x}$ and $\bar{\boldsymbol{q}} \leq \boldsymbol{q}$, belongs to R.
4) Any semi-positive linear combination of observed activities belongs to R (i.e. R is a convex set).

Under these properties, if we observe I DMUs with activities $(\boldsymbol{x}_j, \boldsymbol{q}_j)$, $j = 1, \ldots, I$, then, given the input and output matrices $\boldsymbol{X}$ and $\boldsymbol{Q}$, respectively, and a semi-positive vector $\boldsymbol{\lambda} \in R^I$, we can define the production possibility set as

$$R(\boldsymbol{x},\boldsymbol{q}) = \{(\boldsymbol{x},\boldsymbol{q}) \,|\, \boldsymbol{x} \geq \boldsymbol{X\lambda}, \boldsymbol{q} \leq \boldsymbol{Q\lambda}, \boldsymbol{\lambda} \geq \boldsymbol{0}\}.$$

Looking to the structure of R we see that for the activity $(\boldsymbol{x}, \boldsymbol{q})$, $\boldsymbol{x} \geq \boldsymbol{X\lambda}$ (think of $\boldsymbol{X\lambda}$ as the input vector of the combined activities) or

$$\begin{bmatrix} x_1 \\ x_2 \\ \vdots \\ x_N \end{bmatrix} \geq \begin{bmatrix} x_{11} & x_{12} & \cdots & x_{1I} \\ x_{21} & x_{22} & \cdots & x_{2I} \\ \vdots & \vdots & & \vdots \\ x_{NI} & x_{N2} & \cdots & x_{NI} \end{bmatrix} \begin{bmatrix} \lambda_1 \\ \lambda_2 \\ \vdots \\ \lambda_I \end{bmatrix} = \begin{bmatrix} \sum_{j=1}^{I} x_{1j}\lambda_j \\ \sum_{j=1}^{I} x_{2j}\lambda_j \\ \vdots \\ \sum_{j=1}^{I} x_{Nj}\lambda_j \end{bmatrix}.$$

Here, $\sum_{j=1}^{I} x_{rj}\lambda_{rj}, r = 1,\ldots,N$, represents the sum of weighted inputs that cannot exceed x_r, $r = 1, \ldots, N$. Also, $\boldsymbol{q} \leq \boldsymbol{Q\lambda}$ (we may view $\boldsymbol{Q\lambda}$ as the output vector of the combined activities) or

$$\begin{bmatrix} q_1 \\ q_2 \\ \vdots \\ q_M \end{bmatrix} \leq \begin{bmatrix} q_{11} & q_{12} & \cdots & q_{1I} \\ q_{21} & q_{22} & \cdots & q_{2I} \\ \vdots & \vdots & & \vdots \\ q_{MI} & q_{M2} & \cdots & q_{MI} \end{bmatrix} \begin{bmatrix} \lambda_1 \\ \lambda_2 \\ \vdots \\ \lambda_I \end{bmatrix} = \begin{bmatrix} \sum_{j=1}^{I} q_{1j}\lambda_j \\ \sum_{j=1}^{I} q_{2j}\lambda_j \\ \vdots \\ \sum_{j=1}^{I} q_{Mj}\lambda_j \end{bmatrix}.$$

Thus, $\sum_{j=1}^{I} q_{sj}\lambda_j, s = 1,\ldots,M$, depicts the sum of weighted outputs which cannot fall short of q_s, $s = 1, \ldots, M$. These inequalities stipulate that R cannot contain activities $(\boldsymbol{x}, \boldsymbol{q})$ which are more efficient than linear combinations of observed activities.

14.7 Input-Oriented DEA Model under CRS

(Charnes et al. 1978)

Given the set of input and output observations, let us define the **total factor productivity** of a given DMU, say DMU_o, as

$$\boldsymbol{U}'\boldsymbol{q}_o / \boldsymbol{W}'\boldsymbol{x}_o = \sum_{s=1}^{M} u_s q_{so} \Big/ \sum_{r=1}^{N} u_r x_{ro}, \tag{14.22}$$

where $\boldsymbol{U}'\boldsymbol{q}_o$ is a (virtual) measure of aggregate output produced and $\boldsymbol{W}'\boldsymbol{x}_o$ is a (virtual) measure of aggregate input used. Given (14.22), we can form the following **fractional program** so as to obtain the input and output weights w_r, $r = 1, \ldots, N$, and u_s, $s = 1, \ldots, M$, respectively. Since these weights are the unknowns or variables, the said program is structured as

$$\begin{aligned} &\max_{U,W} \boldsymbol{U}'\boldsymbol{q}_o / \boldsymbol{W}'\boldsymbol{x}_o \text{ s.t.} \\ &\quad \boldsymbol{U}'\boldsymbol{q}_j / \boldsymbol{W}'\boldsymbol{x}_j < 1, j = 1, \ldots, I, \\ &\qquad \boldsymbol{U} \geq \boldsymbol{0}, \boldsymbol{W} \geq \boldsymbol{0}. \end{aligned} \tag{14.23}$$

Clearly this problem involves finding the values of the input and output weights such that the total factor productivity (our efficiency measure) of DMU_o is maximized subject to the **normalizing constraints** that all efficiency measures must be less than or equal to unity. Since $j = 1, \ldots, I$, we need I separate optimizations, one for each DMU to be evaluated, where the particular DMU under consideration receives the "o" designation. Hence each DMU is assigned its own set of optimal input and output weights.

Given that the fractional program (14.23) is nonlinear, it can be converted to a linear program in multiplier form via the following change of variables (see CCR 1978) from $\boldsymbol{W}$, $\boldsymbol{U}$ to $\boldsymbol{\nu}$, $\boldsymbol{\mu}$. That is, for $\boldsymbol{\mu} = t\boldsymbol{U}$, $\boldsymbol{\nu} = t\boldsymbol{W}$ and $t = (\boldsymbol{W}'\boldsymbol{x}_o)^{-1}$, set $\boldsymbol{U} = \frac{1}{t}\boldsymbol{\mu}$ and $\boldsymbol{W} = \frac{1}{t}\boldsymbol{\nu}$ in (14.23) so as to obtain

$$\max_{\mu,\nu} \frac{\frac{1}{t}\mu' q_o}{\frac{1}{t}\nu' x_o} \text{ s.t.}$$
$$\frac{\frac{1}{t}\mu' q_j}{\frac{1}{t}\nu' x_j} \le 1, j = 1, \ldots, I, \tag{14.23.1}$$
$$\frac{1}{t}\mu \ge 0, \frac{1}{t}\nu \ge 0.$$

Since $t = \frac{1}{W' x_o} = \frac{1}{\frac{1}{t}\nu' x_o}$, it follows that $\nu' x_o = 1$. Hence (14.23.1) can ultimately be rewritten as

$$\begin{aligned} &\max_{\mu,\nu} \gamma = \mu' q_o \text{ s.t.} \\ (PLP_o) \qquad & \nu' x_o = 1 \\ & \mu' q_j - \nu' x_j \le 0, j = 1, \ldots, I, \\ & \nu \ge 0, \nu \ge 0, \end{aligned} \tag{14.23.2}$$

the **multiplier form** of the DEA input-oriented problem. Here (14.23.2) involves $M + N$ variables in $I + 1$ structural constraints.

We note briefly that: (i) the fractional program (14.23.1) is equivalent to the linear program (14.23.2); and (ii) the optimal objective function values in (14.23.1) and (14.23.2) are independent of the units in which the inputs and outputs are measured (given that each DMU uses the same units of measurement). Hence our measures of efficiency are *units invariant.*

The optimal solution to (14.23.2) will be denoted as λ^*, μ^*, and ν^*, where $\mu^*$and ν^* represent optimal price vectors. Given this solution, let us consider the concept of CCR-efficiency. Specifically, **CCR-efficiency**: (i) DMU_o is **CCR-efficient** if $\gamma^* = 1$ and there exists at least one optimal set of price vectors (μ^*, ν^*) with $\mu^* > 0$, $\nu^* > 0$; and (ii) otherwise DMU_o is **CCR-inefficient**.

Hence DMU_o is CCR-inefficient if either (i) $\gamma^* < 1$ or (ii) $\gamma^* = 1$ and at least one element of (μ^*, ν^*) is zero for every optimal solution to (PLP_o). Suppose DMU_o is CCR-inefficient with $\gamma^* < 1$. Then there must be at least one DMU whose constraint in $(\mu^*)'Q - (\nu^*)'X \le 0'$ holds as a strict equality since, otherwise, γ^* is suboptimal. Let the set of such $j \in J = \{1, \ldots, I\}$ be denoted as

$$\hat{E}_o = \left\{ j \in J \mid \sum\nolimits_{s=1}^{M} \mu_s^* q_{sj} = \sum\nolimits_{r=1}^{N} \nu_r^* x_{rj} \right\}.$$

Let the subset E_o of $\hat{E}_o$ be composed of CCR-efficient DMUs. It will be termed DMU_o's **reference set** or **peer group**. Clearly, it is the existence of this collection of efficient DMUs that renders DMU_o inefficient. The set spanned by E_o is DMU_o's **efficient frontier**. Thus, all inefficient DMUs are measured relative to their individual reference sets.

The dual of the multiplier problem (14.23.2) is termed the input-oriented **envelopment form** of the DEA problem and appears as

$$\begin{aligned} &\min_{\theta,\lambda} \ \theta \text{ s.t.} \\ (DLP_o) \quad &-\boldsymbol{q}_o + \boldsymbol{Q\lambda} \geq \boldsymbol{0} \\ &\theta \boldsymbol{x}_o - \boldsymbol{X\lambda} \geq \boldsymbol{0} \\ &\theta \,\text{unrestricted}, \boldsymbol{\lambda} \geq \boldsymbol{0}, \end{aligned} \tag{14.24}$$

where θ is a real dual variable and $\boldsymbol{\lambda}$ is an $I+1$ vector of nonnegative dual variables. Here (14.24) involves $I+1$ variables and $M+N$ constraints.

Let us examine the structure of (PLP_o) and (DLP_o) in greater detail. Looking to their objectives:

(PLP_o) – seeks to determine the vectors of input and output multipliers (or normalized shadow prices) that will maximize the value of output $\boldsymbol{q}_o$.

(DLP_o) – seeks to determine a vector $\boldsymbol{\lambda}$ that combines the observed activities of the I firms and a scalar θ that depicts the level of input contraction required to adopt a technically efficient production technology. In this regard, $\theta(\leq 1)$ will be termed the **efficiency score** for the jth DMU. And if $\theta = 1$, then the DMU is on the frontier and consequently is termed **technically efficient**. As indicated earlier, this program must be solved I times (once for each DMU).

Additionally, the usual complementarity between primal variables and dual structural constraints and between dual variables and primal structural constrains holds (see Table 14.3).

Due to the nonzero restriction on the input and output data, the dual constraint $-\boldsymbol{q}_o + \boldsymbol{Q\lambda} \geq \boldsymbol{0}$ requires that $\boldsymbol{\lambda} \neq \boldsymbol{0}$ since $\boldsymbol{q}_o \geq \boldsymbol{0}$ and $\boldsymbol{q}_o \neq \boldsymbol{0}$. Thus, from $\theta \boldsymbol{x}_o - \boldsymbol{X\lambda} \geq 0$, we have $\theta > 0$ so that the optimal value of θ, θ^* must satisfy $0 < \theta^* \leq 1$. Moreover, from linear programming duality (Theorem 6.2), $\boldsymbol{\mu}'\boldsymbol{q}_o \leq \theta$ at a feasible solution and, at the optimum, $(\boldsymbol{\mu}^*)'\boldsymbol{q}_o = \theta^* \leq 1$. Clearly input-oriented radial efficiency requires $(\boldsymbol{\mu}^*)'\boldsymbol{q}_o = \theta^* = 1$.

Table 14.3 Complementarity between (PLP_o) and (DLP_o) variables and constraints.

(PLP_o) Constraints	(DLP_o) Variables	(DLP_o) Constraints	(PLP_o) Variables
$\boldsymbol{\nu}'\boldsymbol{x}_o = 1$	θ unrestricted	$\theta \boldsymbol{x}_o - \boldsymbol{X\lambda} \geq \boldsymbol{0}$	$\boldsymbol{\mu}$
$\boldsymbol{\mu}'\boldsymbol{Q} - \boldsymbol{\nu}'\boldsymbol{X} \leq \boldsymbol{0}'$	$\boldsymbol{\lambda} \geq \boldsymbol{0}$	$-\boldsymbol{q}_o + \boldsymbol{Q\lambda} \geq \boldsymbol{0}$	$\boldsymbol{\nu}$

What is the connection between the production possibility set R and the dual problem (DLP_o)? The constraints of (DLP_o) restrict the activity $(\theta \boldsymbol{x}_o, \boldsymbol{q}_o)$ to R while the objective in (DLP_o) seeks the θ that produces the smallest (radial) contraction of the input vector $\boldsymbol{x}_o$ while remaining in R. Hence, we are looking for an activity in R that produces at least $\boldsymbol{q}_o$ of DMU_o while shrinking $\boldsymbol{x}_o$ proportionately to its smallest feasible level.

14.8 Input and Output Slack Variables

When $\theta^* < 1$, the proportionately reduced input vector $\theta^* \boldsymbol{x}_o$ indicates the fraction of $\boldsymbol{x}_o$ that is needed to produce $\boldsymbol{q}_o$ efficiently, with $(1 - \theta^*)$ serving as the maximal proportional reduction allowed by the production possibility set if input proportions are to be kept invariant. Hence *input excesses* and *output shortfalls* can obviously exist. In this regard, from (DLP_o), let us define the slack variables $\boldsymbol{s}^- \in R^N$ and $\boldsymbol{s}^+ \in R^M$, respectively, as *wasted inputs* $\boldsymbol{s}^- = \theta \boldsymbol{x}_o - \boldsymbol{X}\boldsymbol{\lambda} \geq \boldsymbol{0}$ and *forgone outputs* $\boldsymbol{s}^+ = -\boldsymbol{q}_o + \boldsymbol{Q}\boldsymbol{\lambda} \geq \boldsymbol{0}$. To determine any input excesses or output shortfalls, let us employ the following **two-phase approach**:

Phase I. Solve (DLP_o) in order to obtain θ^*. (Remember that, at optimality, θ^* equals the (PLP_o) objective value and that serves as the Farrell efficient or CCR-efficient value.)

Phase II Extension of (DLP_o). Given θ^*, determine the optimal $\boldsymbol{\lambda}$ that maximizes the input slacks $\boldsymbol{s}^- = \theta^* \boldsymbol{x}_o - \boldsymbol{X}\boldsymbol{\lambda}$ and output slacks $\boldsymbol{s}^+ = \boldsymbol{Q}\boldsymbol{\lambda} - \boldsymbol{q}_o$. Given that $\boldsymbol{\lambda}$, $\boldsymbol{s}^-$, and $\boldsymbol{s}^+$ are all variables, this Phase II linear program is

$$
\begin{aligned}
&\max_{\boldsymbol{\lambda}, \boldsymbol{s}^-, \boldsymbol{s}^+} \ w = \boldsymbol{1}'\boldsymbol{s}^- + \boldsymbol{1}'\boldsymbol{s}^+ = \sum_{s=1}^{N} s_R^- + \sum_{s=1}^{M} s_s^+ \quad \text{s.t.} \\
&(EDLP_o) \qquad \boldsymbol{s}^- = \theta^* \boldsymbol{x}_o - \boldsymbol{X}\boldsymbol{\lambda} \\
&\qquad\qquad\qquad \boldsymbol{s}^+ = -\boldsymbol{q}_o + \boldsymbol{Q}\boldsymbol{\lambda} \\
&\qquad\qquad\qquad \boldsymbol{\lambda} \geq \boldsymbol{0}, \boldsymbol{s}^- \geq \boldsymbol{0}, \boldsymbol{s}^+ \geq \boldsymbol{0}.
\end{aligned}
\tag{14.25}
$$

Clearly, this Phase II problem seeks a solution that maximizes the input excesses and output shortfalls while restricting θ to its Phase I value θ^*.

A couple of related points concerning this two-phase scheme merit our attention. First, an optimal solution $(\boldsymbol{\lambda}^*, (\boldsymbol{S}^-)^*, (\boldsymbol{S}^+)^*)$ to (14.25) is called a **maximal-slack solution**. If any such solution has $(\boldsymbol{S}^-)^* = \boldsymbol{O}^N$ and $(\boldsymbol{S}^+)^* = \boldsymbol{O}^M$, then it is termed a **zero-slack solution**. Second, if an optimal solution $(\theta^*, \boldsymbol{\lambda}^+, (\boldsymbol{S}^-)^*, (\boldsymbol{S}^+)^*)$ to the two-phase routine has $\theta^* = 1$ (it is radially efficient) and is zero-slack, then DMU_o is CCR-efficient. Thus, DMU_o is CCR-efficient if no radial contraction in the input vector $\boldsymbol{x}_o$ is possible and there are no positive slacks. Relative to some of our preceding efficiency concepts, we note briefly that:

i) if $\theta^* = 1$ with $(\boldsymbol{S}^-)^* \geq \boldsymbol{0}$ or $(\boldsymbol{S}^+)^* \geq \boldsymbol{0}$, then DMU_o is Farrell efficient; but
ii) if $\theta^* = 1$ and $(\boldsymbol{S}^-)^* = \boldsymbol{O}^N$, $(\boldsymbol{S}^+)^* = \boldsymbol{O}^M$, then DMU_o is Koopmans efficient.

Example 14.1 Suppose we have $N = 2$ inputs, $M = 1$ output, and $I = 5$ DMUs. Then x_{ij} can be interpreted as "firm i uses input j in the amount x_{ij}," $i = 1, \ldots, 5$, and $j = 1,2$, while q_i, $i = 1, \ldots, 5$, is the quantity of output produced by firm i (see Table 14.4).

Suppose DMU_1 is taken to be DMU_o and let

$$\boldsymbol{\nu} = \begin{bmatrix} \nu_1 \\ \nu_2 \end{bmatrix}, \boldsymbol{x}_i = \begin{bmatrix} x_{i1} \\ x_{i2} \end{bmatrix}, i = 1, \ldots, 5.$$

Then looking to (14.23.2) or (PLP_o):

$$\begin{aligned} &\max_{\mu,\boldsymbol{\nu}} \gamma = \mu q_1 \text{ s.t.} \\ &\boldsymbol{\nu}'\boldsymbol{x}_1 = 1 \\ &\mu' q_i - \boldsymbol{\nu}'\boldsymbol{x}_i \leq 0, i = 1, \ldots, 5 \\ &\mu \geq 0, \boldsymbol{\nu} \geq \boldsymbol{0}, \end{aligned}$$

or

$$\begin{aligned} &\max_{\mu,\nu_1,\nu_2} \gamma = \mu q_1 \text{ s.t.} \\ &\quad \nu_1 x_{11} + \nu_2 x_{12} = 1 \\ &\mu q_1 - \nu_1 x_{11} - \nu_2 x_{12} \leq 0 \\ &\mu q_2 - \nu_1 x_{21} - \nu_2 x_{22} \leq 0 \\ &\mu q_3 - \nu_1 x_{31} - \nu_2 x_{32} \leq 0 \\ &\mu q_4 - \nu_1 x_{41} - \nu_2 x_{42} \leq 0 \\ &\mu q_5 - \nu_1 x_{51} - \nu_2 x_{52} \leq 0 \\ &\quad \mu, \nu_1, \nu_2 \geq 0. \end{aligned}$$

Table 14.4 Input and output quantities for DMU_i, $i = 1, \ldots, 5$.

DMU	Output	Input 1	Input 2
1	q_1	x_{11}	x_{12}
2	q_2	x_{21}	x_{22}
3	q_3	x_{31}	x_{32}
4	q_4	x_{41}	x_{42}
5	q_5	x_{51}	x_{52}

The dual to this multiplier problem is the envelopment problem (14.24) or (DLP_o). To construct this program, let

$$\underset{(2\times 5)}{\boldsymbol{X}} = [\boldsymbol{x}_1, \boldsymbol{x}_2, \boldsymbol{x}_3, \boldsymbol{x}_4, \boldsymbol{x}_5], \quad \underset{(1\times 5)}{\boldsymbol{Q}} = [q_1, q_2, q_3, q_4, q_5], \text{and}$$

$$\underset{(5\times 1)}{\boldsymbol{\lambda}} = \begin{bmatrix} \lambda_1 \\ \lambda_2 \\ \lambda_3 \\ \lambda_4 \\ \lambda_5 \end{bmatrix}.$$

Then

$$\begin{aligned} &\min_{\theta,\lambda} \ \theta \text{ s.t.} \\ &-q_o + \boldsymbol{Q\lambda} \geq 0 \\ &\theta \boldsymbol{x}_1 - \boldsymbol{X\lambda} \geq \boldsymbol{0} \\ &\theta \text{ unrestricted}, \boldsymbol{\lambda} \geq \boldsymbol{0} \end{aligned}$$

or

$$\begin{aligned} &\min_{\theta,\lambda} \ \theta \text{ s.t.} \\ &-q_1 + \textstyle\sum_{i=1}^{5} q_i \lambda_i \geq 0 \\ &\theta x_{11} - \textstyle\sum_{i=1}^{5} x_{i1} \lambda_i \geq 0 \\ &\theta x_{12} - \textstyle\sum_{i=1}^{5} x_{i2} \lambda_i \geq 0 \\ &\theta \text{ unrestricted}, \lambda_1, \ldots, \lambda_5 \geq 0. \end{aligned}$$

■

Example 14.2 Consider the case where $N = 2$, $M = 2$ and $I = 3$. As before, x_{ij} is interpreted as "firm i uses input j in the amount x_{ij}," $i = 1,2,3$, and $j = 1, 2$. Additionally, q_{ik}, means that "firm i produces output k in the amount q_{ik}," $i = 1, 2, 3$, and $k = 1, 2$ (see Table 14.5).

Table 14.5 Input and output quantities for DMU_i, $i = 1$ 2, 3.

DMU	Output 1	Output 2	Input 1	Input 2
1	q_{11}	q_{12}	x_{11}	x_{12}
2	q_{21}	q_{22}	x_{21}	x_{22}
3	q_{31}	q_{32}	x_{31}	x_{32}

Let DMU_1 serve as DMU_o and let

$$\boldsymbol{\nu} = \begin{bmatrix} \nu_1 \\ \nu_2 \end{bmatrix}, \boldsymbol{x}_i = \begin{bmatrix} x_{i1} \\ x_{i2} \end{bmatrix}, i = 1,2,3, \boldsymbol{q}_i = \begin{bmatrix} q_{i1} \\ q_{i2} \end{bmatrix}, i = 1,2,3,$$

$$\text{with}\,\boldsymbol{\mu} = \begin{bmatrix} \mu_1 \\ \mu_2 \end{bmatrix}.$$

Then the multiplier problem (14.23.2) or (PLP_o) appears as

$$\begin{aligned} &\max_{\mu,\nu} \boldsymbol{\mu}'\boldsymbol{q}_1 \text{ s.t.} \\ &\boldsymbol{\nu}'\boldsymbol{x}_1 = 1 \\ &\boldsymbol{\mu}'\boldsymbol{q}_i - \boldsymbol{\nu}'\boldsymbol{x}_i \le 0, i = 1,2,3, \\ &\boldsymbol{\mu} \ge \boldsymbol{0}, \boldsymbol{\nu} \ge \boldsymbol{0}, \end{aligned}$$

or

$$\begin{aligned} &\max_{\mu,\nu} \mu_1 q_{11} + \mu_2 q_{12} \quad \text{s.t.} \\ &\nu_1 x_{11} + \nu_2 x_{12} = 1 \\ &\mu_1 q_{11} + \mu_2 q_{22} - \nu_1 x_{11} - \nu_2 x_{12} \le 0 \\ &\mu_1 q_{21} + \mu_2 q_{22} - \nu_1 x_{21} - \nu_2 x_{22} \le 0 \\ &\mu_1 q_{31} + \mu_2 q_{32} - \nu_1 x_{31} - \nu_2 x_{32} \le 0 \\ &\mu_1, \mu_2, \nu_1, \nu_2 \ge 0. \end{aligned}$$

The dual of the program (14.23.2) is the envelopment form (14.24) or (DLP_o) and can be structured as follows. Let

$$\underset{(2\times 3)}{\boldsymbol{X}} = [\boldsymbol{x}_1, \boldsymbol{x}_2, \boldsymbol{x}_3], \ \underset{(2\times 3)}{\boldsymbol{Q}} = [q_1, q_2, q_3], \text{and } \underset{(3\times 1)}{\boldsymbol{\lambda}} = \begin{bmatrix} \lambda_1 \\ \lambda_2 \\ \lambda_3 \end{bmatrix}.$$

Then we seek to

$$\begin{aligned} &\min_{\theta,\lambda} \ \theta \text{ s.t.} \\ &-\boldsymbol{q}_1 + \boldsymbol{Q}\boldsymbol{\lambda} \ge \boldsymbol{0} \\ &\theta \boldsymbol{x}_1 - \boldsymbol{X}\boldsymbol{\lambda} \ge \boldsymbol{0} \\ &\theta \,\text{unrestricted}, \boldsymbol{\lambda} \ge \boldsymbol{0} \end{aligned}$$

or

$$\min_{\theta,\lambda} \theta \text{ s.t.}$$

$$-q_{11} + q_{11}\lambda_1 + q_{21}\lambda_2 + q_{31}\lambda_3 \geq 0$$

$$-q_{12} + q_{12}\lambda_1 + q_{22}\lambda_2 + q_{32}\lambda_3 \geq 0$$

$$\theta x_{11} - x_{11}\lambda_1 - x_{21}\lambda_2 - x_{31}\lambda_3 \geq 0$$

$$\theta x_{12} - x_{12}\lambda_1 - x_{22}\lambda_2 - x_{31}\lambda_3 \geq 0$$

$$\theta \text{ unrestricted}; \lambda_1, \lambda_2, \lambda_3 \geq 0.$$

■

Example 14.3 Suppose $N = 2$, $M = 1$, and $I = 5$ (Table 14.6).
Looking to the primal multiplier problem (14.23.2) for DMU_1:

$$\max_{\mu_1,\nu_1,\nu_2} \gamma = \mu \quad \text{s.t.}$$

$$5\nu_1 + 4\nu_2 = 1$$

$$\mu - 5\nu_1 - 4\nu_2 \leq 0$$

$$\mu - 8\nu_1 - 3\nu_2 \leq 0$$

$$\mu - 9\nu_1 - 2\nu_2 \leq 0$$

$$\mu - 6\nu_1 - 6\nu_2 \leq 0$$

$$\mu - 3\nu_1 - 6\nu_2 \leq 0$$

$$\mu, \nu_1, \nu_2 \geq 0.$$

Table 14.6 Input and output quantities and optimal values for DMU_i, $i = 1, \ldots, 5$.

DMU	Output	Input x_1	Input x_2	γ^*	ν_1^*	ν_2^*
1	1	5	4	1.000	0.111	0.111
2	1	8	3	0.928	0.071	0.142
3	1	9	2	1.000	0.000	0.500
4	1	6	6	0.750	0.083	0.083
5	1	3	6	1.000	0.333	0.000

Similarly, for DMU_2:

$$\max_{\mu_1,\nu_1,\nu_2} \gamma = \mu \quad \text{s.t.}$$
$$8\nu_1 + 3\nu_2 = 1$$
$$\mu - 5\nu_1 - 4\nu_2 \leq 0$$
$$\mu - 8\nu_1 - 3\nu_2 \leq 0$$
$$\mu - 9\nu_1 - 2\nu_2 \leq 0$$
$$\mu - 6\nu_1 - 6\nu_2 \leq 0$$
$$\mu - 3\nu_1 - 6\nu_2 \leq 0$$
$$\mu, \nu_1, \nu_2 \geq 0.$$

The linear multiplier problems for the remaining three DMUs are structured in a similar fashion.

A glance at the γ^* column in Table 14.6 reveals that DMU_1, DMU_3, and DMU_5 all have efficiency scores of unity while DMU_2 and DMU_4 are inefficient (these two DMUs have an efficiency score less than unity). Since DMU_4 has an efficiency score of 0.750, it follows that this DMU can reduce its input usage by 25%. In addition, DMU_2 has an efficiency score of 0.928 so that it can reduce its input usage by about 7.2%. ■

Example 14.4 Given the input and output data presented in Example 14.3, let us solve program (14.25). We start by determining θ^* via problem (14.2). To this end, for DMU_1:

$$\min_{\theta,\lambda} \theta \quad \text{s.t.}$$
$$\lambda_1 + \lambda_2 + \lambda_3 + \lambda_4 + \lambda_5 \geq 1$$
$$5\theta - 5\lambda_1 - 8\lambda_2 - 9\lambda_3 - 6\lambda_4 - 3\lambda_5 \geq 0$$
$$4\theta - 4\lambda_1 - 3\lambda_2 - 2\lambda_3 - 6\lambda_4 - 6\lambda_5 \geq 0$$
$$\boldsymbol{\lambda} \geq \boldsymbol{0}.$$

And for DMU_2:

$$\min_{\theta,\lambda} \theta \quad \text{s.t.}$$
$$\lambda_1 + \lambda_2 + \lambda_3 + \lambda_4 + \lambda_5 \geq 1$$
$$8\theta - 5\lambda_1 - 8\lambda_2 - 9\lambda_3 - 6\lambda_4 - 3\lambda_5 \geq 0$$
$$3\theta - 4\lambda_1 - 3\lambda_2 - 2\lambda_3 - 6\lambda_4 - 6\lambda_5 \geq 0$$
$$\boldsymbol{\lambda} \geq \boldsymbol{0}.$$

The programs for the remaining DMUs are constructed in a similar fashion. The results for this Phase I problem appear in Table 14.7 and in Figure 14.7.

Table 14.7 Optimal solutions for the Phase I problem.

DMU	θ^*	λ_1^*	λ_2^*	λ_3^*	λ_4^*	λ_5^*
1	1.000	1.000	0.000	0.000	0.000	0.000
2	0.928	0.392	0.000	0.607	0.000	0.000
3	1.000	0.000	0.000	1.000	0.000	0.000
4	0.750	0.750	0.000	0.000	0.000	0.250
5	1.000	0.000	0.000	0.000	0.000	1.000

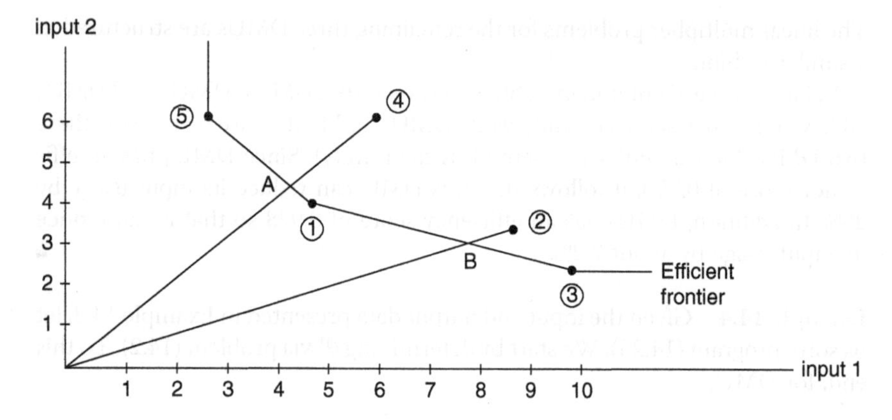

Figure 14.7 The efficient frontier.

Clearly DMU_1, DMU_3, and DMU_5 all lie on the efficient frontier since their efficiency scores (the estimated θs) are each unity while DMU_2 and DMU_4 are inefficient (these two DMUs have efficiency scores less than unity). Since DMU_4 has an efficiency score of 0.750, it follows that it can reduce its input usage by 25%. And since DMU_2 has an efficiency score of 0.928, it can curtail its input usage by 7.2%. Thus: (i) DMUs 1, 3, and 5 are Farrell efficient (and also Koopmans efficient) in that no radial reduction in input usage is possible if their output levels are to be maintained; and (ii) DMUs 2 and 4 are inefficient and thus can proportionately reduce their input amounts while holding output constant.

As Figure 14.7 reveals, the reference set for DMU_4 is $E_4 = \{1, 5\}$ while that of DMU_2 is $E_2 = \{1, 3\}$. That is, DMU_4 can be compared with a **synthetic DMU** on the line segment from DMU_5 to DMU_1 by forming a convex combination of these latter two points (Point A). Hence the synthetic DMU consists of 75% of DMU_5 and 25% of DMU_1 (see Table 14.7) or

$$\lambda_1^* \, 5 + \lambda_5^* \, 1 \; = A,$$

$$0.250\begin{bmatrix}3\\ \\6\end{bmatrix} + 0.750\begin{bmatrix}5\\ \\4\end{bmatrix} = \begin{bmatrix}4.50\\ \\4.50\end{bmatrix}.$$

Similarly, DMU_2 can be compared with a synthetic DMU consisting of a convex combination of DMU_1 and DMU_3 (Point B), i.e. this synthetic DMU consists of 60.7% of DMU_1 and 39.2% of DMU_3 or

$$\lambda_3^* \, 1 + \lambda_1^* \, 3 \; = B,$$

$$0.392\begin{bmatrix}5\\ \\4\end{bmatrix} + 0.607\begin{bmatrix}9\\ \\2\end{bmatrix} = \begin{bmatrix}7.423\\ \\2.782\end{bmatrix}.$$

In sum, DMU_4 can be made efficient by reducing its input by 25% or by forming a convex combination of DMU_5 and DMU_1 (we end up at Point A). Similarly, DMU_2 can be brought into efficiency by curtailing its input usage by 7.2% or by constructing a convex combination of DMU_1 and DMU_3 (Point B).

We next turn to an examination of the input excesses and output shortfalls that may exist at an optimal solution to the envelopment problem (14.24). In particular, we seek to determine the values of the input and output slacks $\boldsymbol{s}^-$ and $\boldsymbol{s}^+$, respectively. To this end, let us look to the two-phase method outlined earlier. From DMU_1, given $\theta^* = 1$ obtained from Phase I, the Phase II extension is

$$\begin{aligned}
&\max_{s_1^+, s_1^-, s_2^-, \lambda} \quad s_1^+ + s_1^- + s_2^- \quad \text{s.t.}\\
&5\lambda_1 + 8\lambda_2 + 9\lambda_3 + 6\lambda_4 + 3\lambda_5 + s_1^- = 5\\
&4\lambda_1 + 3\lambda_2 + 2\lambda_3 + 6\lambda_4 + 6\lambda_5 + s_2^- = 4\\
&\lambda_1 + \lambda_2 + \lambda_3 + \lambda_4 + \lambda_5 - s_1^+ = 1\\
&\qquad s_1^+, s_1^-, s_2^- \geq 0; \; \boldsymbol{\lambda} \geq \boldsymbol{0}.
\end{aligned}$$

And for DMU_2, given $\theta^* = 0.928$, we have $\theta^* \boldsymbol{x}_0' = (7.424, 2.784)$ so that we seek to

$$\begin{aligned}
&\max_{s_1^+, s_1^-, s_2^-, \lambda} \quad s_1^+ + s_1^- + s_2^- \quad \text{s.t.}\\
&5\lambda_1 + 8\lambda_2 + 9\lambda_3 + 6\lambda_4 + 3\lambda_5 + s_1^- = 7.424\\
&4\lambda_1 + 3\lambda_2 + 2\lambda_3 + 6\lambda_4 + 6\lambda_5 + s_2^- = 2.784\\
&\lambda_1 + \lambda_2 + \lambda_3 + \lambda_4 + \lambda_5 - s_1^+ = 1\\
&s_1^+, s_1^-, s_2^- \geq 0; \; \boldsymbol{\lambda} \geq \boldsymbol{0}.
\end{aligned}$$

Table 14.8 Optimal solutions for the Phase II Extension.

DMU	s_1^+	s_1^-	s_2^-	λ_1	λ_2	λ_3	λ_4	λ_5
1	0.000	0.000	0.000	1.000	0.000	0.000	0.000	0.000
2[a]	0.000	−0.008	0.000	0.392	0.000	0.608	0.000	0.000
3	0.000	0.000	0.000	0.000	0.000	1.000	0.000	0.000
4	0.000	0.000	0.000	0.750	0.000	0.000	0.000	0.250
5	0.000	0.000	0.000	0.000	0.000	0.000	0.000	1.000

[a] No feasible solution.

The programs for the remaining DMUs are determined in like fashion. The results for Phase II appear in Table 14.8.

As these results reveal, DMU_1, DMU_3, and DMU_5 are Farrell as well as Koopmans efficient. ■

14.9 Modeling VRS (BCC 1984; Cooper et al. 2007; Banker and Thrall 1992 and Banker et al. 1996)

14.9.1 The Basic BCC (1984) DEA Model

The CCR (1978) DEA model considered earlier was specified under the assumption of CRS. We may modify the CCR (1978) model to accommodate the presence of VRS by looking to the DEA formulation offered by BCC (1984).

To introduce VRS, BCC (1984) add $\boldsymbol{1}'\boldsymbol{\lambda} = 1$ to the constraint system in the envelopment form of the input-oriented DEA problem. This additional constraint (called the **convexification constraint**) admits only convex combinations of DMUs when the efficient production frontier is formed. This convexification constraint essentially shrinks the production possibility set T to a new production possibility set $\bar{T}$ (so that $\bar{T} \subseteq T$) and thus converts a CRS technology to a VRS technology, where the letter production possibility set has the form

$$\bar{T}(\boldsymbol{x},\boldsymbol{q}) = \{(\boldsymbol{x},\boldsymbol{q}) | \boldsymbol{x} \geq \boldsymbol{X\lambda}, \boldsymbol{q} \leq \boldsymbol{Q\lambda}, \boldsymbol{1}'\boldsymbol{\lambda} = 1, \boldsymbol{\lambda} \geq \boldsymbol{O}\}.$$

This revised technology set satisfies strong disposability and convexity but not homogeneity; it amounts to a polyhedral convex free disposal hull. In fact, it is the *smallest* set satisfying the convexity and strong disposability property subject to the condition that each activity $(\boldsymbol{x}_j, \boldsymbol{q}_j)$, $j = 1, \cdots, I$, belongs to $\bar{T}$. As we shall see below, the convexification constraint leads to the definition of a new variable that determines whether production is carried out in regions of increasing, decreasing, or CRS. Hence, we shall examine returns to scale locally at a point on the efficiency frontier, and ultimately relate it to the sign of this new variable.

The **input-oriented BCC (1984) model** is designed to evaluate the efficiency of DMU_0 by generating a solution to the *envelopment form* of the (primal) linear program

$$\begin{aligned} &\min_{\theta_B, \boldsymbol{\lambda}} \theta_B \text{ s.t.} \\ (PBLP_o) \quad &\theta_B \boldsymbol{x}_o - \boldsymbol{X\lambda} \geq \boldsymbol{O} \\ &-\boldsymbol{q}_o + \boldsymbol{Q\lambda} \geq \boldsymbol{O} \\ &\boldsymbol{1}'\boldsymbol{\lambda} = 1 \\ &\boldsymbol{\lambda} \geq \boldsymbol{O}. \end{aligned} \tag{14.26}$$

The dual to $(PBLP_o)$ is its *multiplier form*

$$\begin{aligned} &\max_{\boldsymbol{\mu}, \boldsymbol{\nu}, u_o} z = \boldsymbol{\mu}'\boldsymbol{q}_o - u_o \text{ s.t.} \\ (DBLP_o) \quad &\boldsymbol{\nu}'\boldsymbol{x}_o = \boldsymbol{1} \\ &-\boldsymbol{\nu}'\boldsymbol{x}_j + \boldsymbol{\mu}'\boldsymbol{q}_j - u_o \leq 0, j = 1, \ldots, I, \\ &\boldsymbol{\nu} \geq \boldsymbol{O}, \boldsymbol{\mu} \geq \boldsymbol{O}, u_o \text{ unrestricted,} \end{aligned} \tag{14.27}$$

where u_o is the dual variable associated with the convexification constraint $\boldsymbol{1}'\boldsymbol{\lambda} = 1$ in (14.26).[10]

It is instructive to examine the geometric nature of an optimal solution to the dual problem $(DBLP_o)$. If the I constraint inequalities in (14.27) are replaced by $-\boldsymbol{\nu}'\boldsymbol{X} + \boldsymbol{\mu}'\boldsymbol{Q} - u_o\boldsymbol{1} \leq \boldsymbol{O}$, then we can interpret this inequality as a supporting hyperplane to $\bar{T}$ at the optimum. In this regard, the vector of coefficients $(\boldsymbol{\nu}^*, \boldsymbol{\mu}^*, u_o^*)$ of the supporting hyperplane to $\bar{T}$ is an optimal solution to the dual problem (14.27).

10 The fractional program equivalent to (14.27) is

$$\begin{aligned} &\max_{\boldsymbol{U}, \boldsymbol{W}, u_o} \frac{\boldsymbol{U}'\boldsymbol{q}_o - u_o}{\boldsymbol{W}'\boldsymbol{x}_o} \text{ s.t.} \\ &\frac{\boldsymbol{U}'\boldsymbol{q}_j - u_o}{\boldsymbol{W}'\boldsymbol{x}_j} \leq 1, j = 1, \ldots, I, \\ &\boldsymbol{U} > \boldsymbol{O}, \boldsymbol{W} \geq \boldsymbol{O}, u_o \text{ unrestricted.} \end{aligned} \tag{14.28}$$

14.9.2 Solving the BCC (1984) Model

Let us focus on the envelopment form ($PBLP_o$). A solution to this problem can be obtained by employing a two-phase routine similar to that used to solve the CCR dual program ($DBLP_o$), i.e.

Phase I. Solve ($PBLP_o$) in order to obtain $\theta_B^*(\geq \theta^*)$.

Phase II Extension of (***PBLP***$_o$). Given θ_B^*, determine the optimal $\boldsymbol{\lambda}$ which maximizes the input slacks $\boldsymbol{s}^-$ and the output slacks $\boldsymbol{s}^+$. Thus, the Phase II linear program is

$$
\begin{aligned}
&\max_{\boldsymbol{\lambda},\boldsymbol{s}^-,\boldsymbol{s}^+} \quad w = \boldsymbol{1}'\boldsymbol{s}^- + \boldsymbol{1}'\boldsymbol{s}^+ \quad \text{s.t.}\\
&\qquad \boldsymbol{1}'\boldsymbol{\lambda} = 1\\
&\qquad \boldsymbol{s}^- = \theta_B^*\boldsymbol{x}_o - \boldsymbol{X\lambda}\\
&(EPBLP_o)\ \ \boldsymbol{s}^+ = -\boldsymbol{q}_o + \boldsymbol{Q\lambda}\\
&\qquad \boldsymbol{\lambda} \geq \boldsymbol{O}, \boldsymbol{s}^- \geq \boldsymbol{O}, \boldsymbol{s}^+ \geq \boldsymbol{O}.
\end{aligned}
\tag{14.29}
$$

Let an optimal solution to (14.29) be denoted as $\left(\theta_B^*, \boldsymbol{\lambda}^*, (\boldsymbol{s}^-)^*(\boldsymbol{s}^+)^*\right)$. An important characterization of this solution is provided by the notion of **BCC-Efficiency**: (i) If program (EPBLP_o) generates an optimal solution for (PBLP_o) with $\theta_B^* = 1$ and $(\boldsymbol{s}^-)^* = \boldsymbol{O}^N$, $(\boldsymbol{s}^+)^* = \boldsymbol{O}^M$, then DMU_o is termed BCC-efficient; (ii) otherwise, it is called BCC-inefficient.

Moreover, if DMU_o is BCC-efficient, its **reference set** E_o can be expressed as $E_o = \{j \in \mathcal{J} \,|\, \lambda_j^* > 0,\ \mathcal{J} = \{1, \ldots, I\}\}$.

Given the possibility of multiple optimal solutions to (14.29), any one of them may be selected to determine

$$
\begin{aligned}
\theta_B^*\boldsymbol{x}_o &= \sum_{j \in E_o} \lambda_j^* \boldsymbol{x}_j + (\boldsymbol{s}^-)^*,\\
\boldsymbol{q}_o &= \sum_{j \in E_o} \lambda_j^* \boldsymbol{q}_j - (\boldsymbol{s}^+)^*
\end{aligned}
\tag{14.30}
$$

so that an improved activity can be calculated (via the **BCC-projection**) as

$$
\begin{aligned}
\hat{\boldsymbol{x}}_o &= \theta_B^*\boldsymbol{x}_o - (\boldsymbol{s}^-)^*\\
\hat{\boldsymbol{q}}_o &= \boldsymbol{q}_o - (\boldsymbol{s}^+)^*.
\end{aligned}
\tag{14.31}
$$

In addition, it can be demonstrated that this improved activity $(\hat{\boldsymbol{x}}_o, \hat{\boldsymbol{q}}_o)$ is BCC-efficient (on all this see Cooper et al. 2007).

14.9.3 BCC (1984) Returns to Scale

We noted in the preceding section that u_o is a dual variable associated with the convexification constraint $\boldsymbol{1}'\boldsymbol{\lambda} = 1$ in (14.26). It was also mentioned that the constraint $-\boldsymbol{\nu}'\boldsymbol{X} + \boldsymbol{\mu}'\boldsymbol{Q} - u_o\boldsymbol{1} \leq \boldsymbol{O}$ in the dual problem (DBLP$_o$) is a supporting hyperplane to $\bar{T}$ at an optimal solution $(\boldsymbol{\nu}^*, \boldsymbol{\mu}^*, u_o^*)$. Thus, $(\boldsymbol{\nu}^*, \boldsymbol{\mu}^*, u_o^*)$ is an optimal solution to the dual problem (DBLP$_o$) associated with the primal problem (PBLP$_o$). So if DMU$_o$ is BCC-efficient, the vector of coefficients $(\boldsymbol{\nu}^*, \boldsymbol{\mu}^*, u_o^*)$ of the supporting hyperplane to $\bar{T}$ at $(\boldsymbol{x}_o, \boldsymbol{q}_o)$ renders an optimal solution to (PBLP$_o$), and vice versa. So, when a DMU is BCC-efficient, we can rely on the *sign of u_o^* to describe returns to scale* (specifically, **returns to scale** as described in Banker and Thrall (1992)). Given that $(\boldsymbol{x}_o, \boldsymbol{q}_o)$ is on the efficient frontier:

i) Increasing returns to scale exist at $(\boldsymbol{x}_o, \boldsymbol{q}_o)$ if and only if $u_o^* < 0$ for all optimal solutions.
ii) Decreasing returns to scale exist at $(\boldsymbol{x}_o, \boldsymbol{q}_o)$ if and only if $u_o^* > 0$ for all optimal solutions.
iii) Constant returns to scale exist at $(\boldsymbol{x}_o, \boldsymbol{q}_o)$ if and only if $u_o^* = 0$ in any optimal solution.

It is instructive to examine a simplified application of (i)–(iii). Let us consider the single-input, single-output case (Figure 14.8). In this instance, the

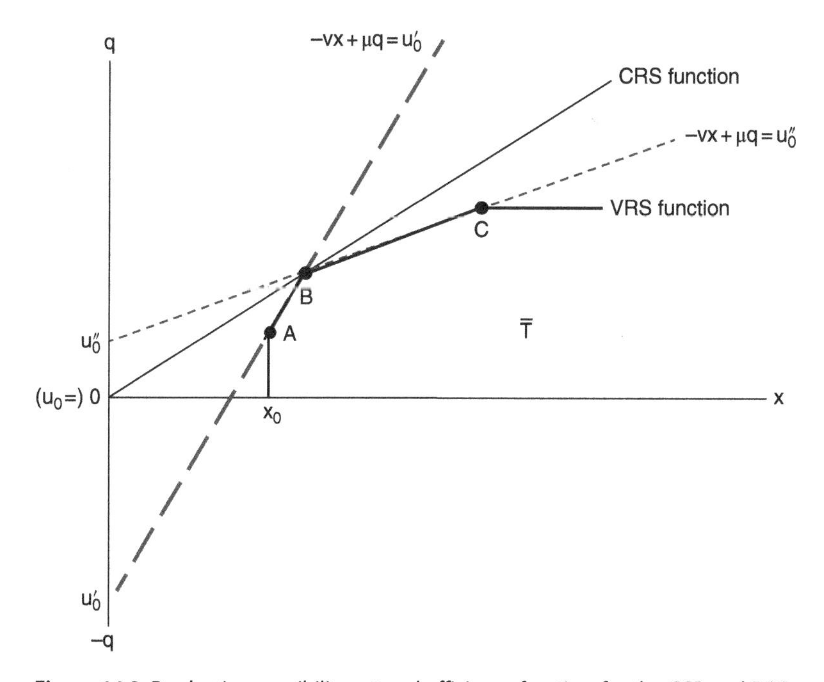

Figure 14.8 Production possibility set and efficiency frontiers for the CCR and BCC models.

supporting hyperplane to $\bar{T}$ has the form $-vx + uq = u_o$. Additionally, we know that the average product $AP = q/x$ is calculated as the slope of a ray from the origin to the VRS frontier. Clearly, AP is increasing as x increases from 0 to x_o (e.g. see the plane of support $-vx + uq = u_o', u_o' < 0$, which is tangent to $\bar{T}$ along edge AB). AP reaches a maximum at Point B (when $x = x_o$ with $u_o = 0$) and then begins to decrease as x is increased beyond x_o (e.g. $-vx + uq = u_o'', u_o'' > 0$, is tangent to $\bar{T}$ along edge BC). Hence we have a range of increasing returns from 0 to x_o (with $u_o' < 0$), constant returns at x_o (where $u_o = 0$), and a range of decreasing returns beyond x_o (with $u_o'' > 0$).

Suppose we have obtained an optimal solution to the ($PBLP_o$) envelopment problem (14.26). Then we may project $(\boldsymbol{x}_o, \boldsymbol{q}_o)$ into a point $(\hat{\boldsymbol{x}}_o, \hat{\boldsymbol{q}}_o)$ on the efficient frontier using the **BCC-projection**

$$\begin{aligned} \hat{x}_{ro} &= \theta_B^* x_{ro} - \left(s_r^-\right)^*, r = 1,\dots,N; \\ \hat{q}_{so} &= q_{so} + \left(s_r^+\right)^*, s = 1,\dots,M. \end{aligned} \tag{14.32}$$

Additional details on the use of (14.32) in forming a modified dual (multiplier) problem can be found in Banker et al. (1996) and in Cooper et al. (2007).

14.10 Output-Oriented DEA Models

We noted at the outset of this chapter that, in general, an input-oriented DEA model focuses on the minimization of input usage while producing at least the observed output level. That is, it identifies TE as the proportionate reduction in input usage, with output levels kept constant. In contrast to this modeling approach, an output-oriented DEA model maximizes output while using no more than the observed inputs. Now TE is measured as the proportionate increase in output, with input levels held fixed. These two efficiency measures are equivalent or render the same objective function value under CRS but display unequal objective values when VRS is assumed.

Let us first consider the **output-oriented DEA model under CRS**. It is structured in **envelopment form** as

$$\begin{aligned} &\max_{\eta,\mu} \ \eta \ \text{ s.t.} \\ (PLP_o) \quad &\boldsymbol{x}_o - \boldsymbol{X}\boldsymbol{\mu} \geq \boldsymbol{O} \\ &\eta \boldsymbol{q}_o + \boldsymbol{Q}\boldsymbol{\mu} \leq \boldsymbol{O} \\ &\boldsymbol{\mu} \geq \boldsymbol{O}, \eta \text{ unrestricted.} \end{aligned} \tag{14.33}$$

As one might have anticipated, there exists a direct correspondence between the input-oriented model (14.24) and the output-oriented model (14.33). That is, the solution to (14.33) can be derived directly from the solution to (14.24). To see this, let

$$\boldsymbol{\lambda} = \boldsymbol{\mu}/\eta \text{ and } \theta = {}^{1}/_{\eta}. \tag{14.34}$$

Then substituting (14.34) into (14.24) results in the output-oriented model (14.33). So at optimal solutions to problems (14.33) and (14.24),

$$\boldsymbol{\lambda}^* = \boldsymbol{\mu}^*/\eta^* \text{ and } \theta^* = {}^{1}/_{\eta^*}. \tag{14.34.1}$$

Since $\theta^* \leq 1$, it follows that $\eta^* \geq 1$, i.e. as η^* increases in value, the efficiency of DMU_o concomitantly decreases. And while θ^* represents the input reduction rate, η^* depicts the output expansion rate, where $\eta^* - 1$ represents the proportionate increase in outputs that is achievable by DMU_o given that input levels are held constant. The TE for DMU_o is given by $0 < {}^{1}/_{\eta^*} \leq 1$. So if DMU_o is efficient in terms of its input-oriented model, then it must also be efficient in terms of its associated output-oriented model. Thus, the solution of the output-oriented model (14.33) can be obtained from the solution to the input-oriented model (14.24). In fact, the improvement in $(\boldsymbol{x}_o, \boldsymbol{q}_o)$ can be obtained from

$$\begin{aligned} \hat{\boldsymbol{x}}_o &= \boldsymbol{x}_o - (\boldsymbol{t}^-)^*, \\ \hat{\boldsymbol{q}}_o &= \eta^* \boldsymbol{q}_o + (\boldsymbol{t}^+)^*, \end{aligned} \tag{14.35}$$

where $\boldsymbol{t}^- = \boldsymbol{x}_o - \boldsymbol{X\mu}$ and $\boldsymbol{t}^+ = \boldsymbol{Q\mu} - \eta \boldsymbol{q}_o$ are the vectors of slack variables for (14.33). Additionally, at optimal solutions to problems (14.33) and (14.24),

$$(t^-)^* = (s^-)^*/\theta^*, (t^+)^* = (s^+)^*/\theta^*. \tag{14.36}$$

Moreover, both the input- and output-oriented DEA models will determine exactly the same efficient frontier, and thus label the same set of DMUs as efficient.

The dual of (14.33) is the **output-oriented multiplier problem**

$$\begin{aligned} &\min_{p,y} \ \boldsymbol{p}'\boldsymbol{x}_o \ \text{ s.t.} \\ (DPLP_o) \quad &\boldsymbol{y}'\boldsymbol{q}_o = 1 \\ &-\boldsymbol{pX} + \boldsymbol{yQ} \leq \boldsymbol{O} \\ &\boldsymbol{p} \geq \boldsymbol{O}, \boldsymbol{y} \geq \boldsymbol{O}. \end{aligned} \tag{14.37}$$

If we impose the assumption of VRS, then the appropriate output-oriented DEA model (in envelopment form) appears as

$$\begin{aligned} &\max_{\eta,\mu} \ \eta \ \text{ s.t.} \\ &\boldsymbol{x}_o - \boldsymbol{X\mu} \geq \boldsymbol{O} \\ &\eta \boldsymbol{q}_o - \boldsymbol{Q\mu} \leq \boldsymbol{O} \\ &\boldsymbol{1}'\boldsymbol{\mu} = \boldsymbol{1} \\ &\boldsymbol{\mu} \geq \boldsymbol{O}, \eta \text{ unrestricted}. \end{aligned} \tag{14.38}$$

Then the dual (multiplier) form of (14.38) is

$$\begin{aligned} &\min_{u,\nu,\nu_o} \; z = \nu' x_o - \nu_o \quad \text{s.t.} \\ &u' q_o = 1 \\ &\nu X - u Q - \nu_o 1 \geq O \qquad u \geq O, \nu \geq O, \nu_o \text{ unrestricted,} \end{aligned} \tag{14.39}$$

where ν_o is the scalar associated with $1'\mu = 1$ in the envelopment form (14.38).

References and Suggested Reading

Allen, R.G.D. (1960). *Mathematical Economics*, 2e. London: Macmillan & Co., Ltd.

Banker, R.D. and Thrall, R.M. (1992). Estimation of returns to scale using data envelopment analysis. *European Journal of Operational Research* **62** (1): 74–84.

Banker, R.D., Charnes, A., and Cooper, W.W. (1984). Some models for estimating technical and scale inefficiencies in data envelopment analysis. *Management Science* **30**: 1078–1092.

Banker, R.D., Bantham, I., and Cooper, W.W. (1996). A note on returns to scale in DEA. *European Journal of Operational Research* **88**: 583–585.

Barankin, E. and Dorfman, R. (1958). *On Quadratic Programming. Berkeley Series in Statistics*, vol. **2**, 285–317. Berkeley, CA: University of California Press.

Baumol, W.J. (1977). *Economic Theory and Operations Analysis*, 4e. New Jersey: Prentice-Hall, Inc.

Beattie, B.R., Taylor, C.R., and Watts, M.J. (2009). *The Economics of Production*, 2e. Florida: Krieger Publ. Co.

Belinski, M. and Baumol, W. (1968). The dual in nonlinear programming and its interpretation. *Review of Economic Studies* **35**: 237–256.

Bitran, G. and Magnanti, T. (1976). Duality and sensitivity analysis for fractional programs. *Operations Research* **24**: 684–686.

Bitran, G. and Novaes, A. (1973). Linear programming with a fractional objective. *Operations Research* **21**: 22–29.

Chadha, S. (1971). A dual fractional program. *ZAMM* **51**: 560–561.

Charnes, A. and Cooper, W. (1962). Programming with linear fractional Functionals. *Naval Research Logistics Quarterly* **9**: 181–186.

Charnes, A., Cooper, E., and Henderson, A. (1953). *Introduction to Linear Programming*. New York: Wiley Blackwell

Charnes, A., Cooper, W.W., and Rhodes, E. (1978). Measuring the efficiency of decision making units. *European Journal of Operational Research* **2**: 429–444.

Coelli, T.J. and Rao, D.S.P. (2005). Total factor productivity growth in agriculture: a Malmquist index analysis of 93 countries, 1980–2000. *Agricultural Economics* **32**: 115–134.

Linear Programming and Resource Allocation Modeling, First Edition. Michael J. Panik.

Coelli, T.J., Rao, D.S.P., O'Donnell, C.J., and Battese, G.E. (2005). *An Introduction to Efficiency and Productivity Analysis*, 2e. New York: Springer.

Cooper, W.W., Seiford, L., and Tone, K. (2007). *Data Envelopment Analysis*. New York: Springer.

Craven, B. (1978). *Mathematical Programming and Control Theory*. New York: Halsted Press.

Craven, B. and Mond, B. (1973). The dual of a linear fractional program. *Journal of Mathematical Analysis and Applications* **42**: 507–512.

Craven, B. and Mond, B. (1976). Erratum, the dual of a linear fractional program. *Journal of Mathematical Analysis and Applications* **55**: 807.

Dano, S. (1966). *Industrial Production Models*. New York: Springer Verlag, Inc.

Dantzig, G., Orden, A., and Wolfe, R. (1955). The general simplex method for minimizing a linear form under inequalitiy restraints. Pacific J. Math. **5**: 183–195.

Dantzig, G. (1951). Maximization of a linear function of variables subject to linear inequalities. In: *Activity Analysis of Production and Allocation* (ed. T. Koopman), 373–359. New York: Wiley Blackwell.

Debreu, G. (1951). The coefficient of resource utilization. *Econometrica* **19**: 273–292.

Dorfman, R. (1951). *Application of Linear Programming to the Theory of the Firm*. Berkeley, CA: Univ. of California Press.

Dorfman, R. (1953). Mathematical, or linear, programming: a nonmathematical exposition. *American Economic Review* **53**: 797–825.

Dorfman, R., Samuelson, P.A., and Solow, R. (1958). *Linear Programming and Economic Analysis*. New York: McGraw-Hill Book Co.

Dorn, W. (1960). Duality in quadratic programming. *Quarterly of Applied Mathematics* **18**: 155–162.

Dreyfus, S. and Freimer, M. (1962). A new approach to the duality theory of linear programming. In: *Applied Dynamic Programming* (ed. R. Bellman and S. Dreyfus). Princeton, NJ: Princeton University Press.

Färe, R. and Lovell, C.A.K. (1978). Measuring the technical efficiency of production. *Journal of Economic Theory* **19**: 150–162.

Färe, R., Grosskopf, C., and Margaritis, D. (2008). Efficiency and productivity: Malmquist and more. In: *The Measurement of Productive Efficiency and Productivity Growth* (ed. H. Fried et al.). New York: Oxford University Press.

Farrell, M.J. (1957). The measurement of productive efficiency. *Journal of the Royal Statistical Society, Series A* **120** (3): 253–290.

Ferguson, C.E. (1971). *The Neoclassical Theory of Production and Distribution*. Cambridge: Cambridge University Press.

Fried, H., Lovell, C.A.K., and Schmidt, S. (2008). *The Measurement of Productive Efficiency and Productivity Growth*. New York: Oxford University Press.

Frisch, R. (1965). *Theory of Production*. Chicago: Rand McNally & Co.

Gale, D. (1960). *The Theory of Linear Economic Models.* Chicago: The University of Chicago Press.

Gass, S. (1975). *Linear Programming: Methods and Applications,* 4e. New York: McGraw-Hill Book, Co.

Goldman, A. and Tucker, A. (1956). Theory of linear programming. In: *Linear Inequalities and Related Systems* (ed. H. Kuhn and A. Tucker), 60–61. Princeton, NJ: Princeton University Press.

Hadar, J. (1971). *Mathematical Theory of Economic Behavior.* Reading, MA: Addison-Wesley Publishing Co.

Hadley, G. (1963). *Linear Programming.* Reading, MA: Addison-Wesley Publishing Co., Inc.

Jehle, G.A. and Reny, P.J. (2011). *Advanced Microeconomic Theory,* 3e. England: Pearson Education Ltd.

Kogiku, K.C. (1971). *Microeconomic Models.* New York: Harper & Row Publishers.

Koopmans, T.C. (1951). Analysis of production as an efficient combination of activities. In: *Activity Analysis of Production and Allocation* (ed. T.C.K.). New York: John Wiley & Sons.

Kornbluth, J. and Salkin, G. (1972). A note on the economic interpretation of the dual variables in linear fractional programming. *Journal of Applied Mathematics and Mechanics (ZAMM)* **52**: 175–178.

Kunzi, H. et al. (1966). *Nonlinear Programming.* Massachusetts: Blaisdell Pub. O.

Kydland, F. (1962). Duality in fractional programming. *Naval Research Logistics Quarterly* **19**: 691–697.

Lasdon, L. (1970). *Optimization Theory for Large Systems.* New York: The Macmillan Co.

Lemke, C.E. (1954). The dual method of solving the linear programming problem. *Naval Research Logistics Quarterly* **1**: 48–55.

Lemke, C.E. (1965). Bimatrix equilbrium points and mathematical programming. *Management Science* **11**: 681–689.

Lemke, C.E. (1968). On complementary pivot theory. In: *Mathematics of the Decision Sciences* (ed. G.B. Dautzig and Venoitt). Providence: American Mathematical Society.

Liao, W. (1979). Effects of learning on resource allocation decisions. *Decision Sciences* **10**: 116–125.

Lovell, C.A.K. (1994). Linear programming approaches to the measurement and analysis of productive efficiency. *TOP* **2**: 175–248.

Martos, B. (1964). Hyperbolic programming. *Naval Research Logistics Quarterly* **11**: 135–155.

Nanda, R. and Adler, G. (eds.) (1982). *Learning Curves: Theory and Application.* Atlanta: Industrial Engineering and Management Pr.

Naylor, T. (1966). The theory of the firm: a comparison of marginal analysis and linear programming. *Search Engine Journal* **32L**: 263–274.

Naylor, T. and Vernon, J. (1969). *Microeconomics and Decision Models of the Firm*. New York: Harcourt, Brace & World, Inc.

Panik, M.J. (1976). *Classical Optimization: Foundations and Extensions*. The Netherlands: North-Holland Publishing Co.

Panik, M.J. (1992). A note on modeling internal technological externalities in linear profit programs. *European Journal of Operational Research* **59**: 147–160.

Panik, M.J. (1993). Constrained sales maximization under a linear technology. *International Review of Economics and Finance* **2**: 97–106.

Panik, M.J. (1994). The fundamental theorem of game theory revisited. *Applied Mathematics Letters* **7**: 77–99.

Panik, M.J. (1996). *Linear Programming: Mathematics, Theory, and Algorithms*. The Netherlands: Kluwer Academic Publ.

Panik, M.J. (1994). Duality and the geometry of linear programming. *International Journal of Mathematical Education in Science and Technology* **25**: 187–192.

Panik, M.J. (1990). Output learning and duality in joint production programs. *ZOR* **34**: 210–217.

Pfouts, R. (1961). The theory of cost and production in the multiproduct firm. *Econometrica* **29**: 650–658.

Reeves, G. and Sweigart, J. (1981). Product-mix models when learning effects are present. *Management Science* **17**: 204–2112.

Reeves, G. (1980). A note of the effects of learning on resource allocation decisions. *Decision Sciences* **11**: 169–170.

Ryan, V. (1962). *Price Theory*. New York: St. Martins Press.

Saaty, T. and Bram, J. (1964). *Nonlinear Mathematics*. New York: McGraw-Hill Book Co.

Schnaible, S. (1976). Fractional programming. I, duality. *Management Science* **22**: 858–867.

Seiford, L.M. and Thrall, L.M. (1990). Recent developments in DEA: the mathematical approach to frontier analysis. *Journal of Econometrics* **46**: 7–38.

Shephard, R.W. (1970). *Theory of Cost and Production Functions*. Princeton, NJ: Princeton University Press.

Shephard, R.W. (1953). *Cost and Production Functions*. Princeton, NJ: Princeton University Press.

Swarup, K. (1965). Linear fractional functionals programming. *Operations Research* **13**: 1029–1036.

Takayama, A. (1997). *Mathematical Economics*, 2e. Cambridge: Cambridge University Press.

Thompson, G.E. (1971). *Linear Programming*. New York: The Macmillan Co.

Vandermullen, D. (1971). *Linear Economic Theory*. Englewood Cliffs, NJ: Prentice-Hall, Inc.

Wagner, H. and Yuan, J. (1968). Algorithmic equivalence in linear fractional programming. *Management Science* **14**: 301–306.

Williams, A. (1963). Marginal values in linear programming. *Journal of the Society for Industrial and Applied Mathematics* **2**: 82–94.

Wolf, P. (1959). The simplex algorithm for quadratic programming. *Econometrica* **27**: 382–398.

Index

Page references to Figures are followed by the letter 'f', references to Tables by the letter 't', while references to Footnotes are followed by the letter 'n'

a

accounting loss 141, 183
  activity levels 150, 189
  optimal accounting loss figures 5, 10
activity levels 8, 124, 284
  interpreted as individual product levels 148–155
  interpreted as individual resource levels 186–193
  optimal/optimal output 4, 9
activity vector
  changing component of 210–215
  multiactivity production function 129
  parameterizing 245–256
activity/activities
  see also activity levels; activity vector
  additive 129
  composite 129, 132–133, 158
  independent 130
  input 1–2
  marginal cost function for 276–284, 282t
  output 6, 7, 183–185, 187, 190, 191
    composite output activity 171
    multiactivity joint-production model 171–174
    output activity mix 189
    output activity vector 165
  production 1, 4, 6, 123
  quadratic programming and activity analysis 335–338
  shadow 181
  simple 123, 130
  supply function for the output of 257–262
additive activity 129
additive identity 20
additive inverse 20
admissible solutions 298
Alder, G. 162
algebra
  matrix 13–20
  vector 20–22
Allen, R.G.D. 123
allocative efficiency (AE) 380, 383–384
almost complementary basic solution 330–332

artificial augmented structural constraint system 72
artificial linear programming problem
  inconsistency and redundancy 78, 81, 83
  M-Penalty method 73, 75–77
artificial objective function 74
artificial variables 73, 93, 111
  complementary pivot method 330, 331
  inconsistency and redundancy 79, 84
  linear fractional functional programming 352
  two-phase method 88
artificial vectors
  inconsistency and redundancy 78, 79, 81, 82
  M-Penalty method 73, 75
  two-phase method 91, 93
associative law
  matrix algebra 14, 15
  vector algebra 20
augmented linear programming problem 36
augmented matrix 23, 24
augmented structural constraint system 36, 39, 53
  artificial 72
  primal 155–156
average cost function
  average (real) resource cost function 168, 169f, 179f
  determining 286–295
average product 136f, 290, 291–292
  average product function 139f
average productivity
  average productivity function 127, 138
  determining 286–295
average profit 2, 354
average variable cost 142, 292

b

back-substitution 25
Banker, R.D. 373–374, 385–386, 398–402
basic feasible solutions *see* feasible solutions
basic solution 28
basic variables 39, 156
  see also nonbasic variables
  complementary pivot method 333
  computational aspects 43–48, 68, 69
    improving basic feasible solutions 50, 51, 53–55, 58–60, 62–66
  duality theory 106, 121
    dual simplex method 113, 115, 117, 118, 120
  inconsistency and redundancy 80, 81, 83, 84
  M-Penalty method 73, 76, 77
  parametric programming 229, 237, 239, 255, 272, 274, 278, 281, 287
  sensitivity analysis 200, 202, 203, 205, 213
  simplex-based optimization methods 331–334, 340–344, 355
  structural changes 223, 226
  two-phase method 87, 90
basis for $\mathcal{E}^m$ 27
basis inverse, updating 256
basis matrix B 39, 49, 201, 203
Baumol, W.J. 123, 146–148, 311–315, 338
BCC (Banker, Charnes and Cooper) model
  basic 398–399

input-oriented 399
projection 400, 402
returns to scale 401–402
solving 400
Belinski, M. 311–315, 338
best-practice extremal frontier 373
binding constraints 37
Bitran, C. 345, 346, 355
boundary point 29
Bram, J. 326

c

canonical forms 35–36
primal problem 95, 97
capacity, excess 4
Cauchy-Schwarz inequality 21
certainty 1, 6
Chadha, S. 347–353
Charnes, A. 338, 373–374, 385–390
closed half-planes 29
coefficient matrix, changing component of 202–208, 209t
Coelli, T.J. 377–379, 383–384
commutative law
matrix algebra 14
vector algebra 20, 21
competition, perfect *see* perfect competition
complementarity, perfect 124
complementarity condition 329
complementary inputs 1, 2
complementary outputs 7
complementary pivot method 329–335
complementary slackness conditions 150–151, 189, 306, 324
complementary slackness theorems 319–320
strong 104–106, 109–111
weak 102–104, 106, 109–111, 116
complementary solutions/ complementary basic solutions 329
composite activity 129, 132–133, 158
composite output activity 171
computational aspects 43–70
degenerate basic feasible solutions 66–69
dual simplex method 114–121
improving basic feasible solutions 48–65
simplex matrix 59–65, 68–70
simplex method 43–48, 69–70
constant product curves 125
see also isoquants
constant returns to scale (CRS) 373
input-oriented DEA model under 387–390
constraint system 35
convex cone 30, 31f, 32
convex hull 33
convex polygons 174
convex polyhedral cone 31, 133
convex polyhedron 33, 34, 339
convex sets
boundary point 29
closed half-planes 29
cones 31–33
convex combination of X_1, X_2, 29
hyperplane 29
interior point 29
linear form 29
and n-dimensional geometry 29–34
open half-planes 30
open or closed 29
quadratic programming 322
set of all convex combinations 33
spherical δ-neighborhood 29
strict separability 30
strictly bounded 29
supporting hyperplane 30
weak separation theorem 30

convexification constraint 398
Cooper, W.W. 338, 385–386, 398–402
cost efficiency (CE) 383–384
cost indifference curves *see* isocost curves
cost minimization 7
  and joint production 180–184
  producing a given output 284–285
costs
  see also cost efficiency (CE); cost indifference curves; cost minimization; marginal cost; total cost
  average cost functions, determining 286–295
  marginal *see* marginal cost
  optimal dollar value of total cost 9
  total imputed cost of firm's minimum output requirements 9
  total potential cost reduction 9
Craven, B. 338, 347–353
critical values
  parametric analysis 228, 236, 237, 242, 246, 249, 250
  parametric programming and theory of the firm 262, 267, 269, 277
CRS *see* constant returns to scale (CRS)

d

Dano, S. 123, 139–146
data envelopment analysis (DEA) 373–404, 374
see also BCC (Banker, Charnes and Cooper) model; decision making units (DMUs)
  allocative efficiency 380
  best-practice extremal frontier 373
  CCR (Charnes, Cooper and Rhodes) model 398, 400
  constant returns to scale (CRS) 373
    input-oriented DEA model under 387–390
  convexification constraint 398
  input and output slack variables 390–398
  input distance function (IDF) 378–379
  input-oriented 373, 387–390
    envelopment form 389
  isoquants 375
  modeling 385–386
  multiplier form 388
  nonparametric 374
  nonstochastic 374
  output correspondence 375
  output distance function (ODF) 377–378
  output-oriented 373, 402–404
  production correspondence 386–387
  projection path 373
  set theoretic representation of a production technology 374–377
  solving the BCC model 400
  strong disposability 374, 375
  technical efficiency 379, 380–383
  technology set 374, 375
  variable returns to scale (VRS) 373, 398–402
Debreu, G. 380–383
decision making units (DMUs) 373, 374, 377, 379
  see also data envelopment analysis (DEA)
degree of input-oriented technical efficiency 380–381

efficient frontier 389
fully efficient 382, 383
input and output slack variables 393–397
peer group 388
reference set 388, 400
synthetic 396, 397
unit isoquant of fully efficient DMUs 380
unit production possibility curve of fully efficient DMUs 381, 384
degenerate basic feasible solutions 66–69
demand function for a variable input 262–269
diagonal matrix 14
diminishing returns 135
direct proportionality 2, 6–7
distance 21
distributive laws
matrix algebra 14, 15
vector algebra 20, 21
divisibility, perfect *see* perfect divisibility
DMUs *see* decision making units (DMUs)
Dorfman, R. 123
Dorn, W. 326
Dreyfus, S. 315–320
dual degeneracy 121
dual feasibility *see* primal optimality (dual feasibility)
dual problem
see also duality theory; primal problem
artificial augmented form 113
dual quadratic programs 326–328
dual solution, constructing 106–113
duality theorems 103, 349, 353
generalized multiactivity profit-maximization model 161
joint production and cost minimization 184
as minimization (maximization) problem 95
optimal solutions 145, 184, 306, 399, 401
reformulation 297–310
simplex matrix 116
single-activity profit maximization model 141
dual quadratic programs 325–328
dual simplex method 113–114
addition of a new structural constraint 221, 222
basic feasible solutions 114, 117, 119
computational aspects 114–121
deletion of a variable 223
as an internalized resource allocation process 157
optimal solutions 114, 121, 122
summary 121–122
dual solution, constructing 106–113
dual structural constraints 183, 184, 338
data envelopment analysis (DEA) 389
dual solution, constructing 107, 108
duality and complementary slackness theorems 320
duality theorems 104, 105
multiactivity profit maximization model 145
reformulation of primal and dual problems 308
simplex method 156
single-activity profit maximization model 141
symmetric duals 96

dual structural constraints (*cont'd*)
  unsymmetrical duals 99
dual support cone 301
duality theory 4, 95–122, 297–370
  see also dual problem; dual simplex method; dual structural constraints
  and complementary slackness theorems 315–320
  constructing the dual solution 106–113
  identity matrix *see* identity matrix
  Lagrangian saddle points 297, 311–315
  in linear fractional functional programming 347–353
  optimal solutions 95, 106, 107, 121, 122, 298, 302, 306, 313–315
    duality theorems 101, 104
  pivot operations 114, 118, 120, 122
  primal 95
  reformulation of primal and dual problems 297–310
  simplex matrix 107–116, 118, 119, 121
  symmetric duals 95–97, 98
  Taylor formula 316, 317
  theorems 100–106, 315–319, 348–353
  unsymmetrical duals 97–100

e

echelon matrix 23, 24
economic efficiency 126, 167
economic rent 338
efficiency
  allocative, cost, and revenue 383–384
  constant returns to scale (CRS) 388
  economic 126, 167
  efficient subsets 375
  fully efficient DMUs 382, 383
  input efficient subsets 377
  output efficient subsets 375
  technical *see* technical efficiency (TE)
  unit isoquant of fully efficient DMUs 380
  unit production possibility curve of fully efficient DMUs 381, 384
elementary row operation 16
excess capacity 4
existence theorem 101
expansion path 126, 127, 294
  joint output 167, 169f
expected payoff 357, 359
extreme point solutions 39–40
  parametric analysis 231, 234
extreme points 33, 43
  linear fractional functional programming 339–340
  parametric analysis 238, 239

f

factor learning
  learning economies 162
  learning index 163
  learning rates 162, 163
  negative exponential 163
  and optimum product mix model 164–165
  progress elasticity 163
factor substitution 130
Färe, R. 374–379
Farrell, M.J. 380–383
feasible directions 298
feasible solutions
  see also optimal solutions; primal optimality (dual feasibility); solutions
  basic/optimal basic

addition of a new structural constraint 220, 221, 222
degenerate 39, 66–69
deletion of a structural constraint 223–224
demand function for a variable input 262, 264–269
determination of marginal productivity, average productivity and marginal cost 287–288, 290, 293
dual simplex method 114, 117, 119
dual solution, constructing 106, 108, 111
improving 48–65
inconsistency and redundancy 79–85
linear fractional functional programming 340, 345, 352
marginal (net) revenue productivity function for an input 271, 273, 274, 278, 279, 281, 284
minimizing cost of producing a given output 285
M-Penalty method 73, 76, 77, 78
new variable, addition of 217–219
nondegenerate 40, 66
parametric analysis 227–229, 231–233, 235–236, 239–243, 245, 248
quadratic programming 324
resource allocation with a fractional objective 354
simplex method 155, 156
supply function for the output of an activity 258–260
two-phase method 90–93
updating the basis inverse 256
definition 36
dual quadratic programs 325, 326
duality theorems 349
extreme points 43
linear complementarity problem (LCP) 329
nonbasic 89
profit indifference curves 148
region of 35, 38, 148, 298
Ferguson, C.E. 123, 139–146
finite cone 30–31
firm
technology of 123–125
theory of *see* theory of the firm
fixed coefficients linear technology 157
fixed inputs 4, 5, 11
see also inputs
data envelopment analysis (DEA) 384
quadratic programming 335
theory of the firm 155, 157, 159, 164
activity levels 150, 151
multiactivity profit maximization model 144, 145n
and parametric programming 257, 262, 269, 271, 275–277, 284, 290
single-activity profit maximization model 140, 142
fixed resources 5, 6
see also resources
foregone profit 4, 5
fractional objective, resource allocation with 353–356
fractional programs 387
Freimer, M. 315–320

Fried, H. 374–379
Frisch, R. 181n
fundamental theorems of linear programming 101, 102

g

game theory 356–363
  defining a game 356
  expected outcome 357
  expected payoff 357, 359
  fundamental theorem 359–360
  generalized saddle point 357
  matrix games 357–360
    transformation to a linear program 361–363
  mixed strategies 358, 359, 360, 363
  normal form 357
  outcome strictly dominated 357
  payoff function 357
  saddle point solution 358
  strategies 356, 358, 359, 360
  two-person games 356, 357
  value of the game 359
  zero-sum game 356
Gauss elimination technique 24
generalized multiactivity profit-maximization model 157–161, 335, 353
generalized saddle point 359
generalized saddle value 360
gross profit 3
  gross profit margin 141, 144, 159, 161, 337, 338
  simplex-based optimization methods 337, 338, 355
  theory of the firm 145, 156, 157, 164
    and parametric programming 257, 262, 263, 269, 276–278, 281, 283–286

h

Hadar, J. 123, 139–146
half-line 124
homogeneity 21, 26, 27
hyperplanes 29, 30, 339

i

identity matrix 14, 45n
  duality theory 108, 112
  M-Penalty method 72–75
  parametric programming 256
  sensitivity analysis 201
inconsistency 78–85
increasing (real resource) opportunity cost 175
indifference curves
  cost 184–186, 190
  production 146, 147
  profit 146–148, 151–156
infeasibility form 88
input distance function (IDF) 378–379
input isoquants 376–377
inputs
  see also input distance function (IDF); input isoquants; outputs
  activities 1, 2
  complementary 1
  demand function for a variable input 262–269
  fixed *see* fixed inputs
  input activities 1–2
  input correspondence 376
  input efficient subsets 377
  input set 376
  input-conserving orientation 379, 380
  input-oriented BCC model 399
  input-oriented DEA model 387–390

limitational 124
marginal (net) revenue productivity function for 269–276
optimal value of 5
shadow 180, 181n, 182
slack variables 390–398
strong disposability 375, 376
interior point 29
isocost curves 184–186, 190
surface 192f, 193
iso-input transformation curve 167, 174
isoquants 125–127, 128f, 138, 375
see also theory of the firm
ABCD 133, 134
input 376–377
joint process 131
multiple processes 133f
output 375
parametric representation 131
unit 131, 132, 380
"well-behaved," 135

j

joint output expansion path 167, 169f
joint output linear production model 172
joint process isoquant 131
joint process linear production model 130
joint process transformation curve 172
joint product transformation curve (iso-input) 167, 174
joint production 6
see also production function; theory of the firm
and cost minimization 180–184
multiactivity joint-production model 171–180, 177f, 179f
processes 165–166

k

Karush-Kuhn-Tucker equivalence theorem 313–315
Karush-Kuhn-Tucker necessary and sufficient conditions 297
complementary pivot method 329–330
quadratic programming 323, 327, 337
Karush-Kuhn-Tucker theorem 303–310
corollaries 306–310
Kogiku, K.C. 123
Kornbluth, J. 347–353
Kuhn-Tucker-Lagrange necessary and sufficient conditions 160
Kydland, F. 347–353, 355

l

Lagrange technique 303
Lagrange multipliers 107, 160, 304, 306
Lagrangian expressions 107
Lagrangian saddle points 297, 311–315
linear fractional functional programming 350
Lasdon, L. 338
learning economies 162
learning index 163
learning rates 162, 163
Lemke, C.E. 113–114, 329–335
Liao, W. 164
limitationality 2, 7, 124, 166
mutual 126, 127
limiting subset 129
linear combinations 26–27
linear complementarity problem (LCP) 329
linear dependence 26–29
and linear independence 27
linear form 29, 321

linear fractional functional programming 338–346
  duality in 347–353
linear model for the firm 1–2
linear programming problem
  artificial 73
  augmented 36
  deletion of a structural constraint 223–224
  graphical solution to 37
  linear fractional programming 338
  new variable, addition of 217
  optimal solution to *see* optimal solutions
  sensitivity analysis 195–196
  surrogate 88, 90, 91, 93
  symmetric duals 95
linear technology 123
Lovell, C.A.K. 374–379, 383–386

m

Magnanti, T. 355
marginal (net) revenue productivity function for an input 269–276
marginal cost
  activity 189, 276–284, 282t
  determining 286–295
  imputed or shadow costs 9, 10
  joint production and cost minimization 183
  marginal (real) resource cost function 168, 169f, 179f
  marginal cost function for an activity 276–284
  marginal cost relationships 142
  multiactivity joint-production model 175n
marginal product 136f, 175, 290, 291f
  see also average product
  marginal product function 170
marginal productivity
  determining 286–295
  marginal productivity function 127, 138
marginal profitability 156
marginal revenue 142
market prices 155
Martos, B. 338, 355
mathematical foundations
  convex sets and n-dimensional geometry 29–34
  linear dependence 26–29
  matrix algebra 13–20
  simultaneous linear equation systems 22–26
  vector algebra 20–22
matrix
  algebra *see* matrix algebra
  augmented 23, 24
  basis matrix B 39, 49, 201, 203
  coefficient, changing component of 202–208, 209t
  defined 13
  diagonal 14
  echelon 23, 24
  identity 14, 73–75, 256
  nth order matrix A 18
  output technology 171
  premultiplier 15
  postmultiplier 15
  rank 23, 24
  simplex *see* simplex matrix
  submatrix 13, 25, 45
  transposition of 14
  triangular 14
matrix algebra 13–20
  see also matrix
  elementary row operation 16
  multiplication 15
  sweep-out process 18
  TYPE I operation 17
  TYPE II operation 17
  TYPE III operation 18
matrix games 357–360

transformation to a linear program 361–363
maximal-slack solution 390
method
complementary pivot 329–335
dual simplex *see* dual simplex method
M-Penalty 71–78, 111, 294
simplex *see* simplex method; simplex-based optimization methods
two-phase 87–94
minimization of the objective function 85–86
Minkowski-Farkas theorem 302–303
mixed structural constraint system 71
Mond, B. 347–353
M-Penalty method 71–78
basic feasible solutions 73, 76, 77, 78
dual solution, constructing 111
identity matrix 72–75
mixed structural constraint system 71
parametric programming 294
slack variables 71, 72, 76
surplus variables 71, 72, 76
multiactivity joint-production model 171–180, 177f, 179f
see also transformation curve
composite output activity 171
increasing (real resource) opportunity cost 176
joint output linear production model 172
joint process transformation curve 172
joint product transformation curve 174
output technology matrix 171
parametric representation of the transformation curve 172
rate of product transformation 175
unit transformation curve 172, 173f
multiactivity production function 129–139, 136f
additive activity 130
composite activity 130
diminishing returns 135
factor substitution 130
joint process linear production model 130
process substitution 130
technical rate of substitution 134–135
multiactivity profit maximization model 143–146
multiplicative identity 15, 20
mutual limitationality 126, 127

n

Nanda, R. 162
Naylor, T. 123
n-dimensional Euclidean space 21
negative exponential 163
nonbasic variables 39, 69, 118, 197, 212, 228
see also basic variables
improving basic feasible solutions 50, 51, 54, 55
simplex method 44, 46, 47
simplex-based optimization methods 332, 340, 341
nondegeneracy assumption 44
nonnegativity conditions 35
nonreversible production activities 4
norm of X 21
normalizing constraints 387
Novaes, A. 345, 346
null vector 20, 27, 30, 98

O

objective function 2, 3, 6, 8, 37, 43
  see also hyperplane; objective function coefficients
  artificial 74
  canonical forms 35
  computational aspects 46, 66, 67, 69, 70
    improving basic feasible solutions 48, 49, 51, 52, 55, 57, 60, 61, 65
  deletion of a variable 223
  dual solution, constructing 110
  duality theorems 100, 102
  interpretation 269n
  minimization 85–86
  optimal 61, 388
  parameterizing 227, 228–236, 257
  primal 100, 106, 107, 223, 299, 307
  quadratic programming 321
  sensitivity analysis 198, 205
  surrogate 87–88
  two-phase method 88, 89
objective function coefficients 63, 164, 195, 217
  see also objective function
  changing 196–199, 209
  parametric programming 229, 231, 259, 260
  theory of the firm 182, 188
operational level 124, 131
opportunity cost 175
optimal (imputed) costs of output quotas 10
optimal (imputed) value of all fixed resources 5, 6
optimal (imputed) value of outputs produced 10
optimal accounting loss figures 5, 10
optimal activity levels 4
optimal criterion 51, 78, 85, 88, 92, 93, 113, 220
  dual solution 107, 108, 112
  inconsistency and redundancy 80, 81, 84
  parametric programming 228, 237, 245, 249, 255
  sensitivity analysis 196–198
optimal dollar value of total cost 9
optimal dollar value of total profit 4
optimal objective function 61, 388
optimal output activity levels 9
optimal output configuration 9
optimal primary-factor/labor-grade mix 8–9
optimal product mix 4, 270
  and factor learning 164–165
optimal shadow price configuration 5
optimal simplex matrix *see* simplex matrix
optimal solutions 4, 5, 10, 36–38, 40, 43, 67, 78, 202
  see also feasible solutions; solutions
  canonical forms 36
  data envelopment analysis (DEA) 388, 390, 397, 399–402
  dual problem 145, 184, 306, 399, 401
  duality theory 95, 106, 107, 298, 313–315
    dual simplex method 114, 121, 122
    reformulation of primal and dual problems 302, 306
    theorems 101, 104
  existence and location 38–39
  parametric programming 241, 246, 265, 271, 272
  quadratic programming 322–324
  simplex-based optimization methods 322, 326, 328, 337, 338, 346, 347, 349, 352, 355, 363
  structural changes 217, 222, 223f

theory of the firm 141, 145, 184
optimal utilization information 4
optimal value of inputs 5
optimality evaluators 48
optimality theorem 48, 51
optimum product-mix model 164–165
output activities 6, 7, 183–185, 187, 190, 191
multiactivity joint-production model 171–174
output activity mix 189
output activity vector 165, 174
output distance function (ODF) 377–378
output efficient subsets 375
output process ray 166
output substitution 172
output technology matrix 171
output transformation curves 174, 175, 176f, 184–186
outputs
see also inputs; output activities; output distance function (ODF); output efficient subsets; output process ray; output substitution; output technology matrix; output transformation curve
cost minimization 284–285
fixed level 131
joint output expansion path 167, 169f
joint output linear production model 172
optimal (imputed) value of outputs produced 10
optimal imputed costs of output quotas 10
optimal output activity levels 9
optimal output configuration 9
output correspondence 375
output set 375
output-augmenting orientation 380
output-oriented DEA 377–378, 402–404
output-oriented multiplier problem 403
quotas 10, 167, 175, 187
slack variables 390–398
supply function for the output of an activity 257–262
total imputed cost of firm's minimum output requirements 9
transformation surface 190, 191f
unit level 132, 354
overproduction 9, 182, 189

p

Panik, M. 123, 351
parallelogram law, vector addition 132
parametric analysis 11, 227–256
see also parametric programming and theory of the firm
basis inverse, updating 256
critical values 228, 236, 237, 242, 246, 249, 250
parameterizing an activity vector 245–256
parameterizing the objective function 227, 228–236, 257
parameterizing the requirement vector 236–245, 277
primal feasibility 228, 236, 237, 245, 249, 250, 253
revised feasibility criterion 237
revised optimality condition 228, 245, 249
parametric programming and theory of the firm 257–295
see also parametric analysis
average cost functions, determining 286–295
average productivity, determining 286–295

parametric programming and theory of the firm (*cont'd*)
  ceteris paribus assumption 257, 262, 269, 276, 290, 292
  critical values 262, 267, 269, 277
  demand function for a variable input 262–269
  marginal (net) revenue productivity function for an input 269–276
  marginal cost, determining 286–295
  marginal cost function for an activity 276–284, 282t
  marginal productivity, determining 286–295
  minimizing cost of producing a given output 284–285
  supply function for the output of an activity 257–262
parametric representation of the isoquant 131
parametric representation of the transformation curve 172
payoff function 357
perfect competition 1, 6
  generalized multiactivity profit-maximization model 158
  multiactivity profit maximization model 143–144
  single-activity profit maximization model 140, 142
perfect complementarity 124, 166
perfect divisibility 1, 6, 124, 166
pivot operations 74, 89
  complementary pivot method 332, 334
  computational aspects 52, 56, 70
  dual simplex 121, 157, 253, 271, 274
  duality theory 114, 118, 120, 122
  parametric programming 239, 253, 259, 260, 264, 265, 271, 274
pivotal term 52
plane of support theorem 30–31
polar support cone 300
postmultiplier matrix 15
post-optimality analysis 10, 195
premultiplier matrix 15
primal feasibility 220, 280
  duality theory 114, 120, 122
  parametric analysis 228, 236, 237, 245, 249, 250, 253
  sensitivity analysis 200, 206–208, 209t, 210, 213, 214
primal objective function 100, 106, 107, 223, 299, 307
primal objective value 121, 325
primal optimality (dual feasibility)
  see also sensitivity analysis
  changes in technology 214, 215
  changing a component of the coefficient matrix 203, 204, 206–208
  changing a component of the requirements vector 200, 202t
  changing objective function coefficient 196, 198, 199t
  changing product and factor prices 212–213
  changing resource requirements 213
  parametric analysis 245, 246, 250
primal problem 4
  see also dual problem; duality theory
  canonical form 95, 97
  dual quadratic programs 325–327
  dual simplex method 114, 115
  dual solution, constructing 107
  duality theorems 100, 102, 103, 306, 348, 349
  generalized short-run fixed-coefficients profit-maximization model 159–160
  Lagrangian saddle points 311
  as maximization (minimization) problem 95

reformulation 297–310
structural constraints 96, 99
symmetric duals 95–97
unsymmetrical duals 98, 99
primal simplex matrix 108, 114, 118, 201
primal simplex method 113
primary-factor, optimal 8–9
principal diagonal 14
problems
artificial 75–77
dual *see* dual problem
linear complementarity problem (LCP) 329
linear programming *see* linear programming problem
optimization 284, 286
output-oriented multiplier 403
parametric 258, 260–262, 265–267, 269, 270
primal *see* primal problem
primal maximum 297
profit maximization 149
reformulation of primal and dual problems 297–310
saddle point 297, 312
symmetrical 284
process ray 124
process substitution 130, 172
product mix, optimal 4, 164–165, 270
product transformation curve 177f
joint product 167, 174
"well-behaved" product 175
product transformation function, single-process 167–170, 169f
production activities 1, 4, 6, 123
production correspondence 386–387
production function
see also joint production; theory of the firm
joint production process 165–166
multiactivity 129–139, 136f
single-process 125–127, 128f, 129
production indifference curves 146, 147
production possibility set 374, 386
production time 354
profit
see also profit indifference curves; profit maximization model; profit maximization problem
average 2, 354
foregone 4, 5
gross *see* gross profit
objective function 3
total *see* total profit
profit indifference curves 146–148, 151–156
profit maximization model
assumptions underlying 1–2
generalized multiactivity 157–161, 335, 353
multiactivity 143–146
short-run fixed-coefficients 140–141
short-run linear technology 144
simplex-based optimization methods 335, 336
single-activity 139–142
profit maximization problem 149
progress elasticity 163
proportionality, direct 2, 6–7

q

quadratic forms 321, 363–371
classification 367–368
definite 367, 368–370
general structure 363–365
indefinite 367
necessary and sufficient conditions for the definiteness and semi-definiteness of 369–370

quadratic forms (*cont'd*)
  necessary conditions for definiteness and semi-definiteness of 368–369
  semi-definite 367, 368–370
  symmetric 366–367
  theorems 368–371
quadratic programming 321–324
  and activity analysis 335–338
  dual programs 325–328
  Karush-Kuhn-Tucker necessary and sufficient conditions 324, 327
  primal programs 325
  theorems 325–328
quasi-rents 338n

r

rank 22, 23
rate of product transformation 175
redundancy 39, 78–85, 89, 126
Reeves, G. 164, 165
reference set 388, 400
requirements space 41, 294
requirements vector 41
  changing component of 200–202, 209–210
  determination of marginal productivity, average productivity and marginal cost 286–290
  marginal (net) revenue productivity function for an input 272–274
  parameterizing 236–245, 277
resources
  activity levels interpreted as individual resource levels 186–193
  allocation process
    dual simplex method 157
    fractional objective, resource allocation with 353–356
    simplex method 155–156
  average (real) resource cost function 168, 179f
  changing resource requirements 213
  fixed 5, 6
  level of utilization 166, 167, 185
  marginal (real) resource cost function 168, 179f
  optimal utilization information 4
  optimal valuation of the firm's fixed resources 5
  resource requirements vector 124, 158
  total (real) resource cost function 168, 169f, 179f
  total imputed value of firm's fixed resources 5
returns to scale
  BCC (Banker, Charnes and Cooper) model 401–402
  constant 373
  data envelopment analysis (DEA) 401–402
  technology of the firm 123, 124
  variable 373
revenue efficiency (RE) 383–384
robustness 10

s

Saaty, T. 326
saddle points
  game theory 358, 359
  generalized 359
  Lagrangian *see* saddle points, Lagrangian
saddle points, Lagrangian 297, 311–315
  problem 297, 312
  theorems 312–315
    necessary and sufficient condition 312–313
    sufficient condition 313
Salkin, G. 347–353
scalar (inner) product 20

Schnaible, S. 347–353
Seiford, L.M. 373–374
sensitivity analysis 10, 195–215
changes in technology 213–215
changing a component of an activity vector 210–215
changing a component of the coefficient matrix 202–208, 209t
changing a component of the requirements vector 200–202, 209–210
changing product and factor prices 211–213
changing resource requirements 213
objective function coefficient, changing 196–199, 209
post-optimality analysis 10, 195
primal feasibility 200, 206–208, 209t, 210, 213, 214
simplex matrix 195, 196, 198, 201, 205, 213, 214
summary of effects 209–215
shadow activities 181
shadow inputs 180, 181n, 182
shadow prices 4
dual simplex method 157
optimal shadow price configuration 5
resource allocation with a fractional objective 355
single-activity profit maximization model 140
Shephard, R.W. 377–379
short run
firm operating in 2
generalized short-run fixed-coefficients profit-maximization model 159
short-run fixed-coefficients profit-maximization model 140–141
short-run linear technology profit-maximization model 144
short-run supply curve 142
simple activity 123, 130
simplex 34
simplex matrix 50–57, 331
computational aspects 59–65, 68–70
duality theory 107–116, 118, 119, 121
optimal
computational aspects 61, 63
duality theory 109, 110, 113, 294
parametric analysis 227, 230–234, 238–245, 247, 248, 251, 253, 255
parametric programming and theory of the firm 258, 259, 263, 270, 277, 278, 285, 286
redundancy 79
sensitivity analysis 195, 196, 198, 201, 205, 213, 214
simplex-based optimization methods 345, 346, 352, 353, 355
structural changes 217, 218, 221, 223–225
theory of the firm 149, 188
parametric programming 227, 263, 270, 285, 286, 289, 294
activity vector 247, 248, 251, 253, 255
marginal cost function for an activity 277, 278
parameterizing the objective function 229–235
parameterizing the requirement vector 238–245
supply function for the output of an activity 258, 259
primal 108, 114, 118, 201
sensitivity analysis 195, 196, 198, 201, 205, 213, 214
simplex-based optimization methods 342, 343, 345, 346, 352, 353, 355

simplex matrix (*cont'd*)
structural changes 217–225
theory of the firm 149, 188
variations of standard simplex routine 72, 74–76, 79–83, 88–93
simplex method 4, 43–48
see also simplex-based optimization methods
dual *see* dual simplex method
dual solution, constructing 111
as an internal resource allocation process 155–156
nondegeneracy assumption 44
primal 113
summary 69–70
symmetric duals 95
variations of standard simplex routine 71–94
simplex-based optimization methods 321–371
see also simplex method
complementary pivot method 329–335
duality in linear fractional functional programming 347–353
game theory 356–363
linear fractional functional programming 338–346
matrix games 357–360
optimal solutions 322, 326, 328, 337, 338, 346, 347, 349, 352, 355, 363
quadratic forms 363–371
quadratic programming 321–324
and activity analysis 335–338
dual quadratic programs 325–328
resource allocation with a fractional objective 353–356
simplex matrix 342, 343, 345, 346, 352, 353, 355
simultaneous linear equation systems 22–26
consistency 22
determinate solutions 26
homogenous 26, 27
n linear equations in *n* unknowns 22
rank 22, 23
theorems 23–26
underdetermined 25
single-activity profit maximization model 139–142
single-process product transformation function 167–170, 169f
single-process production function 125–129, 128f
average productivity function 127
expansion path 126, 127
limiting subset 129
marginal productivity function 127
slack variables 82, 189, 303, 403
see also basic variables; complementary slackness theorems; nonbasic variables; variables
dual solution 112
duality theorems 103, 105
input and output 390–398
M-Penalty method 71, 72, 76
nonnegative 36, 45, 67, 71, 109, 149, 160, 211, 285, 342
dual simplex method 115, 116
duality theory 303, 306
improving basic feasible solutions 57, 59, 62, 64
linear fractional functional programming 352, 353
structural changes 219, 222, 224
primal 103–105, 112, 145, 306, 319, 320
simplex-based optimization methods 320, 340, 353
structural changes 219, 221, 222, 224

theory of the firm 150, 156, 160, 161, 183, 184
solutions
admissible 298
almost complementary basic 330–332
basic 28
canonical forms 36
complementary/complementary basic 329
dual, constructing 106–113
extreme point 39–40, 43
feasible *see* feasible solutions
graphical, to linear programming problem 37
maximal-slack 390
optimal *see* optimal solutions
and requirements spaces 41
saddle point 358
zero-slack 390
solutions space 41
spaces
n-dimensional Euclidean 21
requirements 41, 294
solutions 41
vectors 20, 28
spanning set, vectors 27
spherical δ-neighborhood 29
standard forms 36
static models 1, 6
strategies, game theory 356
maximin 360
minimax 360
mixed 358, 359, 360, 363
strict separability 30
strong complementary slackness theorems 104–106, 109–111
strong disposability 374, 375, 376
structural changes 11, 217–226
addition of a new structural constraint 219–222
addition of a new variable 217–219
deletion of a structural constraint 223–226
deletion of a variable 223
optimal solutions 217, 222, 223f
simplex matrix 217–225
structural constraints 4, 5, 9
activity levels 150
addition of 219–222
artificial augmented structural constraint system 72
augmented structural constraint system 36, 39, 53
primal 155–156
canonical forms 35
complementary pivot method 329
deletion of 223–226
dual 183, 184, 338
data envelopment analysis (DEA) 389
dual solution, constructing 107, 108
duality and complementary slackness theorems 320
duality theorems 104, 105
multiactivity profit maximization model 145
reformulation of primal and dual problems 308
simplex method 156
single-activity profit maximization model 141
symmetric duals 96
unsymmetrical duals 99
generalized multiactivity profit-maximization model 161
inconsistency and redundancy 75
inequality 97, 302
linear fractional functional programming 340, 352
original system 78, 79, 85
primal problem 96, 99
reformulation of primal and dual problems 307
sensitivity analysis 202
single-activity profit maximization model 142

submatrix 13, 25, 45
substitution
  factor 130
  output 172
  process 130, 172
  technical rate of 134–135, 154
sum vector 1 21
supply function for the output of an activity 257–262
supporting hyperplane 30
surplus variables 86, 306, 320, 353
  see also basic variables; nonbasic variables; slack variables; variables
  duality theory 96, 103, 104, 105, 108, 112–114, 119
  inconsistency and redundancy 79, 82
  M-Penalty method 71, 72, 76
  nonnegative 71, 188, 221, 353
  structural changes 221–226
  theory of the firm 141, 145, 150, 157, 182, 184
    activity levels 188, 189
surrogate linear programming problem 88, 90, 91, 93
surrogate objective function 87–88
Swarup, K. 340, 342
Sweigart, J. 164, 165
symmetric duals
  duality theory 95–97, 98
  joint production and cost minimization 183
  theory of the firm 141, 144, 148, 161

t

tangent support cone 298
technical efficiency (TE) 126, 132, 167, 379, 389
  degree of input-oriented technical efficiency 380–381
  degree of radial technical inefficiency 381, 382
  input-oriented measure 381
  measuring 380–383
  output-oriented measure 382
  projection points 382
  radial measures 380, 381, 382
  technically efficient projection point 381
technical rate of substitution 134–135, 154
technological changes 213–215
technological independence 165
technological interdependence 6, 165
technology of the firm 123–125
technology set 374, 375
theorems
  basic feasible solutions 40
  complementary slackness *see* complementary slackness theorems
  convex sets 30–34
  duality 100–106, 315–319, 348–353
  existence 101
  fundamental, of linear programming 101, 102
  game theory 359–360
  Karush-Kuhn-Tucker 303–310
  Karush-Kuhn-Tucker equivalence 313–315
  minimization of the objective function 86
  Minkowski-Farkas 302–303
  necessary and sufficient condition 312–313
  optimal solutions 38–39
  optimality 48, 51
  plane of support 30–31
  quadratic forms 368–371
  quadratic programming 325–328
  reformulation of primal and dual problems 299–310
  saddle points, Lagrangian 312–315

simultaneous linear equation systems 23–26
sufficient condition 313
weak separation 30
theory of the firm 123–193
activity levels interpreted as individual product levels 148–155
activity levels interpreted as individual resource levels 186–193
cost indifference curves 184–186
dual simplex method 157
factor learning and optimum product-mix model 161–165
generalized multiactivity profit-maximization model 157–161
isoquants *see* isoquants
joint production
and cost minimization 180–184
multiactivity joint-production model 171–180, 177f, 179f
processes 165–166
multiactivity production function 129–139, 136f
multiactivity profit maximization model 143–146
optimal solutions 141, 145, 184
and parametric programming *see* parametric programming and theory of the firm
profit indifference curves 146–148, 151–156
simplex method 155–156
single-activity profit maximization model 139–142
single-process product transformation function 167–170, 169f
single-process production function 125–127, 128f, 129
technology of the firm 123–125
Thompson, G.E. 123
Thrall, R.M. 373–374, 385–386, 398–402
total cost
imputed cost of all output requirements 10
imputed cost of firm's minimum output requirements 9
joint production and cost minimization 183
marginal cost function for an activity 276, 281, 283
optimal dollar value of 9
total (real) resource cost function 168, 169f, 179f
total conversion cost 158
total variable cost (TVC) 263, 292–294
total factor productivity 387
total profit 2–6, 211, 295, 338, 354
activity levels 149, 150
optimal dollar value of 4
profit indifference curves 146–148
simplex method 155
single-activity profit maximization model 140
theory of the firm 140, 146–150, 155
total product function 127, 170
transformation curves 167
joint process 172
joint product 167, 174
output 174, 175, 176f, 184–186
parametric representation 172
unit 172, 173f, 174
"well-behaved" product 175
transposition, matrix 14
triangular inequality 21
triangular matrix 14
two-person games 356, 357
two-phase method 87–94
infeasibility form 88
input and output slack variables 390

two-phase method (*cont'd*)
surrogate linear programming problem 88
surrogate objective function 87–88

u
unit column vector 21
unit isoquant 131, 132, 380
unit level of activity 124
unit transformation curve 172, 173f, 174
unrestricted variables 86–87
unsymmetrical duals 97–100

v
Vandermullen, D. 123
variable returns to scale (VRS) 373
modeling 398–402
variables
addition of a new variable 217–219
artificial *see* artificial variables
basic *see* basic variables
deletion of 223
demand function for a variable input 262–269
legitimate 72, 79, 88
nonbasic *see* nonbasic variables
slack *see* slack variables
surplus *see* surplus variables
unrestricted 86–87
vector algebra 20–22
see also activity vector; vectors
vector space 20, 28
vectors
activity *see* activity vector
algebra *see* vector algebra
artificial 73, 75, 78, 79, 81, 82, 91, 93
components 20
definition 20
nonbasic 202, 203, 256
null vector 20, 27, 30, 98
orthogonal 21
output activity 165, 174
requirements *see* requirements vector
resource requirements 124, 158
spanning set 27
sum vector 1 21
unit column 21
Vernon, J. 123
vertex (of a cone) 30
vertex (of a convex set) 33

w
Wagner, H. 347–353
weak complementary slackness theorem 102–104, 106, 109–111, 116
weak disposability 376
weak separation theorem 30

y
Yuan, J. 347–353

z
zero-slack solution 390
zero-sum game 356